Sustainable Development Goals Series

The **Sustainable Development Goals Series** is Springer Nature's inaugural cross-imprint book series that addresses and supports the United Nations' seventeen Sustainable Development Goals. The series fosters comprehensive research focused on these global targets and endeavours to address some of society's greatest grand challenges. The SDGs are inherently multidisciplinary, and they bring people working across different fields together and working towards a common goal. In this spirit, the Sustainable Development Goals series is the first at Springer Nature to publish books under both the Springer and Palgrave Macmillan imprints, bringing the strengths of our imprints together.

The Sustainable Development Goals Series is organized into eighteen subseries: one subseries based around each of the seventeen respective Sustainable Development Goals, and an eighteenth subseries, "Connecting the Goals," which serves as a home for volumes addressing multiple goals or studying the SDGs as a whole. Each subseries is guided by an expert Subseries Advisor with years or decades of experience studying and addressing core components of their respective Goal.

The SDG Series has a remit as broad as the SDGs themselves, and contributions are welcome from scientists, academics, policymakers, and researchers working in fields related to any of the seventeen goals. If you are interested in contributing a monograph or curated volume to the series, please contact the Publishers: Zachary Romano [Springer; zachary.romano@springer.com] and Rachael Ballard [Palgrave Macmillan; rachael.ballard@palgrave.com].

Lira Luz Benites Lazaro
Leandro Luiz Giatti
Laura S. Valente de Macedo
Jose A. Puppim de Oliveira
Editors

Water-Energy-Food Nexus and Climate Change in Cities

Springer

Editors
Lira Luz Benites Lazaro
Department of Environmental Health
School of Public Health
University of São Paulo
São Paulo, Brazil

Leandro Luiz Giatti
Department of Environmental Health
School of Public Health
University of São Paulo
São Paulo, Brazil

Laura S. Valente de Macedo
Fundação Getulio Vargas (FGV)
São Paulo School of Management
(FGV-EAESP)
São Paulo, Brazil

Jose A. Puppim de Oliveira
Fundação Getulio Vargas
FGV EAESP and FGV EBAPE
São Paulo, Brazil

ISSN 2523-3084 ISSN 2523-3092 (electronic)
Sustainable Development Goals Series
ISBN 978-3-031-05471-6 ISBN 978-3-031-05472-3 (eBook)
https://doi.org/10.1007/978-3-031-05472-3

This Springer imprint is published by the registered company Springer Nature Switzerland AG
The registered company address is: Gewerbestrasse 11, 6330 Cham, Switzerland

Foreword

More than ever, cities are at the frontline of global change, and their local government leaders need to rapidly act in nearly every dimension of urban life.

Cities, towns, and their surrounding regions are often seen as centers of opportunity and prosperity, and, indeed, they often act as important hubs for economic opportunity, innovation, co-creation, production, consumption, inclusivity, as well as social and cultural interaction.

At the same time, our current global challenges, including global public health crises such as COVID-19; the need to end systemic inequities and poverty and to better apply technological advancement; the increasing impacts of climate change, biodiversity loss, and land degradation; the effects of rapid urbanization and demographic change; and the lack of resilience in urban service systems, processes, and approaches are all particularly visible in urban environments and require urgent action from local government leaders.

Moreover, as important engines of economic development, cities rely heavily on water, energy, and agricultural produce to sustain development. For this reason, cities can and must play an important catalyzing role in optimizing water, food, and energy synergies and closing the water, energy, and nutrients loops.

Addressing the strong nexus between water, energy, and food at urban level in policy making is therefore essential to enhance sustainability both at local and global levels.

Projections by relevant UN agencies suggest that the demand for freshwater, energy, and food is rising due to demographic changes, economic development, and international trade, among others, and play out particularly in urban areas.

Population growth and increase in economic productivity will require an increase in agricultural production by about 70% by 2050 and about 50% more primary energy by 2035.

Changes in energy usage and types of energy production (for example, fossil fuels replaced by hydropower or biofuels) affect water usage and impact agricultural production.

Climate change is causing additional pressure on water availability and quality as well as on agricultural production. Extreme events have severe socioeconomic and environmental consequences, causing further vulnerability of people and natural resources. Actions to mitigate and adapt to climate change can have strong implications for the surface and ground water system and its users.

The resulting conflicts in the allocation of water and between the water, energy, and food sectors cause additional concerns in regard to the sustainable management of surface and ground water bodies and the interlinkages with urban areas, where a very large proportion of the world's population lives.

Taking a water-energy-food (WEF) nexus approach in urban policy making, governance, and management will help address these complex and interlinked challenges by exploiting available synergies across all policy areas, maximizing coherence, and promoting positive trade-offs between different policies.

A core premise of the WEF nexus approach is that the policy objectives in one sector (water, energy, or food) can interact with those in other sectors because they are either preconditions for the realization of another sector's objectives, or one sector (system) imposes conditions or constraints on what can be achieved in the other sectors. In other words, some policy objectives have synergies across sectors, while others require trade-offs. Objectives in different sectors can also reinforce action towards objectives in other sectors.

Urban policy formulation that is based on a WEF nexus approach will go beyond sectoral solutions and actively address synergies, trade-offs, as well as externalities between the water, energy, and food sectors. And thus it can support the transition to global sustainability.

Literature on the nexus approach follows often three core themes: the nature of the relationships between water, energy, and food; the consequences of change in one sector for change in the other sectors; and the implications for policy making.

The last of these themes is relatively unexplored, and there is a lack of evidence and knowledge about the governance, and institutional and political economy factors that determine the effectiveness of the nexus approach.

It is therefore much welcomed that the authors in this book examine and review in particular the potential, the benefits, and the policy implications of the WEF nexus approach at the urban level and within a context of current global sustainability agendas; through a series of hand-on cases, they present the opportunities of the WEF nexus approach to achieve innovation and transformative change; and they discuss concrete areas of synergy and policy initiative that the WEF nexus approach allow to raise urban resilience.

When a group of forward-thinking cities founded ICLEI in 1990, they took action for sustainability, well before sustainable development was the predominant model for change. The entry point was protecting and conserving the environment. Since then, ICLEI has evolved with a focus on mainstreaming sustainable development in all aspects of life.

Our ICLEI network reaches well over 2500 cities, towns, and regions and is widely recognized as a leading organization with visionary leaders acting worldwide. We represent small, large, and often fast-growing cities, towns, and counties, vast metropolitan areas, capitals, provinces, and regions.

With the start of the Decade of Action in 2021 and through the UN sustainability frameworks in place towards 2030 and 2050, notably the Paris Climate Agreement and new commitments (NDCs), the Sustainable Development Goals (SDGs), and the New Urban Agenda – for which we advocated intensively over the last two decades – cities and their leaders must now, more than ever, implement the global sustainable development agenda in collaboration with all levels of government, academia/research, the private sector, and civil society.

We are grateful to the authors for the guidance presented in this book on the use of the WEF nexus approach to help sustain and scale up action over the span of this decisive decade for both people and planet alike.

Secretary General, ICLEI-Local Governments for Sustainability
Bonn, Germany
Gino Van Begin

Preface

Sustainable and resilient cities are crucial for securing social inclusion and improving the quality of life for their inhabitants, given that the production chains and the supply of water, energy, and food are strongly interdependent. However, the intricate multi-scale relations between natural resources and the urban systems underpinning the water-energy-food nexus are still poorly understood, even as the contexts become increasingly complex.

This book aims to explore and strengthen the transdisciplinary aspects of the water-energy-food nexus in cities in the face of the challenges and threats of climate change. Experts highlight urban nexus issues and/or propose solutions to decision makers globally when dealing with the social, economic, and environmental consequences of the siloed governance of water, energy, and food, and the implications for achieving Sustainable Development Goals (SDGs).

In this book, the nexus approach is conceived as a multidimensional means of scientific investigation that seeks to understand the complex and non-linear interrelationships and interdependencies between water-energy-food under climate change, and to better understand the wider implications of this for urban societies. This approach proposes to reduce trade-offs among development goals and generate co-benefits that help encourage sustainable development. Governing the nexus in cities is one of the greatest resource challenges of our time, but one that can also generate relevant alternatives with which to tackle climate change. The concepts of governance and nexus reflect important challenges in an increasingly interconnected and rapidly changing world. This requires a broad-based investigation of the relationships between disciplines and methods and the co-creation of solutions between different actors in society.

Thus, we conceptualize the dialogue between the nexus and climate change in urban areas, emphasizing its socio-political dimensions while discussing the role of cities in the contemporary global environmental crises. As a result, this linkage is characterized as a necessary perspective for pursuing the Sustainable Development Goals (SDGs), given the recognition of the importance of interconnectivity between traditionally isolated sectors. Furthermore, nexus approaches cannot be restricted to isolated territorial boundaries; it is necessary to understand and act from a territorial perspective in the analysis of open systems, in which cities, due to their disproportionate uses of resources, form the center of the debate concerning sustainability and planetary limits. In addition, socio-political dimensions are considered to be a critical element in the nexus approach used within this book. The linkage is not limited to technical approaches so that the nexus perspective refers to the engagement of different institutions and social actors, interconnecting existing governance structures and seeking effective contributions to address non-sustainability and resource vulnerabilities.

The book presents discussions and case studies covering the range of current international concerns relating to sustainable urban development, the nexus, and climate change dialogue. Figure 1 shows the more prominent words in the textual content of the chapters, while Fig. 2 indicates the most relevant book content topics by frequency, demonstrating the strength of their relationships within the texts of the various chapters. The emphasis of these relevant themes reinforces the correspondence of the entire book's content with its objectives.

The book targets policy experts, academics, and researchers (including PhD and master's students of courses in public policy, development planning, urban and regional studies, public health, environment, climate change, energy studies, environmental planning/engineering, and transportation). We organized this book with the intention of contributing to global discussions concerning how cities can represent possibilities to navigate and manage sustainability from local to global. We believe that the wealth of knowledge presented in these studies will also inspire teachers of undergraduate and graduate courses to use *WEF Nexus and Climate Change in Cities* as a topical and informative classroom textbook.

São Paulo, Brazil

Lira Luz Benites Lazaro
Leandro Luiz Giatti
Laura S. Valente de Macedo
Jose A. Puppim de Oliveira

Fig. 1 Cloud of the 200 most frequent words based on chapter contents

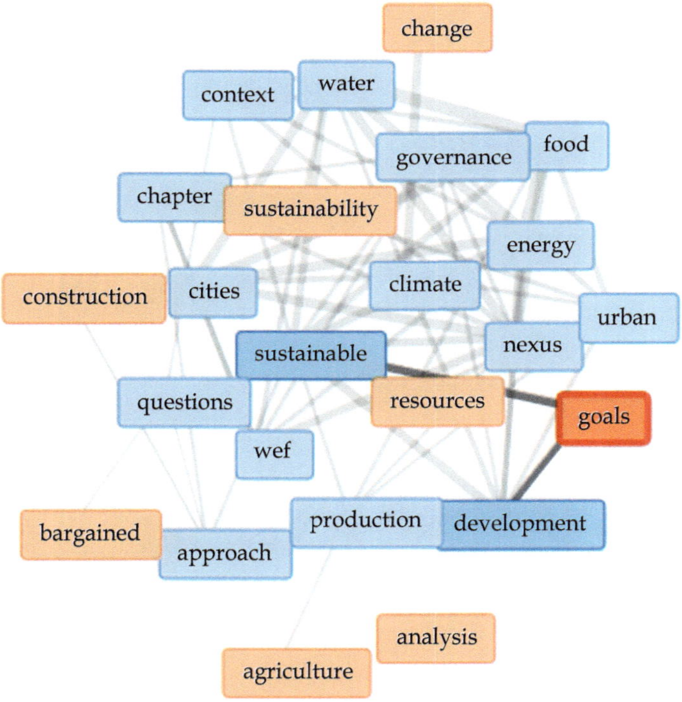

Fig. 2 Links of the most frequent topics based on chapter contents

Acknowledgments

Lira Luz Benites Lazaro acknowledges the financial support received from São Paulo Research Foundation (FAPESP), Grant: 2017/17796-3, and Grant: 2019/24479-0.

Leandro Luiz Giatti acknowledges funding by the National Council for Scientific and Technological Development (CNPq – proc. n. 309840/2018-0) and by the São Paulo Research Foundation (FAPESP – proc. n. 2015/03804-9).

Laura S. Valente de Macedo would like to thanks the São Paulo Research Foundation (FAPESP), Brazil for funding her research through grant number 2018/20714-1.

Jose A. Puppim de Oliveira thanks the support of the Belmont Forum (grant number NEXUS2016:152); the JPI Urban Europe (grant number 11221480); the São Paulo Research Foundation (FAPESP) (grant number 2017/50425-9); Coordination for the Improvement of Higher Education Personnel (CAPES) (grant number 88881.310380/2018-01); and National Council for Scientific and Technological Development (CNPq) (grant number 442472/2020).

Contents

About the Editors

Lira Luz Benites Lazaro is a postdoctoral fellow at the Department of Environmental Health, School of Public Health, University of São Paulo, Brazil. Lira is visiting researcher at Durham Energy Institute and the Department of Anthropology at Durham University, UK. She holds a master's and PhD in Latin America integration from the University of São Paulo, Brazil, and a second PhD in Earth system science at the National Institute for Space Research – INPE, Brazil. Lira develops research on the political and social dimensions of climate change, responses of social actors regarding climate change, corporate social responsibility, climate policy analysis, energy policy analysis, water-energy-food nexus security, and governance. Her areas of research include social sciences and interdisciplinary perspectives. Her most recent research projects examined the governance of water-energy-food nexus with a focus on biofuels.

Leandro Luiz Giatti is an associate professor in the Department of Environmental Health, School of Public Health, University of São Paulo (USP), Brazil. He has international expertise in sustainability and health, health promotion, participatory research approaches, and urban water-energy-food nexus. His experience is characterized through interdisciplinary studies and with the main efforts dedicated to socioenvironmental vulnerable groups, also with studies in the Brazilian Amazon Region. Giatti is a member of the Institute of Advanced Studies at USP and associate editor of the journal *Ambiente & Sociedade* (SCIELO – Brazil). During his academic research projects, he has advised theses and dissertations of graduate students and has also supervised postdoctoral partners.

Laura S. Valente de Macedo is a research fellow at the Global Cities Center of the Institute of Advanced Studies of the University of São Paulo (IEA-USP). She is also a fellow at the Center for Studies on Infrastructure and Environmental Solutions (CEISA) of Fundação Getúlio Vargas (FGV-EAESP) in Brazil, where she also undertook postdoctoral research on urban green and blue infrastructure and the food-water-energy nexus. With an extensive career in environmental policies, she has held senior advisory or executive positions in both state and federal government departments since 1991. From 2002 to 2011, she was ICLEI's regional director for Latin America, leading projects in a wide range of sustainability policies and management issues, such as climate change mitigation, green public procurement, and urban resilience. Ms. Macedo also lectured on sustainable consumption, cities, and climate change and collaborated as co-author and editor in numerous publications. She has a PhD and an MSc in environmental science from the University of São Paulo (USP), and an MSc in environmental change from the University of Oxford as a Chevening scholar. Before her career in the environmental field, she was a practicing architect in São Paulo.

Jose A. Puppim de Oliveira is a faculty member at Fundação Getulio Vargas (FGV/EAESP and FGV/EBAPE) in Brazil and a Visiting Fudan Chair Professor at the Institute for Global Public Policy. He has extensive experience in interdisciplinary research projects, particularly involving the developing world. He has over 20 years of experience in universities, businesses, think tanks, and governmental and international agencies in research, teaching, training, and academic management. His research and policy interests concentrate on themes around sustainable development, studying the patterns of governance, institution building, and policy implementation at different levels, and looking at how global and national institutions are interlinked to governance and action. He holds a PhD in planning from the Massachusetts Institute of Technology, USA.

Contributors

Lincoln M. Alves Climate researcher at the National Institute for Space Research (INPE). He is responsible for providing technical and scientific information quality to guide public policies for adaptation to regional environmental changes. His areas of expertise include Earth System Science and Environmental Sciences, acting on the following themes: atmospheric sciences, climate, meteorology, Amazon and climate modeling, biosphere-atmosphere interaction, climate change. He is the lead author of Atlas of the IPCC Sixth Assessment Report (AR6-WG1).

Paulo Nocera Alves Junior is an assistant professor at the Escuela de Ingeniería de Coquimbo (EIC), at the Universidad Católica del Norte (UCN), Chile. He holds a Ph.D. degree in Industrial Engineering from the University of São Paulo (USP/EESC), a Master's degree in Industrial Engineering from the São Paulo State University (UNESP/FEB), and a Bachelor's degree in Industrial Engineering from the University of Franca (UNIFRAN). He concluded a postdoctoral project at the USP/ESALQ/ESALQ-LOG, about the evaluation of public policies for the development of green corridors to Agro-Logistics in Brazil. His research interests lie in the following subjects: inventory control, logistics, supply chain, sustainability, efficiency analysis, operations research, data envelopment analysis, optimal control theory, multivariate analysis, and quantitative methods.

Mateus Henrique Amaral is a biologist with a master's degree in science, acting during graduation; masters in the areas of urban sustainability and health information systems; and currently, a PhD student in the Department of Environmental Health at the School of Public Health of the University of São Paulo, with the main line of work being the environmental justice aspects related to the food systems.

Gustavo F. B. Arcoverde is a geographer from Brasília University (UnB) and has a master's and a doctorate's degree in Remote Sensing from INPE. He is a public employee in the National Institute for Space Research (INPE) and has worked with indicators of sustainability and climate impacts, hydro-vegetational resilience, and modeling of land use. He is currently part of the coordination of the Laboratory of Analysis and Indicators of Sustainability and is formally working on the following projects: (1) Transition to sustainability and the water-agriculture-energy nexus: Exploring an integrative approach with cases of study in the Cerrado and Caatinga biomes; (2) Water Resources in the Paraíba do Sul Basin: Integrating natural and anthropic aspects; (3) Information and Analysis System on the Impacts of Climate Change – AdaptaBrasil –MCTI; and (4) After the Installation of Hydroelectric Power Plants: Social and Environmental Processes that occur after the construction of Belo Monte, Jirau, and Santo Antônio in the Brazilian Amazon.

Daniela Bacchi Bartholomeu holds a Ph.D. in Applied Economics and a Bachelor's degree in Economics, both from the University of São Paulo, Luiz de Queiroz College of Agriculture (USP/ESALQ). She is a researcher and collaborator on the Group of Research and Extension in Agroindustrial Logistics (ESALQ-LOG). She worked as a researcher at Research Support Center (NAP), Innovation Center in Sustainable Logistics (CALS), and Center for Advanced Studies in Applied Economics (CEPEA). Between 2011 and 2012, she was a professor in USP/ESALQ. She has experience in the areas of transport economics, with emphasis on low carbon logistics, agroindustrial products, solid waste, sustainability, and GHG emissions, among others.

Marcos Rehder Batista is a PhD candidate in Economic Development at the Institute of Economics (UNICAMP); a researcher at the Center for Applied, Agricultural and Environmental Economics at the same Institute; a collaborator in the Waste Food-Energy-Water Urban Living Lab (SP in Natura Lab): Mapping and Reducing Waste in the Food-Energy-Water Nexus; and visiting researcher at the Center for Studies in Public Government Policies (GVceapg) at the São Paulo School of Business Administration of Fundação Getúlio Vargas (FGV EAESP). A Bachelor in Social Sciences and Master in Sociology from Unicamp, he is currently researching governance in state and non-state public policies for sustainable development and innovation economics.

Rodrigo A. Bellezoni is a postdoctoral researcher at the São Paulo Management School (FGV). He investigated the effects of nature-based solutions on the food-water-energy nexus in cities, which inspired him to write this chapter. Having over 10 years of work experience within energy and environmental topics in leading organizations, Rodrigo has developed strong expertise in assessing and reporting in diverse range of industry sectors, mainly energy and agriculture. He has conducted water-energy-food-carbon nexus research in biofuel production chains during his PhD in Energy Planning (UTS and UFRJ). Rodrigo's expertise also includes circular economy, waste management, environmental and energy planning, renewable energy, energy modeling, land-use change, green infrastructure, ecosystem services, carbon accounting, and climate change.

Ana Maria Bertolini is a nutritionist, currently developing a PhD in Global Health and Sustainability at the Department of Environmental Health, School of Public Health of the University of São Paulo (USP), Brazil.

Susanne Börner is currently a Marie Curie Global Fellow at the School of Geography, Earth, and Environmental Sciences at the University of Birmingham in collaboration with the School of Public Health at the University of Sao Paulo.

Evandro A. Branco is a researcher at the Brazilian National Institute for Space Research (INPE), in the Adaptation and Vulnerability Division (DIIAV); co-leader of the Sustainability Indicators Analysis and Development Lab (LADIS); co-leader of the Interinstitutional Research Group on Environmental Planning and Management (PLANGEA); and co-founder and member of the Executive Committee of the Brazilian National Public Policy Articulation for Environmental Education (ANPPEA). His research interests include the science-policy interaction, environmental public policy analysis, adaptive capacity, environmental education, social-environmental science epistemology, and knowledge co-production.

José Eduardo Holler Branco holds a Ph.D. in Transportation Engineering from the University of São Paulo (São Carlos School of Engineering), Master's at Applied Economics from University of São Paulo (Luiz de Queiroz College of Agriculture), and Bachelor's at Agriculture from University of São Paulo (Luiz de Queiroz College of Agriculture). Concluded a postdoctorate about policies for cutting GHG emissions of the Brazilian freight transportation system. Currently is Analyst at Mobility and Transport Department of Piracicaba-SP City Hall and Researcher Associate to Group of Research and Extension in Agro-industrial Logistics (ESALQ-LOG). He has over 10 years of experience in planning, mathematical modeling, and optimization of logistics systems, mainly in agro-industrial supply chains, and worked in more than 30 research and extension projects in the area of transport planning and logistics.

Luciana Londero Brandli holds a degree in Civil Engineering from the Federal University of Santa Maria (1995), a Master's degree in Civil Engineering from the Federal University of Santa Catarina (1998), a Ph.D. in Production Engineering from the Federal University of Santa Catarina (2004), and a Post-Doctorate at the Hamburg University of Applied Sciences (2014). Luciana was Deputy Editor of the *Encyclopedia of Sustainability and Higher Education* and Deputy Editor of the *Encyclopedia of Sustainable Development Goals: Transforming the World We Want*. She serves on the editorial board of several journals; supervises several master's and doctoral students on engineering, environment, and sustainability issues; and has over 300 publications, including books, book chapters, and articles in said journals. His current research interests include sustainability in higher education, the 2030 Agenda for Sustainable Development, climate change, education for sustainability, and sustainable cities. She is currently carrying out the international project "Transforming Universities for the Climate Change," involving the UK, Kenya, Mozambique, and Fiji Island. She is a productivity fellow at CNPQ.

José Vicente Caixeta Filho is a full professor at the Department of Economics, Management, and Sociology of the College of Agriculture "Luiz de Queiroz" - ESALQ, USP (since 1989). He is a Bachelor in Civil Engineering by Polytechnic School, University of São Paulo - USP (1984); Master in Economics by University of New England, Australia (1989); Doctorate in Engineering by Polytechnic School, USP (1993); Creator and Coordinator of the Research and Extension Group in Agribusiness Logistics - ESALQ-LOG (since 2003); and Secretary of Urban Mobility, Traffic, and Transportation of Piracicaba, municipality at the State of São Paulo (2021). His main field of work is logistics (transportation and storage issues, mostly), with the use of operations research models (including mathematical programming).

Clarice Araújo Carvalho has bachelor's and master's degrees in Administration at the Federal University of Bahia (UFBA). Currently, she is a PhD student in the same university. Her master's degree focused on the governance of urban gardens and the strengthening of urban resilience. She started her PhD looking for the connection between food systems, community initiatives, and resilience. She is a FAPESB (Fundação de Amparo à Pesquisa do Estado da Bahia) scholarship holder and member of the research group Governance for Sustainability and Low Carbon Management (GpS) and of the Thematic Chamber of Urban Resilience, initially linked to the City of Salvador for the elaboration of the Plan for Mitigation and Adaptation to Climate Change (PMAMC).

José Cesar Cruz Junior is an associate professor at the Department of Economics, at the Federal University of São Carlos, Brazil. He holds a degree in Economic Sciences from the Federal University of Viçosa (2002), a Master's in Applied Economics from the Federal University of Viçosa (2004), and a PhD in Applied Economics from the University of São Paulo (2009), and he was a visiting professor at the University of Illinois at Urbana-Champaign (2016–2017). He has experience in the field of economics, and his main research interests include risk management in agricultural and financial markets, financial and behavioral economics, agricultural derivatives, and energy markets.

Maria Ester Soares Dal Poz is a Business and Economics Professor at University of Campinas, UNICAMP, Brazil. For the last 25 years, she has been an economics of innovation teacher and researcher, working with research and development market dynamics, especially at bioenergy and pharma market development frontiers. She has been one of the Board of Directors of UNICAMP's innovation agency (INOVA), for 5 years. She is a consultant for national and international institutions, concerning innovation public and private funding systems and its spin offs. From the last 5 years, she has been coordinating sustainable innovation international joint projects, as the Belmont Forum Sustainable Urban Global Initiative, establishing the São Paulo Urban Living Lab (SP *in Natura* Lab) for the State of São Paulo, Brazil. From this year of 2022, she is an UNICAMP's dean office counsellor for innovation funding.

Michele Dalla Fontana is a postdoctoral research fellow at the Environmental Health Department, School of Public Health, University of São Paulo (USP – Brazil). He is a PhD in Regional Planning and Public Policies at the Università IUAV di Venezia. His research areas include climate change adaptation strategies, natural resource management, water-energy-food nexus, systems thinking, and transdisciplinary approaches.

Paulo Antonio de Almeida Sinisgalli is an associate professor 3 at the University of São Paulo. He holds a degree in Civil and Sanitary Engineering from the Instituto Mauá de Tecnologia (1984), a master's degree in Environmental Science from the University of São Paulo (1997), a doctorate in Applied Economics from the State University of Campinas (2005), and a full professorship from the University of São Paulo (2018). He holds a postgraduate degree in Environmental Management from the Technical University of Dresden – Germany (1989/1990) – and was a visiting student at the University of Kent at Canterbury in England (1999/2000). He is accredited in the postgraduate programs at USP: Environmental Science and Modeling of Complex Systems. He has postdoctorate studies at the University of Wageningen in the Netherlands in Participatory Water Resource Management. He works in the areas of ecological economics and water resources management.

José Baltazar Salgueirinho Osório de Andrade Guerra is permanent professor and researcher at the Graduate Programs (Master and Ph.D.) in Administration and Master in Environmental Sciences, at the Universidade do Sul de Santa Catarina (UNISUL, Brazil); fellow of the Cambridge Center for Energy, Environment and Natural Resource Governance (C-EENRG), Department of Land Economy, University of Cambridge, Cambridge, UK; founder of the Center for Sustainable Development/Research Group on Energy Efficiency and Sustainability (Greens), UNISUL; Honorary Citizen of Florianópolis, a title awarded by the City Council of Florianópolis; Citizen of Santa Catarina, a title awarded by the Legislative Assembly of the State of Santa Catarina; and author of 64 scientific articles, 21 books, and 73 book chapters. He is a productivity fellow at the National Council for Scientific and Technological Development (CNPq), Brazil.

Fabiano de Araújo Moreira is a postdoctoral research fellow at the Environmental Health Department, School of Public Health, University of São Paulo (USP – Brazil). He is a PhD in Geography (Environmental Analysis and Territorial Dynamics) at the University of Campinas. His research areas include climate change, adaptation, natural disasters, environment, sustainability, urban development, transdisciplinarity, and water-energy-food nexus.

Carolina Monteiro de Carvalho is a participatory mapping researcher and practitioner at Comunidades Vivas - Mapeamento Participativo and researcher at the Institute of Energy and Environment, University of Sao Paulo.

Ruy de Castro Sobrosa Neto is a Ph.D. student in Business Administration in the area of Innovation and Society (UNISUL); Master in Business Administration in the area of Strategy and Performance (UNISUL), Electrical Engineer and Production Engineer (UFSC), and Specialist in Occupational Safety Engineering (UNISUL); and R&D+I Manager at CGT ELETROSUL, where he has worked as an engineer since 2005. His main publications and research interests are topics related to the energy sector; sustainability and its three dimensions, the water, energy, and food nexus; and industry 4.0. He is a member of the Permanent Ethics Committee of CGT ELETROSUL. He was President of CIPA Headquarters of ELETROSUL in 2010/2011 Management and Member of the Administrative Council of Cepel (Electric Energy Research Center) as representative of ELETROSUL between 2019 and 2022. E-mail: ruydecastrosneto@gmail.com.

Thainara Granero de Melo is Professor of Psychology at the School of Life Sciences, Pontifical Catholic University of Paraná (PUCPR), Brazil, and holds a PhD in Psychology. In the field of Work and Critical Social Psychology, her research areas include rural work, subjectivity, groups, community organizing, development policies, and socio-environmental justice.

Felipe Teixeira Dias is a Master's student at the Graduate Program in Environmental Sciences (technology, environment, and society) at the Universidade do Sul de Santa Catarina (UNISUL), as a CAPES scholarship holder. He is a postgraduate student in Urban Planning and Environmental Management of Cities at Universidade Castelo Branco (UCB); graduated in Law from Centro Universitário FG (UniFG); a researcher at the Center for Sustainable Development/Research Group on Energy Efficiency and Sustainability (GREENS); collaborating researcher at the UniFG Observatory of the Northeastern Semiarid (Research Center on Law to the City and at the Geoprocessing Laboratory (LabGEO); and Editorial Assistant of the Journal of Environmental Management and Sustainability at UNISUL. He is a member of the collegiate and self-assessment and planning committee of the Master's in Environmental Sciences (PPGCA/UNISUL). He has experience in urban and environmental law, real rights and social rights. He works mainly on the following topics: (1) urban policy and the right to sustainable cities; (2) public policies and sustainable development; (3) urban planning and contemporary space production; and (4) geoprocessing applied to environmental, urban, social, and economic analyses. He was a special student in the discipline Law to the City and Interdisciplinary Perspective, in the Postgraduate Program in Social Development (Master's and Doctorate) at the State University of Montes Claros (UNIMONTES). He is a Technologist in Business and Real Estate Transactions.

Gabriela Marques Di Giulio is an associate professor at the Environmental Health Department, School of Public Health, University of São Paulo (USP – Brazil), and a visiting researcher at the Department of Politics, University of York. She is a PhD in Environment and Society. Her research areas include human dimension of climate change; crises, risks, and uncertainties; science-policy-communication interactions; and global health and sustainability.

Ísis Domingues is a PhD candidate in Business Management at the Fundação Getúlio Vargas/EAESP, Brazil, focusing on management of sustainable operations. She holds an MBA from Manchester University; during this time, she studied and researched food security and multistakeholder initiatives in that area. Isis has a master's and bachelor's in Accounting and developed studies in corporate governance and controllership. She also has more than ten years of professional experience in multinational companies of different industries.

Antonio Embid is PhD (University of Zaragoza, Spain, 1977). Professor of Administrative Law at the University of Zaragoza. Scholar of the Juan March (1980) and Alexander von Humboldt Foundations (1981 and 1982). Specialist in water and environmental law, territorial organization of the State, and public rights and freedoms. Ha has directed 23 doctoral theses. Doctor honoris causa by several Argentine universities. He has developed academic and professional activities in numerous countries in Latin America and the Caribbean.

Érika Cristina Francisco has a Degree in Technology of Environmental Sanitation and a Master's and PhD in Chemical Engineering (microalgae biotechnology and biorefinery research areas), from the University of Campinas (UNICAMP). She has worked as a professor in the Department of Environmental Engineering at the University of Passo Fundo and as a consultant at environmental licensing projects. She has a recent postdoctoral degree in Management and Sustainability at the School of Applied Sciences, Unicamp. She is an expert in environmental complex systems analysis by system dynamics and agent-based modelling tools, based on socioeconomic and environmental indicators.

Larissa Miranda Heinisch is a climate change consultant at the Arup's International Development Group based in London. She is trained in architecture and urbanism and holds an MSc City Design and Social Science degree from the London School of Economics Cities Programme. She has experience with projects and studies that promote resilience, sustainability, and nature-based solutions for urban and rural human settlements across Latin America, Asia, Africa, and Europe. Larissa has been engaged with the work of institutions committed to tackling global challenges such as ICLEI – Local Governments for Sustainability – at its South American Secretariat, where she coordinated regional projects and initiatives addressing nature-based solutions and ecosystem services and mitigation and adaptation to climate change. She has also worked with the Article 25 (guidelines for rural health centers in multiple locations in Myanmar), Architecture for Humanity

(Charting Displacement in London project), and UN-Habitat (City Resilience Profiling Programme).

K. M. Nazmul Islam is working as quantitative sustainability scientist at the Commonwealth Scientific and Industrial Research Organisation (CSIRO). His research interests include low-carbon food system, food-water-energy-greenhouse gas emission nexus, climate change mitigation, sustainable food system for cities, urban metabolism, environmental accounting for agriculture, and climate smart food system. His PhD research is focused on water-energy-greenhouse gas emission nexus of food system in Australian cities. As an environmental science and industrial ecology graduate, he has academic, research, and professional experience of more than ten years on strategic and applied research focusing on environmental footprint analysis, greenhouse gas (GHG) inventory, life cycle sustainability assessment (LCSA), material flow analysis (MFA), input-output analysis (IOA), and urban metabolism. He has previously worked in the Food and Agriculture Organization (FAO) of the United Nations, United Nations Development Programme (UNDP), United Nations Office for Project Services (UNOPS), University of Chittagong (Public University), and Stamford University Bangladesh (Private University).

Pedro Roberto Jacobi is a senior full professor at the graduate program of Environmental Science at the Institute of Energy and Environment/University of São Paulo (USP) (Brazil). He is a coordinator of the Research Group on Environmental Governance and of FAPESP project (2017–2022) on the Environmental Governance in Macrometropolitan São Paulo facing climate change and a coordinator of Research Group on Environmental Governance and Research Fellow at the Institute of Energy and Environment at University of São Paulo. He is a research Fellow at the INCLINE/USP Interdisciplinary Climate Investigation Center. He is a member of Regional Network on Climate Change and Decision Making/Unitwin and Coordinator of USP Research Group. Editor of the journal *Ambiente & Sociedade/Anppas/*Brazil. He is a president of the board of the Local Governments for Sustainability – ICLEI – South America.

Steven J. Kenway is a water leader with senior experience in research, industry, and government. He has led water research since 1990 with the University of Queensland, CSIRO, Brisbane Water, Sydney Water, and Private Consulting. He has worked with urban water, wastewater, stormwater, and related energy and greenhouse gas issues. His work addresses urban water security, net-zero carbon water cycle, and circular economy. He creates collaborations, tools, models, and knowledge to address all flows of water – and related energy – into, out of, and within cities. This enables evaluation and management of key concepts such as integrated system performance and sustainable urban design and planning. Steven's work is enhancing performance benchmarking of cities, shaping development, and guiding policy and infrastructure investment locally and internationally.

Tadeu Fabrício Malheiros is a civil and environmental engineer and PhD in Public Health. Currently, Malheiros is an associate professor at the São Carlos School of Engineering at the University of São Paulo and a coordinator of the Postgraduate Professional Master's Program for Teaching Environmental Sciences. Malheiros works mainly on the following topics: sustainability indicators, sustainable development, environment, public health, and solid waste and environmental management.

Liber Martin is PhD (University of Zaragoza, Spain, 2009). Professor of Public and Water Law at the University of Cuyo/Mendoza and Researcher at the National Council of Scientific and Technological Research (CONICET), Argentina. Former visiting Scholar at Université Paris I Panthéon Sorbonne (2011), University of Arizona (2012), and University of California, Berkeley Law (2015). He authored many books, journal articles, and papers on comparative water law and policy, water rights, water public utilities, natural resources, and environmental law and international water law.

Daniele Eckert Matzembacher holds a PhD in Management from the Federal University of Rio Grande do Sul (UFRGS), Brazil, focusing on innovation and sustainability. She was a visiting researcher at the University of Tartu (Estonia) and University of Guelph (Canada), developing collaboration to identify solutions focused on sustainability in food systems. She holds a master's and a bachelor's in Business Administration at UFRGS. She also has over 10 years of experience in the market, working in the public and private sectors in the management area, project coordination, and team management.

Nathália Nascimento is an Earth System Scientist expert in land use dynamics and natural systems modeling. Her research is focused on understanding and modeling the different forms of interaction between man and nature, including social, political, and cultural aspects as a key element of landscape transformation. As a researcher passionate about nature and people, she has extensive field experience in the Amazon, Atlantic Forest, Cerrado, and Gran Chaco biomes. She has recently expanded her studies to the understanding of different types of ecosystems in South America, in particular, the different types of forests and their provision of ecosystem services essential to human well-being. Given the evidence and impacts of climate change on human livelihood, especially in the Amazon region, she has also been dedicated to studying mitigation and adaptation strategies for the most vulnerable human groups in the Brazilian Amazon and Cerrado, Brazil.

Alex Oche is a lecturer in law at the College of Law and an associate research fellow at the Institute for Oil, Gas, Environment and Sustainable Development (OGEES Institute), Afe Babalola University, Ado Ekiti, Nigeria.

Damilola S. Olawuyi SAN, FCIArb., is an associate professor of energy and environmental law at the Hamad Bin Khalifa University (HBKU) College of Law, Doha, Qatar. He is also a chancellor's fellow and director of the

Institute for Oil, Gas, Energy, Environment and Sustainable Development (OGEES Institute), Afe Babalola University, Ado Ekiti, Nigeria. As prolific and highly regarded scholar, his most recent book publications include *Environmental Law in Arab States* (Oxford University Press, 2022), *Local Content and Sustainable Development in Global Energy Markets* (Cambridge University Press, 2021), *The Human Rights Based Approach to Carbon Finance* (Cambridge University Press, 2018), *Extractives Industry Law in Africa* (Springer, 2018) and *Climate Change Law and Policy in the Middle East and North Africa Region* (Routledge, 2021).

Jean P. H. B. Ometto Senior Researcher at the Brazilian Institute of Space Research and Head of Strategic Projects Division (DIPE3) and CNPq Fellow. His academic and research activities are centered on the diagnosis and prognosis of anthropogenic activities and climatic changes, in natural biogeochemical processes, land use and land cover, greenhouse gas emissions, and sustainability indicators, including mitigation and adaptation options. He was appointed as IPCC contributor to WG-II – Impacts, Adaptation, and Vulnerability (for AR5 and AR6 cycles) – and to the IPCC Task Force on GHG Emission Inventories; Coordinate Lead Author on IPBES on Regional Assessment; Member of the American Geophysical Union's Committee on International Participation; Member of the Coordination of the FAPESP Research Programme on Global Climate Change; Member of the Belmont Forum Steering Committee; Vice-Chair of the Brazilian Research Network on Global Climate Change; and Member of the Scientific Steering Committee of the Global Carbon Project.

Sophia B. N. Picarelli is an environmental manager, with a master's degree in Environmental Technology from the Technological Research Institute of the State of São Paulo. She is currently conducting doctoral research focused on nature-based solutions and the potential to climate change adaptation in Brazilian cities, at the Environmental Science Program, University of São Paulo. With more than 15 years of experience in the environmental agenda, working in the public, private, and third sector, she has worked mainly on topics related to biodiversity and climate change. She joined ICLEI South America's team for ten years, concluding this cycle as the regional manager for the Biodiversity and Circular Development. Currently, she is a Climate Solutions director at the Conservation International – Brazil.

Alessandro Luis Piolli is a collaborative researcher at the School of Applied Sciences, in the University of Campinas (UNICAMP), working at the Belmont Forum project "Waste Food-Energy-Water Urban Living Lab", SP *in Natura* Lab. He has a Master's and a PhD in Scientific and Technological Policy at UNICAMP. He is graduated in biological sciences at UNICAMP and post-graduated (lato sensu specialization) in scientific journalism at the Laboratory of Advanced Studies in Journalism (Labjor UNICAMP). For 10 years, he was an associate researcher at Labjor and, for 6 years, has coordinated the project of the Public Centre for Social Technology of Campinas, from the Innovation Policy Analysis Group (GAPI-Unicamp). He has worked as an invited professor in the postgraduate courses at the same university (Scientific and Technological Policy Program) and Fundação Oswaldo Cruz, Fiocruz (Postgraduate Program in Health Information and Communication).

Suzana Más Rosa is currently a postdoctoral fellow at the School of Business Administration at the Federal University of Bahia (UFBA). She holds a bachelor's degree in Veterinary Medicine from the University of Marília (UNIMAR), a master's in Geochemistry and the Environment, and a PhD in Analytical Chemistry from UFBA. She participates in the Research Group on Global Environmental Governance and the Carbon Market (GpS) and in the Urban Resilience and Innovation for Sustainability Thematic Chambers at the Salvador Panel on Climate Change, where she coordinates the Thematic Chamber on Waste.

Stephane Louise Boca Santa is a Ph.D. student in Business Administration at the University of Southern Santa Catarina (UNISUL) and master in Accounting and Bachelor in Accounting Sciences from the Federal University of Santa Catarina (UFSC). She is currently part of the Research Group on Energy Efficiency and Sustainability (GREENS) at the University of Southern Santa Catarina (UNISUL).

Patrícia Marra Sepe is a geologist, master in Geosciences and Environment, and PhD in Geography. She is currently a geologist at the Municipal Department of Urban Development of the City of São Paulo. She has experience in the area of urban environment and urban and regional planning, with emphasis on environmental planning, acting mainly on the following themes: land use and master plan, environmental indicators, instruments to encourage environmental preservation, green infrastructure, agro-ecology and rural areas, and climate change in cities.

Karen C. Seto is the Frederick C. Hixon Professor of Geography and Urbanization Science. An urban and land change scientist, she is one of the world's leading experts on contemporary urbanization and global change. Her research focus is how urbanization will affect the planet. She has pioneered methods to reconstruct urban land use with satellite imagery and has developed novel methods to forecast urban expansion. She has conducted urbanization research in China for 20 years and in India for more than 10. Her research has generated insights on the links between urbanization and land use, food systems, biodiversity, and climate change. Seto is an elected member of the US National Academy of Sciences, the Connecticut Academy of Science and Engineering, and the American Association for the Advancement of Science. She earned a PhD in Geography from Boston University.

Thiago Coelho Soares graduated in Business Administration from Universidade do Sul de Santa Catarina (2004) and holds a Master's degree in Business Administration from Universidade Federal de Santa Catarina (2006) and PhD in Business Administration from Universidade Federal de Santa Catarina (2016). He is currently a professor in the Administration graduate program at the Universidade do Sul de Santa Catarina. He has experience in the area of Administration, with an emphasis on Business Administration, working mainly on the following topics: business costs, organizational performance evaluation, and evaluation of the sustainability of HEIs.

Peter Toledo Peter Mann de Toledo holds a degree in Biological Sciences from the Federal University of Paraná, a master's degree in Geosciences from the Federal University of Rio Grande do Sul, and a Ph.D. in Geology from the University of Colorado at Boulder. He is currently a senior researcher at the National Institute for Space Research (INPE/MCTI). He works at the division of impact, adaptation, and social and environmental vulnerabilities. He has experience in Paleozoology and Paleoecology and in Environmental Sciences, working mainly on the following topics: public policies in Science and Technology in the Amazon, ecosystem evolution, and paleontology and paleoecology of South America. He is an advisor professor of the Master's and Doctoral Course in Environmental Sciences at the Federal University of Pará, Brazil. He is currently developing interdisciplinary research on the Amazon and neotropical biodiversity. He is a member of the Management Board of the AdaptaBrasil MCTI Platform at risk of climate change impacts.

Alberto Matenhauer Urbinatti is a sociologist with a Master's degree in Sociology from the University of Campinas. He holds a PhD in Public Health from the University of São Paulo. He is a postdoctoral researcher at the Department of Science and Technology Policy of the University of Campinas.

Andréa Cardoso Ventura is an adjunct professor at the School of Administration, Federal University of Bahia (UFBA), where she coordinates the Research Group on Governance for Sustainability and Low Carbon Management (GpS), with research focused on innovations and solutions for sustainable development. She holds a master's degree in Business Administration from the same university. She also has a master's degree in Latin American Contemporary Studies from the Universidad Complutense de Madrid – UCM (2009) – and specialized in marketing at the UNIFACS (2003). She graduated in Social Communication/Public Relations (UNESP, 1999) and in Law (ITE, 1996). She is a fellow researcher at the UNESCO Chair in Sustainability and at the Salvador Panel on Climate Change (GpS), where she coordinates the thematic chambers on the Urban Resilience and Innovation for Sustainability.

Luciana M. Vieira is a professor of Management of Sustainable Operations at the Fundação Getulio Vargas/EAESP, Brazil. She holds a PhD in Agricultural Economics from the University of Reading (UK). Luciana is a visiting fellow at the King's College London and Institute of Development Studies (both in the UK). Her research interests are global value chains, multistakeholder initiatives, and agri-food chains.

1 Water-Energy-Food Nexus in Cities: Opportunities for Innovations to Achieve Sustainable Development Goals in the Face of Climate Change

Lira Luz Benites Lazaro, Leandro Luiz Giatti, Laura S. Valente de Macedo, and Jose A. Puppim de Oliveira

Abstract

The urban water-energy-food (WEF) nexus approach in cities represents a pathway for coping with trade-offs in the search for achieving the Sustainable Development Goals (SDGs). The scientific literature on the WEF nexus has grown enormously since 2012. Recently, it has become more diversified with the evolution of new topics and expanded scope, demonstrating the inherent complexity associated with nexus thinking and placing this methodology at the science-0policy-society interface. Cities are central to the sustainability agenda and have been at the core of plans and strategies implementation since United Nations Conference on Environment and Development (UNCED 92). Thus, applying the nexus in urban contexts allows the exploration of local complexities and uncertainties and the engagement of different social actors to produce actions that transcend scales and dialogue with global concerns, such as climate change. Unquestionably, the urban nexus stimulates multilevel and intersectoral governance, contributing to coping with the challenges and contradictions of the SDGs. Innovations, understood in a broad sense as doing things differently, are essential to moving the WEF nexus from theory into practice. Here we explore nexus discussions via cases involving urban-rural relationships, circular economy, institutional perspectives, logistics, urban food production, and food waste reduction and analyze the consequences for urban climate mitigation and adaptation.

L. L. B. Lazaro (✉)
Department of Environmental Health, School of Public Health, University of São Paulo, São Paulo, Brazil

Durham Energy Institute and Department of Anthropology, Durham University, Durham, UK
e-mail: lbenites@usp.br

L. L. Giatti
Department of Environmental Health, School of Public Health, University of São Paulo, São Paulo, Brazil

L. S. Valente de Macedo
Fundação Getulio Vargas (FGV), São Paulo School of Management (FGV EAESP), São Paulo, SP, Brazil

Center for Global Cities, Institute for Advanced Studies, University of São Paulo (IEA-USP), São Paulo, SP, Brazil

J. A. Puppim de Oliveira
Fundação Getulio Vargas, FGV EAESP and FGV EBAPE, São Paulo, Brazil

L. L. B. Lazaro et al. (eds.), *Water-Energy-Food Nexus and Climate Change in Cities*, Sustainable Development Goals Series, https://doi.org/10.1007/978-3-031-05472-3_1

Keywords

Water-energy-food nexus · Urban nexus · Sustainable Development Goals · Science-policy-society interface · Multilevel governance · Nexus innovations

1.1 Introduction

Urbanization and its links with global environmental challenges have made the water-energy-food (WEF) nexus approach crucial in natural resources management, policy, and governance. Responding to the unsustainability challenges ahead becomes ever-more urgent and will require changes in the way societies manage and consume their resources. Decision-making based on classical management theories still emphasizes forecasting, planning, organizing, and controlling in separate sectors. However, this management model no longer provides the solutions required for "wicked problems" inherently systemic and often structural. Such problems demand new multidisciplinary and inter−/multi-sectoral approaches to address complex, interconnected, and interdependent relational challenges, besides different ways of doing business and managing finite resources (Lazaro et al., 2021a). At the same time, a global agenda that includes implementing the Paris Agreement and achieving the Sustainable Development Goals (SDGs) emphasizes the importance of using integrated policies, governance, strategies, plans, assessments, and approaches, thus fostering the WEF nexus approach.

In particular, the 2011 Bonn Conference on the nexus of water, energy, and food security highlighted the need for understanding the nexus approach as a critical prerequisite to attaining sustainable development. It emphasized that understanding the nexus helps to "develop policies, strategies, and investments to explore synergies and mitigate the trade-offs between the objectives of sustainable development, with the active participation of government agencies, the private sector, and civil society" (Bonn, 2011, p. 1). Furthermore, important international organizations such as FAO and the International Renewable Energy Agency (IRENA) emphasized that by 2050, the demand for primary energy will almost double, increasing by 80%, and the need for water and food will increase. Thus, these organizations recommend rethinking how society produces and consumes natural resources and doing so in ways that meet competing demands for limited resources amidst the tremendous effects of climate change.

Diverse challenges in governance, policy consistency, planning, monitoring, and evaluation related to implementing the 2030 Agenda and the Paris Agreement have focused on the WEF nexus approach as a conceptual tool that can contribute to achieving the SDGs (Lazaro & Giatti, 2021a). The SDGs illuminate the nature and scope of the broad-scale challenge of searching for alternatives against the background of a looming global unsustainability crisis in all its diverse dimensions. However, in the Anthropocene, interconnections and interdependencies among sectors and scales add complexity to new challenges. The usual sectoral solutions inevitably generate trade-offs that can compromise and hinder the search for sustainability (Steffen et al., 2015; Whitmee et al., 2015). The complexity of sustainability challenges requires initiatives that bridge scales and sectors and cut across disciplines to unleash transformative potential. An essential driver of transformation consists of adopting urban interventions that enable intersectoral and multi-scale gains, generating co-benefits capable of considering and effectively dealing with the diversity of objectives and goals associated with SDGs (De Andrade Guerra et al., this volume; Doll & Puppim de Oliveira, 2017). Urban interventions, for instance, can address climate change by providing access to water and sanitation, sustainable mobility, more compact and resilient settlements, and urban adaptation strategies. The way to plan and execute these interventions should be conditional upon the search for greater efficiency in diverse fields to reduce the likelihood of potential trade-offs (Sachs et al., 2019).

In an urban age that increasingly concentrates human activities in urbanized environments, cities become the locus of fundamental intersections to confront today's global challenges. Today, urban metabolism's linear model drives most of the pressures and demands associated with production and consumption chains, thus playing a pivotal role in generating humanity's pressure on the biosphere and ecosystems. This model has remained fundamentally unchanged in the history of Western civilization. Like other living organisms, cities continue to feed on natural resources and generate waste, now more than ever at an unsustainable rate. Furthermore, cities harbor profound inequalities related to environmental degradation processes, pollution, and social vulnerability (MEA, 2005; Kjellstrom et al., 2008; Bulkeley, 2010; Seto et al., 2017). Large urban centers with millions of inhabitants drain regional resources substantially and demand ecosystem services, generate pollution and inequalities, and foment environmental injustices in their peripheries (Amaral et al., 2021).

However, cities can also provide solutions to sustainability problems as they are centers of knowledge and innovation and political and economic power (Puppim de Oliveira, 2019). Cities can mobilize human, technological, and financial resources at a large scale, required to address these sustainability challenges. Cities are also critical elements in searching for alternatives to mitigate greenhouse gas emissions and adapt to the risks associated with global climate change. Moreover, there is a vast potential for co-benefits related to climate action in cities, which could also lead to improvements in energy efficiency, waste management, sustainable housing, generation of income, and protection of green areas (Doll & Puppim de Oliveira, 2017; Sethi & Puppim de Oliveira, 2018). Such actions, in turn, can help achieve many of the SDGs. Thus, promoting Agenda 2030 within a multilevel governance perspective places cities as hubs for the convergence of international global agendas, such as the Paris Accord on Climate Change and the Sendai Framework for Disaster Risk Reduction 2015–2030.

This book addresses the water-energy-food nexus in urban contexts (WEF urban nexus) from a systems perspective and explores how the nexus approach can foster innovation. Through its cross-cutting potential, the FWE nexus approach can promote new knowledge about the implementation of synergetic alternatives, connecting scales and sectors that, conventionally, operate in isolation and are managed in silos.

1.2 The Evolution of the WEF Nexus Concept

The WEF nexus is an innovative approach to understanding the interactions, interdependencies, synergies, and trade-offs between natural resources interrelationships, jointly addressing water, energy, and food management (Hoff, 2011). The WEF nexus idea emerged during the World Economic Forum in 2008 when international business leaders called for a better understanding of how water management links to economic growth across the WEF-climate connectivity assemblage. Later, in the 2011 Bonn Conference, international organizations and governments recognized that water, energy, and food policies could not be managed in isolation. This conference was the watershed moment for consolidating the WEF nexus concept (Lazaro et al., 2021a). The 2012 United Nations Conference on Sustainable Development (Rio + 20) promoted the "nexus thinking" approach, which became a central mechanism to achieve the SDGs.

The WEF nexus is a cross-cutting approach that can form the conceptual underpinning to SDGs, thus catalyzing the potential for convergence between the different global sustainability agendas in cities. It is an innovative perspective that furthers intersectoral gains and provides synergies and multiple benefits in pursuing SDGs (Islam & Kenway, this volume; Valencia et al., 2019) while mitigating greenhouse gas emissions and fostering climate change adaptation. Altering the nature of the interconnections between water, energy, and food systems allows us to address other issues such as health, climate change, and biodiversity loss. For example, access to energy (SDG 7) is critical for the economic growth of poor communities (SDG 8), improves health services (SDG 3), reduces poverty (SDG 1), and provides food and clean water (SDGs 2 and 6); if well managed, such resource-use patterns can help mitigate climate change (SDG 13).

One of the main problems that the nexus approach seeks to overcome is uncoordinated traditional decision-making that manages an urban sector without considering its synergies and trade-offs with other sectors, overlooking how the policies of one industry can negatively or positively affect another. This fragmented approach brings about the need for transformation through more integrative approaches that balance public policies, social achievements, and the sustainable management of natural resources and ecosystems (Lazaro et. al. 2020).

Until 2012, there were few publications on the three-element interactions in the WEF nexus, as Fig. 1.1 shows. Our search in Scopus, conducted in November 2021 without limiting the period, found 653 articles with the keyword "water-energy-food nexus." The first papers on the water-energy-food nexus appeared in 2012, and since then, their numbers have increased year on year (Fig. 1.1). According to Cairns and Krzywoszynska (2016), this development is due to the nexus concept gaining prominence as a framework for funding research. For instance, the European Union's Horizon 2020 program refers explicitly to the nexus and food security, low-carbon energy, sustainable water management, and climate change mitigation. Other examples include collaborative research initiatives, such as the United Kingdom-Brazil "Healthy Urban Life and the Social Sciences of the Food-Water-Energy Nexus" (ESRC, 2015) and the Belmont Forum/Joint Programming Initiative (JPI), sustainable urbanization global initiative on the food, water, and energy nexus—SUGI-FWE Nexus (2017). Furthermore, numerous international institutions highlight the nexus concept as a new perspective defining sustainable development. The World Bank, the Food and Agriculture Organization of the United Nations (FAO), the United Nations World Water Resources Assessment Program, the OECD, the UN Economic Commission for Latin America (ECLAC), and the Global Water Partnership have all produced policy documents and perspectives on the nexus approach.

However, despite being promoted by influential global policy leaders and embraced by many researchers, holistic and integrated approaches to resource management and sectoral planning such as the nexus concept still need to be adopted beyond the academic domain. Although its initial conception occurred within the World Economic Forum, little is known about the extent to which companies practice WEF nexus thinking. Lazaro et al. (2021a) argue that while governments play a critical role in designing and implementing inclusive policies to minimize impacts and trade-offs between nexus sectors, businesses play an equally important role in driving positive change toward sustainable management.

We have found that existing literature has emphasized computational and mathematical models to analyze and quantify the relationships between the WEF nexus components. Nexus assessments frequently rely on quantitative approaches, and mixed methods do not predomi-

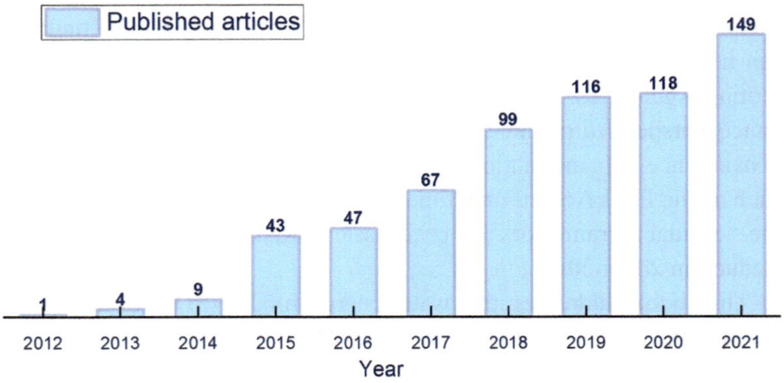

Fig. 1.1 Annual number of scientific publications concerning the water-energy-food nexus

nate; nexus studies are usually confined to disciplinary silos such as energy or water efficiency (Lazaro et al., 2021b). Nexus thinking is posed very strongly as a technological issue related to sectoral performance in mitigating trade-offs, which strongly contributes to the predominance of quantitative approaches (Dalla Fontana et al., 2020). However, when measures from one sector operate to influence another, problems arise, generating uncertainties about the real cross-sector possibilities. It also raises doubts about how specific nexus thinking measures may or may not function concerning inequalities among different social actors in accessing water, energy, and food resources (Di Felice et al., 2021). These findings reinforce the understanding that the sociopolitical dimensions of the nexus should be further explored using qualitative approaches.

Both quantitative and qualitative methods can be critical to a nexus assessment, such as identifying the interlinkages among sectors, which includes data analysis. This analysis can help understand the constraints to review sectoral planning and implementation of policy measures and legislation. At the same time, monitoring the governance, policy decisions, and the dialogue and discourses of the decision-makers and experts can help identify critical interlinkages and analyze how decision-making focused only in one sector can cause impacts in other sectors (Flammini et al., 2014).

We performed a bibliometric analysis of the 653 articles' abstracts using the Bibliometrix software (Aria & Cuccurullo, 2021) and identified the topic evolution; we divided the sample into four periods (2012–2015, 2016–2017, 2018–2019, 2020–2021) and considered the number of articles published in each period. Besides publications increasing, we verified an evolution in the topics covered, as shown in Fig. 1.2.

We noted that the number of publications on sustainable development issues related to the WEF nexus in the first period (2012–2015) was not significant compared to the third and fourth periods, in which it steadily increased. In subsequent periods, new themes were incorporated, such as land use, urban nexus, circular economy, and WEF and waste nexus. Studies have gone beyond the three original nexus sectors and have included other dimensions, such as climate, land, and environmental impact. Figure 1.2 also shows clusters of the topics. In the lower left quadrant, the set in brown color covers issues such as climate change, resilience, adaptation, and water scarcity. In the upper left quadrant, the cluster in pink color shows topics like urban nexus, environmental justice, governance, and water-energy-food nexus. In particular, the WEF nexus in urban areas has gained prominence as an approach that can be useful for managing urban metabolism and the resilience of cities (Bellezoni et al., this volume; de Macedo et al., 2021). The relationship between waste and the dynamics of WEF nexus sectors in cities characterizes these elements' interconnected generation, allocation, and consumption processes. It indicates the need to integrate resource management processes that increase the efficiency of natural resource use and consumption, transform infrastructural systems, and better urban planning to reduce GHG emissions and waste generation (Artioli et al., 2017; Lazaro & Giatti, 2021b).

The burden of urbanization on ecosystem services in peri-urban areas, public health, and climate change has inevitably increased in the Anthropocene. Therefore, the nature of urban territories and their growing consumption pressure significantly affect the availability and interdependence between water, energy, and food resources. The WEF urban nexus can provide fundamental guidelines for sustainable solutions to the expected sharp growth in water, energy, and food demand accompanying rapid urbanization (Lazaro & Giatti, 2021a). Estimates are that by 2030 the overall global population will reach 8.6 billion people, 68% in urban areas, and demands for energy, water, and food are expected to increase significantly (OECD/FAO, 2013). The urban scenario represents both a challenge and an opportunity to understand and direct resources toward more sustainable configurations. In particular, mobilizing action and understanding that the nexus approach in cities can benefit urban populations by dealing with some of the most complex global problems, including climate change, while helping achieve sustainable

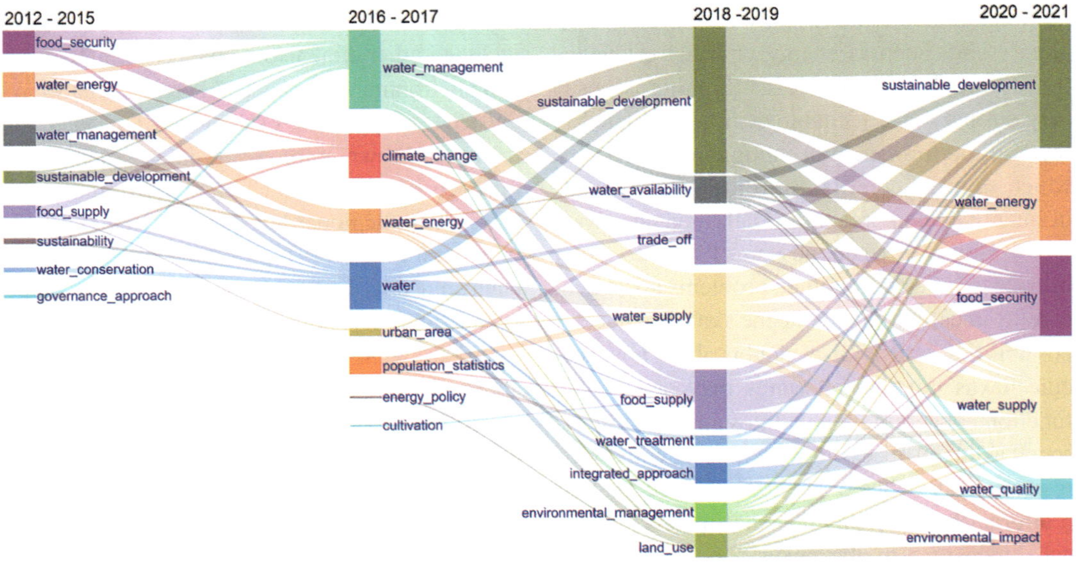

Fig. 1.2 Sankey of the thematic evolution on the WEF nexus research from the bibliometric analysis

development goals (Covarrubias, 2019; Lazaro & Giatti, 2021a).

Some authors have criticized the nexus approach as a novelty (Simpson & Jewitt, 2019; Muller, 2015). However, instead of being a cause for concern, this approach can inspire discussions about a "paradigm shift" to respond to global environmental challenges. For instance, the nexus can foster using traditional knowledge (Lazaro et al., 2021a), such as nature-based solutions (NbS) that support the WEF nexus to tackle climate change impacts (Bellezoni et al., this volume). Because finite natural resources continue to be managed conventionally, in isolation and prioritizing segmentation, we now have to deal with the shortcomings of silo thinking, linearity, and fragmentation. Therefore, the main challenge becomes getting key players on board to embark on this paradigm shift, moving away from metrics of success, profitability, and market share based on a development model that relies on depreciating the base of natural resources (Lazaro & Giatti, 2021b).

At the closing plenary session of the 2011 Bonn Conference, Klaus Topfer, executive director of the Institute for Advanced Sustainability Studies in Germany, called for a return to the nexus process for problem-solving. He reminded the audience that "the Rio 92 Conference had discussed environment and development through a logic of nexus," noting that this interconnectedness perspective had been undermined and added that after 20 years, "[…] this sustainable development message should be reinvigorated through a nexus approach" (Martin-Nagle et al., 2011, p. 5). The alert fostered the internationalization of the WEF nexus research.

Figure 1.3a, b, respectively, show the scientific production and collaboration map per country. The countries with the highest published articles are the USA and the UK, followed by China and Brazil. The country with the most significant number of collaborations was the UK, as depicted in the network in purple (Fig. 1.3b). The UK's centrality in boosting and contributing research and action on nexus issues has been enhanced since a 2009 speech by Professor Sir John Beddington—then Chief Scientific Advisor to the UK government. On that occasion, Beddington raised the concern about a "perfect storm" of interlinked problems challenging humanity.

Furthermore, since 2012, the nexus has become the focus of several research programs, funding calls, and other initiatives in the UK, such as the research council initiatives and the

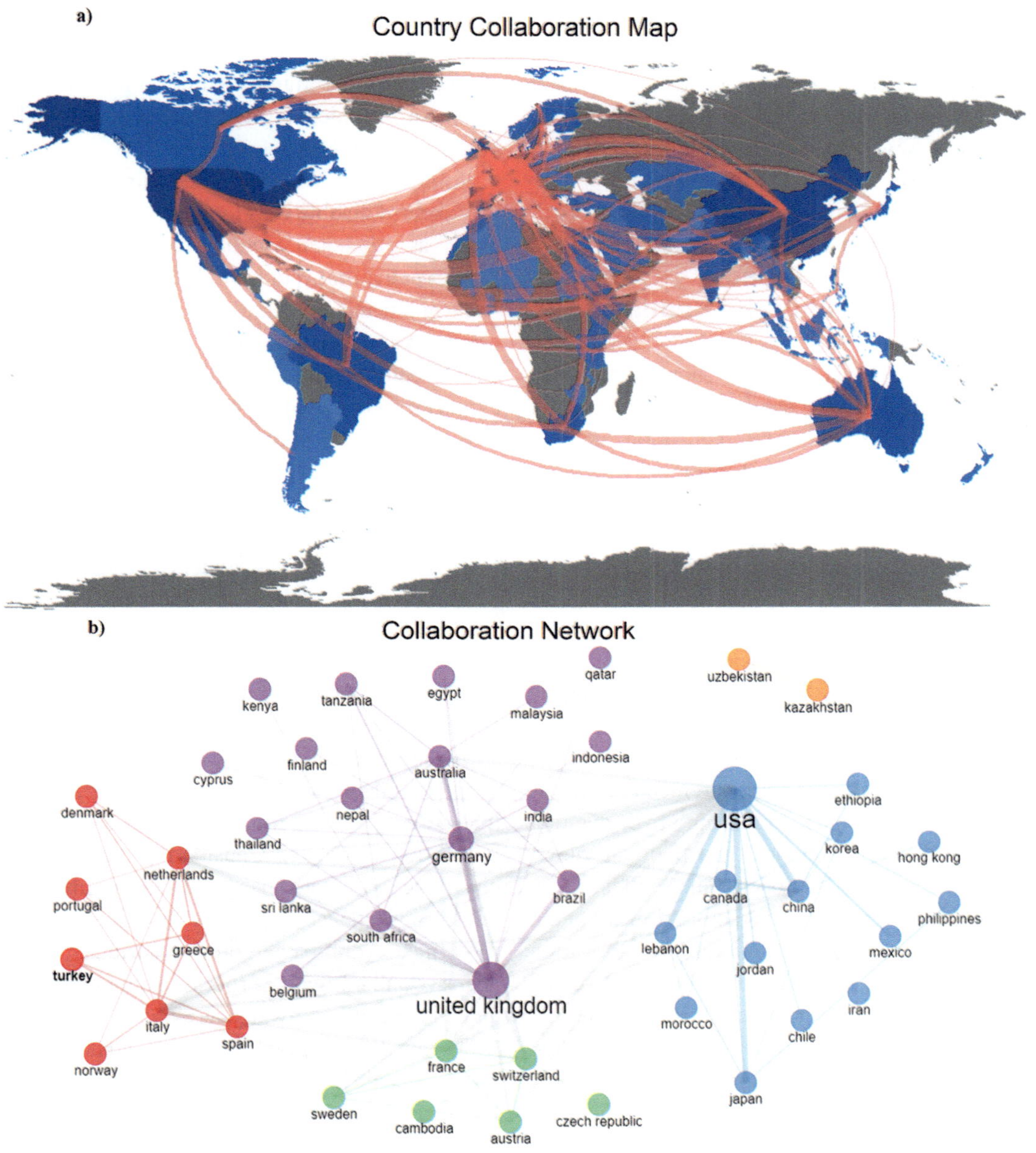

Fig. 1.3 (**a**) Scientific production per country. (**b**) Collaboration between countries from the bibliometric analysis

Newton Fund. These circumstances have fostered an increase in the use of nexus-related terminology in academic and governmental circles in the UK, ratcheting with the rise in research and policy involvement (Cairns & Krzywoszynska, 2016). Figure 1.3b also shows a close relationship of the collaboration between the UK and Germany. The latter organized the Bonn Conference "The Water, Energy and Food Security Nexus: Solutions for the Green Economy," which aimed to explore the WEF nexus as an innovative method for achieving sustainability. It presented initial evidence concerning how a nexus approach can improve water, energy, and food security in a green economy, increasing efficiency, reducing trade-offs, and creating cross-sector synergies. The results of this conference served as a specific German contribution to the United Nations Conference on Sustainable Development "Rio + 20," held in Brazil in 2012. There has since been significant progress in the science linked to the WEF nexus, along with several calls to bring it to practice.

1.3 Science-Policy-Society Interfaces Based on Hybridity and Humility

Our bibliographical research supports the idea that work on the WEF nexus does not limit itself to analyzing technological innovation and the search for efficiency and optimization, which has been criticized as a reductionist and technical/managerial approach. The breadth of current nexus-related research also underscores that understanding the sociopolitical dimensions associated with the interconnections between sectors (Cairns & Krzywoszynska, 2016; Dalla Fontana et al., 2020) is critical to advance the incorporation of the WEF nexus in urban practice. On the other hand, nexus thinking and its associated innovations imply tremendous transformational changes, requiring hybridization between different technical and political views, diverse scales of approach, and collaborative involvement of many social actors (Lazaro et al., 2021a). Thus, the understanding of innovation goes beyond the purely technical perspective and does not necessarily involve R&D or something new to the world. It includes changes that allow us to do things differently in the context stakeholders operate.

Hybridization is rooted in complex interdependencies and reflects knowledge gaps and uncertainties around outcomes, disputes, and trade-offs. In this sense, it is impossible to predict nexus outcomes, and there is no single vision that embraces the interrelated causes, effects, and consequences. As there is no single causal direction for complex systems, nexus innovations relate to different actors, values, knowledge, and imagined outcomes within nexus relationships. Therefore, nexus innovations go beyond regular sectoral decision-making and the conventional challenges of academic knowledge production when assisting policymakers. It requires framing conversations and narratives between actors around desirable futures while simultaneously expanding the debate beyond technicalities (Di Felice et al., 2021).

Applying the WEF nexus thinking as a pathway for sustainable development allows it to be used as a science-policy-society interface. Additionally, innovation production must evolve from analyzing potential applications and integrating knowledge to promote transformation, which requires humility by researchers and policymakers in knowledge co-production processes. This opening-up process can be implemented through a humility-based nexus framework. Knowledge co-production consists of constantly and collaboratively framing the full extent of the problem; exploring and understanding the vulnerabilities; invoking distribution concerns related to the nexus innovations and continuous learning; and securing accessible, transparent, accountable, and actively shared knowledge (Urbinatti et al., 2020b).

As a result, it becomes necessary to constantly promote reflection on the benefits and trade-offs that are most likely to occur due to each nexus innovation initiative. Considering the massive challenge of climate change and its possible systemic consequences, applying the water-energy-food nexus approach could contribute as an adaptive and mitigative process to climate change and enhance urban resilience.

The nexus approach in urban contexts can boost sustainable urban development while focusing on the global climate crisis. Furthermore, it can spark reflection, contributing to the debate in science-policy-society relations. For instance, "Why must specific nexus solutions be implemented?" leads to a necessary politicization of the issue. The question "Which interactions among nexus dimensions are we considering and what resources are we evaluating?" addresses the extent to which interrelationships can be optimized. For example, promoting urban agriculture connects to demands for local water resources. However, it can save water from other regions, conserve energy for food transport and trade, and develop synergies by exploiting local organic compost production. This example also encourages reflecting on another question: "Where are the nexus interactions operating?"—this is relevant as nexus innovations can relate to flows that transcend spatial scales, creating opportunities to contextualize urban measures that connect to global concerns (Dalla Fontana et al., 2021).

The current search for resilience within a sustainable city perspective also links to targeting the global climatic crisis from local contexts. In this sense, nexus thinking applied in the search for urban resilience also links to other reflections such as "Which resilience are we talking about?"

Cities depend on external resources, so it is necessary to consider synergies at regional scales in tackling resource scarcity to foster systemic urban resilience. Another question is "resilience for whom?". In this case, the nexus approach also helps recognize the necessary politicization of nexus and resilience issues while underscoring the need to address deep urban and regional inequities among social groups inside or outside cities (Meerow & Newell, 2016).

These are legitimate questions that emphasize the need for constant learning about actions and potentialities and, most importantly, promote dialogue involving several social actors. In short, the nature of the science-policy-society interface for applying the nexus requires constant hybridization on the one hand and receptivity to new knowledge on the other, which should be obtained through opening up the dialogue with society on behalf of policymakers and researchers. These successive rounds of questioning are also legitimately posed as ways to understand the limits of knowledge and the limits of the approach toward the water-energy-food nexus.

Overall, urbanizing the nexus agenda means interplaying integrative policy frameworks and urban governance to address global challenges (Artioli et al., 2017). On the other hand, the implementation of nexus thinking and innovations in urban contexts also connects to the imperative of establishing socio-material flows of water, energy, and food. It represents incorporating what is missing in the urban nexus perspective since the recommendation for hybridization corresponds to the dynamics of daily practices, policies, ideologies, networks, or cultural meanings that can shape the provision of resources (Covarrubias, 2019).

Vulnerable urban communities and marginalized social groups, for instance, can provide actual examples of pathways for nexus innovations related to social practices and local commonsense knowledge. Although these social groups live under conditions of severe resource scarcity, their daily routines can teach academics and policymakers about legitimate alternatives building on nexus synergies (Giatti et al., this volume). In addition, analyzing flows and potentialities of concurrent multidimensional urban problems can trigger nexus innovations. For example, planners and policymakers can apply nexus thinking to implement food banks in addressing the convergence between food loss and waste and urban poverty. Such an initiative can contribute via different sectors and benefits, reducing demand for waste management, reducing hunger and vulnerability, and mitigating pressures for water and energy in food production (Vieira et al., this volume). Moreover, nexus innovations can be implemented through governance structures involving local government practitioners, nongovernmental organizations, and representatives of production sectors and local communities.

Food systems surrounding urban settlements can also contribute to implementing the urban nexus. Rural communities face the challenge of resource scarcity and the threats of social vulnerability and poverty. This reflection contributes to the multiscaling of the questions around nexus innovations. The contestation of inequalities in these rural peri-urban territories can underpin the political motivation to engage actors capable of synergizing the nexus. Thus, participatory and socially produced alternatives, such as managerial adjustments based on agroforestry systems, can enhance food production with resulting ecological gains and mitigation of water resource impacts and scarcity (de Melo, this volume). These rural innovations and synergies can integrate urban nexus processes. For example, when urban policymakers recognize these potentialities, local government measures can encourage food acquisition from farmers who adopt ecological and nexus innovations.

The integration of diversified experiences and possibilities can demonstrate the need for better communicative processes, making dialogue possible between diverse knowledge holders (Urbinatti, this volume). This integration characterizes the perspective of humility by researchers and decision-makers, recognizing that laypersons produce such relevant knowledge for nexus innovations. The creative capacity of different social actors connects to a myriad of alternatives. Also, opening up to such diversity represents legitimate alternatives grounded in the local context and provides relevant political insertions in the debate. This expansive process is also desirable to integrate the conventional siloed management of water, energy, and food production chains, leading to more sophisticated, robust, and adaptive governance systems (Pahl-Wostl & Hare, 2004; Pahl-Wostl, 2019). Therefore, the governance of processes to incorporate the WEF nexus in urban initiatives is fundamental to move from theory to practice.

1.4 Multilevel and Intersectoral Governance for the Nexus

Governance has become a household expression for scholars and practitioners in social and political sciences, gaining particular relevance in international relations studies to address the interplay between institutions and actors engaged in global environmental policies. Rosenau defined governance as activities that include institutions and informal mechanisms alongside other societal actors (1992: 4). But in summary, the term "governance" implies an intricate set of structures, actors, and processes, both public and private. As a result, it became a beacon for multilateralism. In terms of the WEF nexus, strengthening governance and building institutions to better connect the three elements (WEF) is critical to improve their interconnections and to generate co-benefits.

Multilevel governance has gained increasing academic attention to analyze regionalism and European studies, further to other applications and definitions that include "network governance," "community governance," and "new modes of governance." The multilevel governance (MLG) concept emerged in the early 1990s, challenging then prevailing theories of neo-functionalism and intergovernmentalism to describe decision-making processes beyond the state-centric representation of the European Union (Piattoni, 2010).

In his seminal article, Marks (1992) proposed that when considering cohesion policies, multilevel governance would transcend the nation-state and include continuously integrated negotiation, both vertically and horizontally, involving governments from different territorial levels—supranational, national, regional, and local. When analytical elements of federative regimes and decentralization were integrated into this idea, the concept was used to describe the increasingly complex and dense network of connections between players from different spheres. Led by national governments, these networks are often transboundary (Svedin et al., 2001).

Thus, the MLG approach has crossed the EU-focused analytical boundaries and evolved into a more complex and encompassing theory for examining situations involving the participation of multiple actors. The MLG framework has been widely adopted in addressing global environmental issues such as climate change to include subnational governments and non-state actors from different sectors in civil society. This lens has been beneficial for analyzing subnational participation in the global environmental change polity. Moreover, the concept's classification in Type I and Type II (Hooghe & Marks, 2003) is particularly appropriate for the analyses of federative systems (see Piattoni, 2010: 24), worthy of further exploration in countries such as Brazil.

As societies become more complex, environmental challenges demand governance approaches that address interconnectivity and interdependence. The 17 Sustainable Development Goals (SDGs) adopted by signatories of Agenda 2030 under the auspices of the United Nations (UN, 2015) mainly involve intertwined issues requiring problem-solving strategies based on multilevel and intersectoral governance. Indeed, implementing SDGs 16 (peace, justice, and strong institutions) and 17 (partnerships for the goals) will enable the other 15 SDGs; without institutions and collaborative efforts, other objectives hang in the balance. As Meuleman postulates, "[...] governance is about the quality of institutions, processes, tools, skills, etc., which should enable effective policymaking and policy implementation. [...] the SDGs will not be attained by 2030 without well-functioning public institutions and effective governance at all levels" (2021: 5).

Managing water, energy, and food production for growing urbanized populations will become increasingly challenging, and planning is crucial to achieving its effectiveness. The nexus between demand and supply of finite natural resources requires a planning perspective that engages social systems in their entirety. Governance must include all sectors across all levels of government and even between countries sharing natural resources, as in the Latin America and the Caribbean (LAC) region (Embid & Martin, this volume).

Multilevel governance shows that local and regional governments play a crucial role in implementing relevant public policies for a sustainable transition (Lazaro et al., 2022). The case of a severe drought in southeast Brazil between 2014 and 2015 exemplifies the interdependence on resources. The sizeable territorial scale and different jurisdictions involved demand integrated management within a multilevel and intersectoral governance approach (Alves et al., this volume). The authors describe a unique region encompassing the two largest metropolises in Brazil, São Paulo (RMSP) and Rio de Janeiro (RMRJ). They are joined by a hydrological system comprised of connected river basins and reservoirs that supply millions of households along that Paraíba do Sul River. It also provides energy via nine hydroelectric power plants that serve 52 municipalities in two states and a robust agricultural sector. Governance of the whole system involves the three spheres of government: the national Water and Basic Sanitation Agency (ANA); water and power state departments in Minas Gerais, Rio de Janeiro, and São Paulo; and representatives from municipal governments and civil society gathered in different water basin committees acting in collaboration.

There is much room for developing research in nexus thinking and multilevel governance. Urbinatti et al.'s (2020a) literature assessment demonstrated that attention on nexus governance was concentrated between 2012 and 2016, focusing on water management. In contrast, scholarship seldomly mentioned MLG in conjunction with the WEF nexus between 2007 and 2018.

In developing countries, where the pressures of urgent needs and social issues commonly act as barriers to planning and effective policy implementation, there is much to learn, not only from successful experiences and methodologies in developed countries. Finding adequate and creative solutions is possible in surprisingly difficult and unexpected conditions, such as those faced by vulnerable communities neglected by the state (Carvalho et al., this volume; Giatti et al., this volume), not only in Brazil but in the Amazon basin in general (Nascimento et al., this volume). However, studies that measure and document results and processes to learn from experiences and improve their fungibility are still lacking.

Therefore, applying the MLG concept combined with other analytical approaches such as the actor-network theory can help better understand interactions between different actors and jurisdictions in managing natural resources and climate change challenges (Colding et al., 2020). Particularly in developing nations and cities, a multilevel governance approach to the WEF nexus should provide an appropriate analytical framework for a more balanced management and distribution of natural resources, besides fostering opportunities to push for innovations through the connections between the water, energy, and food systems.

1.5 Innovations: Moving the WEF Nexus in Cities Forward

The WEF nexus provides many opportunities to address sustainability challenges in cities. However, as discussed above, the potential for improvement in theory and the academic literature on nexus is not automatically translated into practice. The chapters in this book bring a series of new perspectives on how the WEF nexus can advance the sustainability agenda in urban areas and beyond. Nevertheless, to get those ideas together and apply them on the ground, we need to innovate in understanding city-based resource flows and make them more sustainable. For example, cities are experimenting and innovating with a series of initiatives to incorporate climate change into their policy agendas, which requires capacity and connection with the broader urban system beyond governments (Roppongi et al., 2017; Puppim de Oliveira, 2011).

Cities need to innovate to realize the benefits of the WEF nexus, not only in technologies or restricted to R&D but also in a combination of different innovative initiatives to expand the outreach of the nexus concept and its application at a variety of social levels. The nexus perspective includes new social and organizational approaches to deploy existing opportunities that improve sustainability through its adoption. Cities concentrate the demand for food, water, and energy, and in 2020, their 4.4 billion inhabitants accounted for over 75% of the global energy consumption (UN-HABITAT, 2020). Since they all come together in cities, this concentration provides opportunities for synergies in managing these three resources. At the same time, cities have the human and financial capacities to develop innovations to improve the WEF nexus. Combining those resources could generate new solutions to urban problems based on the nexus. For example, the city of São Paulo has innovated with the Link the Dots (*Ligue os Pontos*) program that brings together urban farmers to improve the sustainability of their practices and connect their products to markets (Moreira et al., this volume). The joint work of several municipal departments that previously had not cooperated in this particular aspect generated a new perspective on urban agriculture by safeguarding the city's water resources, reducing transportation needs, and improving product quality.

Cities have tremendous potential for improving urban sustainability and implementing innovations in nexus-based initiatives to deal with global environmental changes. In particular, climate changes can lead to water scarcity in several regions, consequently affecting food production and energy generation, which rely on the water systems in many places. Climate change should heavily influence Brazil's Cerrado and Caatinga biomes in the following decades. However, they offer extensive opportunities for agriculture expansion and production of renewable energy if innovative policies and management practices for the region are in place anticipating these changes, which will also mitigate climate change (Alves et al., this volume). In the Amazon, cities are vulnerable to changing weather patterns that have already been detected in the region (Nascimento et al., this volume). They could worsen the precarious condition of existing urban infrastructure, as many municipalities lack proper sanitation and waste management. Improving the WEF nexus could be achieved by employing circular economy approaches to reduce waste and eliminate the pollution of water bodies caused by inadequate waste disposal practices (Paes et al., 2021).

New management and institutional approaches are vital in boosting WEF nexus innovation. Such methods allow the connection between WEF silos and improve the nexus among the three components. The WEF nexus approach could accelerate sustainability transitions in cities (Dal Poz et al., this volume). On the other hand, academic literature and implementation measures have developed many ways to address sustainability challenges applicable to the nexus during past decades, such as circular economy, participatory governance, energy transition conceptual models, participatory and integrated planning, and ecosystem-based solutions (Bellezoni et al., this volume). New technologies, such as digital tools and artificial intelligence (AI), also emerged to collect and process information for improving management and policy.

Alves Junior et al. (this volume) suggest that innovations in logistics between Brazilian cities and producers adopting planning models that consider improving WEF nexus opportunities could significantly increase grain production efficiency and sustainability in the country, one of the largest exporters of agricultural products. These innovations would include greenhouse gas reduction at levels that have worldwide consequences. Moreover, bringing supply and demand closer in the food sector could significantly reduce food losses and waste with environmental, economic, social, and health (nutritional) consequences, addressing many SDGs. Better food bank operation could be an opportunity for improvements in the food supply chain with implications for waste, water, and energy (Vieira et al., this volume).

To allow those kinds of initiatives to emerge, new institutional approaches need to be developed, such as the "institutional bricolage" in the case of urban farmers in São Paulo (de Melo, this volume). The bricolage strategy to integrate policies at the different levels of government has allowed farmers to improve agriculture practices to make production more organic. Innovations in the WEF nexus will come not only from governments but from a variety of actors' initiatives. Using participatory planning, civil society groups can bring new resources and knowledge to innovate urban practices. In Salvador, Brazil, the Urban Vegetable Gardens Project (*Projeto Hortas Urbanas*), a community garden initiative, succeeded because the city government and civil society organizations worked together (Carvalho et al., this volume). Organic agriculture gardens can provide fresh vegetables to vulnerable communities and reduce artificial fertilizers and transportation.

A more integrated innovative approach to WEF in cities could also catalyze sustainable investments from the public and private sectors, particularly in rapidly urbanizing developing countries (Olawuyi & Oche, this volume). Increasing demand for food, water, and energy in cities in developing countries requires the significant deployment of investments to guarantee and secure supply. However, the lack of integration among the sectors that generate inefficiency and uncertainties caused by climate change, human rights, and political instability prevent many investors from deploying their financial resources. In addition, new approaches to climate change adaptation and mitigation can also identify opportunities to improve WEF flows (Jacobi et al., this volume). Many cities have already adopted adaptation plans that could be used to identify opportunities to strengthen the WEF nexus. Nevertheless, there is a long way to go until WEF nexus integration is a reality for such cities, as it is still little understood by many policymakers and the public. Many challenges exist, particularly in the communication and political processes to integrate WEF nexus in public policies, resulting in more sustainable outcomes (Urbinatti, this volume).

1.6 Concluding Remarks

Nexus thinking and related concepts have evolved since 2012 and aggregated new perspectives in their application to the urban context. This evolution has been accompanied by a complexification of the necessary knowledge production, demanding actions on problems still filled with uncertainties and disputes. Nexus issues require constantly analyzing outcomes and implications of their implementation in different sectors and scales by the actors involved. Nevertheless, while nexus innovations are opportunities for cross-sectoral gains, they cannot be reduced to technical issues, given the complexity of intrinsic factors, the cross-scalar issues involved, and uncertainties concerning the potential outcomes and consequences. Therefore, nexus innovations demand constant reflections and debates from different social actors, stimulating more robust and adaptive multilevel and intersectoral governance structures. We foresee exciting opportunities to promote synergies from the urban nexus perspective through the cases explored in this volume and founded on our bibliographic review. However, we caution that there is much to be done to successfully integrate the WEF nexus in the urban agenda and encourage engagement among the public and policymakers. The search for sustainable development can benefit enormously from nexus innovations. It must, nevertheless, be the subject of constantly questioning, reflecting, and exploring its multidimensionality, bringing together all sectors, scales, and diverse social actors.

Acknowledgment The authors acknowledge the financial support received from São Paulo Research Foundation (FAPESP), Process: 2015/03804–9. LLBL also acknowledges the financial support received from FAPESP (Grant: 2017/17796–3 and 2019-24479–0). LLG acknowledges National Council for Scientific and Technological Development, Process: 309840/2018-0, Brazil. LVM would like to thank the São Paulo Research Foundation (FAPESP) in Brazil for funding her research through grant number 2018/20714-1. JAPO acknowledges the financial support received from FAPESP, Process: 2017/50425-9, Coordination for the Improvement of Higher Education Personnel (CAPES) Grant Number 88881.310380/2018-01; and National Council for Scientific and Technological Development (CNPq) Grant Number 442472/2020.

References

Alves Junior, P. N., Branco, J. E. H., Bartholomeu, D. B., Cruz Júnior, J. C., & Caixeta Filho, J. V. (this volume). *Methods for evaluating food-energy-water nexus: Data envelopment analysis and network equilibrium model approaches*. Springer.

Alves, L. M., Ometto, J. P. H. B., Arcoverde, G. F. B., & Branco, E. A. (this volume). *Water–energy–food nexus under climate change: Analyzing different regional socio-ecological contexts in Brazil*. Springer.

Amaral, M. H., Lazaro, L. L. B., de Almeida Sinisgalli, P. A., da Fonseca Alves, H. P., & Giatti, L. L. (2021). Environmental injustices on green and blue infrastructure: Urban nexus in a macrometropolitan territory. *Journal of Cleaner Production, 125829*.

Aria, M., & Cuccurullo, C. (2021). *Bibliometrix R Package* [WWW Document]. https://www.bibliometrix.org/. Accessed 11.11.21.

Artioli, F., Acuto, M., & Mcarthur, J. (2017). The water-energy-food nexus: An integration agenda and implications for urban governance. *Political Geography, 61*, 215–223.

Bellezoni, R. A., Seto, K. C., & Puppim de Oliveira, J. A. (this volume). *What can cities do to enhance water-energy-food nexus as a sustainable development strategy?* Springer.

Bonn 2011 Conference. The Water, Energy and Food Security Nexus – Solutions for a Green Economy. (Accessed 11 August 2021) https://www.iucn.org/sites/dev/files/import/downloads/bonn_conference_nexus_messages.pdf

Bulkeley, H. (2010). Cities and the governing of climate change. *Annual Review of Environment and Resources, 35*.

Cairns, R., & Krzywoszynska, A. (2016). Anatomy of a buzzword: The emergence of 'the water-energy-food nexus' in UK natural resource debates. *Environmental Science & Policy, 64*, 164–170.

Carvalho, C. A., Rosa, S. M., & Ventura, A. C. (this volume). *Urban gardens and composting: Effective governance for strengthening urban resilience and community waste management*. Springer.

Colding, J., Barthel, S., & Samuelsson, K. (2020). Supporting bottom-up human agency for adapting to climate change. *One Earth, 3*(4), 392–395. ISSN:2590-3322.

Covarrubias, M. (2019). The nexus between water, energy and food in cities: Towards conceptualizing socio-material interconnections. *Sustainability Science, 14*, 277–287.

Dal Poz, M. E., Piolli, A., Francisco, E. C., & Batista, M. R. (this volume). *WEF nexus innovations: The institutional agenda for sustainability*. Springer.

Dalla Fontana, M., de Araújo, M. F., Di Giulio, G. M., & Malheiros, T. F. (2020). The water-energy-food nexus research in the Brazilian context: What are we missing? *Environmental Science & Policy, 112*, 172–180.

Dalla Fontana, M., Wahl, D., de Araujo Moreira, F., Offermans, A., Ness, B., Malheiros, T. F., & Di Giulio, G. M. (2021). The five Ws of the water-energy-food nexus: A reflexive approach to enable the production of actionable knowledge. *Frontiers in Water, 3*, 124. https://doi.org/10.3389/frwa.2021.729722

de Andrade Guerra, J. B. O., Santa, S. L. B., Sobrosa-Neto, R. C., Brandli, L. L., & Soares, T. C. (this volume). *The importance of water-energy and food nexus in the promotion of sustainable cities in the perspective of the sustainable development goals*. Springer.

de Macedo, L. S. V., Picavet, M. E. B., Puppim de Oliveira, J. A., & Shih, W. Y. (2021). Urban green and blue infrastructure: A critical analysis of research on developing countries. *Journal of Cleaner Production, 313*, 127898.

de Melo, T. G. (this volume). *The water-energy-food nexus and the micro-politics of everyday: A view from institutional bricolage*. Springer.

Di Felice, L. J., Cabello, V., Ripa, M., & Madrid-Lopez, C. (2021). Quantitative storytelling: Science, narratives, and uncertainty in nexus innovations. *Science, Technology, & Human Values*, 01622439211053819.

Doll, C. N. H., & Puppim de Oliveira, J. A. (Eds.). (2017). *Urbanization and climate co-benefits: Implementation of win-win interventions in cities*. Routledge.

Embid, A., & Martin, L. (this volume). *The water-energy-food nexus in Latin America and the Caribbean: Priority interconnections*. Springer.

Flammini, A., Puri, M., Pluschke, L., & Dubois, O. (2014). *Walking the nexus talk: Assessing the water-energy-food nexus in the context of the sustainable energy for all initiative*. FAO.

Giatti, L. L., Börner, S., & Carvalho, C. M. (this volume). *Learning about the nexus from vulnerable urban communities*. Springer.

Hoff, H. (2011). Understanding the Nexus. Background Paper for the Bonn 2011 Nexus conference Stock. Environ. Inst. (2011), pp. 1–52.

Hooghe, L., & Marks, G. (2003). Unraveling the central state, but how? Types of multilevel governance. *American Political Science Review, 97*(2), 233–243.

Islam, K. M. N., & Kenway, S. K. (this volume). *The food-water-renewable energy nexus resource-security examples for Asian-Pacific cities*. Springer.

Jacobi, P. R., Picarelli, S. B. N., Heinisch, L. M., & Sinisgalli, P. A. A. (this volume). *The challenges of the food, water, and energy nexus and potential interlinkages with instruments to tackle climate change: Cases of Brazilian cities*. Springer.

Kjellstrom, T., Friel, S., Mercado, S., Havemann, K., & Satterthwaite, D. (2008). *Our cities, our health, our future. Acting on social determinants for health equity in urban settings*. Report to the WHO Commission on Social Determinants of Health from the Knowledge Network on Urban Settings.

Lazaro, L. L. B., & Giatti, L. L. (2021a). O nexo água-energia-alimentos – uma abordagem para cidades sustentáveis e o desenvolvimento sustentável. In P. Jacobi & L. Giatti (Eds.), *Inovação para governança da macrometrópole paulista face à emergência climática* (pp. 79–97).

Lazaro, L. L. B., & Giatti, L. L. (2021b). Doing business in a climate change context – business case on water-energy-food nexus of biofuels. In M. Lackner et al. (Eds.), *Handbook of climate change mitigation and adaptation*. Springer.

Lazaro, L. L. B., Giatti, L. L., Sousa Junior, W. C., & Giarolla, A. (2020). Land-water-food nexus of biofuels: Discourse and policy debates in Brazil. *Environment and Development, 33*, 100491.

Lazaro, L. L. B., Giatti, L. L., & de Oliveira, J. A. P. (2021a). Water-energy-food nexus approach at the core of businesses – How businesses in the bioenergy sector in Brazil are responding to integrated challenges? *Journal of Cleaner Production, 303*, 127102.

Lazaro, L. L. B., Giatti, L. L., Bermann, C., Giarolla, A., & Ometto, J. (2021b). Policy and governance dynamics in the water-energy-food-land nexus of biofuels: Proposing a qualitative analysis model. *Renewable and Sustainable Energy Reviews, 149*, 111384.

Lazaro, L. L. B., Soares, R. S., Bermann, C., Collaço, F. M. A., Giatti, L. L., & Abram, S. (2022). Energy Research & Social Science Energy transition in Brazil: Is there a role for multilevel governance in a centralized energy regime ? *Energy Research and Social Science, 85*, 102404.

Marks, G. (1992). Structural policy in the European Community. In A. Sbragia (Ed.), *Euro-politics, Institutions and policymaking in the "new" European Community* (pp. 191–225). The Brookings Institution.

Martin-Nagle, R., et al. (2011). *The water, energy and food security nexus – Solutions for the green economy*. Conference Synopsis. Available at: www.ooskanews.com. Accessed 5 Oct 2021.

Meerow, S., & Newell, J. P. (2016). Urban resilience for whom, what, when, where, and why? *Urban Geography*, 1–21.

Meuleman, L. (2021). Public administration and governance for the SDGs: Navigating between change and stability. *Sustainability, 2021*(13), 5914. https://doi.org/10.3390/su13115914

Millennium Ecosystem Assessment. (2005). *Ecosystems and human well-being*. Island Press.

Moreira, F. A., Fontana, M. D., Bertolini, A. N., Sepe, P. M., Malheiros, T. F., & Di Giulio, G. M. (this volume). *Urban living labs and the water-energy-food nexus: Experiences from the GLOCULL project in São Paulo, Brazil*. Springer.

Muller, M. (2015). The 'Nexus' as a step back towards a more coherent water resource management paradigm. *Water Alternatives, 8*, 675–694.

Nascimento, N., Lazaro, L. L. B., & Amaral, M. (this volume). *How can the water-energy-food nexus approach contribute to enhancing the resilience of Amazonian cities to climate change?* Springer.

OECD/FAOAgricultural Outlook 2013–2022: Highlights (2013) (Accessed 8 September 2021) https://www.oecd.org/tad/agricultural-policies/OECD-FAO_Outlook_2013-2022.pdf

Olawuyi, D. S., & Oche, A. (this volume). *WEF nexus and sustainable Investments in West Africa: The case of Nigeria*. Springer.

Paes, M. X., Campos-Silva, J. V., & Puppim de Oliveira, J. A. (2021). Integrating circular economy in urban Amazon. *npj Urban Sustainability, 1*, 29.

Pahl-Wostl, C. (2019). Governance of the water-energy-food security nexus: A multilevel coordination challenge. *Environmental Science & Policy, 92*, 356–367.

Pahl-Wostl, C., & Hare, M. (2004). Processes of social learning in integrated resources management. *Journal of Community & Applied Social Psychology, 14*, 193–206.

Piattoni, S. (2010). *The theory of multilevel governance: Conceptual, empirical and normative challenges*. Oxford University Press.

Puppim de Oliveira, J. A. (2011). Why an air pollution achiever lags on climate policy? The case of local policy implementation in Mie, Japan. *Environment and Planning A, 43*(8), 1894–1909.

Puppim de Oliveira, J. A. (2019). Sustainability challenges in an urban century: Can we change urbanization paths to make cities the solutions for rather than the drivers of global problems? *Challenges in Sustainability, 7*(1), 1–4.

Roppongi, H., Suwa, A., & Puppim de Oliveira, J. A. (2017). Innovating in sub-national climate policy: The mandatory emissions reduction scheme in Tokyo. *Climate Policy, 17*(4), 516–532.

Rosenau, J. N. (1992). Governance, order and change in world politics. In J. N. Rosenau & O. Czempiel (Eds.), *Governance without government: Order and change in world politics* (pp. 1–29). Cambridge University Press.

Sachs, J. D., Schmidt-Traub, G., Mazzucato, M., Messner, D., Nakicenovic, N., & Rockström, J. (2019). Six transformations to achieve the sustainable development goals. *Nature Sustainability, 2*, 805–814.

Sethi, M., & Puppim de Oliveira, J. A. (Eds.). (2018). *Mainstreaming climate co-benefits in Indian cities: Post-Habitat III innovations and reforms*. Springer.

Seto, K. C., Golden, J. S., Alberti, M., & Turner, B. L. (2017). Sustainability in an urbanizing planet. *Proceedings of the National Academy of Sciences, 114*, 8935–8938.

Simpson, G. B., & Jewitt, G. P. (2019). The development of the water-energy-food nexus as a framework for achieving resource security: A review. *Frontiers in Environmental Science, 7*, 8.

Steffen, W., Broadgate, W., Deutsch, L., Gaffney, O., & Ludwig, C. (2015). The trajectory of the Anthropocene: The great acceleration. *The Anthropocene Review, 2*, 81–98.

Svedin, U., O'Riordan, T., & Jordan, A. (2001). Multilevel governance for the sustainability transition. In T. O'Riordan (Ed.), *Globalism, localism and identity: Fresh perspectives on the transition to sustainability* (pp. 43–60). Earthscan.

UN-HABITAT. (2020). *World cities report 2020 – The value of sustainable urbanization*. UN-HABITAT. 418 p. ISBN:978-92-1-132872-1.

United General Assembly, Transforming our world: the 2030 Agenda for Sustainable Development, 21 October 2015. A/RES/70/1. Available at: https://www.refworld.org/docid/57b6e3e44.html

Urbinatti, A. M. (this volume). *Innovations towards 'the nexus' in the science-politics-society interface: What transformations do we really seek?* Springer.

Urbinatti, A. M., Lazaro, L. L. B., Carvalho, C. M., & Giatti, L. L. (2020a). The conceptual basis of water-energy-food nexus governance: Systematic literature review using network and discourse analysis. *Journal of Integrative Environmental Sciences, 17*(2), 21–43. https://doi.org/10.1080/1943815X.2020.1749086

Urbinatti, A. M., Dalla Fontana, M., Stirling, A., & Giatti, L. L. (2020b). 'Opening up' the governance of water-energy-food nexus: Towards a science-policy-society interface based on hybridity and humility. *Science of The Total Environment*, 140945.

Valencia, S. C., Simon, D., Croese, S., Nordqvist, J., Oloko, M., Sharma, T., Taylor Buck, N., & Versace, I. (2019). Adapting the sustainable development goals and the new urban agenda to the city level: Initial reflections from a comparative research project. *International Journal of Urban Sustainable Development, 11*, 4–23.

Vieira, L. M., Domingues, I., & Matzembacher, D. E. (this volume). *Food waste redistribution and implications for achieving the sustainable development goals: The case of a food bank in Sao Paulo Municipality, Brazil*. Springer.

Whitmee, S., Haines, A., Beyrer, C., Boltz, F., Capon, A. G., de Souza Dias, B. F., Ezeh, A., Frumkin, H., Gong, P., Head, P., et al. (2015). Safeguarding human health in the Anthropocene epoch: Report of the Rockefeller Foundation–Lancet Commission on planetary health. *The Lancet, 386*, 1973–2028.

Part I

Sustainable Development: Cities and the Big Picture

2 The Importance of Water-Energy-Food Nexus in the Promotion of Sustainable Cities in the Perspective of the Sustainable Development Goals

José Baltazar Salgueirinho Osório de Andrade Guerra, Stephane Louise Boca Santa, Ruy de Castro Sobrosa Neto, Luciana Londero Brandli, Thiago Coelho Soares, and Felipe Teixeira Dias

Abstract

The water-energy-food nexus is directly connected with the implementation of the 17 Sustainable Development Goals (SDGs) that make up the United Nations 2030 Agenda. Therefore, the water-energy-food nexus is directly related to the objective of building sustainable cities, to achieve the 17 SDGs. In this research, the nexus composed by the triad—water-energy-food—is evidenced, based on *the following*: SDG 2: zero hunger and sustainable agriculture; SDG 6: clean water and sanitation; SDG 7: clean and

J. B. S. O. de Andrade Guerra (✉)
Centre for Sustainable Development (GREENS), The University of Southern Santa Catarina, Florianópolis, Brazil

Cambridge Centre for Environment, Energy and Natural Resource Governance, (CEENRG), University of Cambridge, Cambridge, UK

S. L. B. Santa
Administrations, University of Southern Santa Catarina (UNISUL), Florianópolis, Brazil

Accounting, University of Federal Santa Catarina (UFSC), Florianópolis, Brazil

Centre for Sustainable Development (GREENS), Florianópolis, Brazil

R. de Castro Sobrosa Neto
Centre for Sustainable Development (GREENS), The University of Southern Santa Catarina, Florianópolis, Brazil

L. L. Brandli
University of Passo Fundo, Passo Fundo, Brazil

T. C. Soares
Business Administration, University of Southern Santa Catarina (UNISUL), Florianópolis, Brazil

Business Administration, University of Federal Santa Catarina (UFSC), Florianópolis, Brazil

Centre for Sustainable Development (GREENS), Florianópolis, Brazil
e-mail: thiago.soares@unisul.br

F. T. Dias
Centre for Sustainable Development (GREENS), Florianópolis, Brazil

Environmental Science, University of Southern Santa Catarina, Florianópolis, Brazil

L. L. B. Lazaro et al. (eds.), *Water-Energy-Food Nexus and Climate Change in Cities*, Sustainable Development Goals Series, https://doi.org/10.1007/978-3-031-05472-3_2

affordable energy; connected to SDG 11: sustainable cities and communities. As a result, water quality, access to clean and renewable energy, and food security play and exercise a central role in the cities' sustainability. Thus, to make more sustainable cities, they must adopt energy efficiency strategies and promote renewable and clean energy technologies. In this connection, this research aims to analyze the importance of the water, energy, and food nexus for the promotion of sustainable cities in the framework of the Sustainable Development Goals. To that effect, we used a methodology suggestion that considered the need for an integrative literature review, as well as an analysis of field practical data. Furthermore, a theoretical analysis was structured, thus proposing three levels of hypotheses described and discussed in the text. As a result, this article has a double aspect: it contributes to broadening the literature, and at the same time it points out mechanisms for managers and municipal policymakers to develop better policies and strategies that enable the achievement of the SDGs, especially those listed above, having as a consequence strategies to mitigate the impacts of climate change.

Keywords

Water-energy-food nexus · Sustainable cities · Sustainable Development Goals · Sustainable development

2.1 Introduction

The phenomenon of urbanization is undoubtedly an issue that, although not emerging from the contemporary catastrophic process, is a topic that has been discussed for some time, taking into account both the humanistic condition and the direction that will be given in the use of urban spaces and their resources (Lefebvre, 2016). With the process of industrialization and migration of populations from rural to urban spaces, several issues emerged, such as health crises, housing deficit, urban swelling, and environmental deterioration among others (Alfonsin, 2001).

In this perspective, themes focused on the issue of resources inserted in the framework of urban space show a latent gap and thus suggest studies and mechanisms for the present-future projection of these resources. To validate the perspectives of sustainable cities, it is important to consider the rational use of spaces, resources and their fair distribution.

Thus, the water, energy, and food nexus, or simply nexus, first presented at the Bonn Conference in 2011, is perhaps the UN's greatest contribution to postmodern society as a tool for building a world in which the needs of the largest number of inhabitants on our planet are met. The nexus provides resources for improving water, energy, and food security for the communities that use it. The logic of the nexus combines three strategic resources in an integrated way. It is through that logic that the developed and developing nations can be presented with an intelligent view capable of guiding technicians, scientists, managers, and society in general, in the optimization of solutions, improving governance in all sectors (Islam et al., 2020; Schlör et al., 2020).

The concept of sustainability has been introduced, considering the well-being of the planet with continued growth and human development. The definition offered by the World Commission on Environment and Development has been seen as the guiding principle for long-term global development and is based on economic development, social development, and environmental protection: "meeting the needs of the present without compromising the ability of future generations to meet their own needs" (WCED, 1987).

In the last decades, many significant international events and commitments have impacted the evolution of sustainable development (Beynaghi et al., 2014; Lozano et al., 2014). In 2015, the 2030 Agenda for Sustainable Development was launched. This global development agenda includes 17 UN Sustainable Development Goals, or SDGs, and 169 milestones. The agenda addresses critical issues facing the world today, including the eradication of extreme poverty, tackling global inequality and climate change, promoting sustainable urbanization and industrial development, protecting natural ecosystems, and fostering the even growth of peaceful communities and governing institutions. A set of 232 indicators has also been developed to measure progress on SDG goals and targets, within and across countries (UN, 2018a). After 5 years, some progress is visible according to the last report of the UN (2020a): the share of children and youth out of school had fallen; the incidence of many communicable diseases was in decline; access to safely managed drinking water had improved; and women's representation in leadership roles was increasing. At the same time, the number of people suffering from food insecurity was on the rise, the natural environment continued to deteriorate at an alarming rate, and dramatic levels of inequality persisted in all regions.

The goals call for action "for people, planet, prosperity, peace, and partnership" to be implemented by "all countries and all stakeholders, acting in collaborative partnership."

With the adoption of the SDGs, countries and cities around the world are turning to the question of SDG implementation. How to make practicable these ambitious global goals? How can the national and local governments contribute? And why must cities and human settlements play a crucial role in their implementation (Kanuri et al., 2016)?

In this agenda, there is a specific goal with a focus on cities. The SDG 11 calls for "making cities and human settlements including, safe, resilient, and sustainable." That means creating opportunities for all, safe and affordable housing, better conditions in public transport, creating green spaces, improving urban planning, preparing cities with resilient societies and economies, and evenly increasing management with participatory ways.

Cities are critical to achieving the SDGs. As the world continues to urbanize, sustainable development depends increasingly on the successful management of urban growth for meeting the needs of the populations (UN, 2019). According to the United Nations' report, the world's population is expected to increase by 2 billion persons in the next 30 years, from 7.7 billion currently to 9.7 billion in 2050. That means that many cities and countries will face challenges including infrastructures such as housing, transportation, energy systems, and basic services such as education, health care, and food (UN, 2018b). Cities encompass some of the main issues to sustainable urban development, such as urban planning, transport systems, water, sanitation, waste management, disaster risk reduction, access to information, education, and capacity building (UN, 2020b).

The data shows that cities produce more than 50% of the world's greenhouse gas emissions (Hoornweg et al., 2011), are responsible for the use of 60% of the world's energy, and contribute to 70% of global waste (UN-Habitat, 2020). For this reason, cities are frequently considered the optimum strategic scale for action to mitigate climate change, in terms of land use, carbon control policies, and transitions toward a green economy.

In a scenery where climate change is accelerating, the world crisis is due to the scarcity of resources, and their poor distribution is becoming more worrying. One of the approaches is the water-energy-food nexus (Urbinatti, 2020). Toboso-Chavero et al. (2018) argue that cities are rapidly growing; they need to look for ways to optimize resource consumption, especially the food, energy, and water nexus.

Due to that, cities are increasingly paying attention to the sustainable development drivers. A sustainable city maximizes the benefits in economic and social dimensions under relevant constraints on environmental limitations (Mori & Yamashita, 2014). The authors emphasize the need for a balance between social and economic living standards and environmental sustainability.

Since 2016, the progress of the SDGs has been monitored globally, regionally, and subnationally. The SDSN (Sustainable Development Solutions Network) and Bertelsmann Stiftung annually publish the SDG Index and Dashboards (sdgindex.org).

In terms of the SDG 11 progress, the US Cities Sustainable Development Report (Lynch et al., 2019) provides an entry point into the United Nations' Sustainable Development Goals at the city level in the United States. The report shows that city performance isn't as closely tied to size, location, or goal; solutions and furthering progress require localized understanding and interventions. The last report published by SDSN discusses the impact of COVID-19 on SDG progress (Sachs et al., 2020). It is indisputable that COVID-19 will have severe negative impacts on most SDGs. In terms of sustainable cities and communities (SDG 11), the negative impact has been mixed or moderate; there is a rise in urban poverty and vulnerability; shut down of public transports, lower access to public/green spaces; movements of population that vary across countries; and sharp short-term reduction in pollution levels.

There is no doubt that there are still many challenges ahead for cities to be inclusive, productive, and environmentally friendly. New challenges have arisen due to COVID-19, an unprecedented crisis that is threatening lives and livelihoods, impacting health, the economy, and society in all countries (Leal Filho et al., 2020).

Thus, in the scope of sustainable urban spaces, a relevant aspect that even guided the theme selection, as well as the way to explore the bibliography, is about urban sustainability. The urban sustainability proposed by SDG 11 connected with the resources distributed and used in the urban setting, that is, water, energy, and food nexus, highlights the need to think and rethink policies that are more in line with the socio-environmental cities' outlook.

Given the aspects addressed, this article aims to analyze the importance of the water, energy, and food nexus for the promotion of sustainable cities in the context of the objectives of sustainable development.

2.2 Theoretical Framework

The literature review is divided into two themes: (1) the energy-water-food nexus as graphene for the SDGs and (2) the energy-water-food nexus in the context of sustainable cities.

2.2.1 The Energy-Water-Food Nexus as Graphene for the SDGs

Graphene is a derivative of carbon (one of its crystalline forms), as are the carbon nanotubes (diamond and graphite), and when added to other materials has remarkable thermal, mechanical, and electrical properties. Most applications in energy storage devices revolve around the application of graphene, as it can improve the performance, functionality, and durability of many applications. It is the strongest, lightest, and thinnest material that exists, being much more resistant than steel even though it is a two-dimensional material, that is, it has the thickness of an atom (Alam et al., 2019; Olabi et al., 2021).

And just like graphene, which comes from a simple, elementary, and necessary raw material for the formation of life, as we know it on Earth, the nexus is also a simple concept, whose very name clarifies its purpose and which establishes a relationship among three important resources for life on Earth (Hua et al., 2020). When investigating its potential, one can think of the nexus as a kind of graphene for the achievement of the Sustainable Development Goals (SDGs) and sustainability. Just as graphene has the potential to transform the world (Bandala & Berli, 2019; Makgabutlane et al., 2020), so does the nexus, which has incredible potential with much to be researched, being able to enhance the service of all SDGs.

The nexus is aligned not only with the concept of sustainability but also with the concept of industry 4.0. In this sense, we are already beginning to observe the emergence of concepts such as sustainability 4.0. If for some authors, industry 4.0 is seen more for its negative aspects, such as being responsible for the loss of countless jobs and for fostering the concentration of income; on the other hand, it can be an excellent ally of developed nations in meeting the SDGs and in the development of circular economy models (Dantas et al., 2020; Bai et al., 2020).

The techniques of automated irrigation and automation in planting are examples of how to optimize food production and also to reintroduce forests, thereby meeting various SDGs and avoiding diplomatic friction in the intervention in the natural wealth of other nations (USW, 2010; Goh et al., 2020). The automatic irrigation system allows the plants to receive water in the exact amount that the plant needs, and it is even possible to schedule irrigation. Thus, it eases the monitoring and irrigation time, which generally needs an extended working time, especially if the gardens have plants of different species, thus requiring irrigation in different quantities. This process is especially useful for greenhouses, including vertical gardens, thus allowing a large amount of food to be produced in a reduced space. For automated planting, specific equipment is used, which can use sensors for temperature and humidity of the air and soil and sensors for solar radiation.

It is observed the importance of interventions and support from governments in supporting research and incubators to address market failures so that the SDGs are achieved, remembering that sustainable development manages to meet the needs of the current generation without compromising the existence of future generations, at the same time that the best education is that which takes place through example (Surana et al., 2020).

Thus, it could be stated that if the concept of the nexus was already known and widely spread worldwide, it could have prevented the majority of developed countries from destroying their forests and from now being so concerned with forests and resources preserved by underdeveloped or developing countries. Even though Brazil has about 2/3 of its original forests preserved, the cleanest energy matrix in the world, passenger cars that do not use diesel, and is a pioneer in biofuels such as ethanol, it suffers from harsh environmental legislation that slows down the updating of its road and rail network, among other things, just like other developing countries. It is above all in the social dimension of sustainability that the ability to minimize conflicts is identified in the nexus, presenting evidence of becoming a central element in the development of a new era. Thus, while the SDGs can be seen as a set of guiding concepts for actions capable of developing the actions, the nexus contributes to the achievement of the SDGs by presenting concepts and strategic visions of resource management indispensable to human life, and in this sense the industry 4.0 also contributes to the achievement of several SDGs, highlighting two notable characteristics, the first is the potential to reforest large areas of deforested land and the automation associated with tools, software, irrigation systems, and machinery, which allow jumps in the volume of food production in the world, making it possible to feed the more than 7 billion human beings that currently live on our planet (Bai et al., 2020; Dantas et al., 2020).

For Islam et al. (2020), most studies related to the nexus were carried out by the United States, China, and Australia, representing 23%, 17%, and 15% of the total, respectively. Engineering thought from the perspective of the nexus is capable of turning professionals from the so-called hard sciences into social scientists, who, by performing engineering thinking with the logic of the nexus, sophisticate engineering projects and processes and thus meet important concepts and theories defended by social scientists, creating synergies and increasing efficiency (Proctor et al., 2020; Schlör et al., 2020).

Since the nexus relates three irreplaceable resources for human survival, at the same time that it guides managers in the efficient treatment of resources, the nexus also demonstrates their role as science and practice in international relations between countries and in their domestic markets. This art of orienting human relations in the face of strategic resources is what is identified in the nexus as an important instrument of diplomacy. Wars originate by obtaining resources, and then there are strategic management and governance tools that raise awareness and guide the importance of resources, the responsibility of all parties interested in these resources, and the encouragement of management tools capable of optimizing the use of resources prove to be diplomatic elements and, in essence, aligned with the UN objective, in creating the nexus in 2011 and the SDGs in 2015 (Salmoral et al., 2019; Hua et al., 2020; Schlör et al., 2020).

If, on the one hand, diplomacy can be understood as the art of preserving the rights and interests of the State in negotiation with foreign governments, on the other hand, through diplomatic relations, countries can bargain the resource of other countries, avoiding military wars and practicing wars of narratives and media in the defense of their interests. In this sense, international law recognizes the ability of States to exercise diplomatic protection over their national interests, and thus, again, the nexus is capable of structuring diplomatic deals on a rationalist basis capable of carrying out historical rescues (Hassan et al., 2017; Klimes et al., 2019; Proctor et al., 2020).

By drawing upon present and future scenarios of the planet while at the same time placing on the table, in numbers, what each country has done for the conservation of global nature throughout its history, reveals which countries have done little in this direction. Similarly, this points out which countries have advanced technology and how they plan to share these technologies so that the least developed countries reach the minimum acceptable Human Development Index (HDI) indicators that enable communities to optimize and reach the level of the most important SDGs. At the same time, we should avoid exacerbating the gap between the under-developed and developed countries, which mostly have advanced technology but have destroyed most of their forests. These are important challenges to be overcome in reaching the SDGs (Castillo et al., 2019; Santos, 2020).

The high potential for reforestation made possible by industry 4.0 is greater in the most technologically developed countries, which are precisely the countries that most polluted their waters and destroyed their original forests and which today charge countries that have preserved high percentages of their forests, to keep them untouchable. Industry 4.0 can help countries such as the United States, Canada, Russia, China, and the countries of the European Union to carry out reforestation of their original areas, avoiding diplomatic wear. And so it is clear that industry 4.0 can be an integrating element of countries, as can the nexus (Klimes et al., 2019).

Although cities can employ bioclimatic architecture solutions in their projects, aligned to the nexus, enabling water collection, treatment, and reuse, the use of energy efficiency and IoT concepts, along with renewable sources of energy generation, and also privilege the production of part of the food to be consumed in the urban environment, the fact is that large centers have become conflicting zones, as they do not support the demands and expectations of their inhabitants. Therefore, concepts such as those of sustainable cities and/or healthy people are more easily served in smaller cities, with a better population distribution from large centers to small centers, enabling direct service to important SDGs (del Río Castro et al., 2020; Liu et al., 2021).

2.2.2 The Energy-Water-Food Nexus in the Context of Sustainable Cities

Since the second half of the last century, discussions have been intensified about the urbanization process, which took place simultaneously in different parts of the world. In this connection, Lefebvre (2016) emphasized that this process occurred simultaneously with the advance of industrialization, which caused cities to spread into rural spaces and, consequently, led to the invasion of these spaces.

In this way, also according to Lefebvre (2016), that reckless urbanization driven by industrialization, whether from the perspective of rural-urban migration of people seeking jobs and economic improvement or from the perspective of enjoying the sociable spaces, resulted in the disorderly use of natural resources existing in the spaces that made up the cities.

As a result, the Industrial Revolution showed humanity how much the human being is capable of producing but also how much the human being is capable of consuming. Meadows (1972) projections impacted the world because they demonstrated that there is a limit to growth, imposed by nature itself. From this point on, many actions were taken to preserve the environment. However, where there are factories, there is employment. And a movement of people leaving the countryside searching for a better quality of life in cities could be seen all over the world.

New projections have been made; according to the UN (2019), 55% of the world's population lives in urban areas, and the population in urban areas is expected to be 70% of the world population by 2050. Therefore, cities need to prepare and transform themselves, becoming sustainable, healthy, intelligent, and resilient. Governments need to consider the goals for 2030 of the Sustainable Development Goals (SDGs) as a whole and contribute to solving so many problems that many cities still have.

Conceptualizing sustainable, healthy, intelligent, and resilient cities can be complex, as it requires urban transformation with economic and systemic changes. In general, it is a process that will make cities efficient, livable, inclusive, and environmentally friendly (Chehri & Mouftah, 2019). Much research is being carried out on urban sustainability to contribute and conceptualize urban sustainability.

A sustainable city can be defined as cities that develop responsibly in terms of social, environmental, and economic dimensions (Brito et al., 2019). These cities must reduce impact, waste, and emissions and expand recycling and encourage local businesses (El Ghorab & Shalaby, 2016), in addition to promoting environmental education. An active intervention aimed at improving the network of green spaces (Haase et al., 2017), the quality of human life (Li & Yi, 2020), and economic growth (Jing & Wang, 2020), is an essential driver to this aim.

We sought to define characteristics, indicators, and sub-indicators for sustainable, healthy, intelligent, and resilient cities. Anand et al. (2017) proposed a series of criteria and indicators; for this research, the following stand out: food security, use of renewable energy, and water quality/availability. Brito et al (2019) mentioned, among others, the following criteria: people, water, energy efficiency, and waste.

Deng et al. (2019), when proposing an evaluation method for urban sustainability, divided the evaluation points into primary dimensions, namely, building and installation, natural environment, people's satisfaction, and transportation system. Each primary dimension is made up of subdimensions. Regarding the subdimensions, the following stand out: energy efficiency, consumption efficiency, energetic environment, and health care.

Meerow (2020) cites the danger of rainwater. Su et al. (2019) present the indicators: energy consumption by gross domestic product (GDP) and water consumption by GDP. Langellier et al. (2019), with a view to a healthy city, present the indicators: changes in food preferences and nutritional literacy, among others. Based on this information, the importance of issues related to energy, water, and food is noticeable.

As highlighted by Sachs (2015) in the book *The Age for Sustainable Development*, the definition of sustainable cities is threefold, involving the three dimensions of sustainable development. Sustainable cities are economically productive, socially and politically inclusive, and environmentally sustainable. It means that cities must promote an inclusive and efficient economic activity and preserve the biodiversity, air, and water and physical health and safety of the citizens, especially in an age of increasing vulnerability to extreme climate catastrophes from climate change.

However, even though there are many sustainable city initiatives worldwide, there are also many difficulties and challenges.

The official road map for achieving the SDGs in urban environments, *Getting Started on the SDGs in Cities* was proposed by Kanuri et al. (2016) with collaboration from SDSN's Thematic Network on Sustainable Cities. The guide underly four steps: Step 1: Initiate an inclusive and participatory process of SDG localization. This includes raising awareness of the SDGs, engaging stakeholder collaboration, and potentializing political leadership. Step 2: Set the local SDG agenda: equipping the SDGs with ambitious but realistic local agendas, evidence-based decision-making, and public involvement. Step 3: Plan for SDG implementation: using goal-based planning, both long-term and multi-sectoral, and supporting it with financial resources and partnerships. Step 4: Monitor SDG progress: developing local monitoring and evaluation systems that are affordable, comprehensive, and effective in reliably capturing progress on local goals and targets.

In Brazil, the first connotation given to the idea of sustainable cities occurred through the normative scope of Federal Law 10.257 of 2001, also called the Statute of Cities. This law, in addition to establishing and regulating the Brazilian Urban Policy, aims to establish basic guidelines to govern the common use of urban spaces and also to establish mechanisms to conceptualize and seek the construction of apparatuses for urban sustainability (Brasil, 2001).

From this perspective, later in Brazil, the Sustainable Cities Program, an initiative of organized civil society in 2011, aims to contribute to the sustainability of cities through the involvement of the teams responsible in the city halls. The program proposes a process based on guidelines, indicators, and targets, organized into 12 thematic areas and linked to SDGs: (1) governance; (2) common natural resources; (3) equity, social justice, and a culture of peace; (4) local management for sustainability; (5) urban planning and design; (6) culture for sustainability; (7) dynamic, creative, and sustainable local economy; (8) education for sustainability and quality of life; (9) better mobility; (10) local action for health; (11) from local to global; and (120 responsible consumption and lifestyle options. The relevance of this program is that it is based on the engagement of local governments with goals for action during their management time (Programa Cidades Sustentáveis, 2017, 2020).

Although many collaborations between politics, research, business, and civil society have been developed in cities worldwide, Khair et al. (2020) emphasize the importance of local community active participation in promoting a city more sustainable, especially regarding their monitoring process.

Weymouth and Hartz (2018) also argue that to achieve the implementation of SDGs in cities, collaborative problem-solving and decision-making are required. According to them, the role of new partnerships in governance and the empowered participation of everyday people to develop public wisdom can help cities to integrate and implement SDGs and bring transformative change.

Dalla Fontana et al. (2020) researched the water-energy-food nexus in the Brazilian context. The authors mentioned that when implementing policies aimed at a specific sector, objectives in other areas may be impaired since the nexus requires an integrated approach. The research by Yuan et al. (2020) aims to measure the sustainability of the urban food-energy-water nexus in cities. The model used is composed of nine indicators, which are food production, malnutrition, agricultural area, energy consumption, electrical access, renewable energy, water consumption, access to water, and wastewater collection.

Bazzana et al. (2020) studied the impact of the water-energy-food nexus infrastructure on local well-being. According to the authors, the nexus proposes the use of natural resources in the social and economic context; however, the well-being of the population about investments in infrastructure is controversial.

When analyzing the effects of the construction of a specific hydroelectric plant, the results showed that there was a competition for land between rural communities and the water and energy sectors. In addition, the population's economic constraints can reduce access to new services (Bazzana et al., 2020).

However, thinking about the water-energy-food nexus in the context of sustainable cities proposes environmentally, economically, and socially viable policies but also intelligent, resilient, inclusive, and healthy. It is therefore important to assess whether the indicators contribute to the Sustainable Development Goals (SDGs). With that, we have the hypotheses of this research:

H1. There is an influence of sustainable city energy indicators in the SDGs.

H2. There is an influence of water indicators of sustainable cities in the SDGs.

H3. There is an influence of food indicators from sustainable cities in the SDGs.

2.3 Methodology

The theoretical-methodological structure proposed by this research followed two pathways: (1) selection of the methods used and (2) division of the steps and procedures used in the collection and treatment of data. For the first stage, the choice of methods and guidelines, their nature, way of obtaining data, and way of approach are highlighted.

Regarding its nature, it is proposed as applied research, since it aims to investigate and correlate data obtained directly, with the findings in the literature in the search for answers to the proposed hypotheses. To help with this type of research, there is the direct research method, which, through its mechanisms, seeks evidence in the delimited field for analysis and sampling (Martins & Theóphilo, 2018). On the other hand, regarding the approach, this research is based on a quantitative approach, since, from the perspective of Saunders et al. (2016), the quantitative approach allows the formulation of hypotheses, which can be tested (contributing to the development of the theory) or examined in future research.

About the steps of procedures used in this research, three major basic cores stand out, namely, (1) literature review, using systematic mechanisms for an integrative review; (2) procedures for preparing indicators and sub-indicators; and (3) structuring of data collection and analysis mechanisms and definition of the application area (Florianópolis).

Thus, in the first stage, for the integrative review, databases were not specifically defined, since the objective of integrative reviews is to obtain the main elements to support a given topic. However, in the searches, we used the main repositories/publishers such as Elsevier, Periódicos CAPES, Springer, and Google Scholar, searching keywords "water-energy-food nexus"; "sustainable cities"; "water-energy"; and "food nexus," defined to obtain more relevant and quoted results, compatible with the thematic proposal.

For the second stage, regarding the procedures for the development of indicators and sub-indicators, the literature review was essential, as it allowed the extraction of the most cited indicators. Subsequently, in the third stage, the data collection instrument was elaborated, using the questionnaires through the Google Forms tool, proposing a data collection instrument connected with the findings of the literature review.

In a subsidiary way, an analysis was carried out using descriptive statistics and the technique of modeling structural equations of partial least squares. For that, the real data were collected through direct research applying questionnaires. The dimensions listed in the present study were identified, translated, analyzed, and adapted from existing models in the literature mentioned in this work, aiming at maintaining compatibility with the theme and context of this research.

Regarding the procedures of the third stage, we highlight (1) the structuring of the research instrument and (2) data collection. Thus, the first version of the instrument had 72 questions on the 4 dimensions. On the other hand, about data collection, the period used was between June, 27 and August, 18, 2020, online, through the Google Forms tool with dissemination on social networks and email distribution.

Regarding data collection, initially, the completion and validation of the questionnaires received were verified. A total of 75 questionnaires were collected and validated. The criterion used was the selection of respondents living in the metropolitan region of the city of Florianópolis. The delimitation of this region followed the sampling method; despite the non-probabilistic sampling, this can be considered a homogeneous group, with at least one characteristic in common, residents of Florianópolis, as recommended by Flynn et al. (1990) and Hourneaux Jr et al. (2018).

And as the last procedure, the data collected were entered into Excel spreadsheets and analyzed using descriptive statistics and the modeling of the technique partial least squares structural equations (partial least squares), supported by the SmartPLS software, version 3. Subsequently, there was a compatibility comparison of the data obtained with the literature review, culminating in the discussion established in this research.

2.4 Presentation of Results

The sustainable city model is composed of 3 indicators, which are (1) energy, with 9 sub-indicators; (2) water, with 13 sub-indicators; and (3) food, with 3 sub-indicators.

Table 2.1 Strategic map of the energy indicator

1 Indicator	Energy indicator
Description	Actions aimed at energy efficiency
Authors	Bao and Toivonen (2014), Taecharungroj et al. (2018), Sokolov et al. (2019), Brito et al. (2019), Anand et al. (2017), Alyami (2019), Subadyo et al. (2019), Su et al. (2019), Li and Yi (2020), Ruan et al. (2020), Brilhante and Klaas (2018), Sokolov et al. (2019), Deng et al. (2019) and Vukovic et al. (2019)
1.1 Sub-indicator	Energy consumption by GDP
Description	Energy consumed for a population
Authors	Su et al. (2019), Li and Yi (2020), Ruan et al. (2020) and Brilhante and Klaas (2018)
1.2 Sub-indicator	Solar energy and energy from waste plants
Description	Clean energy sources
Authors	Subadyo et al. (2019) and Sokolov et al. (2019)
1.3 Sub-indicator	Wind force
Description	Wind power source
Authors	Subadyo et al. (2019)
1.4 Sub-indicator	Smart housing
Description	Residences with clean technologies
Authors	Anand et al. (2017) and Deng et al. (2019)
1.5 Sub-indicator	Hydroelectric plants
Description	Amount of energy generated in hydroelectric
Authors	Subadyo et al. (2019)
1.6 Sub-indicator	Energy measurement or sub-measurement
Description	Energy spent on a population
Authors	Alyami (2019)
1.7 Sub-indicator	Potential for renewable energy and passivity
Description	Amount of energy that can be generated
Authors	Bao and Toivonen (2014), Anand et al. (2017), Brilhante and Klaas (2018) and Alyami (2019)
1.8 Sub-indicator	Energy-efficient building systems
Description	Clean technologies for buildings
Authors	Bao and Toivonen (2014), Alyami (2019), Subadyo et al. (2019) and Vukovic et al. (2019)
1.9 Sub-indicator	Energy-efficient technologies
Description	Investment in clean technologies
Authors	Anand et al. (2017) and Sokolov et al. (2019)

Table 2.1 shows the strategic map of the energy indicator, including the description of the indicator and the sub-indicators, as well as the authors who supported the indicator and the sub-indicators.

Yuan et al. (2020) observed that renewable energy plays an essential role in the nexus and that future work should focus on the technological innovation associated with the nexus. State-of-the-art technology, which in recent years has become accessible on some scale to remote regions, enhanced the quality of life in the countryside, minimizing rural exodus and enabling reverse movement, especially after the COVID-19 world pandemic in which people have been forced to isolate at home to avoid crowds, which is uncommon in rural regions.

Table 2.2 shows the strategic map of the water indicator, including the description of the indicator and the sub-indicators, as well as the authors who supported the indicator and the sub-indicators.

The nexus also provides resources where there is a regional shortage, for example, through the so-called virtual water imports. Whether through

Table 2.2 Strategic map of the water indicator

2	Indicator	Water indicator
Description		Actions aimed at water efficiency
Authors		Bao and Toivonen (2014), Taecharungroj et al. (2018), Sokolov et al. (2019), Brito et al. (2019), Alyami (2019), Subadyo et al., (2019), Su et al. (2019), Meerow (2020), Brilhante and Klaas (2018), Giles-Corti et al. (2019), Steiniger et al. (2020), Jing and Wang (2020) and Li and Yi (2020)
2.1	Sub-indicator	Access to drinking water
Description		Access to drinking water in a population
Authors		Bao and Toivonen (2014), Brilhante and Klaas (2018), Alyami (2019), Giles-Corti et al. (2019) and Steiniger et al. (2020)
2.2	Sub-indicator	Wastewater capacity
Description		The capacity of wastewater in a city
Authors		Brilhante and Klaas (2018), Alyami (2019) and Jing and Wang (2020)
2.3	Sub-indicator	Water intake (rain, runoff)
Description		Investment in water intake
Authors		Alyami (2019) and Sokolov et al. (2019)
2.4	Sub-indicator	Per capita water consumption
Description		Water consumption of a population
Authors		Brilhante and Klaas (2018) and Su et al. (2019)
2.5	Sub-indicator	Industrial wastewater produced
Description		Quantity of wastewater produced from industries
Authors		Jing and Wang (2020) and Li and Yi (2020)
2.6	Sub-indicator	Renewable water source
Description		Investment in renewable water
Authors		Alyami (2019)
2.7	Sub-indicator	Quality of wastewater treatment service
Description		Water quality
Authors		Steiniger et al. (2020)
2.8	Sub-indicator	Water reuse (recycled)
Description		Investment in water recycling
Authors		Alyami (2019)
2.9	Sub-indicator	The leak detection system, meter system
Description		Monitoring of water leakage
Authors		Alyami (2019)
2.10	Sub-indicator	Drainage, sewage, and water systems
Description		Drainage, sewage, and water actions
Authors		Giles-Corti et al. (2019)
2.11	Sub-indicator	Water conservation technology and accessories
Description		Investment in clean water technologies
Authors		Alyami (2019)
2.12	Sub-indicator	Sewage and garbage treatment
Description		Actions aimed at the treatment of sewage and garbage
Authors		Bao and Toivonen (2014), Brilhante and Klaas (2018), Alyami (2019), Steiniger et al. (2020), Su et al. (2019) and Li and Yi (2020)
2.13	Sub-indicator	Surface runoff production
Description		Estimation of runoff production
Authors		Meerow (2020)

land degradation, water scarcity, or food crises, the main challenge under these various levels of restrictions is to reconcile long-term and global objectives, not limited to an immediate vision of economic benefits, privileging the guarantee of local livelihoods and nonnegotiable human rights to water and food, as well as energy, resources directly related to the HDI of nations (Xu, 2019; Islam et al., 2020; Liu et al., 2021).

Table 2.3 shows the strategic map of the food indicator, including the description of the indicator and the sub-indicators, as well as the authors who supported the indicator and the sub-indicators.

One of the issues that have become a concern with estimates of population growth is how to feed everyone. Today, malnutrition is still an unsolved

Table 2.3 Strategic map of the food indicator

3	Indicator	Food indicator
Description		Actions focused on food
Authors		Bao and Toivonen (2014), Langellier et al. (2019), Anand et al. (2017), Rosales (2011), Giles-Corti et al. (2019), He et al. (2020) and Steiniger et al. (2020)
3.1	**Sub-indicator**	Nutritional literacy
Description		Provision of nutrition professionals
Authors		Langellier et al. (2019)
3.2	**Sub-indicator**	Taxation of ultra-processed foods
Description		Control of ultra-processed foods
Authors		Langellier et al. (2019)
3.3	**Sub-indicator**	Access to food with sustainable and quality production
Description		Providing healthy food
Authors		Bao and Toivonen (2014), Langellier et al. (2019), Anand et al. (2017), Rosales (2011), Giles-Corti et al. (2019), He et al. (2020) and Steiniger et al. (2020)

problem, affecting 40% of the world's population. Another problem that often goes unnoticed is the deficiency of micronutrients, such as vitamins. There is also a third problem, obesity, common mainly in rich countries; about 15% of the world population are obese (Sachs, 2017).

Many people are not well informed about the nutrients they need to have a healthy life, so nutritional literacy and the availability of nutrition professionals are important. Many people do not have a varied menu; thus, they are deficient in nutrients.

2.4.1 PLS (Partial Least Squares) Analysis

In this subsection, the analysis of partial least squares will be carried out, through two steps: evaluation of the measurement model and analysis of the structural model, both of which are broken down in sequence.

2.4.2 Evaluation of the Model: Validity and Reliability

From the exportation of the collected primary data to the SmartPLS software, version 3, and its configurations having been made, the report of the preliminary data obtained was generated. The evaluation of the model started through its convergent validity, reliability, and discriminating validity, as recommended by Hair Junior et al. (2017).

It should be noted, however, that the sequence of analyses occurred by the recommendation of Bido et al. (2019), namely, (1) convergent validity; (2) discriminant validity (DV); (3) reliability; one of the assumptions for assessing reliability is that its convergent and discriminant validity are adequate. Thus, if any problem in the convergent or discriminant validity is diagnosed, it is suggested not to proceed to the reliability assessment.

Thus, it was found that the AVE of the latent variable (LV) values greater than 0.50 are acceptable, according to Ringle et al. (2014), and values greater than 0.40 may be acceptable in applied social sciences.

Once the convergent validity is assured, the next step was to assess the model's discriminant validity, which indicates whether the constructs or variables are independent of each other (Hair Junior et al., 2017). According to Ringle et al. (2014), there are two ways to ascertain this indicator: (a) by observing the cross loads, that is, indicators with higher factor loads in their respective LV than in the others, as recommended by Chin (1998) and (b) Fornell-Lacker criterion, in which the square roots of the AVE must be greater than the correlations between the constructs (Fornell & Larcker, 1981).

First, cross factorial loads were evaluated, according to the Chin criterion (1998), which proved to be adequate.

Subsequently, the RV was evaluated according to the Fornell-Larcker criteria, which according to Hair Junior et al. (2017) is considered more conservative. Table 2.4 shows the values of the

Table 2.4 Values of the correlations between LV and square roots of the AVE values on the main diagonal (highlighted)

	Energy	Food	SDG	Water
Energy	0.779			
Food	0.571	0.808		
SDG	0.632	0.661	0.791	
Water	0.762	0.672	0.716	0.842

Source: Made by the author, based on research data (SMARTPLS3®, 2020)

Table 2.5 Values related to the internal consistency of the model

Dimension	Cronbach's alpha	Composite reliability	AVE
Energy	**0.870**	**0.902**	**0.607**
Food	**0.695**	**0.840**	**0.653**
SDG	**0.956**	**0.961**	**0.625**
Water	**0.954**	**0.960**	**0.709**

Source: Made by the author, based on research data (SMARTPLS3®, 2020)

correlations between LV and square roots of the AVE values on the main diagonal (highlighted).

Through the analysis of Table 2.4, it appears that all the values of the correlations between the LV are less than the square roots of their AVE; therefore, the Fornell-Larcker criterion was met.

Finally, the values of internal consistency were evaluated using Cronbach's alpha and composite reliability. Table 2.5 shows the referred values, together with the values related to the AVE.

Table 2.5 shows that the Cronbach's alpha of the constructs is greater than 0.60. Also, the reliability criterion met was considered, through the composite reliability indexes, which were higher than the minimum limit of 0.7 (Hair Junior et al., 2017).

Therefore, by validating the measurement model, based on the criteria described above, the next subsection will be dedicated to the analysis of the structural model.

2.4.3 Evaluation of the Structural Model

The first evaluation carried out consisted of the analysis of collinearity, which is the inflation variation factor (IVF). According to Hair Junior et al. (2009), failure to comply with this assumption may make inferences based on the model that is erroneous or unreliable. It is emphasized that, in the context of PLS-SEM, IVF value equal to or greater than five indicates a potential collinearity problem (Hair Junior et al. 2011). However, according to Hair Junior et al. (2009), one should consider removing one of the corresponding indicators if the level of collinearity is very high, as indicated by an IVF value equal to or greater than ten.

As all values are less than ten, it was decided to keep all variables. Subsequently, Pearson's determination coefficients (R^2) were evaluated. According to Ringle et al. (2014, p. 67), R^2 "assesses the portion of the variance of endogenous

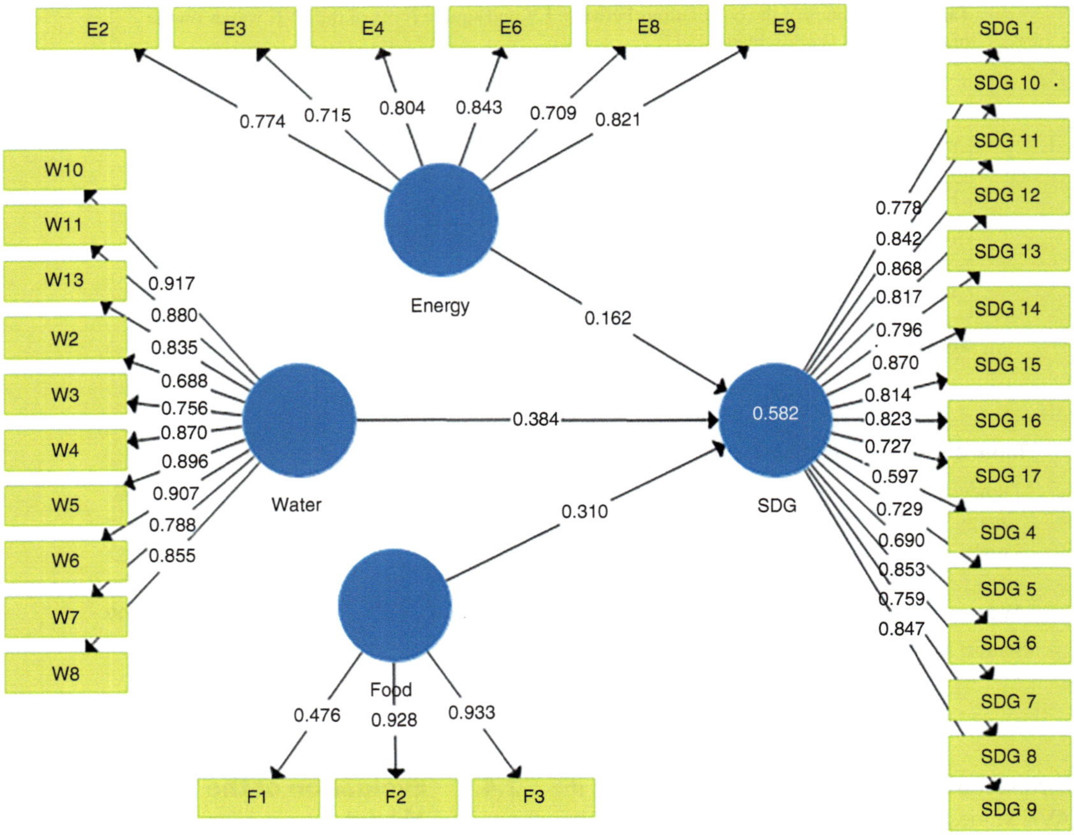

Source: SmartPLS3® (2020).

Fig. 2.1 Proposed model, R^2, and path coefficients. (Source: SmartPLS3® 2020)

variables, which is explained by the structural model." Fig. 2.1 shows the structure of the measurement model, with the values of R^2 and path coefficients.

According to Cohen (1988), for the area of social and behavioral sciences, the coefficient usually varies between 2% and 26%, with R^2 = 2% considered a small effect; R^2 = 13% medium effect; and R^2 = 26% large effect. Hair Junior et al. (2011) consider that R^2 results above 0.20 are considered high in disciplines such as consumer behavior.

It appears that the endogenous LV ODS has an R^2 of 0.582, above the percentage suggested as large/high; according to the classifications of Cohen (1988) and Hair Junior et al. (2011), all have a large effect on the model. The model explained a substantial part of the variation of endogenous variables, specifically, 58.20%.

To test the significance of the pointed relationships, the bootstrapping technique was used, which, according to Ringle et al. (2014), is a resampling technique used to assess the significance (p-value) of the correlations (measurement models) and the regressions (structural model). Thus, a bootstrapping resampling procedure and analysis were performed with 5000 bootstrap samples per group. As shown in Table 2.6, only in Hypothesis 1 is it above the reference value (1.96). In this case, H_0 was rejected, and it can be said that the correlations and regression coefficients are significant, providing support for this part of the proposed model.

Then, it is observed in the nexus that the interdisciplinarity and multidisciplinarity that this science possesses are capable of meeting the urgent need to improve water, energy, and food security for the poorest. And it is capable to guide the

Table 2.6 Hypothesis testing

Hypothesis	Path	T statistics	P values	Results
H1	Energy → SDG	1404	0.161	Not supported
H2	Food → SDG	2670	0.008	Supported
H3	Water → SDG	2362	0.019	Supported

Source: Made by the author (2020)

richest not to waste, managing responsibly and intelligently necessary and limited resources. This can be stated especially if indices and trends of population growth on the planet are noted (Proctor et al. 2020; Schlör et al., 2020).

Yuan et al. (2020) state that cities are the places where the consumption of food, energy, and water occurs and that consumption creates challenges that have a strong impact on natural sources. Just as the nexus can guide the design of engineering and architecture projects, it can also assist in the analysis of sustainable, smart, and healthy cities and their governance mechanisms.

2.5 Conclusion

With this study, we got the understanding that cities, in their most unique forms, are inextricably connected with the existing nexus in the triad: water-energy-food, resulting from the rapid urbanization process, which during the last decades has proved to be of extreme concern.

In this light, it appears that the main problems involving the topic of rapid urbanization; unregulated and unbridled use of natural resources; and their resulting social, economic, and environmental problems have resulted in a repetitive logic in different parts of the world.

Thus, as a result of this reflection and the research carried out between the literature and the practice, this article aimed to analyze the importance of the water-energy-food nexus for the promotion of sustainable cities in the context of sustainable development objectives.

As a result, we found the most cited indicators and sub-indicators in the literature, which make up a sustainable city model regarding the water-energy-food nexus. The model had a practical application with the population of a city, thus validating the relevant indicators and sub-indicators for this population.

Cities are the places where the consumption of food, energy, and water occurs, and this consumption creates challenges that have a strong impact on natural sources. Just as the nexus can guide the design of engineering and architecture projects, it can also assist in the analysis of sustainable, smart, and healthy cities and their governance mechanisms (Yuan et al., 2020).

However, given the positive aspects of industry 4.0, it is clear that it allows for an opposite movement, from cities to the countryside. This movement helps to reach the quality-of-life indicators, with the advantage that small cities have better indicators about urban violence, level of air and noise pollution. The consequences of these actions can be reflected in lower levels of stress, making it even easier to access land, which makes green development paths feasible, stimulating circular and family economy models (Dantas et al., 2020; Liu et al., 2021).

Supported by industry 4.0, the nexus establishes new engineering concepts and production standards, capable of guiding also the commercial relations that occur both between countries and in their domestic markets.

The results help municipal managers and policymakers to formulate policies and strategies that enable the SDGs to be met. Furthermore, the research contributes to the understanding of the interrelationships between the water-energy-food nexus and sustainable cities in the context of the Sustainable Development Goals and strategies to mitigate climate change.

Acknowledgments This study was conducted by the Centre for Sustainable Development (Greens), from the University of Southern Santa Catarina (UNISUL) and Ânima Institute (AI), in the context of the project BRIDGE—Building Resilience in a Dynamic Global

Economy: Complexity Across Scales in the Brazilian Food-Water-Energy Nexus—funded by the Newton Fund, Fundação de Amparo à Pesquisa e Inovação do Estado de Santa Catarina (FAPESC), Coordenação de Aperfeiçoamento de Pessoal de Nível Superior (CAPES), National Council for Scientific and Technological Development (CNPq), and the Research Councils United Kingdom (RCUK).

References

Agenda 2030. Os 17 Objetivos de Desenvolvimento Sustentável. Disponível em: http://www.agenda2030.org.br/ods/17 /. Acesso em 30 Nov 2020.

Alfonsin, B. (2001). O Estatuto da Cidade e a construção de cidades sustentáveis, justas e democráticas. *Revista Direito e Democracia*. Available in: http://www.periodicos.ulbra.br/index.php/direito/article/view/2405

Alam, F., Choosri, M., Gupta, T. K., Varadarajan, K. M., Choi, D., & Kumar, S. (2019). Electrical, mechanical, and thermal properties of graphene nano platelets reinforced UHMWPE nanocomposites. *Materials Science and Engineering: B, 241*, 82–91.

Alyami, S. H. (2019). Opportunities and challenges of embracing Green City principles in Saudi Arabia future cities. *Access, 7*, 178584–178595. https://doi.org/10.1109/access.2019.2959026

Anand, A., Rufuss, D. D. W., Rajkumar, V., & Suganthi, L. (2017). Evaluation of sustainability indicators in smart cities for India using MCDM approach. *Energy Procedia, 141*, 211–215. https://doi.org/10.1016/j.egypro.2017.11.094

Bai, C., Dallasega, P., Orzes, G., & Sarkis, J. (2020). Industry 4.0 technologies assessment: A sustainability perspective. *International Journal Of Production Economics*, 107776.

Bandala, E. R., & Berli, M. (2019). Engineered nanomaterials (ENMs) and their role at the nexus of food, energy, and water. *Materials Science for Energy Technologies, 2*(1), 29–40.

Bao, S., & Toivonen, M. (2014). The specificities and practical applications of Chinese eco-cities. *Journal of Science and Technology Policy Management, 5*(2), 162–176. https://doi.org/10.1108/JSTPM-05-2014-0020

Bazzana, D., Zaitchik, B., & Gilioli, G. (2020). Impact of water and energy infrastructure on local Well-being: An agent-based analysis of the water-energy-food nexus. *Structural Change and Economic Dynamics, 55*, 165–176. https://doi.org/10.1016/j.strueco.2020.08.003

Brasil. (2001). *L10257 Estatuto da Cidade*. Available in: http://www.planalto.gov.br/ccivil_03/leis/leis_2001/l10257.htm. Accessed 25 Jan 2021.

Beynaghi, A., Moztarzadeh, F., Maknoon, R., Waas, T., Mozafari, M., Hugé, & Leal Filho, W. (2014, November). Towards an orientation of higher education in the post Rio + 20 process: How is the game changing? *Futures, 63*, 49–67.

Bido, D., De, S., & Da Silva, D. (2019). Smart PLS 3: Especificação, estimação, avaliação e relato. *Administração: Ensino e Pesquisa, Rio de Janeiro, 20*(2), 488–536 mai/ago. 2019.

Brilhante, O., & Klaas, J. (2018). Green City concept and a method to measure Green City performance over time applied to fifty cities globally: Influence of GDP, population size and energy efficiency. *Sustainability, 10*(6), 23. https://doi.org/10.3390/su10062031

Brito, V. T. F., Ferreira, F. A. F., Perez-Gladish, B., Govindan, K., & Meidute-Kavaliauskiene, I. (2019). Developing a green city assessment system using cognitive maps and the Choquet integral. *Journal of Cleaner Production, 218*, 486–497. https://doi.org/10.1016/j.jclepro.2019.01.060

Castillo, R. M., Feng, K., Sun, L., Guilhoto, J., Pfister, S., Miralles-Wilhelm, F., & Hubacek, K. (2019). The land-water nexus of biofuel production in Brazil: Analysis of synergies and trade-offs using a multiregional input-output model. *Journal of Cleaner Production, 214*, 52–61.

Chehri, A., & Mouftah, H. T. (2019). Autonomous vehicles in the sustainable cities, the beginning of a green adventure. *Sustainable cities and society, 51*, 101751. https://doi.org/10.1016/j.scs.2019.101751

Chin, W. W. (1998). The partial least squares approach for structural equation modeling. In G. A. Marcoulides (Ed.), *Modern methods for business research* (pp. 295–336). Lawrence Erlbaum Associates.

Cohen, J. (1988). *Statistical power analysis for the Behavioral Sciences* (2nd ed.). Psychology Press.

Dalla Fontana, M., Moreira, F., Di Giulio, G., & Malheiros, T. (2020). The water-energy-food nexus research in the Brazilian context: What are we missing? *Environmental Science & Policy, 112*, 172–180. https://doi.org/10.1016/j.envsci.2020.06.021

Dantas, T. E. T., & de-Souza, E. D., Destro, I. R., Hammes, G., Rodriguez, C. M. T., & Soares, S. R. (2020). How the combination of circular economy and industry 4.0 can contribute towards achieving the sustainable development goals. *Sustainable Production and Consumption*.

del Río Castro, G., Fernández, M. C. G., & Colsa, Á. U. (2020). Unleashing the convergence amid digitalization and sustainability towards pursuing the sustainable development goals (SDGs): A holistic review. *Journal of Cleaner Production*, 122204.

Deng, W., Peng, Z., & Tang, Y.-t. (2019). A quick assessment method to evaluate sustainability of urban built environment: Case studies of four large-sized Chinese cities. *Cities, 89*, 57–69. https://doi.org/10.1016/j.cities.2019.01.028

El Ghorab, H. K., & Shalaby, H. A. (2016). Eco and green cities as new approaches for planning and developing cities in Egypt. *Alexandria engineering journal, 55*(1), 495–503. https://doi.org/10.1016/j.aej.2015.12.018

Flynn, B. B., Kakibara, S. S., Schroeder, R. G., Bates, K. A., & Flynn, E. J. (1990). Empirical research methods in operations management. *Journal of Operations Management, 9*(2), 250–284.

Fornell, C., & Larcker, D. F. (1981). Evaluating structural equation models with unobservable variables and measurement error. *Journal of Marketing Research, 18*(1), 39–50.

Giles-Corti, B., Lowe, M., & Arundel, J. (2019). Achieving the SDGs: Evaluating indicators to be used to benchmark and monitor progress towards creating healthy and sustainable cities. *Health Policy*. https://doi.org/10.1016/j.healthpol.2019.03.001

Goh, C. S., Ahl, A., & Woo, W. T. (2020). Sustainable transformation of land-based economic development in the era of digital revolution. *Trends in Biotechnology*.

Haase, D., Kabisch, S., Haase, A., Andersson, E., Banzhaf, E., Baró, F., & Wolff, M. (2017). Greening cities – To be socially inclusive? About the alleged paradox of society and ecology in cities. *Habitat international, 64*, 41–48. https://doi.org/10.1016/j.habitatint.2017.04.005

Hair Junior, J. F., Ringle, C. M., & Sarstedt, M. (2011). Pls-Sem: Indeed a silver bullet. *The Journal of Marketing Theory and Practice, 19*(2), 139–152.

Hair Junior, J. F., et al. (2009). *Análise multivariada de dados* (6th ed.). Bookman.

Hair Junior, J. F., et al. (2017). *A primer on partial least squares structural equation Modeling (PLS-SEM)* (2nd ed.). SAGE.

Hassan, M., Afridi, M. K., & Khan, M. I. (2017). Environmental diplomacy in South Asia: Considering the environmental security, conflict and development nexus. *Geoforum, 82*, 127–130.

He, X., Lin, M., Chen, T. L., Liu, B., Tseng, P. C., Cao, W., & Chiang, P. C. (2020). Implementation plan for low-carbon resilient city towards sustainable development goals: Challenges and perspectives. *Aerosol and Air Quality Research, 20*(3), 444–464. https://doi.org/10.4209/aaqr.2019.11.0568

Hoornweg, D., Sugar, L., & Gómez, C. L. T. (2011). Cities and greenhouse gas emissions: Moving forward. Environment & Urbanization. International Institute for Environment and Development (IIED). 207. *Vol, 23*(1), 207–227.

Hourneaux, Jr., Gabriel, & Gallardo-Vázquez. (2018). Triple bottom line and sustainable performance measurement in industrial companies. *Revista de Gestão, 25*(4).

Hua, T., Zhao, W., Wang, S., Fu, B., & Pereira, P. (2020). Identifying priority biophysical indicators for promoting food-energy-water nexus within planetary boundaries. *Resources, Conservation and Recycling, 163*, 105102.

Islam, K. N., Kenway, S. J., Renouf, M. A., Lam, K. L., & Wiedmann, T. (2020). A review of the water-related energy consumption of the food system in nexus studies. *Journal of Cleaner Production*, 123414.

Jing, Z., & Wang, J. (2020). Sustainable development evaluation of the society–economy–environment in a resource-based city of China: A complex network approach. *Journal of cleaner production*, 121510. https://doi.org/10.1016/j.jclepro.2020.121510

Kanuri, C., Revi, A., Espey, J., & Kuhle, H. (2016). *Getting started with the SDGs in cities: A guide for stakeholders*.

Khair, N. K. M., Lee, K. E., & Mokhtar, M. (2020). Sustainable city and community empowerment through the implementation of community-based monitoring: A conceptual approach sustainability 2020, 12, 9583.

Klimes, M., Michel, D., Yaari, E., & Restiani, P. (2019). *Water diplomacy: The intersect of science, policy and practice*.

Langellier, B. A., Kuhlberg, J. A., Ballard, E. A., Slesinski, S. C., Stankov, I., Gouveia, N., & Diez Roux, A. V. (2019). Using community-based system dynamics modeling to understand the complex systems that influence health in cities: The SALURBAL study. *Health & Place, 60*, 102215. https://doi.org/10.1016/j.healthplace.2019.102215

Leal Filho, W., Brandli, L. L., Salvia, A. L., Rayman-Bacchus, L., & Platje, J. (2020). COVID-19 and the UN sustainable development goals: Threat to solidarity or an opportunity? *Sustainability, 12*, 5343.

Lefebvre, H. (2016). *O Direito à Cidade* (1st ed.). Nebli.

Li, W., & Yi, P. (2020). Assessment of city sustainability—Coupling coordinated development among economy, society and environment. *Journal of cleaner production, 256*, 120453. https://doi.org/10.1016/j.jclepro.2020.120453

Liu, Z., Huang, Q., He, C., Wang, C., Wang, Y., & Li, K. (2021). Water-energy nexus within urban agglomeration: An assessment framework combining the multiregional input-output model, virtual water, and embodied energy. *Resources, Conservation and Recycling, 164*, 105113.

Lozano, R., et al. (2014). A review of commitment and implementation of sustainable development in higher education: Results from a worldwide survey. *Journal of Cleaner Production*.

Lynch, A., Lo Presti, A., & Fox, C. (2019). *The 2019 US cities sustainable development report*. Sustainable Development Solutions Network (SDSN).

Makgabutlane, B., Nthunya, L. N., Maubane-Nkadimeng, M. S., & Mhlanga, S. D. (2020). Green synthesis of carbon nanotubes to address the water-energy-food nexus: A critical review. *Journal of Environmental Chemical Engineering, 104736*.

Martins, G. D. A., & Theóphilo, C. R. (2018). *Metodologia da Investigação Científica Para Ciências Sociais Aplicadas* (Atlas, 3 ed.). Atlas.

Meadows, et al. (1972). *The Limits to Growth*. Potomac Associates – Universe Books.

Meerow, S. (2020). The politics of multifunctional green infrastructure planning in new York City. *Cities, 100*, 102621. https://doi.org/10.1016/j.cities.2020.102621

Mori, K., & Yamashita, T. (2014). Methodological framework of sustainability assessment in City sustainability index (CSI): A concept of constraint and maximisation indicators. *Habitat International*.

Olabi, A. G., Abdelkareem, M. A., Wilberforce, T., & Sayed, E. T. (2021). Application of graphene in energy storage device–a review. *Renewable and Sustainable Energy Reviews, 135*, 110026.

Plan International. *Conheça os 17 Objetivos de Desenvolvimento Sustentável.* Disponível em: https://plan.org.br/conheca-os-17-objetivos-de-desenvolvimento-sustentavel/. Acesso em 29 Nov 2020.

Proctor, K., Tabatabaie, S. M., & Murthy, G. S. (2020). Gateway to the perspectives of the food-energy-water nexus. *Science of the Total Environment, 142852.*

Programa Cidades Sustentáveis. (2020). *Gps – Guia para Elaboração do Plano de Metas*. São Paulo. Disponível em: https://www.cidadessustentaveis.org.br/arquivos/Publicacoes/Guia_para_Elaboracao_do_Plano_de_Metas.pdf.

Programa Cidades Sustentáveis. (2017). *Guia de Uso do Sistema de Indicadores para a Construção de Observatórios*. São Paulo. Disponível em: https://www.cidadessustentaveis.org.br

Ringle, C. M., Silva, D. da., & Bido, D. (2014). Modelagem de equações estruturais com utilização do Smart PLS. *Revista Brasileira de Marketing*, 13(2), 56–73, maio/2014.

Rosales, N. (2011). Towards the Modeling of sustainability into urban planning: Using indicators to build sustainable cities. *Procedia Engineering, 21*, 641–647. https://doi.org/10.1016/j.proeng.2011.11.2060

Ruan, F., Yan, L., & Wang, D. (2020). The complexity for the resource-based cities in China on creating sustainable development. *Cities, 97*, 102571. https://doi.org/10.1016/j.cities.2019.102571

Sachs, J. (2015). *The age of sustainable development* (542p). Columbia University Press.

Sachs, J. D. (2017). *A era do desenvolvimento sustentável* (J. Araújo, Trans. Vol. 1). Actual.

Sachs, J., Schmidt-Traub, G., Kroll, C., Lafortune, G., Fuller, G., & Woelm, F. (2020). *The sustainable development goals and COVID-19. Sustainable development report 2020.* Cambridge University Press.

Salmoral, G., Schaap, N. C., Walschebauer, J., & Alhajaj, A. (2019). Water diplomacy and nexus governance in a transboundary context: In the search for complementarities. *Science of the Total Environment, 690*, 85–96.

Santos, I. T. (2020). Confronting governance challenges of the resource nexus through reflexivity: A cross-case comparison of biofuels policies in Germany and Brazil. *Energy Research & Social Science, 65*, 101464.

Saunders, M., Lewis, P., & Thornhill, A. (2016). *Research methods for business students* (7th ed.). Pearson Education.

Schlör, H., Märker, C., & Venghaus, S. (2020). Developing a nexus systems thinking test–A qualitative multi-and mixed methods analysis. *Renewable and Sustainable Energy Reviews*, 110543.

Sokolov, A., Veselitskaya, N., Carabias, V., & Yildirim, O. (2019). Scenario-based identification of key factors for smart cities development policies. *Technological Forecasting and Social Change, 148*, 119729. https://doi.org/10.1016/j.techfore.2019.119729

Steiniger, S., Wagemann, E., de la Barrera, F., Molinos-Senante, M., Villegas, R., de la Fuente, H., & Barton, J. R. (2020). Localizing urban sustainability indicators: The CEDEUS indicator set, and lessons from an expert-driven process. *Cities, 101*. https://doi.org/10.1016/j.cities.2020.102683

Su, M., Xie, H., Yue, W., Zhang, L., Yang, Z., & Chen, S. (2019). Urban ecosystem health evaluation for typical Chinese cities along the belt and road. *Ecological Indicators, 101*, 572–582. https://doi.org/10.1016/j.ecolind.2019.01.070

Subadyo, A. T., Tutuko, P., & Jati, R. M. B. (2019). Implementation analysis of Green City concept in Malang - Indonesia. *International Review for Spatial Planning and Sustainable Development, 7*(2), 36–52. https://doi.org/10.14246/irspsd.7.2_36

Surana, K., Singh, A., & Sagar, A. D. (2020). Strengthening science, technology, and innovation-based incubators to help achieve sustainable development goals: Lessons from India. *Technological Forecasting and Social Change, 157*, 120057.

Sustainable Development Report – Sustainable Development Report (sdgindex.org). Accessed 21 Nov 2020.

Taecharungroj, V., Tachapattaworakul Suksaroj, T., & Rattanapan, C. (2018). The place sustainability scale: Measuring residents' perceptions of the sustainability of a town. *Journal of Place Management and Development, 11*(4), 370–390. https://doi.org/10.1108/JPMD-04-2017-0037

Toboso-Chavero, S., Nadal, A., Petit-Boix, A., Pons, O., Villalba, G., Gabarrell, X., Josa, A., & Rieradevall, J. (2018). Towards productive cities environmental assessment of the food-energy-water nexus of the urban roof mosaic. *Journal of Industrial Ecology, 23*(4), 767–780.

UN-Habitat. UN-Habitat – A Better Urban Future | UN-Habitat|UN-Habitat – A Better Urban Future | UN-Habitat. Accessed 21 Nov 2020.

United Nations. (2018a). *Global indicator framework for the sustainable development goals and targets of the 2030 agenda for sustainable development.* Available online: https://unstats.un.org/sdgs/indicators/. Accessed 21 Nov 2020.

United Nations. (2018b). *Department of Economic and Social Affairs 68% of the world population projected to live in urban areas by 2050, says UN.* https://www.un.org/development/desa/en/news/population/2018-revision-of-world-urbanization-prospects.html. Accessed 21 Nov 2020.

United Nations. (2019). *Department of economic and social affairs, population division. World population prospects 2019: Highlights*. ST/ESA/SER.A/423.

United Nations. (2020a). *The sustainable development goals report 2020.*

United Nations. (2020b). *Sustainable cities and human settlements*. Available online: https://sdgs.un.org/ar/node/91

Urbinatti, A. M. (2020). *Nexos de sustentabilidade e intersetorialidade: Políticas públicas em periferias urbanas* (108p). Tese doutorado USP.

USW. Farms Here, Forests There (2010). Disponível em: http://assets.usw.org/our-union/pulp-paper-forestry/farms-here-forests-there-report-5-26-10.pdf. Acessoem 29 Nov 2020.

Vukovic, N., Rzhavtsev, A., & Shmyrev, V. (2019). Smart City: The case study of Saint-Peterburg 2019. *International Review*, (1–2), 15–20.

Weymouth, R., & Hartz, J. (2018). Principles for integrating the implementation of the sustainable development goals in cities Karp. *Urban Science, 2*, 77.

World Commission on Environment and Development. (1987). *Our common future*. Oxford University Press.

Xu, Z., Li, Y., Herzberger, A., Chen, X., Gong, M., Kapsar, K., & Liu, J. (2019). Interactive national virtual water-energy nexus networks. *Science of the Total Environment, 673*, 128–135.

Yuan, M. H., Chiueh, P. T., & Lo, S. L. (2020). Measuring urban food-energy-water nexus sustainability: Finding solutions for cities. *Science of the Total Environment, 752*, 141954.

3 What Can Cities Do to Enhance Water-Energy-Food Nexus as a Sustainable Development Strategy?

Rodrigo A. Bellezoni, Karen C. Seto, and José A. Puppim de Oliveira

Abstract

Cities are dependent on hinterlands – whether local or global – for water, energy, and food (WEF) to sustain urban activities. With the projected growth of urban population and consumption, the demand for natural resources tends to increase. Moreover, climate change will potentially increase the insecurity of the availability of WEF in cities. Decision-makers in cities are often faced with the very challenging issue of resource management due to scarcity of resources that generates conflicts among stakeholders. Therefore, the risks associated with rapid urbanization and climate change have highlighted the need to reconfigure the development of cities to optimize and reduce the use of resources in order to achieve the Sustainable Development Goals (SDGs). Nevertheless, various approaches have been developed in the last decades to improve the WEFN. Thus, this chapter presents challenges and opportunities for improving the governance of cities over WEF systems and the nexus among them. Using the WEF nexus framework, cities would benefit from a transition toward a circular economy that uses renewable resources and designs cyclical and efficient systems. This would encourage innovative responses and effective partnerships toward smarter cities able to tackle climate change.

Keywords

Resilient cities · Integrated planning · Resource flows · Teleconnections · Innovation · Governance

3.1 Introduction

Most of the urban population and consumption are concentrated in cities (UN, 2019). In 2020, 55% of the world's population lived in urban areas (TWB, 2020). In developing countries, one-third of their population is living in informal settlements (UN-HABITAT, 2021). The UN estimates that the global urban population will increase by 2.5 billion by 2050, with 90% of this growth in developing countries (UN, 2019).

R. A. Bellezoni (✉)
São Paulo School of Management (FGV EAESP), Fundação Getúlio Vargas (FGV), São Paulo, SP, Brazil
e-mail: rodrigo.bellezoni@fgv.edu.br

K. C. Seto
Yale School of the Environment, FGV EAESP and FGV EBAPE, New Haven, CT, USA

J. A. Puppim de Oliveira
Fundação Getulio Vargas, FGV EAESP and FGV EBAPE, São Paulo, Brazil

L. L. B. Lazaro et al. (eds.), *Water-Energy-Food Nexus and Climate Change in Cities*, Sustainable Development Goals Series, https://doi.org/10.1007/978-3-031-05472-3_3

Urban areas take up just 3% of the world's surface area, but they account for 60–80% of global energy consumption and 75% of global carbon emissions (IPCC, 2014). Urbanization rate of a country is more correlated to its carbon dioxide emissions than GDP per capita, showing that urbanization per se has a multiplying effect on climate change (Sethi & Puppim de Oliveira, 2015).

This projected urban population growth, in turn, tends to stress water and energy systems (TWB, 2020). Consumption and waste are further generated from growth in urban population, which have environmental consequences. As such, there is the dilemma of building new developments on existing agriculture lands or redeveloping existing unutilized or underutilized properties (Bren d'Amour et al., 2017). The transformation of nonurban landscapes also poses significant challenges for reducing the rate of biodiversity loss and related ecosystem functionality (Güneralp et al., 2017). These interlinks between components create a complex framework for understanding the development of cities.

Cities define societies. We were used to thinking of cities with clear boundaries, with suburbs far from central regions, with industrial parks far from residential areas. Towns grow into small cities and small cities become larger, and some larger cities reach the scale of conurbation known as megacities. Thus, the world is comprised of thousands of cities and towns which, in turn, are connected by supply chains, labor, and capital and require infrastructure connections (Fig. 3.1). In fact, cities are supported by resources coming mainly from rural regions located around the world far distant from the centers of consumption. Therefore, the sustainability of a city can no longer be considered in isolation from the sustainability of human and natural resources it uses from proximal or distant regions, or the combined resource use and impacts of cities globally (Seitzinger et al., 2012).

Every city is connected to its rural hinterland in complex ways. The production of energy requires water, while the supply of water requires energy, and the production of food requires both water and energy (Hoff, 2011; Bazilian et al., 2011; Howells et al., 2013). Within city boundaries, the cross-sectoral WEF interactions are expected to occur in all cities, although variations in magnitudes exist (Haase et al., 2017; SEI, 2018; Newell et al., 2019). Cities are almost entirely dependent on surrounding (or global) hinterlands for WEF to sustain urban population and activities. Beyond city boundaries, we have transboundary effects that should be addressed by regional policies. However, these effects are not only limited to neighboring cities. The concept of land teleconnections (changes that correlate over large geographic distances) applies precisely to the processes that intertwine land use and urbanization, connecting places through their processes and depicting changes in nonurban places that affect urban places and vice versa (Seto et al., 2012). The urban pull for resources results in tension with neighboring communities and with the development of megacities, unwanted environmental impacts arise, especially climate change. In this context, the development of a global system of cities that develop sustainable processes and policies together with nonurban areas becomes fundamental. Therefore, a better understanding of the connection potential between cities and rural areas could increase capacities to make changes and foster governance on a global scale, thus increasing the WEF securities (Seitzinger et al., 2012).

Indeed, population growth, consumption patterns, and rapid urbanization have been taking their toll on the natural resources that support our living conditions. The resultant effect of this growth has increased the demand for different kinds of energy, such as for generating electricity, transportation, and heating. Water quality and quantity challenges, for example, have typically been met with the addition of more gray infrastructure, such as reservoirs and conventional treatment plants, which require more energy and materials for their construction. Thus, some of humanity's biggest concerns, such as energy and water provision, poverty, hunger, and climate change, are, in some way, city problems. Improving WEF nexus in cities can be an alternative to mitigate some of the increasing demand for resources.

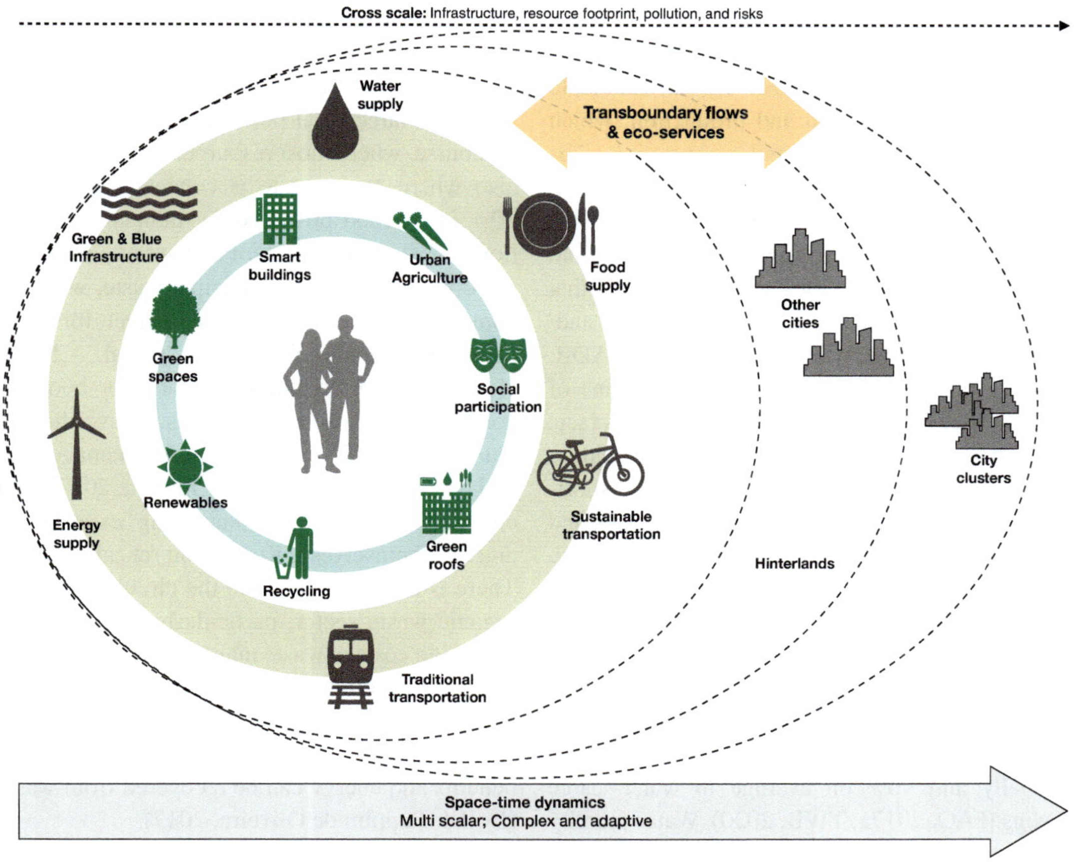

Fig. 3.1 Local, regional, and global impacts from infrastructure. Site-specific actions for more sustainable WEF provision (inner circle). Municipal infrastructure (within a city, light green) and its transboundary effects on other infrastructure, ecosystems, and regions. Governance: Moving from the inner to the outer circle increases the number of stakeholders and complexity and scope of the potential benefits. (Adapted from Suzuki et al. (2010), Ramaswami et al. (2016), and Bellezoni et al. (2021))

In this chapter, we present some challenges that rapid urbanization imposes on cities. We seek to bring together various elements affecting WEF systems into a holistic framework to support multilevel, multi-sectoral, and multi-stakeholder system approaches to cities, highlighting their role in tackling climate change. The aim of this chapter is to present more details on how the concepts of the WEF nexus approach can benefit the integrated planning of greener and smarter cities in promoting synergies and identifying trade-offs in the demand and supply of scarce resources. Thus, it highlights what cities can do to become increasingly sustainable and responsible for a global shift toward sustainability.

3.2 Interlinkages Between Urban Water, Energy, and Food Systems

Cities have three different systems for water, energy, and food, which are not often explicitly connected. Their governance can also be disconnected from the city governance system. The WEFN approach tries to analyze those systems in a more holistic way and finds ways to connect them.

3.2.1 Water System

Cities are often characterized by unsustainable patterns of consumption and production, which contributes to environmental degradation. The actual rate of urbanization is exerting pressure on fresh water supplies, landscapes, food systems, energy commodities, the living environment, and public health. Many cities rely on supplies that lack resilience to seasonal weather patterns and/or the future impacts of climate change (ADB, 2012; FAO, 2020). Drought is probably one of the most severe existential threats that cities face. It can have many effects in urban areas, including increases in water shortages, electricity shortages, water-related diseases, and food prices and food insecurity from reduced supplies (UCCRN, 2018).

Water crises are also drivers of other risks. Perhaps most prominent are the impacts of water on food security. For instance, agricultural production is responsible, on average, for around 70% of surface and groundwater withdrawals globally and 90% on average in water-scarce basins (FAO, 2017a; TWB, 2020). Water scarcity may contribute to negative economic impacts and increased rural to urban migration and is identified by the World Economic Forum (2016) as the risk of greatest concern to the global economy in the coming years. And water scarcity is exacerbated by pollution. Urban planning failures result in waste(water) reaching water bodies, often used for city water supply. In addition to impacting ecosystems and their associated biodiversity, it also harms human health and increases the demand for energy for water treatment. The effects of these water challenges are already in stark display. And so are food resources.

3.2.2 Food System

Urbanization affects every aspect of our food systems, from the way food is produced and priced, to the way it is processed, consumed, recycled, and wasted (Seto & Ramankutty, 2016). Food chains require lots of water and energy inputs for production, being one of the largest sectors in terms of consumption-based GHG emissions per capita (FAO, 2019). In urban areas, dwellers consume up to 70% of the food supply, even in countries with large rural populations (FAO, 2017b). Of course, where most resources are consumed is also where most waste is concentrated. About 30% of all food produced for human consumption is lost (FAO, 2011). Food waste comprises more than 50% of all municipal waste, which is commonly the single highest budget for most local administrations (Silpa et al., 2018). However, there are limited efforts for boosting composting capacity in cities. Nevertheless, many cities are innovating in waste management and composting (Puppim de Oliveira, 2017), and cities are exchanging composting experiences among themselves (Kurniawan et al., 2013). There is a need to upgrade the circularity of the organic waste sector, particularly in places that have high costs in waste management (Paes et al., 2021). Reduction in food waste also means reduction in GHG emissions (Paes et al., 2020), as cities have immense potential for climate co-benefits and energy can be recovered from waste (Doll & Puppim de Oliveira, 2017).

More efficient management of food systems is, therefore, not only strategic from an environmental point of view due to high water and energy consumption and associated emissions. Achieving food security in an era of rapid urbanization will require considerably more understanding about how urban and food systems are intertwined (Seto & Ramankutty, 2016). Thus, food production and consumption are crucial for the sustainable socioeconomic development of cities and peri-urban areas. In low-income countries, food expenditure in cities may be as high as two-third of the total household expenditure, while agro-industry accounts for more than 50% of value added in manufacturing (Crush & Frayne, 2011; FAO, 2017b). And for each type of food, there are different production requirements. The produce can be organic, without the use of fertilizers, or it can come from a monoculture with major land use requirements with landscape impacts. They can be fast food or come from the wet market. For each type of food, there is a related environmental, social, and economic impact, with different effects on WEF systems. Fresh and healthier organic products, produced on a smaller scale are, for example, more expensive. Fast food is more convenient, accessible, and cheaper. Therefore, the access, price, and choice of food also have direct effects on both human diets and health, and the search for solutions to the WEFN also demands reflections on the quality of food and health prevention.

Providing cities with high-quality food today and tomorrow is critical. Cities are increasingly more important actors in terms of eradicating hunger also because all forms of malnutrition are presented in cities. In cities we have both hunger, extreme hunger, obesity, and overweight. In the last three decades, the worldwide incidence of obesity has steadily increased. Today more than 2 billion adults are either overweight or obese (GNR, 2018); in urban areas, this is evidenced by an increasing number of overweight and obese pregnant women (Chen et al., 2018). Thus social, economic, and environmental sustainability of food systems and the evolution of urban diets will be largely dependent on the management of food systems in urban and peri-urban areas. Moreover, bringing food production closer to the consumption hubs, i.e., cities, can also avoid energy consumption, pollution, and GHG emissions as more local production will require less transportation and refrigeration.

3.2.3 Energy System

Another key issue driving GHG gases in cities is energy, which is used in the water and food sector. Today's food system is built upon refrigeration. Refrigeration dependency is the result of several socioeconomic and technological developments. The more lifestyle habits change toward a more modern, technological, and practical society, the greater the changes in the diet of populations, especially urban ones. This shift toward urban living is changing our relationship with food. Higher household incomes and less time to shop for food and cooking at home tend to increase the consumption of fast food, street food, processed and frozen meals, and for foods not locally produced. This boosted the development of supermarkets, house refrigeration, and cold storage facilities. These facilities are the prime infrastructural component for perishables such as fruits, vegetables, fish, and meat. Besides the role of stabilizing market prices and evenly distributing both on-demand and time bases, the cold storage industry renders other advantages and benefits to both farmers and consumers. Again, there is no modern life without increased consumption of energy, especially electricity.

The energy use for space cooling in residential and commercial buildings worldwide has more than tripled between 1990 and 2016 (GCCA, 2019). Just refrigerated storage can account for up to 10% of the total carbon footprint for some products when considering electricity inputs, the manufacturing of cooling equipment, and emissions from lost refrigerants. Electricity for space cooling is growing faster than for any other end use in buildings, and this demand is expected to triple by 2050 (IEA, 2018). And this is not only because temperatures are rising but also because people's incomes are growing, especially in hottest countries. Take the cases of China, India, and Indonesia, some of the countries most affected by climate change. Projections are that they will account for half of all the growth in energy consumption for cooling over the next 30 years (IEA, 2018, 2020).

However, the local socioeconomic and socio-ecological context of a particular urban area should be considered to analyze the urban energy sector. Different factors such as demography, climate, and economics govern electricity usage patterns at local levels. Cities require an uninterrupted supply of energy for transport, industrial and commercial activities, buildings and infrastructure, water distribution, and food production, and one differs from another. In addition, urban sprawl, increasing distances between destinations, and inefficient public transport systems prompt overall reliance on private motorized transport, which has a high-energy consumption, mostly of petroleum products. Urban form and distances from rural areas also affect the way food is transported and consumed. The agro-industrial production of ultra-processed foods requires long distances for transportation, with consequent fossil fuel consumption and GHG emissions, mainly related to the demands of the cold chain. In this regard, urban agriculture advocates argue that some energy can be saved from the avoided food mileage, as well as from avoided packaging and cooling activities, when food is produced locally.

The fast-paced metabolism of cities is putting pressure on limited energy which has an annual demand growth of around 7% in developing countries, while the supply remains stable (UN-HABITAT, 2021). Hence a mismatch between the supply and demand and frequent power rationing in cities. An additional problem to the equation. Although a large share of global energy consumption is centered in cities, energy accounting systems remain predominantly concentrated at the national and regional scales (Grubler et al., 2012; Chowdhury et al., 2020). This lack of energy statistics at the urban level severely hinders planning processes and reliable decision-making.

WEF systems are, therefore, fully intertwined and interdependent. Impacts on one resource, whether from the demand or supply side, affect all the others and, hence, the entire production or consumption chain. The collection, treatment, and distribution of water use energy, and reservoirs are usually located far from urban centers, requiring a lot of energy for pumping. The production of energy requires large amounts of water. Usually, power plants are installed close to the consuming center or to the resource source. Therefore, some water or energy will be lost during the transportation steps. Food production requires a lot of land and water and is part of the peri-urban landscape. Energy is essential for irrigation, fertilizer production, transportation, and food storage. All of these steps emit large amounts of CO_2 each. Although cities are not characterized as producing regions for WEF resources, they are where much of the demand is concentrated.

Therefore, urban problems will not be solved unless urban institutions and governance begin to commit resources to technical assessments of the cities' needs and potential, with a focus on developing strategic plans that improve access to essential resources, through the redesign of blocks, streets, and public services aimed at lower water and energy consumption and with potential for food production within cities themselves. From an integrated analysis of WEF systems, opportunities may arise for reducing external dependence on resources, as well as contributing to more sustainable local actions that can have impacts on broader scales.

3.3 Urban Water-Energy-Food Nexus and the Sustainable Development Goals

WEF resources are crucial for socioeconomic development particularly in urban areas, as they import most of the WEF. Thus, the WEF systems, and the nexus among them, are fundamental to determining the Sustainable Development Goals (SDGs) in cities. How the urban WEF is governed affects the well-being of the population. Thus, urban dwellers, particularly decision-makers in cities, should be concerned about the governance of those resources, and this governance can boost WEFN to avoid present and future risks of scarcity. The lack of proper management of WEF systems may lead to negative trade-offs, which are affected by technological choices. Conventional sector-specific approaches for managing urban WEF miss out on possibilities to address human needs, community priorities, and local, regional, and global resource challenges more efficiently. This is also known as "silo thinking."

Take the example of water management, which is often conducted without considering the other systems involved. Cities share water with other users that are located within a single river basin. While cities are important users of water, for most river basins, they are by no means its major consumers, as agriculture is the largest water consumer in most countries. As a result, security-of-supply issues should always be set in the context of the overall river basin, rather than in the context of the city in question (Abell et al., 2017). So, a robust basin-level management is required for those who share water resources. In the absence of that, socioeconomic advancement in a particular city may be constrained since water shortages impact both a city's level of industrial productivity and local production of food and energy. Thus, the WEFN functions as a shorthand for the links between an urban center's water, energy, and food supplies.

The WEFN concept was firstly proposed in the background paper for the Bonn 2011 Conference, highlighting the "need to secure local livelihoods and the human rights to water and food" (Hoff, 2011). It provides a framework for cities to understand and tackle the challenges that urbanization poses for these critical resources while developing resilient cities. In short, the logic behind the WEFN is that it shifts attention from a one-sector view to a more integrated one (Al-Saidi & Elagib, 2017). However, it is not that easy to implement in practice because cities usually develop themselves without major concerns about resource interlinkages and how their governance can be improved. Cities are organic systems; they just grow and will usually only worry about problems when they become too big. The first step to an integrated planning would be recognizing that WEF resources are interdependent and the city could benefit with a better integration among them (Fig. 3.2). This small step would already represent an important opportunity to provide services to human communities and to optimize cities toward addressing local, national, and global concerns regarding SDGs. However, science also lacks a consensus on which approach to WEFN is most appropriate (Newell et al., 2019; Simpson & Jewitt, 2019). This is because each specific situation requires an understanding of its conditioning factors. Thus, there is no one-size-fits-all solution. Each city has its own challenges, resources, limitations, characteristics, technology, fears, concerns, perils, political will, and potentials.

Therefore, cities will have to better plan their future, or they risk failing in securing WEF in the medium and long term. Their inhabitants and governments have the power to do this. For instance, when the USA pulled out of the Paris Climate Agreement, several American cities doubled down on their climate commitments. Cities can play a key role in adopting integrated approaches for managing water, energy, and food systems. However, few cities have full control of the WEF resources. Advancing WEFN is not so much about the individual projects, as it is about linking and evaluating the activities through systemic and interconnected implications. It creates a framework for valuing trade-offs and making informed decisions about achieving multiple benefits for WEF services.

Thus, sustainable initiatives aimed at adaptation and mitigation of the impacts of climate change must begin in the urban environment and involve people at smaller scales. Many cities are leading the way in changing attitudes toward this, proving to be part of the solution and not just the problem. So far, we have highlighted how fundamental cities are to the world we are building for tomorrow. And how important they are for planning water, energy, and food systems. Next, we will present how integrated urban planning, ecological approaches, digital tools, and political will can be intertwined to make cities a better place.

3.4 Approaches to Address WEF Nexus in Cities

Cities can act in improving WEFN, but there is much more of it, especially in the Global South, which are more vulnerable to WEF constraints. A recent study from OECD (2020) shows that a number of governments are using the post-COVID-19 measures to roll back existing environmental regulations and taxes and increase fossil fuel-intensive infrastructure and electricity. This would reinforce perhaps even more unsustainable development paths. Despite that, we still have a golden opportunity to start designing principles of resilience into our cities (Fig. 3.3). In the last decades, many approaches have allowed cities to govern their resources and be more sustainable in their use and consequently reduce their impacts on climate change.

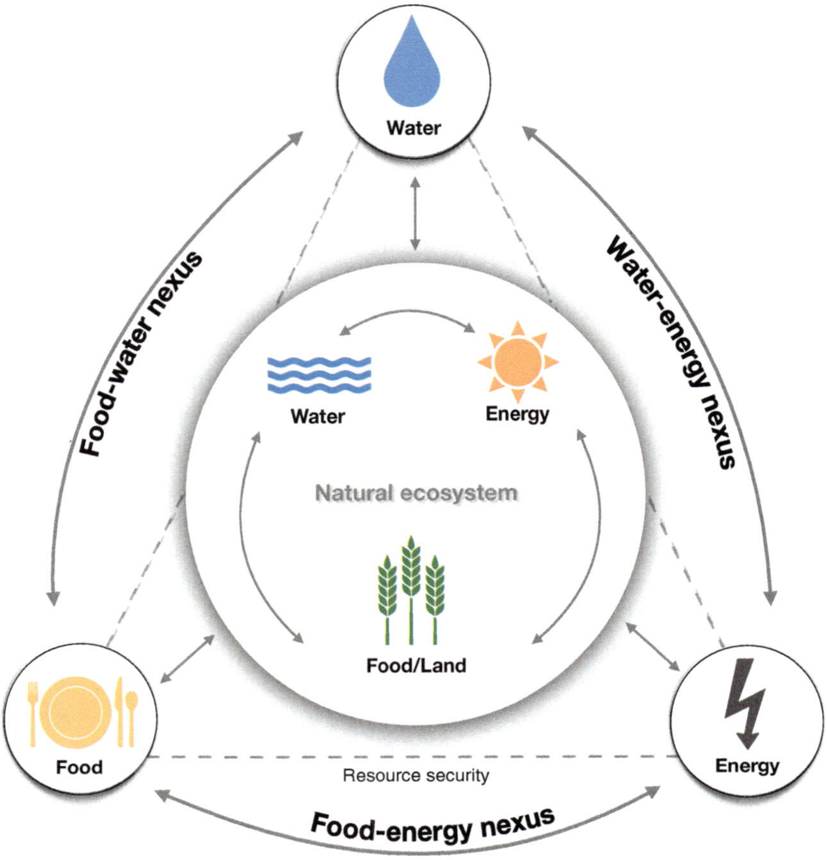

Fig. 3.2 Water-energy-food nexus approach. Integrating policies for WEF can promote more efficient infrastructure and improve resource supply security by identifying synergies between systems in contribution to more sustainable cities. (Adapted from Hoff (2011), Bazilian (2011), and UN-ESCAP (2019))

Policy and governance should focus on providing and innovating basic infrastructure for all, planning based on urban form and identifying multisector synergies for resource efficiency. Self-sufficiency of cities can be stimulated by the use of new technologies and concepts of the circular economy and energy transition, which can benefit local green finance mechanisms. Reductions in inequality and increased social participation can encourage fundamental community processes toward more sustainable cities, through dynamic multisector urban improvements. Some of these approaches will be discussed in more detail below, from the perspective of what cities can do to enhance WEFN as a sustainable development strategy.

3.4.1 Integrated WEF Planning Supported by Ecosystem-Based Approaches

Policies developed for one goal inadvertently influence design solutions in unforeseen areas. This silo thinking is increasingly pushing WEF systems into major problems. The existing WEF-related policies settings are typified by a narrow sectoral, issue-specific, and fragmented viewpoint. These isolated policies are therefore incapable of addressing long-term WEF challenges facing cities and countries, especially for future infrastructure and more severe climate change. To overcome these shortcomings and to provide meaningful insights into policy issues, researchers have been developing analytical frameworks that integrate the technical, economic, and social

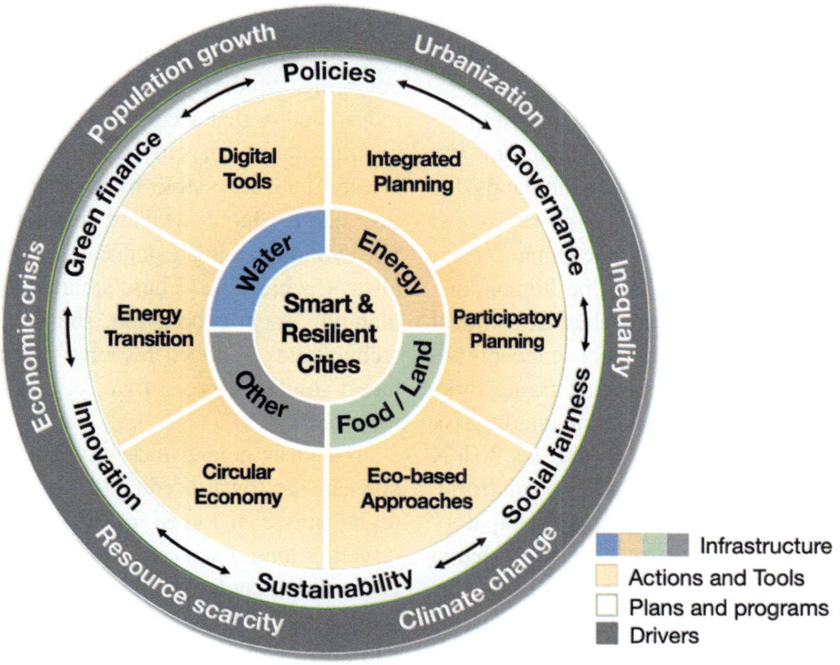

Fig. 3.3 Transition to smart and resilient cities. A green recovery will significantly enhance the resilience of economies and cities in the face of both the severe recession and accelerating environmental challenges. (Adapted from UN-ESCAP (2019) and IEED (2020))

dimensions of WEF policies (Newell et al., 2019; Simpson & Jewitt, 2019). For this to happen at the city level, knowledge of urban morphology, combined with temporal and spatial cross-sectoral infrastructure data, is essential.

Indeed, numerous data sets, from census data to aerial and satellite photographs and remote sensing information, are being integrated to enable planners to characterize urban form. However, regardless of the tool or methodology used, the most important change in WEF policy approach resides in integrating different policies with distinct objectives into a single more embracing and local-specific WEFN policy. The sustainable management of resources implies action at various levels; national management requires resource-use planning policies that consider WEF use and its social, environmental, and economic impacts from an integrated standpoint. Therefore, greater coherence between policies related to water, energy, food, land use, waste, and emissions is needed at local, regional, and national levels to improve the sustainability of WEF. In addition, these policies should consider integration of different government levels. The big question therefore is how to integrate these issues into a comprehensive framework. From a broad review of WEFN projects worldwide, Endo et al. (2017) have pointed that developing methods to integrate interdisciplinary, multi-sectoral, and multidimensional research results is essential to analyze and understand interrelationships and trade-offs among WEF. Therefore, the WEFN approach itself may be used as tools for decision-makers to collectively address economic, social, environmental, and energy goals.

In fact, employing an integrated approach increases the complexity of solutions but is the only way to make real progress toward meeting multiple objectives. Greater integration between WEF demand and supply should also be explored. Overall, infrastructure projects are predominately of a supply nature, i.e., they are focused on the supply of a service rather than a reduction in the demand for a service. This may encourage excessive use rather than efficient use of services, which is counter to SDGs. Therefore, efficiency and conservation should be assessed before supply-side investments.

Considering that climate change will amplify the challenges in balancing elements of the WEFN, the expected increase in demands for WEF will put more pressure on governments, since the environmental impacts estimated from integrated policy scenarios may be higher than those from sector-driven approaches. Thus, local governments should use WEFN concepts to develop shared long-term planning frameworks and tools for achieving SDGs and managing risk for greater resilience. Cities can use the WEFN approach to create a transdisciplinary platform, with integrated design teams using participatory WEFN frameworks. At every stage, a shared WEF framework would help communicate and work together in a coordinated fashion, reducing efforts to micromanage city departments and stakeholders involved in different project stages.

Infrastructure such as WEF, ideally, should be integrated managed through participatory and ecosystem-based approaches. The smart city approach can be an asset in fostering these approaches to overcome urban challenges. Smart cities are systems of people interacting and using energy, materials, services, and finance to catalyze economic development and improved quality of life (Albino et al., 2015). It is based on the concept of a creative and sustainable city, which makes use of technology in its planning process with citizen participation. Its concepts are in line with the growing recognition that cities are now on the front lines in managing change and leading an integrated approach. Smart cities are the future of cities that is happening now, and they are key to better management of urban WEF systems (Alzaabi et al,. 2019; Sukhwani et al., 2020).

Innovative cities have demonstrated that, supported by the appropriate strategic approach, they may greatly enhance WEF resource efficiency by realizing the same value from a much smaller and renewable resource base, while decreasing pollution and waste (Covarrubias, 2019). Cities also need to know how to build resiliently but in a sustainable, fair, equitable and inclusive way. As the switch to home working makes us balk at the back-and-forth of commuting, smart cities are emerging. Modern cities must be designed or redesigned so that access to work, housing, food, health, education, culture, and leisure is co-located and accessible by multiple modes, not just motor vehicles. Having destinations and origins within walking distance is ideal although not always achievable. Thus, smart cities integrate planning, local production, social participation, technologies, reduced resource demand, nature-based solutions, green finance, and tackling climate change.

The ideas behind these "cities of the future" concepts go toward reducing the number of cars on the streets, the need for public transportation, traveling long distances to work, using bicycles, and walking outdoors while enjoying interactions with people and local businesses, where everyone naturally contributes to reducing WEF consumption. It is a win-win situation. Some examples of retrofitting cities into smart cities include implementing end-use efficiency in the energy and water sector; reducing, reusing, and recycling waste; and adapting existing transportation infrastructure to more efficient uses (e.g., routes for bus rapid transit and lanes for bicycles). In fact, poorly planned cities represent a constant drain on resources, and the implementation of integrated planning with smart city concepts can bring many advantages for the joint sustainability of WEF systems (UN-HABITAT, 2021).

While the best design for each city is case-specific, smart city development can contribute to the strategic use of infrastructure resources with urban planning and management to meet the social, environmental, and economic needs of society. Due to more planning and information about demand and supply of services and the needs of people and businesses, governments and society will have better conditions to define sustainable strategies for WEF systems. Dwellers are increasingly willing to pay to live and work in suburbs with more green infrastructure, with green actions and social activities, with less inequality, and more social and environmental responsibility (Zalejska-Jonsson et al., 2020; Ando et al., 2020).

Aware of this and concerned about the increasing WEF demands, many cities across the world have been deploying engineering projects for providing ecosystem services such as rainwater harvesting and flood mitigation, local microclimate control, air purification and CO_2 storage, and wastewater treatment, among others. These built structures are called green and blue infrastructure (GBI) or nature-based solutions (NbS). GBI refers to the city's naturescape, i.e., the mix of trees, gardens, green roofs, living walls, parkland, squares, and reserves (Haase et al., 2014; Andersson et al., 2014; Pitman et al., 2015). Ultimately, NbS are essential for providing urban ecosystem services to urban dwellers and is closed linked to the provision of resources in urban territories. These infrastructures are components of the ecosystem-based adaptation (EbA) approach, which entails incorporating biodiversity and eco-services into the broader adaptation strategy to mitigate climate change effects (Chong, 2014). EbA has the immense potential to contribute to the developmental and environmental goals at local, national, and global levels. Its practices aim to achieve balanced and secure systems that maximize potential synergies and reduce potential conflicts, providing multiple benefits and a wide range of cost-effective applications in different landscapes (Nguyen et al., 2017; Muthee et al., 2021).

Historically, wealthy nations have been investing in gray infrastructure[1] to reduce flooding and water security risks, while less developed nations unable to afford expensive engineering solutions remain at high risk (Abell et al., 2017). This seems like a clear opportunity for developing countries, but NbS also need knowledge, political will, and investment. Indeed, there have been many cities in developing countries introducing innovative approaches to GBI (Macedo et al., 2021). With a combination of growing water demand for agriculture, energy production, domestic and industrial use, and decreased water reliability due to climate change, even developed countries may find that gray solutions alone are insufficient (IEA, 2020). Sustainable water, energy, and food security will require, therefore, an integration of traditional engineering with NbS, given their potential to provide local resources to urban territories.

Indeed, the mosaic of land uses in and around cities includes both the built and natural environment. More recently, open-space design has incorporated principles of ecosystem planning and biodiversity conservation, including spaces for the growing of food crops, tree crops, and small or even vertical farms. In addition to food production, urban agriculture initiatives reduce food mileage by saving energy and emissions at the storage and transport stages (Martin-Moreau & Ménascé, 2019). They also encourage the consumption of fresh, healthier food, generating jobs and boosting the local economy. Urban areas provide a range of benefits to sustain and improve human livelihood and the quality of life through urban ecosystem services, with direct and indirect contributions from ecosystems to human well-being. These contributions are framed in terms of "what ecosystems do" for people (Haines-Young & Potschin, 2018).

On the other hand, it takes a lot of blue to stay green. GBI can also produce ecosystem disservices, defined as functions of ecosystems that are perceived as negative for human well-being (Lyytimäki & Sipilä, 2009; Gómez-Baggethun et al., 2013). Some concerns are related to water and energy security (increased demand), waste and food safety (contamination from sewage usage, water contamination from pesticides), health (mosquito-borne diseases proliferation), and others (Bellezoni et al., 2021). These trade-offs, therefore, need to be understood and evaluated through an integrated framework within a local context and with a variety of stakeholders (Haase et al., 2014, 2017, Kremer et al., 2016). Thus, the management of urban GBI must be connected to the socioecological dynamics of each location, with their cultural, economic, and political specificities jointly guiding the development of a goal-based ecosystem-based-WEFN framework (Bellezoni et al., 2021).

[1]Conventional infrastructure. It refers to structures such as dams, seawalls, roads, pipes or water treatment plants, etc.

3.4.2 Bottom-Up Decision- Making Through Participatory Planning

Cities are multicultural. They should be participatory, promote civic engagement, and engender a sense of belonging and ownership. According to the New Urban Agenda (UN, 2017), cities of tomorrow should provide universal access to water and sanitation, as well as equal access to public goods and quality services in areas such as food security, health, education, infrastructure, and livelihoods. Thus, if the population in cities want to change things, they will have to find their political voice, a voice that is participative, diverse, and collective. And urban governance plays a key role in promoting greater representativeness, popular participation in planning and decision-making, and bottom-up policy design (Fig. 3.4). City governments have a prominent role in overseeing urban WEF systems. They need the power to be able to issue debt, to raise taxes, to zone effectively, and to build affordable housing, but they also need top-down support through enabling policies, capacity provision, and funding schemes.

Participation can strengthen public policy process by bringing resources, ideas, and legitimacy to solve local issues (Puppim de Oliveira, 2005). Participation could also help to strengthen the transdisciplinary nature of WEFN. Participatory process could bring diversity of solutions to WEF problems and maximize urban sustainable potentialities, by bringing new voices to the decision-making process. Importantly, public participation can legitimate and integrate social practices and creativity to find solutions to WEF problems.

And because cities play a key role in people's lives and countries' economies, they grow not only in size and complexity but also in organization, within and between cities. There are hundreds of inner-city coalitions in the world today. Just take a look at ICLEI (Local Governments for Sustainability), a large network that brings together thousands of people in thousands of member cities to contribute to urban gardens, clean water, waste management, and renewable energy.[2] Or look at the World Economic Forum, which is developing smart city and net-zero carbon cities protocols.[3] Or the C40 initiative, which is taking bold climate action, leading the way toward a healthier and more sustainable future.[4] City networks are the movements of the future. They are all showing the world that cities can and must work together to amplify their voice, not only nationally but also globally. Cities need a revolution that is up to their standards that demands the renegotiation of agreements in an independent way, acting as a nation-state, defending the interests of millions of inhabitants, and shaping policies and initiatives from the bottom-up. Many

[2] See more at https://iclei.org

[3] Visit https://www.weforum.org/projects/smart-cities-centre-of-emerging-technologies, https://www.weforum.org/projects/systemic-efficiency

[4] Visit https://www.c40.org

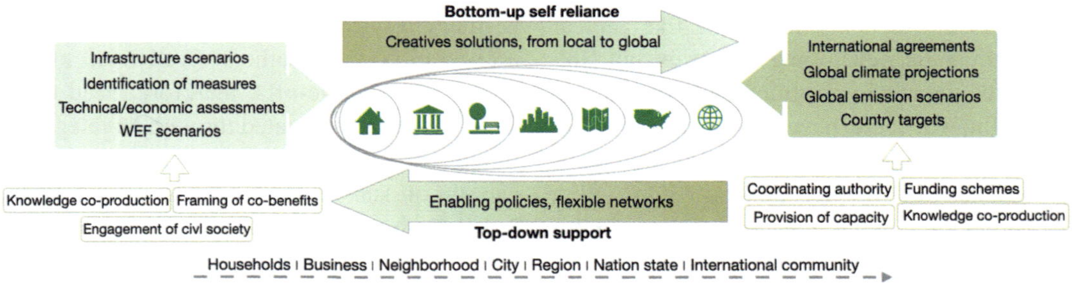

Fig. 3.4 Integration of bottom-up and top-down perspectives. Bottom-up actions are those that begin at the most local level: the city or its neighborhoods and buildings. Top-down support increases the capacity of cities to solve their own problems by providing them with policies, targets, financial mechanisms, etc. (Adapted from Suzuki et al. (2010) and Homsy et al. (2019))

countries are smaller than megacities, and we need to understand and accept this new reality.

If we want to change what our cities look like and how they negotiate, then we have to change the decision-making processes that have given us the results that we have witnessed. Citizens have a key role, as they have the ability to change behaviors and patterns of consuming WEF. In most cases, it is ultimately individual consumers who decide what food they buy. In some cities, the choice to buy only renewable electricity is a reality. The role of citizens is also different depending on region and often income, with globally wealthier cities tending to have greater supply of and access to fundamental resources. Socioeconomic disparities often shape exposure to various risk factors, bringing effects on the quality of life and health of the population (Ramaswami et al., 2016). Therefore, addressing these diverse social, environmental, and infrastructural risk factors represents a new paradigm for urban public services. In this context, many studies argue that participation should actively involve citizens in the transformation of their neighborhood, having substantial effects on WEF systems (Ramaswami et al., 2016; Haldane et al., 2019; IEED, 2020).

Participatory approaches should, therefore, be emphasized to promote better connection with and among decision-makers for WEF policies. This contributes directly to the development of institutional, political, and community capabilities that are fundamental for knowledge transfer on issues that require participation of groups with different interests. For instance, ICLEI has been developing toolkits together with stakeholders to enable local governments to assess their climate risks in the context of urbanization, poverty, and vulnerability to formulate corresponding resilience strategies.[5] Thus, from more inclusive discussions, possibilities for conflict resolution and win-win situations may arise. Complementary perspectives for WEF-integrated planning can contribute to the identification of trade-offs and synergies, highlighting the importance of local actions and policies (bottom-up) focused on local potentials (and demands) to inform the planning of supply on a larger scale (top-down).

3.4.3 Energy and Circular Economy Transitions Toward Urban Sustainability Using New Technologies

As large consumers, cities also discard huge amounts of material that, in a way, could be better used in the activities that power cities, especially through industrial symbiosis. Think of materials not as problems or waste for disposal, but as resources and energy, as opportunities. Cities might benefit from a transition toward a circular economy that uses renewable resources and designs cyclical and efficient systems for provisioning WEF. The circular economy is a rapidly maturing concept that is already seeing real-world applications (Winans et al., 2017; Bjørnbet et al., 2021; Kee et al., 2021).

Solid waste management represents a great opportunity for circular economy. First, cities should follow the well-known waste hierarchy: reduce, reuse, recycle, recover, and dispose. The first three are better developed in the global south. However, recover and dispose options are really underdeveloped degree. For instance, incineration is a nonstarter in developing countries mainly due to wet organic nature of waste and high investment costs. Because of this, composting and biogas are now emerging as viable solutions in the global south. From the application of the very first step, cities would be realizing circular economy through any kind of public awareness campaign to enable waste reduction and segregation and to come to a preferred waste strategy. Waste segregation can be done at source, and wet waste can be disposed through small-scale composting boxes or neighborhood-level biogas plants. Waste can also be taken to a composting yard based near local fresh markets. Household slurry can be disposed in household gardens, promoted by schemes for the promotion of medicine and kitchen gardens. Such activities may be associated with the promotion of GBI by bottom-up

[5] Visit http://southasia.iclei.org/resources/tools.html

initiatives at the household and neighborhood scales.

Decentralizing waste management saves energy, water, land, and emissions and reduces material waste. Thus, curbing food and water waste will therefore contribute to lowering overall energy use. Cities can further reduce their energy consumption by promoting urban agriculture, such as rooftop farming (it is estimated that 30% of urban spaces could be covered) (UN-HABITAT, 2021). Many urban areas are now producing over 20% of their vegetable needs from within city boundaries (Walters & Midden, 2018). Consumption habits need to change, and residents should be encouraged to produce their own food. The food supply chain must be shortened, with a focus on adapting supply channels to redistribute surplus food and reduce food waste (C40, 2019). From the governance side, cities need to ensure that industries pool their resources to create synergy effects. This can be achieved by establishing eco-industrial parks, where waste and by-products of one industry serve as the raw material of another, thereby improving material and energy efficiency and reducing externalities. From an economic perspective, this would also make companies more competitive, as better water/energy/waste management results in cost savings and higher environmental and business performance. In developing countries, governance of urban agriculture (UA) is important to address the challenges UA face with the growth of the urban areas, as most of UA is informal and lack protections in terms of land tenure (Puppim de Olveira & Ahmed, 2021).

Circular economy is well suited to dealing with issues of material supply risks, which are particularly pertinent to the energy transition through renewable energies. For instance, the currently applied treatment of industrial wastewater in Belgium has a high CO_2 footprint. The European Biogas Association has mapped the opportunities for producing biogas from industrial wastewaters from the pulp and paper, biofuels, and food beverage sectors. Results show that biogas has a high potential to mitigate methane emissions from wastewater while at the same time providing a huge potential source of renewable energy (The Digest, 2021). This would reduce energy consumption at wastewater treatment installations, provide a solution for the management of sludge, and create additional green jobs. When producing renewable gas from industrial wastewater, emissions are saved in different ways. First, due to the reduced energy consumption in wastewater treatment installations. Second, by the replacement of fossil energy sources. Last, by bringing the wastewater in a closed and controlled environment, preventing methane releases.

The strong connection between circular economy and energy transition based on previously available renewable resources is therefore clear. It is interesting to note that these are not top-down initiatives, but rather were taken from the bottom up by stakeholders, with a focus on site-specific potentialities. Clearly, the political, fiscal, and regulatory environments need to be in place for both energy and circular economy transitions to occur. Cities therefore need to shift from the current unsustainable development toward using renewable energy sources, not only because of looming resource depletion but also to curb the negative externalities. Sustainable urban WEF systems will need low-carbon technologies on the supply side and efficient distribution infrastructure as well as lowered consumption on the end-user side through WEF "saving techniques."

Buildings have huge water and energy savings potential if they embrace green or low-energy concepts, especially when coupled with GBI. Savings can be made by integrating efficient heating, cooling, insulation, lighting, and water distribution systems in buildings, increasing systems' overall efficiency. Likewise, on-site alternative energy sources such as rooftop solar panels can supplement power from the grid. The use of recycled, reused, or low-energy building materials also contribute to a better energy balance. Similarly, powerlines also need to be improved, calling for a new electrical grid.

Smart grids have been developed to avoid energy losses and improve efficiency in the production, transmission, distribution, and consumption of electricity. Smart grids harmonize supply and demand, providing a solution for the intermittent power supply by helping to balance variable power generation and end-user needs (Khuffash, 2018). Smart electronic meters are at the heart of the system, allowing consumers to track their consumption in real time (Zhang & Li, 2020). Smart grids can be paired with smart appliances or even a whole smart building, which respond to varying electricity supply and prices. Households, offices, and factories can program smart meters to operate certain appliances when power supplies are plentiful and cheaper. Once connected to a smart grid, consumers are able to move from being recipients of energy to being small producers (or prosumers), thus promoting greater connectivity with renewables. Another advantage of smart grids is that they help utilities map the consumption characteristics of customers, facilitating system planning to a smarter energy transition.

There are numerous opportunities to rethink the use of information technology in promoting sustainability not only for the energy sector but also to provide inclusion, credibility, transparency, traceability, and compliance in the use of natural or financial resources by cities. And this will ultimately affect WEF systems. Digitalization is the way to connect millions of decentralized initiatives in megacities, contributing to greater consumer freedom, tariff fairness, land zoning, natural resource management, resource allocation, etc. Decisions of this sort can really impact the global problems that manifest themselves in cities. On the other hand, many city governments have been effective in using technology to turn mobiles in every citizen's hand into a public agent that collects data about the city. During the COVID-19 pandemic, many authorities used data from mobile networks to monitor social distancing. Studies have been investigating big data to understand residents' behavior regarding the use of urban green areas, estimating the ideal distance people are willing to travel to enjoy nature (Zhang & Li, 2020).

Emerging technologies, including the Internet of Things (IoT), virtual reality, artificial intelligence, and blockchain, are enabling social changes that systematically affect economies, values, identities, and possibilities for future generations. Blockchain involves a decentralized global computing database that is intermediary-free, immutable, resilient, and public. The blockchain principle is trust in the system as a whole, not in the issuer. It translates many existing processes into business, governance, environmental, and societal benefits. Blockchain has enormous potential applications for sustainable actions and green finance for climate change. As only 11% of climate funds went to cities (IEED, 2020), there is a need to explore innovative mechanisms to finance sustainable development and urban resilience.

Blockchain technology can provide means for higher transparency in terms of climate finance flows, ensuring while funding reaches its intended recipients and that its impact is tracked. Smart contract-based allocation of climate finance can assure donors that their contributions are applied for their purposes. Multiple studies recognize blockchain's potential to address the present barriers to green finance (CLI, 2018; Marke & Sylvester, 2018; Sanderson, 2018). In combination with IoT sensors, it can reduce transaction costs and enhance transparency and the availability of credible data, bringing greater reliability to investments in urban green projects. Thus, there are many ways in which blockchain can support the instruments established under the Paris Agreement and therefore support the fight against climate change.

Cities could benefit the most from the widespread use of these technologies, as most of the target projects of green bonds are somehow related to them. Cities are learning from each other. How to install zero-emission buildings, how to implement electric car sharing systems and ecosystem-based adaptations for climate risk reduction. Just like in London and Paris, there are quotas on the number of cars on the streets in major Chinese cities. Many young people do not even think about driving. This is all about behavioral changes, long-term integrated planning for infrastructure projects, how to benefit from international schemes by making inclusive, bottom-up decisions that consider local characteristics. Transparent and adaptable governance arrangements, open to public input and scientific study, will empower cities to learn by doing (Ramaswami et al., 2016). Cities have been part of the problem; now they are part of the solution. They have the power to make the world more sustainable, more equitable, and fair. And there will be no equality and sustainability in urban centers, no fighting climate change on a global scale, without a more integrated and humane rethinking of the management of scarce WEF resources. Cities should build sufficient and accessible public infrastructure and invest in smart, inclusive, and sustainable transportation networks that allow people to connect physically and digitally.

3.5 Final Remarks

This chapter presents some arguments for progress toward smarter, more resilient cities and the role of WEF nexus in this process. To do so, city inhabitants and governments need a deep understanding of the institutional, financial, political, social, and behavioral factors that inhibit the necessary changes. Barriers hindering this transition can be explored by researchers aiming to overcome some difficulties in implementing the proposed suggestions.

Urban areas represent great risks and opportunities for mitigation and adaptation to climate change. The emergence of city clusters highlights the importance of managing WEF resources, essential to their success. Improved information about transboundary environmental effects and local well-being impacts are critical to unravel synergies and trade-offs between local versus larger-scale infrastructure networks. Therefore, it is critical to understand the flows of key resources (physical, technological, cultural, and financial) that enable cities to thrive.

Cities must develop with a focus on resilience to lessen their dependence on external resources and the impacts of extreme weather. Best practices exist for long-term planning and regional growth management, and the emergence of new tools for systems analysis and mapping offers great potential for WEF-integrated planning. There are also effective methods for collaborative design and decision-making. Additionally, creative, practical, and cost-effective solutions simultaneously achieve greater benefits than business-as-usual initiatives.

With greater community participation, more creative design solutions are likely to be achieved at local scales. However, part of the challenge in cities implementing new projects is ensuring that all stakeholders have aligned their existing policies and programs and are using their strengths to support the project goals and strategies. Thus, policymakers and authorities need to revise urban risk management practices and adopt new types of data, collaboration, finance, innovation models, and decision-making.

Urban governance should also focus on international mechanisms for long-term urban investments. Many cities are initiating public-private partnerships and/or special financing for smart city development. However, discussions on smart cities focus on high technology, disregarding more basic but innovative and equitable solutions. In this sense, there is a governance paradox with cities in the Global South. Cities in emerging economies and developing countries are where carbon lock-in has not set in and where infrastructure and behaviors are not yet entrenched. Yet, it is often these same places that have limited financial and institutional capacities to leverage regulations and strategies to take advantage of the WEFN framework. Therefore, increasing the capacity of social, political, and governance networks in global cities is key to sustainable changes toward more resilient, just, and sustainable cities.

References

Abell, R., et al. (2017). *Beyond the source: The environmental, economic and community benefits of source water protection*. The Nature Conservancy.

ADB (Asian development Bank). (2012). Green cities. In M. Lindfield & F. Steinberg (Eds.), *Urban development series*. ADB.

Albino, V., Berardi, U., & Dangelico, R. M. (2015). Smart cities: Definitions, dimensions, performance, and initiatives. *Journal of Urban Technology, 22*(1), 3–21. https://doi.org/10.1080/10630732.2014.942092

Al-Saidi, M., & Elagib, N. A. (2017). Towards understanding the integrative approach of the water, energy and food nexus. *Science of the Total Environment, 574*, 1131–1139.

Alzaabi, M., Rizk, Z., & Mezher, T. (2019). Linking smart cities concept to energy-water-food nexus: The case of Masdar City in Abu Dhabi, UAE. In W. Samad & E. Azar (Eds.), *Smart cities in the Gulf*. Palgrave Macmillan. https://doi.org/10.1007/978-981-13-2011-8_6

Andersson, E., Barthel, S., Borgstrom, S., Colding, J., Elmqvist, T., et al. (2014). Reconnecting cities to the biosphere: Stewardship of green infrastructure and urban ecosystem services. *Ambio, 43*, 445–453.

Ando, A. W., Cadavid, C. L., Netusil, N. R., & Parthum, B. (2020). Willingness-to-volunteer and stability of preferences between cities: Estimating the benefits of stormwater management. *Journal of Environmental Economics and Management, 99*, 102274. https://doi.org/10.1016/j.jeem.2019.102274

Bazilian, M., et al. (2011). Considering the energy, water and food nexus: Towards an integrated modelling approach. *Energy Policy, 39*, 7896–7906.

Bellezoni, R. A., Meng, F., He, P., & Seto, K. C. (2021). Understanding and conceptualizing how urban green and blue infrastructure affects the food, water, and energy nexus: A synthesis of the literature. *Journal of Cleaner Production, 289*, 125825. https://doi.org/10.1016/j.jclepro.2021.125825

Bjørnbet, M. M., Skaar, C., Fet, A. M., & Schulte, K. Ø. (2021). Circular economy in manufacturing companies: A review of case study literature. *Journal of Cleaner Production, 294*, 126268. https://doi.org/10.1016/j.jclepro.2021.126268

Bren d'Amour, C., Reitsma, F., Baiocchi, G., Barthel, S., et al. (2017). Future urban expansion and global croplands. *Proceedings of the National Academy of Sciences, 114*, 8939–8944. https://doi.org/10.1073/pnas.1606036114

C40. (2019). *Addressing food-related consumption-based emissions in C40 cities*. Available at: https://c40-production-images.s3.amazonaws.com/other_uploads/images/2269_C40_CBE_Food_250719.original.pdf?1564075020

Chen, C., Xu, Y., & Yan, Y. (2018). Estimated global overweight and obesity burden in pregnant women based on panel data model. *PLoS One, 13*, e0202183.

Chong, J. (2014). Ecosystem-based approaches to climate change adaptation: Progress and challenges. *International Environmental Agreements: Politics, Law and Economics, 14*, 391–405.

Chowdhury, P. K. R., Weaver, J. E., Weber, E. M., Lunga, D., LeDoux, S. T. M., Rose, A. N., Bhaduri, B L. (2020). Electricity consumption patterns within cities: application of a data-driven settlement characterization method. *International Journal of Digital Earth, 13*(1)119–135. https://doi.org/10.1080/17538947.2018.1556355

CLI (Climate Ledger Initiative). (2018). *Navigating Blockchain and climate action*. Available at: https://www.climateledger.org/resources/CLI_Report_Dec181.pdf

Covarrubias, M. (2019). The nexus between water, energy and food in cities: Towards conceptualizing socio-material interconnections. *Sustainability Science, 14*, 277–287. https://doi.org/10.1007/s11625-018-0591-0

Crush, J. S., & Frayne, G. B. (2011). Supermarket expansion and the informal food Economy in Southern African cities: Implications for urban food security. *Journal of Southern African Studies, 37*(4), 781–807.

Doll, C. N. H., & Puppim de Oliveira, J. A. (Eds.). (2017). *Urbanization and climate co-benefits: Implementation of win-win interventions in cities*. Routledge.

Endo, A., Tsurita, I., Burnett, K., Orencio, P. M. (2017). A review of the current state of research on the water, energy, and food nexus. *Journal of Hydrology: Regional Studies, 11*, 20–30. https://doi.org/10.1016/j.ejrh.2015.11.010

FAO. (2011). *Global food losses and food waste – Extent, causes and prevention*. Rome. Available at: http://www.fao.org/3/a-i2697e.pdf

FAO. (2017a). *Water for sustainable food and agriculture. A report produced for the G20 Presidency of Germany*. Rome. Available at: http://www.fao.org/3/i7959e/i7959e.pdf

FAO. (2017b). *The state of food and agriculture. Leveraging food systems for inclusive rural transformation*. Rome. Available at: http://www.fao.org/3/a--i7658e.pdf

FAO. (2019). *FAO framework for the urban food agenda*. Rome. https://doi.org/10.4060/ca3151en.

FAO. (2020). *Green cities initiative*. Available at: http://www.fao.org/green-cities-initiative

GCCA (Global Cold Chain Alliance). (2019). *Cold connection: Cooling and energy consumption – How sustainable is it?* Available at: https://www.gcca.org/resources/news-publications/blogs/cold-connection/cooling-and-energy-consumption---how-sustainable

GNR (Global Nutrition Report). (2018). Available at: https://globalnutritionreport.org/reports/global-nutrition-report-2018/

Gómez-Baggethun, E., et al. (2013). Urban ecosystem services. In T. Elmqvist et al. (Eds.), *Urbanization, biodiversity and ecosystem services: Challenges and opportunities*. Springer.

Grubler, A., Bai, X., Buettner, T., et al. (2012). *Urban energy systems* (Global energy assessment – Toward a sustainable future) (pp. 1307–1400). Cambridge University Press and IIASA.

Güneralp, B., et al. (2017). Urbanization in Africa: Challenges and opportunities for conservation. *Environmental Research Letters, 13*, 015002. Available at: https://iopscience.iop.org/article/10.1088/1748-9326/aa94fe/meta

Haase, D., Larondelle, N., Andersson, E., et al. (2014). A quantitative review of urban ecosystem service

assessments: Concepts, models, and implementation. *Ambio, 43*, 413–433.
Haase, D., Kabisch, S., Haase, A., Andersson, E., et al. (2017). Greening cities – To be socially inclusive? About the alleged paradox of society and ecology in cities. *Habitat International, 64*, 41–48.
Haines-Young, R., & Potschin, M. B. (2018). *Common international classification of ecosystem services (CICES) V5.1 and guidance on the application of the revised structure*. Available at: www.cices.eu
Haldane, V., Chuah, F. L. H., Srivastava, A., et al. (2019). Community participation in health services development, implementation, and evaluation: A systematic review of empowerment, health, community, and process outcomes. *PLoS One, 14*(5), e0216112. https://doi.org/10.1371/journal.pone.0216112
Hoff, H. (2011). *Understanding the nexus (background paper for the Bonn 2011 conference: The water, energy and food security nexus)*. Stockholm.
Homsy, G. C., Liu, Z., & Warner, M. E. (2019). Multilevel governance: Framing the integration of top-down and bottom-up policymaking. *International Journal of Public Administration, 42*, 572–582.
Howells, M., et al. (2013). Integrated analysis of climate change, land-use, energy and water strategies. *Nature Climate Change, 3*, 621–626.
UN-HABITAT. (2021). *Energy*. Available at: https://unhabitat.org/topic/energy
IEA (International Energy Agency). (2018). *IEA the future of cooling: Opportunities for energy-efficient air conditioning (IEA, Paris)*. Available at: https://doi.org/10.1787/9789264301993-en
IEA (International Energy Agency). (2020). *World energy outlook 2020*. IEA. Available at: https://www.iea.org/reports/world-energy-outlook-2020
IEED (International Institute for Environment and Development). (2020). *Disruptive resilience: An agenda for the new normal in cities of the global south*. IEED Briefing (Ed.) Bahadur, A.V. and Dodman, D. Available at: http://pubs.iied.org/17766IIED
IPCC. (2014). *Climate change 2014: Synthesis report. Contribution of working groups I, II and III to the fifth assessment report of the intergovernmental panel on climate change* [Core Writing Team, R.K. Pachauri and L.A. Meyer (Eds.)]. IPCC, Geneva, Switzerland, 151 pp.
Kee, S. H., Chiongson, J. B. V., Saludes, J. P., et al. (2021). Bioconversion of agro-industry sourced biowaste into biomaterials via microbial factories: A viable domain of circular economy. *Environmental Pollution, 271*, 116311. https://doi.org/10.1016/j.envpol.2020.116311
Khuffash, K. A. (2018). Smart grids: Overview and background information. In L. A. Lamont & A. Sayigh (Eds.), Available at *Application of smart grid technologies: Case studies in saving electricity in different parts of the world*. https://doi.org/10.1016/C2014-0-03960-3
Kremer, P., Hamstead, Z., Haase, D., McPhearson, T., et al. (2016). Key insights for the future of urban ecosystem services research. *Ecology and Society, 21*, 29.
Kurniawan, T. A., Puppim de Oliveira, J. A., Premakumara, D. G. J., & Nagaishi, M. (2013). City-to-city level cooperation for generating urban co-benefits: The case of technological cooperation in the waste sector between Surabaya (Indonesia) and Kitakyushu (Japan). *Journal of Cleaner Production, 58*(1), 43–50. https://doi.org/10.1016/j.jclepro.2013.08.002
Lyytimäki, J., & Sipilä, M. (2009). Hopping on one leg – The challenge of ecosystem disservices for urban green management. *Urban for Urban Gree, 8*, 309–315.
Macedo, L. S. V., Picavet, M. E. B., de Oliveira, J. A. P., & Shih, W. Y. (2021). Urban green and blue infrastructure: A critical analysis of research on developing countries. *Journal of Cleaner Production, 127898*.
Marke, A., & Sylvester, B. (2018). Decoding the current global climate finance architecture. In A. Marke (Ed.), *Transforming climate finance and green investment with blockchains* (1st ed., p. 368). Academic. Available at:. https://doi.org/10.1016/b978-0-12-814447-3.00004-5
Martin-Moreau, M., & Ménascé, D. (dir.) (2019). Urban agriculture: Another way to feed cities. *Field Actions Science Reports*, Special Issue 20. Available at: http://journals.openedition.org/factsreports/5536
Muthee, K., Duguma, L., Nzyoka, J., & Minang, P. (2021). Ecosystem-based adaptation practices as a nature-based solution to promote water-energy-food nexus balance. *Sustainability, 13*, 1142. Available at:. https://doi.org/10.3390/su13031142
Newell, J. P., Goldstein, B., & Foster, A. (2019). A 40-year review of food–energy–water nexus literature and its application to the urban scale. *Environmental Research Letters, 14*, 073003.
Nguyen, T. T., Pittock, J., & Nguyen, B. H. (2017). Integration of ecosystem-based adaptation to climate change policies in Vietnam. *Climatic Change, 142*, 97–111.
OECD (Organisation for Economic Co-operation and Development). (2020). *Making the green recovery work for jobs, income and growth. Policy brief: Tackling Coronavirus (COVID-19): Contributing to a global effort*. Available at: https://read.oecd-ilibrary.org/view/?ref=136_136201-ctwt8p7qs5&title=Making-the-Green-Recovery-Work-for-Jobs-Income-and-Growth_
Paes, M. X., de Medeiros, G. A., Mancini, S. D., Bortoleto, A. P., de Oliveira, J. A. P., & Kulay, L. A. (2020). Municipal solid waste management: Integrated analysis of environmental and economic indicators based on life cycle assessment. *Journal of Cleaner Production, 254*, 119848.
Paes, M.X., Campos-Silva, J.V., Puppim de Oliveira, J.A. (2021). Integrating Circular Economy in Urban Amazon. *npj Urban Sustainability, 1*, 1–6. https://doi.org/10.1038/s42949-021-00031-z
Pitman, S. D., Daniels, C. B., & Ely, M. E. (2015). Green infrastructure as life support: Urban nature and climate change. *Transactions of the Royal Society of South Australia, 139*, 97–112.

Puppim de Oliveira, J. A. (2005). Enforcing protected area guidelines in Brazil: What explains participation in the implementation process? *Journal of Planning Education and Research, 24*(4), 420–436.

Puppim de Oliveira, J. A. (2017). Breaking resilience in the urban system for improving resource efficiency: The case of the waste sector in Penang, Malaysia. *International Journal of Urban Sustainable Development., 9*(2), 170–183. https://doi.org/10.1080/19463138.2016.1236027

Puppim de Oliveira, J. A., & Ahmed, A. (2021). Governance of urban agriculture in African cities: Gaps and opportunities for innovation in Accra, Ghana. *Journal of Cleaner Production, 312*, 127730.

Ramaswami, A., Russel, A. G., Patricia, J., Culligan, P. J., et al. (2016). Meta-principles for developing smart, sustainable, and healthy cities. *Science, Special section: Urban planet, 352*(6288), 940–943.

Sanderson, O. (2018). How to trust green bonds. In A. Marke (Ed.), *Transforming Climate Finance and Green Investment with Blockchains* (Vol. 368, 1st ed.). Academic. https://doi.org/10.1016/B978-0-12-814447-3.00020-3

SEI (Stockholm Environment Institute). (2018). *Where is the added value? A review of the water-energy-food nexus literature (Stockholm Environment Institute, Sweden)* SEI working paper.

Seitzinger, S. P., Svedin, U., Crumley, C. L., et al. (2012). Planetary stewardship in an urbanizing world: Beyond City limits. *AMBIO, 41*, 787–794. https://doi.org/10.1007/s13280-012-0353-7

Sethi, M., & Puppim de Oliveira, J. A. (2015). From global 'north-south' to local 'urban-rural': A shifting paradigm in climate governance? *Urban Climate (Elsevier)., 14*(4), 529–543.

Seto, K. C., & Ramankutty, N. (2016). Hidden linkages between urbanization and food systems. *Science – Perspective*, 943–945. Available at: https://science.sciencemag.org/content/352/6288/943/tab-pdf

Seto, K. C., Reenberg, A., Boone, C. G., Fragkias, M., Haase, D., et al. (2012). Urban land teleconnections. *Proceedings of the National Academy of Sciences, 109*(20), 7687–7692. https://doi.org/10.1073/pnas.1117622109

Silpa, K., Yao, L., Bhada-Tata, P., & Van Woerden, F. (2018). *What a waste 2.0: A global snapshot of solid waste management to 2050* (Urban Development Series). World Bank. Available at: https://openknowledge.worldbank.org/handle/10986/30317

Simpson, G. B., & Jewitt, G. P. W. (2019). The development of the water-energy-food nexus as a framework for achieving resource security: A review. *Frontiers in Environmental Science, 7*, 1–9.

Sukhwani, V., Shaw, R., Deshkar, S., Mitra, B. K., & Yan, W. (2020). Role of smart cities in optimizing water-energy-food nexus: Opportunities in Nagpur, India. *Smart Cities, 3*, 1266–1292. https://doi.org/10.3390/smartcities3040062

Suzuki, H., Dastur, A., Moffatt, S., et al. (2010). *Eco2 cities: Ecological cities as economic cities*. World Bank Group. Available at: https://documents.worldbank.org/en/publication/documents-reports/documentdetail/634471468244553955/eco2-cities-ecological-cities-as-economic-cities

The Digest. (2021). *European Biogas Association maps opportunities for wastewater-based biogas*. Available at: https://www.biofuelsdigest.com/bdigest/2021/04/13/european-biogas-association-maps-opportunities-for-wastewater-based-biogas/

TWB (The World Bank). (2020). *Urban development*. Overview. Available at: https://www.worldbank.org/en/topic/urbandevelopment/overview#1

UCCRN (Urban Climate Change Research Network). (2018). *Climate change and cities: Second assessment report of the urban climate change research network* (Rosenzweig, C., Solecki, W., Romero-Lankao, P., et al., Eds.). Cambridge University Press.

UN. (2017). *New urban agenda. United Nations conference on housing and sustainable urban development (Habitat III)*. A/RES/71/256. ISBN: 978-92-1-132731-1.

UN (United Nations). (2019). World urbanization prospects 2018: Highlights (Department of Economic and Social Affairs, Population Division 2019, UN, New York). Available at: https://population.un.org/wpp/Publications/Files/WPP2019_Highlights.pdf

UN-ESCAP (United Nations Economic and Social Commission for Asia and the Pacific). (2019). *The Urban Nexus: Integrating resources for sustainable cities (UN, Thailand)*.

Walters, S. A., & Midden, K. S. (2018). Sustainability of urban agriculture: Vegetable production on green roofs. *Agriculture, 8*, 168. https://doi.org/10.3390/agriculture8110168

Winans, K., Kenda, A., & Deng, H. (2017). The history and current applications of the circular economy concept. *Renewable and Sustainable Energy Reviews, 68*(part 1), 825–833. https://doi.org/10.1016/j.rser.2016.09.123

Zalejska-Jonsson, A., Wilkinson, S. J., & Wahlund, R. (2020). Willingness to pay for green infrastructure in residential development: A consumer perspective. *Atmosphere, 11*, 152. https://doi.org/10.3390/atmos11020152

Zhang, Z., & Li, J. (2020). Big-data-driven carbon emissions reduction. In *Big Data mining for climate change*. Elsevier. https://doi.org/10.1016/B978-0-12-818703-6.00014-3

4 Water-Energy-Food Nexus Under Climate Change: Analyzing Different Regional Socio-ecological Contexts in Brazil

Lincoln M. Alves, Jean P. H. B. Ometto, Gustavo F. B. Arcoverde, Evandro A. Branco, and Peter Toledo

Abstract

Climate change, which has been associated with emerging and even elevated risks on the safety and well-being of societies, has already proved to be a significant challenge in today's world. Furthermore, a new set of IPCC climate projections reveals that rainfall and high-temperature extremes are expected to exceed the extreme conditions already experienced in Brazil. Such conditions are expressed by several events such as the drought and water crisis in the southeastern region in 2014–2015; drought in the northeast region during 2010–2016; and droughts in Amazonia in 2005, 2010, and 2016 and floods in 2009. These extreme events place several challenges on health, infrastructure, agriculture, biodiversity, and water resources, and therefore with significant environmental and socioeconomic implications. A major focus of this chapter is to provide elements to better inform users – from government, organizations, professionals, and individuals – on how climate change is likely to impact the water-energy-food (WEF) nexus in Brazil. For that purpose, two case studies will be discussed: (i) why a nexus approach to resource management is important for Cerrado and Caatinga biome, Brazil's largest available stock of land for agricultural expansion and renewable energy production (NEXUS Project), and (ii) we analyzed the Paraíba do Sul river basin, which encloses one of the most developed and industrialized regions in Brazil. The first case refers to the development of analyses on the transformation for sustainability in the Cerrado and Caatinga biomes, considering the construction of scenarios and indicators, built through a participatory approach. The identification and construction of indicators were carried out in a collaborative way with experts using the 2030 Agenda target framework.

Keywords

Climate change · Nexus · Security · Natural resources · Policies · Sustainability

4.1 Introduction

Currently, a substantial part of the Earth's surface has passed through transformation in its natural ecosystems' cover and biodiversity, by the increased demand of land for agriculture and expansion of urban areas (Ellis, 2015; Foley et al., 2011). As well, the population growth trend poses pressure on the future demands for energy

L. M. Alves (✉) · J. P. H. B. Ometto · G. F. B. Arcoverde · E. A. Branco · P. Toledo
National Institute for Space Research (INPE), São José dos Campos, SP, Brazil
e-mail: lincoln.alves@inpe.br

L. L. B. Lazaro et al. (eds.), *Water-Energy-Food Nexus and Climate Change in Cities*, Sustainable Development Goals Series, https://doi.org/10.1007/978-3-031-05472-3_4

and water, as well as rates and country-based emissions of greenhouse gases (GHGs). The most recent global scientific assessment (e.g., Intergovernmental Panel on Climate Change – IPCC) conveys a quite clear message – global environmental changes are serious, disastrous, and transversal in the global society; thus, we need an urgent and integrated reaction. Central to this are the pillars of the sustainable development goals: (a) human well-being (related to social and political equality); (b) social-financial inclusiveness; and (c) environmental conservation.

Brazil is a continental country, with more than 8 million square kilometers and many biomes, which have permanent preservation areas and legal reserves protected by public policies. On the other hand, Brazil is also an agricultural country that increasingly needs agricultural land. Climate change, which has been associated with emerging and even elevated risks on the safety and well-being of societies, has already proved to be a significant challenge of today's world. The magnitude of the impact of climate change upon Brazilian ecosystems could lead to profound transformations in important drivers of biodiversity, such temperature and precipitation patterns. That will affect the integrity of the ecosystems and the services they provide, compromising the livelihood of millions of people, besides posing a serious risk to biosphere integrity.

In this context, it is essential to understand the range of climate change scenarios to assist in the formulation of sound policies to increase adaptation, mitigation, and resilience of socio-ecological systems in the region. Assessing the impacts of climate change on socio-ecological systems, however, is a challenging task. Part of this challenge resides in the complexity of the climate system, which means that there is no such thing as the "single right trajectory" but an ensemble of possible realizations, which usually we associate to different modelling approaches.

Furthermore, a new set of IPCC climate projections reveals that rainfall and high temperature extremes are expected to exceed the extreme conditions already experienced in Brazil. Such conditions are expressed by several events such as the drought and water crisis in southeastern in 2001 and 2014–2015; drought in northeast during 2010–2016; and droughts in Amazonia in 2005, 2010, and 2016 and floods in 2009. These extreme events place several challenges on health, infrastructure, agriculture, biodiversity, and water resources, and therefore with significant environmental and socioeconomic implications.

The chapter is presenting an outline of current trends in climate and extreme events over Brazil from the beginning of the twentieth century onward based on observations and assessment of their projections up to the end of the twenty-first century. The focus will be on temperature and precipitation that are robust indicators of conditions that are related to agricultural production impacts and food security considerations, possibly with regard to crop tolerances and thresholds. We also bring your attention to two case study areas, in which climate change is likely to impact the water-energy-food (WEF) nexus in Brazil. For that purpose, two case studies will be discussed: (i) we discuss why a nexus approach to resource management is important for Cerrado and Caatinga biome, Brazil's largest available stock of land for agricultural expansion and renewable energy production (NEXUS Project) (Sect. 4.3.1), and (ii) we analyzed the Paraíba do Sul river basin, which encloses one of the most developed and industrialized regions in Brazil (Sect. 4.3.2). The first case refers to the development of analyses on the transformation for sustainability in the Cerrado and Caatinga biomes, considering the construction of scenarios and indicators, built through a participatory approach. The identification and construction of indicators were carried out in a collaborative way with experts using the 2030 Agenda target framework. To date, 149 sustainability indicators were identified, which are predominantly located in SDGs 6, 13, 15, 11, and 1. Final remarks are in Sect. 4.4.

4.2 Climate Change and Its Impacts

Climate change is defined by the IPCC as a significant statistical variation in a given parameter mean state or in its variability that persisted for a period of time (a decade or more), which can be caused by natural phenomena or anthropogenic influence. Both cases can influence the climate. Processes affecting climate can exhibit considerable natural variability. Even in the absence of external forcing, periodic and chaotic variations on a vast range of spatial and temporal scales are observed. Much of this variability can be represented by simple distributions, but many components of the climate system also exhibit multiple states (e.g., El Niño-Southern Oscillation (ENSO) – Cai et al. (2020)). Anthropogenic influence is associated with the increasing emission of GHGs (greenhouse gases). Changes between states can happen as a result of natural variability or as a response to external forcing.

Evidencing climate change is based on careful analysis of observational records of the atmosphere, land, ocean, and cryosphere systems. There is incontrovertible evidence from in situ observations and ice core records that the atmospheric concentrations of greenhouse gases (GHGs) such as carbon dioxide (CO_2), methane (CH_4), and nitrous oxide (N_2O) have increased substantially since the pre-industrial era (Myhre et al., 2014; Ciais et al., 2014). Recent measurements show that carbon dioxide in the atmosphere is now reaching levels 50% higher than when humanity began large-scale burning of fossil fuels during the industrial revolution, despite a growing number of climate change mitigation policies.

In addition, the last IPCC report states unequivocally that warming of the climate system is unequivocal, and most of the observed increase in global average temperatures since the mid-twentieth century is very likely due to the observed increase in anthropogenic greenhouse gas concentrations. Anthropogenic influences have likely affected the global water cycle since 1960 (Hartmann et al., 2013; Hoegh-Guldberg et al., 2018).

The warmer it gets, the greater the risk for more severe changes to the climate and the Earth's system. Small changes in the climate will bring changes that can affect our water supplies, agriculture, power and transportation systems, the natural environment, and even our own health and safety. Although it's difficult to predict the exact impacts of climate change, what's clear is that the climate we are accustomed to is no longer a reliable guide for what to expect in the future.

Over the past decade, the impacts resulting from global change have emerged as a major priority for governments and research organizations around the world. Based on it this chapter uses a set of regional climate model by the Coordinated Regional Downscaling Experiment (CORDEX) (Giorgi & Gutowski, 2015) to provide in a clear and concise manner the estimates of projected future climate change over Brazil with a focus on case study areas (Paraíba do Sul river basin, located in southeastern Brazil, and Cerrado and Caatinga biomes). Despite these regions being located entirely within the tropics, the distributions of climate variables such as temperature, rainfall, and winds are often influenced by local effects of topography, vegetation type, and ocean thermal contrasts, which have a significant effect on the climate. In addition, these regions face numerous environmental and socioeconomic challenges, such as erosion and unregulated land use, desertification, deforestation, overexploitation of resources, pollution, poverty, health problems, and poverty, which make them extremely vulnerable to a changing climate. Then, a challenge is to provide better information about the likely future climate for the benefit of poor rural and urban populations in a changing environment.

Rising global temperatures have been accompanied by changes in weather and climate. Many places around the world including Brazil have seen changes in rainfall, resulting in more floods, droughts, or intense rain, as well as more frequent and severe heat waves causing economic and life losses. Some examples are the droughts in Amazonia in 2005, 2010, and 2016 and floods in 2009 (Marengo & Espinoza, 2015); the droughts in Southern Brazil in 2005–2006 and 2014–2015 (Nobre et al., 2016; Marengo & Alves, 2015); and drought in Northeast Brazil in 2012–2014 (Cunha et al., 2019; Marengo et al., 2018). If these and other changes become more pronounced in the coming decades, they will likely present challenges to our society and our environment.

Regional studies of temperature and precipitation trends for South America show changes that are in line with expected warming, most notably warmer days, heavy precipitation, and consecutive number of dry days (Alexander et al., 2006; Vincent et al., 2005; Haylock et al., 2006; Marengo et al., 2010; Donat et al., 2016; Bezerra et al., 2019; Skansi et al., 2013).

Advances in climate modelling now enable best estimates and likely assessed uncertainty ranges to be given for projected warming for different emission scenarios. Climate projections provide critical information for decision-makers about our future climate, and it helps evaluate risk and opportunity from regional climate change.

In Brazil, the projected average surface air temperature and precipitation for the medium term (2041–2060) relative to the present day (1986–2005) under the Representative Concentration Pathway 8.5 (RCP8.5) scenario (a "business-as-usual" scenario) (Taylor et al., 2012) is shown in Fig. 4.1. The first immediate conclusion from those figures is that it provides better spatially detailed distribution of change compared to the driving general circulation models (GCMs) across Brazil. A number of other key messages can be drawn from Fig. 4.1a: The mean temperature change in central Brazil is much higher from the global mean temperature change, especially for Cerrado and Northeast and Southern Brazil, with potentially more than 3 °C of change by 2041–2060 compared with the baseline period. It also varies considerably with season and magnitude, which means that warming is more pronounced in austral winter (JJA) than summer (DJF). Nonetheless, substantial warming is also expected in all areas and season under a mitigation scenario (RCP4.5) in excess of 1.5 °C of change projected by medium term, relative to the baseline period (figure not shown). This, in itself, is extremely important in the policy context: The discussion of limiting global average temperature change to 2 °C (from preindustrial) will not mean that changes in Brazil are also limited to this level of change.

Precipitation involves more complex local processes than temperature. In this regard, the greater levels of disaggregation in the RCMs potentially provide important information for impact and economic analysis. However, in general, precipitation is less well simulated than temperature, and projections are less robust than those for temperature. Figure 4.1b shows the changes in mean precipitation over the mid-term for the RCP8.5 scenario, and it leads to the following key messages: The results confirm the general conclusions based on the GCMs (IPCC, 2014) but provide much greater information on the spatial distribution of changes, which are particularly important in the case of precipitation; there is a strong spatial pattern of change across Brazil, which is similar in all periods, and there are large differences between RCM and CMIP5 GCM results even in terms of the direction of change. The results also show major differences in precipitation change within Brazil (i.e., between Southern and Northern Brazil). The largest changes projected are for land areas of Northeast Brazil, southern Amazonia, and parts of Cerrado (lower rainfall). The results show the likely decrease in precipitation in those regions in the austral winter. However, there is an increase in precipitation in austral winter over northwest Amazonia and Southern Brazil. The results show a strong drying trend for the RCP8.5 scenario in winter in Northeast Brazil and Cerrado.

Extreme weather and climate events are also an integral aspect of climate variability, and their frequency and intensity may vary with the prospects of climate change. One of the most important questions regarding short-term extreme events is whether their occurrence is increasing or decreasing over time, that is, whether there is a trend for the envelopes within which these events preferentially occur.

Figure 4.2 shows the extreme precipitation indices, for mid-term (2041–2060) modeled cli-

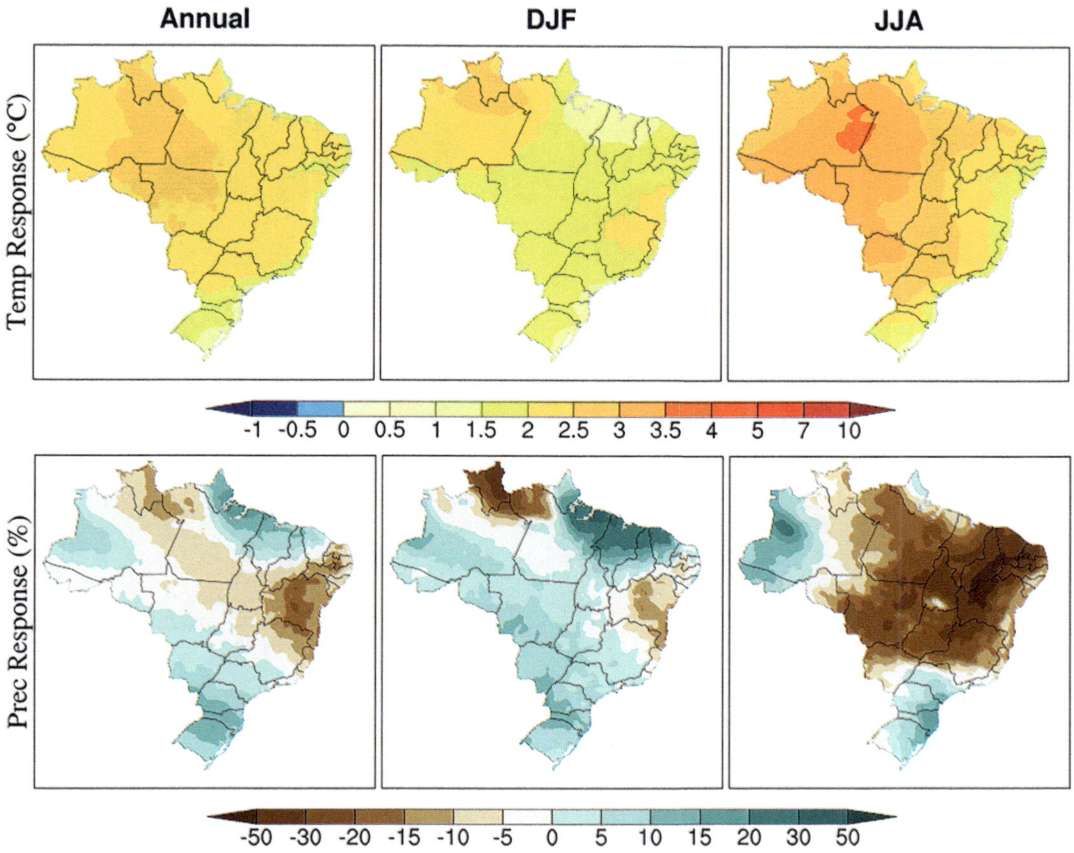

Fig. 4.1 Changes in air temperature (°C) (**a**) and precipitation (**b**) in Brazil for annual mean, austral summer (DJF), and austral winter (JJA) in 2041–2060 relative to 1986–2005 under the RCP8.5 scenario derived from the CORDEX ensemble mean

mates over Brazil from the CORDEX ensemble mean. Future projections show significant changes in most extreme indices. R95P (very-wet-day precipitation) generally tends to increase over most of north and central Brazil. The frequency of consecutive dry days (CDD) tends to increase mainly over Northeast Brazil and Cerrado and Paraiba do Sul basin. Models show future increases in extreme events over most Brazilian regions with warming, i.e., several regions will experience both drier or more frequent dry periods and wetter wet periods in the future (Alves et al., 2021; Coppola et al., 2021). This would augment the risk and vulnerability to fires, and dry conditions would have a negative impact on the population, biodiversity, and human activities on the other hand. Extreme rainfall events result in significant societal impacts, including flooding and landslides, leading to major socioeconomic damages (Marengo et al., 2021).

Potential climate change impacts on Brazil can be extracted from global assessments. For example, a composite index of climate change (comprising temperature and precipitation extremes) shows a strong latitudinal gradient across Brazil with the worst-affected areas in the north (Baettig et al., 2007). The northeast also emerges as a region of severe water stress by the 2050s in global assessments of water resources (Alcamo et al., 2007; Arnell, 2004) and river flow (Milly et al., 2005). More frequent great floods (Fowler et al., 2021), wildfire, and forest loss are expected for Amazonia (Fonseca et al., 2019).

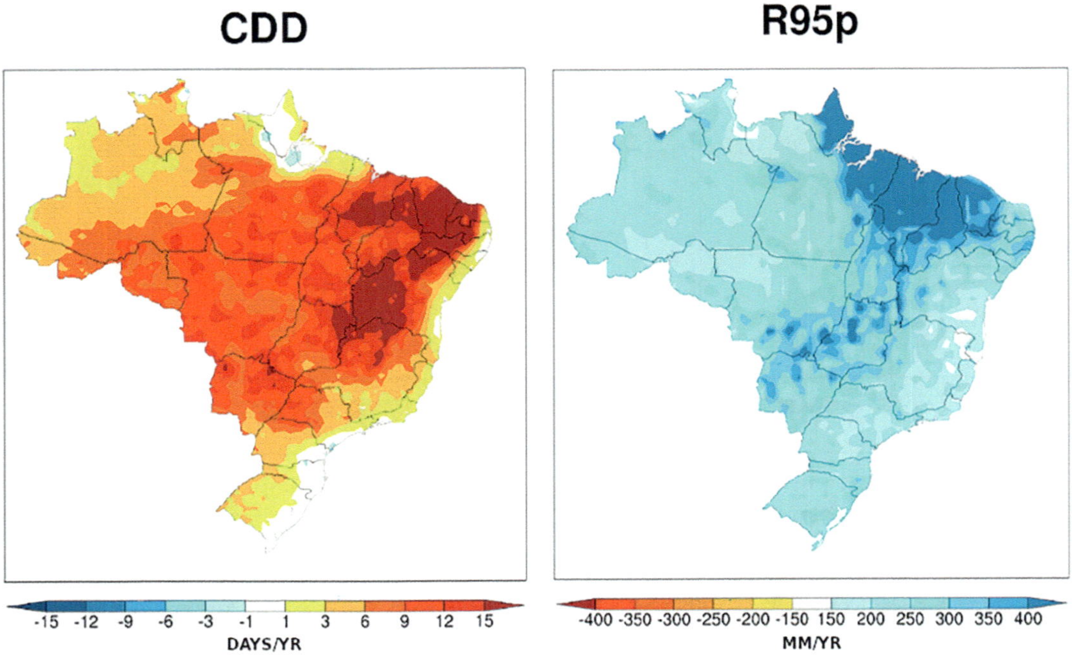

Fig. 4.2 Indices of extremes derived from the CORDEX ensemble mean climate change projections for mid-term (2041–2060), for the RCP8.5 scenario. Indices are defined in Frich et al. (2002). Figures show changes for the timeslice 2041–2060 relative to the present climate (1986–2005)

4.3 Discussion

4.3.1 Nexus Approach to Resource Management in the Cerrado and Caatinga Biomes

Cerrado and Caatinga are unique biomes in the Brazilian territory. Several features could be listed on supporting this statement. Cerrado is a global biodiversity hotspot, the most biodiversity tropical savanna in the world, and encompasses the head water of several important river systems in the country (Klink & Machado, 2005). Caatinga is a biome wholly enclosed within the Brazilian territory, the cradle of endemic species unique in comparation to other semiarid regions in the world. Still, it holds a minor extended area for conservation in the country (ICMBio, 2021). Figure 4.3 shows the location of both biomes. Like Cerrado, Caatinga is home to unique cultural expressions and singular indigenous and traditional communities. Caatinga is the less urbanized region in Brazil, and an important portion of its 27 million inhabitants lives in rural smallholder properties and based on family agricultural production. Similar patterns of smallholder agriculture are found in both biomes, although the expansion of commercial and large-scale agriculture is relatively more important in the Cerrado region (Myers et al., 2000), which has been recognized as a region with high potential to contribute to provide food for increasing global demand (Pires, 2020; Soterroni et al., 2019). The conversion of native vegetation and the lack of broadly distributed conservation areas make these two biomes among the most threatened biomes in the world (Strassburg et al., 2017; Resende et al., 2021), of which 50% of their original vegetation was already replaced by agriculture or pastures (Grande et al., 2020). The Brazilian National Agricultural Census (IBGE, 2017) considers that basically 60% of Cerrado and Caatinga are adequate for agriculture and feedstock production; however, climate change is seen as an emergent threat in the region, with

Fig. 4.3 Location of Cerrado and Caatinga biomes

increasing temperatures and reducing precipitation (Feng, Porporato & Rodriguez-Iturbe, 2013; Soterroni et al., 2019). Climate change impacts in these regions can be expressed by profound changes in the water cycle, increasing losses by evaporation and the frequency of extreme events, for both precipitation and droughts (Alves et al., 2021), affecting the capability for agricultural production (e.g., about 90% of agriculture and almost 100% of the cattle production in Cerrado are rain fed (EMBRAPA, 2019)). Climate change might also affect water availability for human consumption and energy production, as well as the frequency and intensity of fires, which in turn, shall strongly impact the biodiversity, as well as the energy production and transmission (Lapola et al., 2018). Impacts of extreme drought events are seen in the Caatinga biome on the level of soil degradation (Vieira et al., 2021) and losses of soil nutrients (Tôsto et al., 2019), which directly affect the potential for food production and human health and might lead to human migration to the metropolitan areas in the region or to other regions. Despite potential limitation to agricultural production in the next decades, Dalla-Nora et al. (2014) suggest that the increasing demand for agricultural commodities, both internal and international, shall burst the land use change expansion in the Cerrado region, particularly in the Matopiba area. Under the global notion that land is a scarce resource (Lambin and Meyfroidt 2011), the most likely expansion of the agriculture frontiers lays on the tropical belt (Gibbs et al., 2010), and the Brazilian Cerrado is an obvious subject (Soterroni et al., 2019). In legal terms, the Cerrado region has approximately 27 million ha of native vegetation prone to the legally deforested area based on the Brazilian Forest Code (Azevedo et al., 2017; Vieira et al., 2021).

The socio-environmental landscape briefly described for the Cerrado and Caatinga biomes stresses the complex environmental analysis and planning for a sustainable use of resources and development path. About 50 million people live in those regions, under a broad socioeconomic diversity, different opportunities, and access to resources (Almeida et al., 2020).

The urban population represents 70% (Caatinga) to more than 80% (Cerrado) of the total, highlighting the challenges related to the demand and use of natural resources. Some of the main drivers for urban vulnerability in these regions are high population densities, poverty, weak governance, overdemand on urban ecosystems (particularly in the cities located close to the ocean), and poor infrastructure. Those aspects pose strong pressure on the population's basic needs, like water, energy, food, and ecosystems and the already depleted diversity associated with them. The risks on food, water, and energy security are high and exacerbated by the enormous social inequalities (in bigger cities and state capitals, poverty can reach more than 40% of the population economically active (https://cidades.ibge.gov.br/)). Climate change is also an emerging threat. Processes such as sea level rise and extreme precipitation events reach stronger the cities on the coast, and extreme and persistent drought are, in general, more dramatic to small settlements and smallholders in the countryside. Therefore, an integrative analysis of resource vulnerability is critical in these regions.

The heterogeneity of the resources, the important biological diversity, the vulnerability of the natural systems, and the potential for sustainable and renewable energy are threats and opportunities that require a transdisciplinary analysis of the economic sectors itself and the feedback among them. The nexus approach (food, water, energy, and society) is for the integrative management alternative in regions that are thus rich and vulnerable to the available resources. Araujo et al. (2019) state that, under the different and under different governance regimes related to the water, food, and energy systems, the implementation of a nexus approach is challenging but urgently needed. Two dimensions are considered in this approach, one being an interdisciplinary dealing with the resources and the other a transdisciplinary with governance across sectors, different stakeholders, and the policy space (Endo et al., 2020).

Thus, the logic behind the nexus approach requires that the decision taken in one specific sector resounds in a systematic, transparent, and feed forward manner. The food-water-energy requires innovative and multiscale strategy, blending quantitative and qualitative methods (FAO, 2014). Among the challenges of integrative planning in such heterogeneous regions such as Cerrado and Caatinga, understanding the processes driving land use and land use change is critical. Transition from natural vegetation to other uses can be considered one of the primary drivers impacting ecosystem services (Resende et al., 2021), exacerbating climate change impacts on hydrological systems (Lopes et al., 2021) with potentially negative feedback to the climate system itself (Salazar et al., 2015). Richey et al. (2015) observed a decline in the deep aquifer water stocks in the Cerrado region as an effect of increased irrigation in agricultural areas. Unplanned land use changes can also pose strong pressure on the local socio-ecological arrangements, e.g., traditional communities and indigenous communities (Resende et al., 2021; Almeida et al., 2020) and smallholders, and lead to rural exodus to marginal areas on urban settlements (Lapola et al., 2018).

The methodological framework for land use under the FEW approach, for the Cerrado and Caatinga biomes, needs to consider the interconnection among the sectors including the quantification of resource environmental footprint, assessment and systematic simulation, managing trade-offs, and capturing life cycle cross-sector feedbacks and spatial and temporal scale synchronization (Khan et al., 2018; Zhang and Zhou 2019). The scientific literature presents several integrative modelling and tools to apply the nexus concept to inform the policy and resource management processes (Endo et al., 2020). Araujo et al. (2019), working with the nexus+ concept (which includes the socio-ecological security dimension) in the São Francisco river basin (culturally emblematic river system that flows in the heart of the Cerrado and Caatinga biomes), point several challenges to demonstrate the ability of a nexus+ framework to be effective on informing decision-making, even under clearly identified impact of the "no use" of this integrative approach. For instance, river impoundment by large-scale infrastructure with shallow articulation with other key sectors, particularly indigenous territory, local community, and smallholder farming, led to extra pressure on the scarce water resource or inadequate irrigation strategies and soil salinization or lack of sanitation, leading to serious impact on food security (Araujo et al., 2019).

4.3.2 Paraíba do Sul River Basin

The integrated management of large urban centers is one of the significant challenges of the twenty-first century. In contrast, it represents the locus of complex dynamics, both from economic and political strength and several social and environmental issues (NOBRE et al., 2011). In the Brazilian context, São Paulo (RMSP) and Rio de Janeiro Metropolitan Regions (RMRJ) – both relevant from the social, economic, and political point of view in the national context – together hold more than 34 million inhabitants or about 28% of the population residing in urban areas of Brazil (FNEM, 2021). In this situation, the availability and affordability of water resources are a susceptible and strategic element to the viability of essential functions in metropolitan regions.

Understanding the relationship between the two largest metropolitan regions in Brazil and the Vale do Paraíba e Litoral Norte metropolitan region requires an expanded look at the integrated management of water resources as an element of connection between territories throughout Paraiba do Sul river.

The complex basin interconnection systems, implemented since the 1950s and intensified after the so-called water crisis (2013–2015), establish the basis for the creation of a new, even larger, more relevant, sensitive, and complex region called the "hidromegaregião" of São Paulo (SP) and Rio de Janeiro (RJ) (Ribeiro, 2020). The São Paulo Macrometropolis is formed by three bases – Paraíba do Sul river; Guandu river and to Piracicaba, Capivari, and Jundiaí rivers; and, from 2018, connects the Alto Tietê river basin (Fig. 4.4).

The Paraíba do Sul river basin, in this context, is essential because it connects the other regions through the transposition of its waters. The basin comprises around 62,074 km^2, covering 184 municipalities, 57 in the state of RJ, 88 in Minas Gerais, and 39 in São Paulo (Cavalcanti & Marques, 2016). Due to its peculiar characteristics, this basin can be considered a singular connection element, revealing the conflicts and tensions arising from the different levels of decision-making, therefore, fundamental for the study and analysis of complex planning systems and management of waters.

The basin brings together different uses of water in three other states of the Federation. Currently, the primary services are linked to the production of hydropower; flood control; urban, industrial, and agricultural supply; fishing; and recreation (PSR Engenharia & AGEVAP, 2013) and the volumes excepted in the transposition for the Guandu basin in the Rio de Janeiro state. The Paraíba do Sul River is essential for the water supply in the state of Rio de Janeiro and for the economic development of its drainage area. The Paraíba do Sul river is essential for water supply in the Rio de Janeiro state. Since 1952, a 120 m^3/s water volume has been transposed to the Guandu basin, which corresponds to about 80% of RMRJ's water supply, serving around 9 million people (ANA – Agência Nacional de Águas, 2015; CEIVAP, AGEVAP, and COHIDRO, 2014; PROFILL, AGEVAP, and CEIVAP, 2020), besides the hydropower units.

In May 2018, It was concluded the works to the basin's water transposition. Connecting the Jaguarí reservoir upstream of the basin, in the State of São Paulo, to the Atibainha reservoir, in the Piracicaba, Capivari, and Jundiaí (PCJ) river basins. This intervention was quickly approved because it was considered a public utility, emergency, and strategy to deal with the water crisis and increase water security in the São Paulo

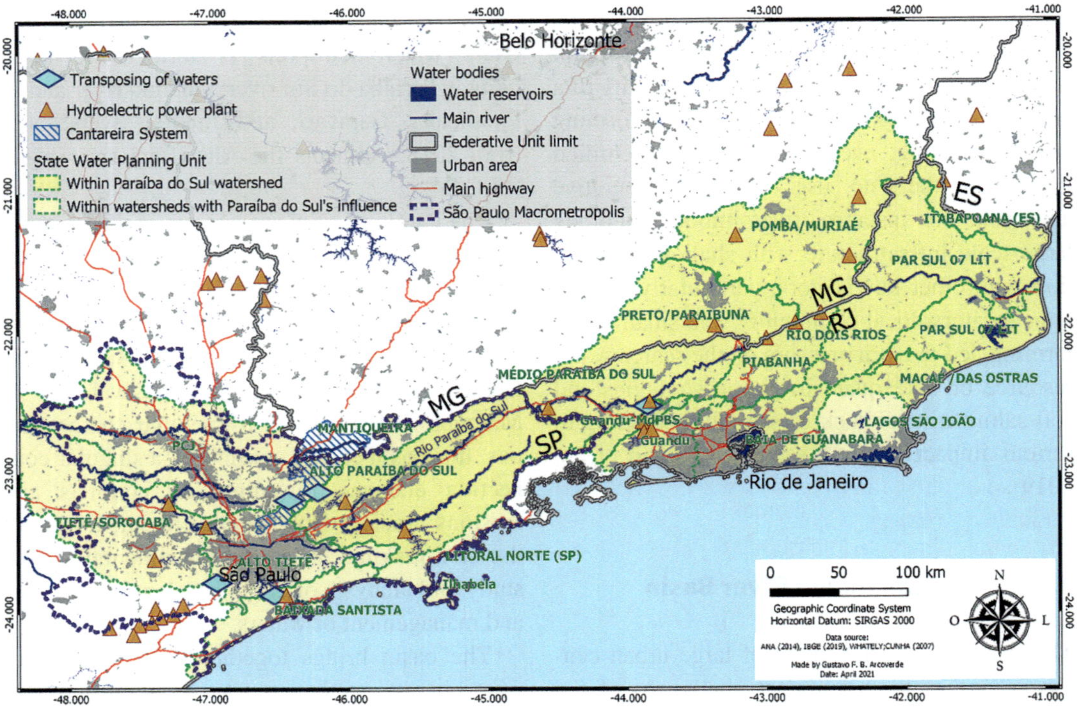

Fig. 4.4 "Hidromegaregião" of São Paulo (SP) and Rio de Janeiro (RJ), with the location of the Paraíba do Sul river basin and its planning units, the federative units, São Paulo Macrometropolis, and water transpositions

Macrometropolis and the PCJ basins (Carmo & Anazawa, 2017). Operationally, this interconnection is carried out through structures to capture an average flow of about 5.13 m^3/s and a maximum flow of 8.5 m^3/s (Engle et al., 2011). The Atibainha reservoir is part of the so-called Cantareira System, which consists of six interconnected dams. The Cantareira System reverts waters to the Alto Tietê basin in compliance with the RMSP. This complex structure aimed to reduce the systemic risk in supplies to the RMSP and the PCJ basin (SABESP, 2015) in response to the severe drought that hit the southeastern region of Brazil between the years 2013 and 2015. Although at first, the volume expected for transposition is small (about 5% of the water volume in the Paraíba do Sul river basin), this decision gives new thresholds to the scenarios of conflicts over water use, vulnerability, and complexity in water management in the Paraíba do Sul river basin (Giatti et al., 2016). The impact dimensions of this decision on the resilience of the entire water system in conditions of significant environmental changes are not known, mainly as related to regional and global climate changes. Given this scenario, an integrated assessment of the water system/demand on the RMSP-RMRJ axis is fundamental, focusing on the Paraíba do Sul administration (ANA – Agência Nacional de Águas, 2015) under various climatic scenarios.

The basin's complexity is also reflected in aspects related to the water allocation within the basin, considering different uses and demands in various territories. In the basin, the agricultural sector stands out, mainly with the temporary crops of rice, corn, and sugarcane and the permanent ones, with the production of coffee, orange, lemon, banana, and tangerine, in addition to the practice of extensive dairy farming (CEIVAP, AGEVAP, and COHIDRO, 2014).

Hydroelectric power generation, although setting up as a nonconsumptive use, is relevant in the basin. The basin has nine hydroelectric plants (with a generation capacity above 30 MW) in operation. They are Paraibuna, Santa Branca, Funil, Picada, Sobragi, Simplício, Ilha dos

Pombos, Nova Maurício, and Barra do Braúna. Moreover, the basin has 37 small hydroelectric plants (PCH) and 29 hydraulic-generating plants (CGH) in operation (PROFILL, AGEVAP, and CEIVAP, 2020).

Besides the demands of the agricultural and energy sectors and the current transposition of part of its waters to the Guandu river basin and the Cantareira System, it is important to highlight the dependence on local communities. Fifty-two municipalities are located along Paraíba do Sul river, and 28 collect water for public supply directly from the river (CEIVAP, AGEVAP, and COHIDRO, 2014).

In addition to this scenario, it is important to summarize a recent event that revealed the vulnerability of the whole region. Between 2014 and 2015, Southeastern Brazil, and specifically the area covered by the São Paulo Macrometropolis, was severely impacted by one of the most significant droughts in history. Marengo and Alves (2015) already warned that events of this magnitude are precisely the type of extreme climate phenomenon projected by the reports of the Intergovernmental Panel on Climate Change (IPCC). The 2014–2015 water crisis impact ranged from water rationing, the interruption of supply in some municipalities and some segments of water users, reflecting the deepening of environmental injustice (Fracalanza & Freire, 2015; Jacobi et al., 2015a).

According to the global projections of the IPCC and numerical modeling experiments, the climatic scenarios for the region suggest a decrease in rainfall, warming in the Amazon region with anomalies in the transport of moisture to southeastern Latin America. The resulting climatic disturbances can be materialized in extreme rainfall events, concentrated in a few days, and periods of more intense and prolonged droughts, such as those of 2014 (Marengo & Alves, 2015). In other words, the region needs to prepare to deal with the possible occurrence of extreme weather events, with direct impacts on the components of the water-energy-food (WEF) nexus.

The understanding of the so-called water crisis, as well as of new events of prolonged droughts, needs to be evaluated in an integrated way, and not only from the climatic point of view. Issues such as the responsibility of stakeholders and institutions involved in governance (Jacobi et al., 2015b; ANA – Agência Nacional de Águas, 2015) and transparency about information and the decision-making process (Martins, 2014; Tadeu, 2016) are essential elements in the identification of the adaptive capacity of the basins in crisis scenarios.

The seek for a balance between the nexus elements in the Paraíba do Sul basin, regarding the characteristics mentioned, demands an essentially multiscaled approach, based on both territorial and temporal scales (Giatti et al., 2016). Pressure from forces in the two largest and most influential metropolitan regions of the country and climate change scenarios leaves the basin into uncertainty and vulnerability from the point of view of the water balance. The governance dimension takes on an even more strategic role to consider such complexity concerning the management of water resources, considering the perspective of the nexus.

Therefore, the territorial coverage and diversity of the basin imply a multilevel water governance model, which can be defined as the sharing – explicit or implicit – of the responsibility for the exercise of the form and to implement as water resource policies by the different administrative levels and territorial (OECD, 2011). Currently, there is a federal river basin committee that comprises the whole river basin and seven other committees for affluent rivers, one in SP, two in the state of Minas Gerais (MG), and another four in RJ. Besides the basin committees, the management structure also includes state management bodies such as ANA (National Water and Basic Sanitation Agency) (on a federal level), INEA (State Institute of the Environment of the State of RJ), IGAM (Instituto Mineiro of Water Management) for the State of MG, and DAEE- Department of Water and Electricity in the State of São Paulo. These institutions, acting together, should converge to a water management model attentive to the challenges of allocating conflicting uses.

This complexity inherent to water governance requires structures and procedures aligned to the search for equality regarding the multilevel and multiscale social stakeholders' participation. That is a challenge for the nexus governance, which demands a strengthening of the governance basin arrangement even more complex, whose alignment between multilevel and multisectoral arrangements is desirable (Giatti et al., 2016).

Therefore, the strengthening of river basin committees emerges as central, legitimate spaces for conflict management, negotiation, and decision-making. Although the basin has a robust and well-structured management arrangement, such a system is still not recognized by the population (Baião et al., 2020). One of the recommendations for such strengthening is actions aimed at the continuous institutional development of spaces, innovative strategies for communication, engagement, and social mobilization to make such areas more diverse and aligned with the concrete challenges of the territories. However, it needs to adopt models based on adaptive management and resilience theory (Engle et al., 2011). Adaptive management of natural resources can be understood as a systematic process of improving management policies and practices by considering the policies and procedures already implemented. Its objective is to increase the adaptive capacity of the adopted approach through the inclusion of social learning processes that take into account the uncertainties associated with the management process and the constant analysis of the results obtained (Pahl-Wostl et al., 2007a).

Human being's ability to predict changes in the ecosystem is limited. The proposed management model incorporates new information and consensus into the management process based on the policies set in practice. Besides past experiences, features are considered essential for the process of adapting to changes (Pahl-Wostl et al., 2007b). In adaptive management, policies are viewed as hypotheses that must be tested and tried, which imply continuous monitoring and evaluation of the effectiveness of the implemented actions (Folke et al., 2005). This concept presupposes the recognition that strategies, objectives, goals, and even policies are continually reviewed and adapted during the management process. Eventual changes may occur due to the emergence of new information and feedback from the process itself. The requirement for this approach is the presence of a dynamic, flexible, and open interface in relation to the process, stakeholders, and institutions involved, as a way of managing natural resources from a holistic and participatory perspective (Pahl-Wostl et al., 2007a; Resilience Alliance, 2010). This setting-up, however, comes up against barriers and bureaucratic procedures, typical of the current water management model adopted by the basin.

4.4 Final Remarks

This chapter aimed to provide elements to better inform users – from government, organizations, professionals, and individuals – how climate change will likely impact the water-energy-food (WEF) nexus in Brazil. In addition to motivating the debate, focus on this new approach stimulates research and interventions, particularly for Cerrado and Caatinga biome and Paraíba do Sul river basin. Both ecosystems and human infrastructure face risks from climate change.

Significant changes in climate and their impacts are already visible over the Brazilian territory. Unlike the average annual precipitation, climate models suggest that most of Brazil will have increased extreme events, such as heavy rainfall intensity and droughts by the mid-twenty-first century. Changes in frequency and intensity of extreme events will cause significant financial and human losses throughout Brazil. For instance, temperature rise and changing precipitation patterns are expected to exacerbate the already acute water shortage problem in the northeast and southern regions. In addition, warming is likely to increase energy demand for air conditioning in the summer, particularly in Southern Brazil. Such extra power demand, compounded by climate change, induced a reduction in hydro-production and could disrupt energy supplies. Climate change represents an added layer of complexity in the WEF nexus approach.

WEF governance requires an understanding of socio-ecological complexity, which implies multilevel, multiscale, and multisectoral policies and stakeholders. Due to high urbanization in Brazil, cities are key systems for advancing the governance of the resources in the country. Faced with this scenario, adaptive management, innovative communication, engagement, social mobilization strategies, and transparency concerning information and the decision-making process lead to a balance of interest and demands on WEF and increase adaptive capacity in the face of compounding effects of climate change and the public policies implemented.

In urban regions or under the influence of large cities, territorial and environmental planning and management structures must deal, simultaneously and systemically, with the search for more efficient, inclusive, and innovative models for the integrated management of natural resources and, on the one hand, due to concerns about environmental and climate equity and justice, through the pursuit of increasing the quality of life of the inhabitants, with a focus on the most vulnerable populations.

Reforms in environmental and territorial management are necessary to overcome bureaucratic barriers, allowing for greater systematization and assessment dynamics, allowing a path for adaptive public planning and greater participation of diverse stakeholders.

Finally, the subjacent premise to the scientific challenge is that the transition to sustainability requires much more than technical solutions (Park et al., 2008). For that, profound and structural sociopolitical changes are needed, including a rethinking of current institutions, dominant standards (e.g., as in science, the valuation of natural sciences about humanities and arts, when looking to environmental issues), cultural options, and the involvement of society in public decisions (Dryzek, 2016).

Acknowledgments This work was supported by the National Institute of Science and Technology for Climate Change Phase 2 under CNPq Grant 465501/2014-1, FAPESP 15/50122-0 and 17/22269-2, and Programa CAPES-ANA 2180/2017 Processo 88881.144932/2017-01.

References

Alcamo, J., Flörke, M., & Märker, M. (2007). Future long-term changes in global water resources driven by socio-economic and climatic changes. *Hydrological Sciences Journal, 52*(2), 247–275. https://doi.org/10.1623/hysj.52.2.247

Alexander, L. V., Zhang, X., Peterson, T. C., Caesar, J., Gleason, B., Klein Tank, A. M. G., Haylock, M., et al. (2006). Global observed changes in daily climate extremes of temperature and precipitation. *Journal of Geophysical Research, 111*(D5), D05109. https://doi.org/10.1029/2005JD006290

Almeida, A. C., Barbieri, A., & Rodrigues, S. (2020). Linking migration, climate and social protection in Brazilian Semiarid: Case studies of Submédio São Francisco and Seridó Potiguar. *Sustentabilidade Em Debate, 11*(2), 238–251. https://doi.org/10.18472/SustDeb.v11n2.2020.31293

Alves, L. M., Chadwick, R., Moise, A., Brown, J., & Marengo, J. A. (2021). Assessment of rainfall variability and future change in Brazil across multiple timescales. *International Journal of Climatology, 41*(S1). https://doi.org/10.1002/joc.6818

ANA – Agência Nacional de Águas. (2015). *Encarte Especial Sobre a Crise Hídrica – Conjuntura Recursos Hídricos No Brasil*. Brasília.

Araujo, M., Ometto, J., Rodrigues-Filho, S., Bursztyn, M., Lindoso, D. P., Litre, G., Gaivizzo, L., Ferreira, J. L., Reis, R. M., & Assad, E. (2019). The socio-ecological nexus+ approach used by the Brazilian research network on global climate change. *Current Opinion in Environmental Sustainability, 39*(August), 62–70. https://doi.org/10.1016/j.cosust.2019.08.005

Arnell, N. W. (2004). Climate change and global water resources: SRES emissions and socio-economic scenarios. *Global Environmental Change, 14*(1), 31–52. https://doi.org/10.1016/j.gloenvcha.2003.10.006

Azevedo, A. A., Rajão, R., Costa, M. A., Stabile, M. C. C., Macedo, M. N., dos Reis, T. N. P., Alencar, A., Soares-Filho, B. S., & Pacheco, R. (2017). Limits of Brazil's forest code as a means to end illegal deforestation. *Proceedings of the National Academy of Sciences, 114*(29), 7653–7658. https://doi.org/10.1073/pnas.1604768114

Baettig, M. B., Wild, M., & Imboden, D. M. (2007). A climate change index: Where climate change may be most prominent in the 21st century. *Geophysical Research Letters, 34*(1), L01705. https://doi.org/10.1029/2006GL028159

Baião, C. F. P., Almeida, F. P., Machado, M. A., Roman, M. T. M., Trovarelli, P., Lourenço, S. A. G., Marques, A. R., et al. (2020). Percepção Ambiental Dos

Moradores Da Bacia Do Rio Paraíba Do Sul. In *Simpósio Da Pós-Graduação Em Ciência Do Sistema Terrestre*. São José dos Campos, Brazil.

Bezerra, B. G., Silva, L. L., Santos e Silva, C. M., & de Carvalho, G. G. (2019). Changes of precipitation extremes indices in São Francisco River Basin, Brazil from 1947 to 2012. *Theoretical and Applied Climatology, 135*(1–2), 565–576. https://doi.org/10.1007/s00704-018-2396-6

Cai, W., McPhaden, M. J., Grimm, A. M., Rodrigues, R. R., Taschetto, A. S., Garreaud, R. D., Dewitte, B., et al. (2020). Climate Impacts of the El Niño–Southern Oscillation on South America. *Nature Reviews Earth & Environment, 1*(4), 215–231. https://doi.org/10.1038/s43017-020-0040-3

Cavalcanti, B. S., & Marques, G. R. G. (2016). Recursos Hídricos e Gestão de Conflitos: A Bacia Hidrográfica Do Rio Paraíba Do Sul a Partir Da Crise Hídrica de 2014–2015. *Revista de Gestão Dos Países de Língua Portuguesa, 15*(1), 4. https://doi.org/10.12660/rgplp.v15n1.2016.78411

CEIVAP, AGEVAP, and COHIDRO. (2014). *Plano Integrado De Recursos Hídricos – Pirh Da Bacia Hidrográfica Do Rio Paraíba Do Sul E Planos De Ação De Recursos Hídricos – Parh Das Bacias Afluentes*. Resende. https://ceivap.org.br/conteudo/REL-Produto-C-versao-final-dez16.pdf

Ciais, P., Sabine, C., Bala, G., Bopp, L., Brovkin, V., Canadell, J., Chhabra, A., et al. (2014). Carbon and other biogeochemical cycles. In C. Change (Ed.), *2013 – The physical science basis: Working group I contribution to the fifth assessment report of the intergovernmental panel on climate Change, edited by intergovernmental panel on climate change* (pp. 465–570). Cambridge University Press. https://doi.org/10.1017/CBO9781107415324.015

Coppola, E., Raffaele, F., Giorgi, F., Giuliani, G., Gao, X., Ciarlo, J. M., Sines, T. R., et al. (2021, March). Climate Hazard indices projections based on CORDEX-CORE, CMIP5 and CMIP6 ensemble. *Climate Dynamics*. https://doi.org/10.1007/s00382-021-05640-z

Cunha, A. P. M. A., Zeri, M., Leal, K. D., Costa, L., Cuartas, L. A., Marengo, J. A., Tomasella, J., et al. (2019). Extreme drought events over Brazil from 2011 to 2019. *Atmosphere, 10*(11), 642. https://doi.org/10.3390/atmos10110642

Dalla-Nora, E. L., de Aguiar, A. P. D., Lapola, D. M., & Woltjer, G. (2014). Why have land use Change models for the Amazon failed to capture the amount of deforestation over the last decade? *Land Use Policy, 39*(July), 403–411. https://doi.org/10.1016/j.landusepol.2014.02.004

de los Skansi, M. M., Brunet, M., Sigró, J., Aguilar, E., Groening, J. A. A., Bentancur, O. J., Geier, Y. R. C., et al. (2013). Warming and wetting signals emerging from analysis of changes in climate extreme indices over South America. *Global and Planetary Change, 100*(January), 295–307. https://doi.org/10.1016/j.gloplacha.2012.11.004

da Vieira, R. M. S. P., Tomasella, J., Barbosa, A. A., Polizel, S. P., Ometto, J. P. H. B., Santos, F. C., da Cruz Ferreira, Y., & de Toledo, P. M. (2021). Land degradation mapping in the MATOPIBA region (Brazil) using remote sensing data and decision-tree analysis. *Science of the Total Environment, 782*(August), 146900. https://doi.org/10.1016/j.scitotenv.2021.146900

do Carmo, R. L., & Anazawa, T. M. (2017). Hidromegalópole São Paulo – Rio de Janeiro: Uma Nova Dinâmica Regional? In *Anais Do XVII ENANPUR* (pp. 1–18). São Paulo. http://anpur.org.br/xviienanpur/principal/publicacoes/XVII.ENANPUR_Anais/ST_Sessoes_Tematicas/ST3/ST 3.13/ST 3.13-14.pdf.

Donat, M. G., Alexander, L. V., Herold, N., & Dittus, A. J. (2016). Temperature and precipitation extremes in century-long gridded observations, reanalyses, and atmospheric model simulations. *Journal of Geophysical Research: Atmospheres, 121*(19), 11,174–11,189. https://doi.org/10.1002/2016JD025480

Dryzek, J. S. (2016). Institutions for the Anthropocene: Governance in a changing earth system. *British Journal of Political Science, 46*(4), 937–956. https://doi.org/10.1017/S0007123414000453

Ellis, E. C. (2015). Ecology in an anthropogenic biosphere. *Ecological Monographs, 85*(3), 287–331. https://doi.org/10.1890/14-2274.1

Endo, A., Yamada, M., Miyashita, Y., Sugimoto, R., Ishii, A., Nishijima, J., Fujii, M., et al. (2020). Dynamics of water–energy–food nexus methodology, methods, and tools. *Current Opinion in Environmental Science & Health, 13*(February), 46–60. https://doi.org/10.1016/j.coesh.2019.10.004

Environment and Natural Resources Working Paper No. 58 – FAO (Food and Agriculture Organization). (2014). Rome.

Engle, N. L., Johns, O. R., Lemos, M., & Nelson, D. R. (2011). Integrated and adaptive management of water resources: Tensions, legacies, and the next best thing. *Ecology and Society, 16*(1) http://www.ecologyandsociety.org/vol16/iss1/art19/

Empresa Brasileira de Pesquisa Agropecuária (Embrapa). (2019). Integração Lavoura-Pecuária-Floresta. Available online: https://www.embrapa.br/temaintegracao-lavoura-pecuaria-floresta-ilpf

Feng, X., Porporato, A., & Rodriguez-Iturbe, I. (2013). Changes in rainfall seasonality in the tropics. *Nature Climate Change, 3*(9), 811–815. https://doi.org/10.1038/nclimate1907

FNEM. (2021). *Fórum Nacional de Entidades Metropolitanas*. 2021. https://fnembrasil.org/entidades-metropolitanas/

Foley, J. A., Ramankutty, N., Brauman, K. A., Cassidy, E. S., Gerber, J. S., Johnston, M., Mueller, N. D., et al. (2011). Solutions for a cultivated planet. *Nature, 478*(7369), 337–342. https://doi.org/10.1038/nature10452

Folke, C., Hahn, T., Olsson, P., & Norberg, J. (2005). Adaptive governance of social-ecological systems.

Annual Review of Environment and Resources, 30(1), 441–473. https://doi.org/10.1146/annurev.energy.30.050504.144511

Fonseca, M. G., Alves, L. M., Aguiar, A. P. D., Arai, E., Anderson, L. O., Rosan, T. M., Shimabukuro, Y. E., & Luiz Eduardo Oliveira e Cruz Aragão. (2019). Effects of climate and land-use change scenarios on fire probability during the 21st century in the Brazilian Amazon. *Global Change Biology, 25*(9), 2931–2946. https://doi.org/10.1111/gcb.14709

Fowler, H. J., Wasko, C., & Prein, A. F. (2021). Intensification of short-duration rainfall extremes and implications for flood risk: Current state of the art and future directions. *Philosophical Transactions of the Royal Society A: Mathematical, Physical and Engineering Sciences, 379*(2195), 20190541. https://doi.org/10.1098/rsta.2019.0541

Fracalanza, A. P., & Freire, T. M. (2015). Crise Da Água Na Região Metropolitana de São Paulo: A Injustiça Ambiental e a Privatização de Um Bem Comum. *GEOUSP: Espaço e Tempo (Online), 19*(3), 464. https://doi.org/10.11606/issn.2179-0892.geousp.2015.103064

Frich, P., Alexander, L. V., Della-Marta, P., Gleason, B., Haylock, M., Klein Tank, A. M. G., & Peterson, T. (2002). Observed coherent changes in climatic extremes during the second half of the twentieth century. *Climate Research, 19*, 193–212. https://doi.org/10.3354/cr019193

Giatti, L. L., Jacobi, P. R., Favaro, A. K. M. D. I., & Empinotti, V. L. (2016). O Nexo Água, Energia e Alimentos No Contexto Da Metrópole Paulista. *Estudos Avançados, 30*(88), 43–61. https://doi.org/10.1590/s0103-40142016.30880005

Gibbs, H. K., Ruesch, A. S., Achard, F., Clayton, M. K., Holmgren, P., Ramankutty, N., & Foley, J. A. (2010). Tropical forests were the primary sources of new agricultural land in the 1980s and 1990s. *Proceedings of the National Academy of Sciences, 107*(38), 16732–16737. https://doi.org/10.1073/pnas.0910275107

Giorgi, F., & Gutowski, W. J. (2015). Regional dynamical downscaling and the CORDEX initiative. *Annual Review of Environment and Resources, 40*(1), 467–490. https://doi.org/10.1146/annurev-environ-102014-021217

Grande, T. O., Aguiar, L. M. S., & Machado, R. B. (2020). Heating a biodiversity hotspot: Connectivity is more important than remaining habitat. *Landscape Ecology, 35*(3), 639–657. https://doi.org/10.1007/s10980-020-00968-z

Hartmann, D. L., Klein Tank, A. M. G., Rusticucci, M., Alexander, L. V., Brönnimann, S., Charabi, Y., Dentener, F. J., et al. (2013). Observations: Atmosphere and surface. In T. F. Stocker, D. Qin, G.-K. Plattner, M. Tignor, S. K. Allen, J. Boschung, A. Nauels, Y. Xia, V. Bex, & P. M. Midgley (Eds.), *Climate change 2013: The physical science basis. Contribution of working group i to the fifth assessment report of the intergovernmental panel on climate change*. Cambridge University Press.

Haylock, M. R., Peterson, T. C., Alves, L. M., Ambrizzi, T., Anunciação, Y. M. T., Baez, J., Barros, V. R., et al. (2006). Trends in total and extreme south American rainfall in 1960–2000 and links with sea surface temperature. *Journal of Climate, 19*(8), 1490–1512. https://doi.org/10.1175/JCLI3695.1

Hoegh-Guldberg, O., D. Jacob, M. Taylor, M. Bindi, S. Brown, I. Camilloni, A. Diedhiou, et al. 2018. "Impacts of 1.5°C global warming on natural and human systems." In *Global warming of 1.5°C. An IPCC special report on the impacts of global warming of 1.5°C above pre-industrial levels and related global greenhouse gas emission pathways, in the context of strengthening the global response to the threat of climate change,* V. Masson-Delmotte, P. Zhai, H.-O. Pörtner, D. Roberts, J. Skea, P.R. Shukla, A. Pirani, et al., Translated by Rt13, 175–311. In Press. https://www.ipcc.ch/sr15/chapter/chapter-3.

IBGE - Instituto Brasileiro De Geografia e Estatística. Censo Agropecuário. (2017). Disponível em: https://censos.ibge.gov.br/agro/2017/

ICMBio. (2021). Painel Dinâmico de Informações. Instituto Chico Mendes de Conservação da Biodiversidade, Brasília, Brasil. http://qv.icmbio.gov.br/

IPCC. (2014). *Climate change 2013 – The physical science basis working group I contribution to the fifth assessment report of the intergovernmental panel on climate change*. Cambridge University Press. http://www.cambridge.org/ba/academic/subjects/earth-and-environmental-science/climatology-and-climate-change/climate-change-2013-physical-science-basis-working-group-i-contribution-fifth-assessment-report-intergovernmental-panel-climate-change?format=PB&i

Jacobi, P. R., Cibim, J. C., & Souza, A. N. (2015a). Crise Da Água Na Região Metropolitana de São Paulo – 2013/2015. *GEOUSP: Espaço e Tempo (Online), 19*(3), 422. https://doi.org/10.11606/issn.2179-0892.geousp.2015.104114

Jacobi, P. R., Silva-Sánchez, S., & Fracalanza, A. P. (2015b). Gobernanza Del Agua En La Ciudad de São Paulo: Actores Sociales, Degradación Socioambiental y Acciones Públicas Innovadoras. *América Latina Hoy, 69*(April), 35–51. https://doi.org/10.14201/alh2015693551

Khan, Z., Linares, P., Rutten, M., Parkinson, S., Johnson, N., & García-González, J. (2018). Spatial and temporal synchronization of water and energy systems: Towards a single integrated optimization model for long-term resource planning. *Applied Energy, 210*(January), 499–517. https://doi.org/10.1016/j.apenergy.2017.05.003

Klink, C. A., & Machado, R. B. (2005). Conservation of the Brazilian Cerrado. *Conservation Biology, 19*(3), 707–713. https://doi.org/10.1111/j.1523-1739.2005.00702.x

Lambin, E. F., & Meyfroidt, P. (2011). Global land use change, economic globalization, and the looming land scarcity. *Proceedings of the National Academy of*

Sciences, 108(9), 3465–3472. https://doi.org/10.1073/pnas.1100480108

Lapola, D. M., Pinho, P., Quesada, C. A., Strassburg, B. B. N., Rammig, A., Kruijt, B., Brown, F., et al. (2018). Limiting the high impacts of Amazon forest dieback with no-regrets science and policy action. *Proceedings of the National Academy of Sciences, 115*(46), 11671–11679. https://doi.org/10.1073/pnas.1721770115

Lopes, T. R., Zolin, C. A., Mingoti, R., Vendrusculo, L. G., de Almeida, F. T., de Souza, A. P., de Oliveira, R. F., Paulino, J., & Uliana, E. M. (2021). Hydrological regime, water availability and land use/land cover Change impact on the water balance in a large agriculture basin in the southern Brazilian Amazon. *Journal of South American Earth Sciences, 108*(June), 103224. https://doi.org/10.1016/j.jsames.2021.103224

Marengo, J., & Alves, L. M. (2015). Crise Hídrica Em São Paulo Em 2014: Seca e Desmatamento. *GEOUSP: Espaço e Tempo (Online), 19*(3), 485. https://doi.org/10.11606/issn.2179-0892.geousp.2015.100879

Marengo, J. A., & Espinoza, J. C. (2015, July). Extreme seasonal droughts and floods in Amazonia: Causes, trends and impacts. *International Journal of Climatology*, n/a–n/a. https://doi.org/10.1002/joc.4420

Marengo, J. A., Ambrizzi, T., da Rocha, R. P., Alves, L. M., Cuadra, S. V., Valverde, M. C., Torres, R. R., Santos, D. C., & Ferraz, S. E. T. (2010). Future Change of climate in South America in the late twenty-first century: Intercomparison of scenarios from three regional climate models. *Climate Dynamics, 35*(6), 1073–1097. https://doi.org/10.1007/s00382-009-0721-6

Marengo, J. A., Alves, L. M., Regina, C. S., Alvala, A. N. A. P. A. U. L. A. C. U. N. H. A., Brito, S., & Moraes, O. L. L. (2018). Climatic characteristics of the 2010–2016 drought in the semiarid Northeast Brazil region. *Anais da Academia Brasileira de Ciências, 90*(2 suppl 1), 1973–1985. https://doi.org/10.1590/0001-3765201720170206

Marengo, J. A., Camarinha, P. I., Alves, L. M., Diniz, F., & Betts, R. A. (2021). Extreme rainfall and hydro-geo-meteorological disaster risk in 1.5, 2.0, and 4.0°C global warming scenarios: An analysis for Brazil. *Frontiers in Climate, 3*(March). https://doi.org/10.3389/fclim.2021.610433

Martins, A. (2014). *Sistema Cantareira e a Crise Da Água Em São Paulo: A Falta de Transparência No Acesso à Informação*. São Paulo.

Milly, P. C. D., Dunne, K. A., & Vecchia, A. V. (2005). Global pattern of trends in streamflow and water availability in a changing climate. *Nature, 438*(7066), 347–350. https://doi.org/10.1038/nature04312

Myers, N., Mittermeier, R. A., Mittermeier, C. G., da Fonseca, G. A. B., & Kent, J. (2000). Biodiversity hotspots for conservation priorities. *Nature, 403*(6772), 853–858. https://doi.org/10.1038/35002501

Myhre, G., Shindell, D., Bréon, F.-M., Collins, W., Fuglestvedt, J., Huang, J., Koch, D., et al. (2014). Anthropogenic and natural radiative forcing. In Intergovernmental Panel on Climate Change (Ed.), *Climate change 2013 – The physical science basis: Working group I contribution to the fifth assessment report of the intergovernmental panel on climate change* (pp. 659–740). Cambridge University Press. https://doi.org/10.1017/CBO9781107415324.018

Nobre, C. A., Young, A. F., Saldiva, P., Marengo, J. A., Nobre, A. D., Alves, S., Jr., da Silva, G. C. M., & Lombardo, M. (2011). *Vulnerabilidades Das Megacidades Brasileiras Às Mudanças Climáticas: Região Metropolitana de São Paulo*. São Paulo.

Nobre, C. A., Marengo, J. A., Seluchi, M. E., Adriana Cuartas, L., & Alves, L. M. (2016). Some characteristics and impacts of the drought and water crisis in Southeastern Brazil during 2014 and 2015. *Journal of Water Resource and Protection, 08*(02), 252–262. https://doi.org/10.4236/jwarp.2016.82022

OECD. (2011). *Water governance in OECD countries – A multi-level approach* (OECD Studies on Water). OECD. https://doi.org/10.1787/9789264119284-en

Pahl-Wostl, C., Craps, M., Dewulf, A., Mostert, E., Tabara, D., & Taillieu, T. (2007a). Social learning and water resources management. *Ecology and Society, 12*(2) http://www.ecologyandsociety.org/vol12/iss2/art5/

Pahl-Wostl, C., Sendzimir, J., Jeffrey, P., Aerts, J., Berkamp, G., & Cross, K. (2007b). Managing Change toward adaptive water management through social learning. *Ecology and Society, 12*(2) http://www.ecologyandsociety.org/vol12/iss2/art30/

Park, J., Conca, K., & Finger, M. (2008). *The crisis of global environmental governance* (J. Park, K. Conca, & M. Finger, Eds., 1st ed.). Routledge. https://doi.org/10.4324/9780203929100

Pires, M. O. (2020). Cerrado', old and new agricultural frontiers. *Brazilian Political Science Review, 14*(3). https://doi.org/10.1590/1981-3821202000030006

PROFILL, AGEVAP, and CEIVAP. (2020). *Produto Parcial 06 (PP-06): Consolidação Do Plano Integrado de Recursos Hídricos Da Bacia Hidrográfica Do Rio Paraíba Do Sul (PIRH-PS)*. http://18.229.168.129:8080/publicacoesArquivos/ceivap/arq_pubMidia_AGVP_PS_PIRH_PP-06_REV03_FINAL.pdf

PSR Engenharia, and AGEVAP. (2013). *Avaliação Dos Impactos de Novas Transposições de Vazão No Rio Paraíba Do Sul*. Rio de Janeiro.

Resende, F. M., Cimon-Morin, J., Poulin, M., Meyer, L., Joner, D. C., & Loyola, R. (2021). The importance of protected areas and indigenous lands in securing ecosystem services and biodiversity in the Cerrado. *Ecosystem Services, 49*(June), 101282. https://doi.org/10.1016/j.ecoser.2021.101282

Resilience Alliance. (2010). *Key concepts: Adaptive management. 2010*. https://www.resalliance.org/key-concepts

Ribeiro, C. R. (2020). Entre Bacias, Malhas E Redes: Das Metrópoles Sedentas À Hidromegarregião. *GEOgraphia, 22*(49). https://doi.org/10.22409/GEOgraphia2020.v22i49.a28625

Richey, A. S., Thomas, B. F., Lo, M.-H., Reager, J. T., Famiglietti, J. S., Voss, K., Swenson, S., & Rodell, M. (2015). Quantifying renewable groundwater stress with Grace. *Water Resources Research, 51*(7), 5217–5238. https://doi.org/10.1002/2015WR017349

SABESP. (2015). *EIA/Rima Para a Obra de Interligação Entre as Represas Jaguari (Bacia Do Paraíba Do Sul) e Atibainha (Bacia Do PCJ)*. São Paulo.

Salazar, A., Baldi, G., Hirota, M., Syktus, J., & McAlpine, C. (2015). Land use and land cover change impacts on the regional climate of non-Amazonian South America: A review. *Global and Planetary Change, 128*(May), 103–119. https://doi.org/10.1016/j.gloplacha.2015.02.009

Soterroni, A. C., Ramos, F. M., Mosnier, A., Fargione, J., Andrade, P. R., Baumgarten, L., Pirker, J., et al. (2019). Expanding the Soy Moratorium to Brazil's Cerrado. *Science Advances, 5*(7), eaav7336. https://doi.org/10.1126/sciadv.aav7336

Strassburg, B. B. N., Brooks, T., Feltran-Barbieri, R., Iribarrem, A., Crouzeilles, R., Loyola, R., Latawiec, A. E., et al. (2017). Moment of truth for the Cerrado hotspot. *Nature Ecology & Evolution, 1*(4), 0099. https://doi.org/10.1038/s41559-017-0099

Tadeu, N. D. (2016). *O Sistema Cantareira e a Crise Da Água Em São Paulo: Falta de Transparência, Um Problema Que Persiste*. São Paulo.

Taylor, K. E., Stouffer, R. J., & Meehl, G. A. (2012). An overview of CMIP5 and the experiment design. *Bulletin of the American Meteorological Society, 93*(4), 485–498. https://doi.org/10.1175/BAMS-D-11-00094.1

Tôsto, K. L., Pacheco, F. S., Gomes, L., Miranda, M., Oliveira, S. M. L., Pessi, D. D., Cox, C., & Ometto, J. P. H. B. (2019). Nitrogen use efficiency: A local and regional approach for Brazilian agriculture. *Raega – O Espaço Geográfico Em Análise, 46*(3), 125. https://doi.org/10.5380/raega.v46i3.67095

Vincent, L. A., Peterson, T. C., Barros, V. R., Marino, M. B., Rusticucci, M., Carrasco, G., Ramirez, E., et al. (2005). Observed trends in indices of daily temperature extremes in South America 1960–2000. *Journal of Climate, 18*(23). https://doi.org/10.1175/JCLI3589.1

Zhang, X., & Zhou, G. (2019). Scientometric analysis of ecological footprint of water resources from 2006–2018. *Ekoloji, 28*(107), 1539–1549. http://ekolojidergisi.com/download/a-scientometric-analysis-of-ecological-footprint-of-water-resources-from-2006-2018-5778.pdf

5 How Can the Water-Energy-Food Nexus Approach Contribute to Enhancing the Resilience of Amazonian Cities to Climate Change?

Nathália Nascimento, Lira Luz Benites Lazaro, and Mateus Henrique Amaral

Abstract

Climate change is expected to affect the most diverse regions of the world in diverse ways, posing additional challenges to managers and populations in the countryside and in the cities. In this chapter, we adopt climate anomaly scenarios considering the variables such as maximum temperature, consecutive days of rain, and number of dry days, to select municipalities in the Brazilian Amazon that are likely to face great climate changes in the region. We then analyzed socioeconomic data, producing clusters for groups of municipalities based on the neural network self-organizing maps. Our findings reveal that an analysis of the cities from a nexus perspective shows the impact of climate change in urban development and, at the same time, urban development impacts on the natural resources. The results depict Brazilian Amazon municipalities' vulnerability – they have the lowest level of basic sanitation, waste management, adequate storm drainage, and human development index that makes their population particularly vulnerable to face the climate crisis. Furthermore, impacts can be particularly disastrous for 30 Amazonian municipalities by their critical condition due to climate change and their socioeconomic and water demand index. Our results can be useful for managers of municipalities that may reach critical states due to climate change and serve as an alert to the urgency of adaptation and management strategies.

Keywords

Water-energy-food nexus · WEF nexus · Urban nexus · Amazon cities · Climate change · Climate adaptation · Neural network

N. Nascimento (✉)
Institute of Advanced Studies of University of São Paulo, São Paulo, Brazil

L. L. B. Lazaro
Department of Environmental Health, School of Public Health, University of São Paulo, São Paulo, Brazil

Durham Energy Institute and Department of Anthropology, Durham University, Durham, UK
e-mail: lbenites@usp.br

M. H. Amaral
Department of Environmental Health, School of Public Health, University of São Paulo, São Paulo, Brazil
e-mail: mateusamaral@usp.br

L. L. B. Lazaro et al. (eds.), *Water-Energy-Food Nexus and Climate Change in Cities*, Sustainable Development Goals Series, https://doi.org/10.1007/978-3-031-05472-3_5

5.1 Introduction

Around the world, the increased frequency of extreme climatic events has pointed to the urgency of measures to mitigate climate change impacts, design of adaptation strategies to a new climatological normal, and the need for a plan in different sectors to guarantee human survival and well-being. In this context, water, food, and energy security are recognized as one of the biggest challenges for the near future, given the projections of population growth and the expected change in consumer patterns (United Nations, 2017). It is estimated that the world population exceeds 9 billion people in 2050 and 10 billion in 2070, consequently increasing the pressure on natural resources and the demand for water, food, and energy (United Nations, 2017; OECD-FAO, 2019). The response to this demand can be severely affected by the impacts of climate change, especially the instability in seasons and in the precipitation regime, and the occurrence of extreme weather events such as droughts, storms, and heatwaves (Bouwer, 2019; Zscheischler et al., 2020). These events can affect crop plantations, limit land available for agriculture, and modify river dynamics affecting hydropower systems and the availability and access to freshwater and food (Gephart et al., 2020; Scott et al., 2020; Yalew et al., 2020).

In this sense, the Amazon region stands out as a key player in the local, regional, and global climate system. Globally, the Amazon carbon sequestration and storage activities influence the global carbon balance (Baccini et al., 2017; Müller et al., 2015). Regionally, evapotranspiration and aerosols released by the forest are essential for the formation of clouds and rainfall in the region and the moisture transportation to other regions of South America (Van Der Ent et al., 2010). Locally, shading, evapotranspiration, and moisture retention produce milder microclimates, promoting thermal comfort (Coe et al., 2017; Longo et al., 2020). However, despite this important functioning of Amazonian ecosystems, climate change scenarios for the region suggest major impacts for its ecosystems and population (Marengo & Souza, 2018).

Climate models indicate that different regions of the Amazon will suffer from an increase in temperature, changes in rain patterns that may become sparser and more intense in some regions, and an increase in dry days, in addition to drought events (Marengo & Souza, 2018). It is estimated that precipitation in the Amazon decreased by 1.8% (Spracklen et al., 2018) and that changes in rainfall patterns between the wet and dry seasons have occurred, especially in the southeast of the region (Leite-Filho et al., 2019). Models also have pointed out that if deforestation in the Amazon basin reaches 20% of the total area in a context of constant emissions and increased temperature, the Amazon forest may reach a tipping point, initiating a transition from rainforest to vegetation like a degraded savannah (Nobre et al., 2016; Gatti et al., 2021). This transition is irreversible and will impact on the regional climate and dynamics of vegetation, species, and biogeochemical cycles. These impacts have been extensively studied (Esquivel-Muelbert et al., 2017; Gomes et al., 2019); however, even less is known about the socioeconomic impacts of climate change in the Amazon (Brondízio et al., 2016).

Climate change is expected to impact different regions of the planet unevenly and may be more intense in some regions, bringing dire consequences for the most vulnerable and exposed populations (Ardila et al., 2021). It is important to emphasize that a natural hazard is only considered a disaster when it affects human populations (Wisner et al., 2015). In this context, the impacted populations have their vulnerability determined by the social context, whose structure was socially constructed over time and guided by marginalization and social inequality. The most vulnerable populations inhabit areas at greater risk of natural hazards, with little infrastructure, low income and education, and limited accessibility to services, such as health and sanitation (Raju et al., 2022).

Approximately 30 million people live in the Brazilian Amazon, of which 75% are concentrated in urban areas (IBGE, 2015). The great diversity of Amazonian ecosystems is reflected in the huge diversity of its population, which despite the great concentration of people in the cities, holds hundreds of communities of riverine, fishers, extractives, Indigenous, Afro-descendants, and small and large rural producers, all of which will be directly or indirectly affected by climate change (Brondízio et al., 2016). In addition to thermal discomfort in humans, and uncertainties in the agricultural production, increasing temperatures can affect terrestrial and aquatic species, compromising food security of local communities and the availability of regional food to Amazonian cities (Giannini et al., 2020; Prado-Lima & Val, 2016).

Changes in precipitation and temperature patterns can have a major impact on Amazonian municipalities not prepared for extreme weather events, that is, in less resilient municipalities (Marengo et al., 2021). The concept of resilience has become increasingly common, but no less complex. In this work, we approach it in the context of resilient cities, where resilience is understood as "the capacity of a city or community to prepare for, respond to, and adapt from dangerous and disruptive events, such as natural disasters, economic crises, demographic changes, health epidemics, and others" proposed by Figueiredo et al. (2018). As it is a comprehensive and multidisciplinary topic, the creation of resilience indicators or monitoring tools is not simple and has demanded a solid base in local and regional realities (Feldmeyer et al., 2019). Studying the characteristics of Amazonian municipalities in the face of climate change scenarios is highly relevant, as it exposes how a region with a crucial role in the global climate system can also be affected by climate change, and the urgency of discussing adaptation strategies for the region's municipalities, which historically present among the lowest human development indices in the country. Amazonian municipalities have the lowest level of basic sanitation, waste management, adequate storm drainage, human development index (HDI), and green area per inhabitant of Brazil (IBGE, 2015). In addition, they have high levels of social vulnerability, which makes their population particularly vulnerable to changes in the climate (IPEA, 2017).

The urgent need to find solutions for mitigating impacts and adaptation strategies to climate changes is challenging, but currently, it can count on approaches that facilitate planning actions uniting different sectors and spheres of decision-making. The water-energy-food (WEF) nexus approach emerges as an approach that integrates three fundamental sectors for human survival, seeking joint solutions and envisioning its adoption by different spheres of government. The nexus approach provides an innovative framework to change from conventional sector silos to integrated managing and policy decision. In fact, this approach can be useful for both WEF nexus analysis and policy-making to address coherent policies and the integration of natural resource management across sectors (Benites-Lazaro et al., 2020; Lazaro et al., 2021). Additionally, the nexus approach can be potentially useful as an effective tool in the management and preparation of cities in the context of climate change (Rasul & Sharma, 2016). Water, energy, and food are understood from this perspective as essential goods for human well-being, which need to be managed and planned in an interconnected way so that their production is guaranteed. Furthermore, the nexus approach has a solid foundation in the governance of these resources, which involves issues related to their access and distribution.

Resource management in many cities around the world still depends on traditional planning modes. Particularly in Brazil, the consequences of climate change have been faced in its various expressions, such as floods, droughts, and water, energy, and food scarcity. However, many Brazilian cities continue to employ traditional approaches to urban planning and management that are unable to assess the complex nature of the phenomena, impacts, and possible effects of climate change on the environment and society (Benites-Lazaro & Giatti, 2021). Thus, in this chapter, we examine some climatic scenarios and socioeconomic data and index to discuss how Brazilian Amazon municipalities could be affected by climate changes, the vulnerability of its population, and infrastructure condition to respond to the climate crisis, and we discuss the potential of the nexus approach as an important tool to help governments in the management of municipalities in the Amazon.

5.2 Methodology

5.2.1 Study Area

Due to the extension of the Amazon region, we prioritized analyzing the municipalities that fall under the most alarming projections of climate anomalies in the region. To select the municipalities for this analysis, we used data from the models for the IPCC-AR5, for the climate scenario RCP (Representative Concentration Pathways) 4.5. The RCP4.5 scenario assumes that the radiative forcing should stabilize around 4.5Wm-2 and 6.0Wm-2 after 2100, without exceeding the long-term radiation level of 4.5Wm-2. This is considered the most positive scenario (compared to RCP8.5) and is based on the premise that emissions should stabilize, while policies and reforestation can contribute to increased carbon sequestration (IPCC, 2013).

The regional model data used in this work were obtained from the climate projection platform for Brazil (http://pclima.inpe.br/), considering the data from the Eta Regional Model of the National Institute for Space Research (INPE), HadGEM2-ES Experiment for two periods: next (2011–2040) and medium (2041–2070). We considered the projected anomalies for four variables: maximum temperature, total precipitation, consecutive days with rain above 20 mm, and the number of days without rain. Data were crossed with the municipal limits of the Amazon biome, considering the highest indices for each variable. Thus, we selected 244 municipalities that must present one or more of the following anomalies: temperature above 3 °C, municipalities with three consecutive days of rain with volume above 20 mm, and municipalities with 20 days without rain (Fig. 5.1).

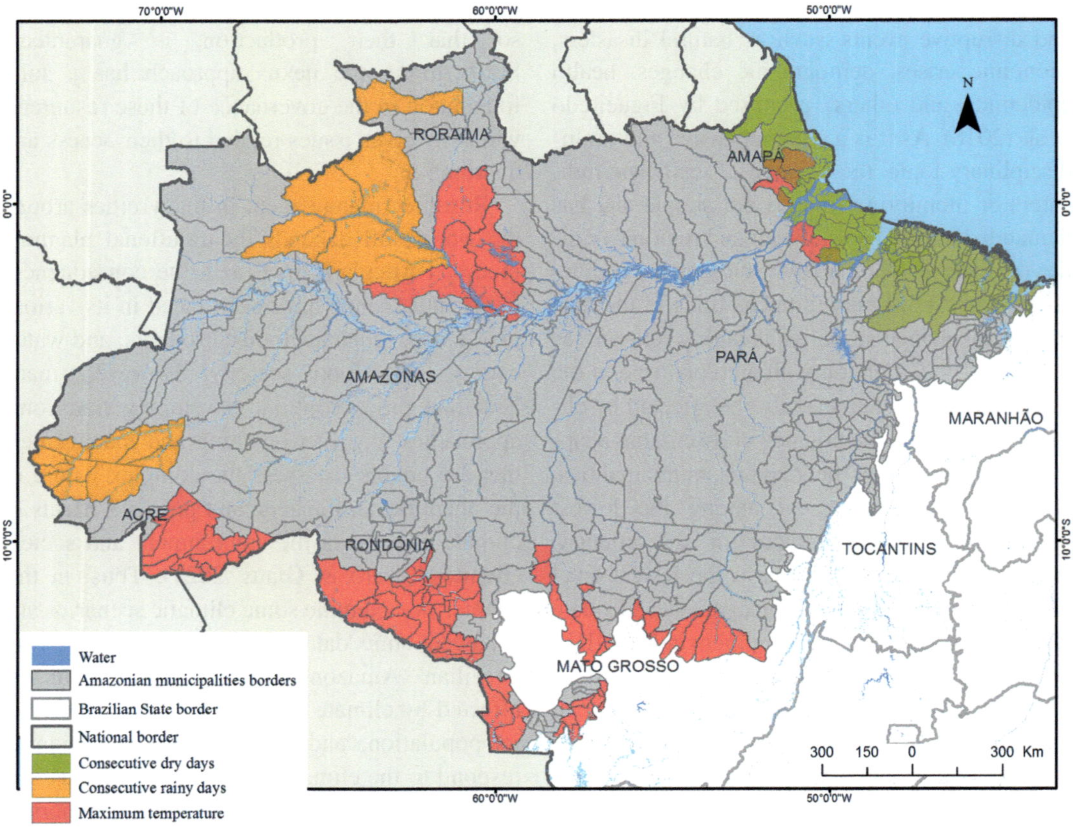

Fig. 5.1 Municipalities selected for analysis

5.2.2 Data Source and Analysis

5.2.2.1 Data Description

For this study, we considered data from ten indicators, referring to the year 2016. All the data utilized for this chapter are from Brazilian public domain information systems. To measure the relations among nexus sectors, we considered the proportion of the flux of water consumed for irrigation and animal grazing. We also included the three dimensions of human development, which are education, health, and employment and income (Table 5.1). However, we considered the Firjan Municipal Development Index as a human development measurement instead of the traditional Human Development Index (HDI) developed by the United Nations. This indicator, elaborated by Firjan System (Industry Federation of the State of Rio de Janeiro), is a study that annually follows the development of the more than 5000 Brazilian municipalities with available public data. The index ranges from 0 to 1 point, which when closer to 1, indicates the greater socioeconomic development of the municipality.

To better focus on Amazônia cities' reality in facing climate change and health challenges, we considered some indicators such as hospital admission rate by diseases that could be related to water quality and environmental conditions related to the heat waves and criminal burn. Thus, we considered the concentration of heat sources as a key element associated with hospital admissions by respiratory and cardiovascular diseases, due the increased human exposure to fine particulate matter from fires (Nunes et al., 2013; Machado-Silva et al., 2020; Alencar et al., 2020). In the same way, we considered acute diarrheal diseases as an important variable to represent diseases associated with poor water quality (Aguiar et al., 2020; Duarte et al., 2019).

5.2.2.2 Self-Organizing Maps with SOMbrero

For data analysis we used self-organizing maps to cluster a numeric dataset in the R package, SOMbrero. SOMbrero implements different variants of the self-organizing map algorithm (also called Kohonen algorithm) (Vialaneix et al., 2020). The Kohonen algorithm represents a class of neural network algorithms in the unsupervised learning category, which permits a nonlinear projection of a large-dimensional data manifold on a

Table 5.1 Description of indicators and their sources

Code	Indicators	Description	Source
F1	Fimd	Firjan Municipal Development Index	Firjan System (1)
F2	fimd_empl_income	Firjan Municipal Development Index (Employment and income)	
F3	fimd_edu	Firjan Municipal Development Index (education)	
F4	fimd_health	Firjan Municipal Development Index (health)	
F5	hr_cd	Hospital admission rate for cardiovascular diseases by 10,000 inhabitants	Department of Informatics of the Unified Health System (2)
F6	hr_rd	Hospital admission rate for respiratory diseases by 10,000 inhabitants	
F7	hr_add	Hospital admission rate for acute diarrheal diseases by 10,000 inhabitants	
F8	hot_spots	Concentration of heat sources	Atlas of Human Development in Brazil (3)
F9	cons_an_m3_s	Proportion of cubic meters of water consumed for animal grazing	Agência Nacional de Águas e Saneamento Básico (4)
F10	cons_ir_m3_s	Proportion of cubic meters of water consumed for irrigation	

(1) Firjan
(2) DATASUS
(3) Atlas Brasil
(4) Agência Nacional de Águas e Saneamento Básico

low (two)-dimensional grid, which can visualize common dependencies between variables, similarity relationships, and cluster structures within larger datasets (Amaral et al., 2021). SOMbrero offers a wide range of plots aimed at giving a comprehensive overview of the resulting clusters. SOMbrero comes with a user-friendly graphical interface, which makes most of its options available in a few clicks, without resorting to the command line, and the interface is programmed using the R package Shiny (Boelaert et al., 2014). For this analysis, we considered 212 municipalities and 10 indicators; this is because both the municipalities and the 2 discarded indicators did not have data for all the variables.

5.3 Results

5.3.1 Municipalities Characteristics

Through analysis of climate scenarios for the Amazon, we select three groups of municipalities where climate anomalies should be more intense (Fig. 5.1). In the northeast of the Amazon, which includes the coasts of the states of Maranhão, Pará, and Amapá, is the group of municipalities where an increase in the number of days without rain is expected. This number can reach 20 days without rain. In addition, this group covers overcrowded regional capitals, such as the capital of the state of Amapá, Macapá, and the capital of the state of Pará, Belém. Municipalities in the center of Amapá (Santana, Ferreira Gomes, and Tartarugalzinho) and the Marajó archipelago (Curralinho) are included in this group and will also face temperature rise above 3 °C. This can pose additional challenges for these municipalities, which may suffer from increased water demand due to heat and, at the same time, suffer from the absence of rain.

Temperature increase is also estimated for the central region of Amazon and the east of the state of Acre, also including overcrowded regional capitals, such as Manaus and Rio Branco. The same is expected to the south of the state of Rondônia and the central region of Mato Grosso. The inclusion of the main capitals and metropolitan areas of the Amazon region in the groups of municipalities with scenarios of major climate anomalies may represent that a substantial portion of the Amazon population will be affected by the consequences of climate change. In addition, the municipalities affected in the states of Rondônia and Mato Grosso are characterized by intense agricultural activities, mainly soy and cattle ranch, which can generate great economic impacts in the region.

An increase of three consecutive rainy days with volume above 20 mm is foreseen for municipalities in western Amazonas and Acre. One particularity of these municipalities is that they are close to important rivers in the region, such as the Rio Negro and Rio Juruá, but they have poor urban sanitation infrastructure. This means that these municipalities can have major social impacts, especially in the health sector, due to the large volume of rainfall and their inability to deal with possible flooding problems.

5.3.2 SOMbrero Analysis

Figure 5.2 shows the cluster from the 212 municipalities; this is the representation of the 36 micro-cluster classification according to the similarities from the 10 variables. The mini clusters 6 and 31 grouped more municipalities, while clusters like 11, for example, did not group any municipality. The clustering component contains the final classification from our dataset. Cluster profile overviews can be plotted either with lines or bar plot, and both provide information similar to that given by names; in this case, we numbered the municipalities (see Appendix). Figure 5.3a shows the ten variables used in this study, one variable is represented, respectively, with a bar that permits to see easily which variable influences which cluster, and Fig. 5.3b shows the cluster of municipalities by the eight federal units. Figure 5.4 presents the supercluster that groups the municipalities into four large clusters.

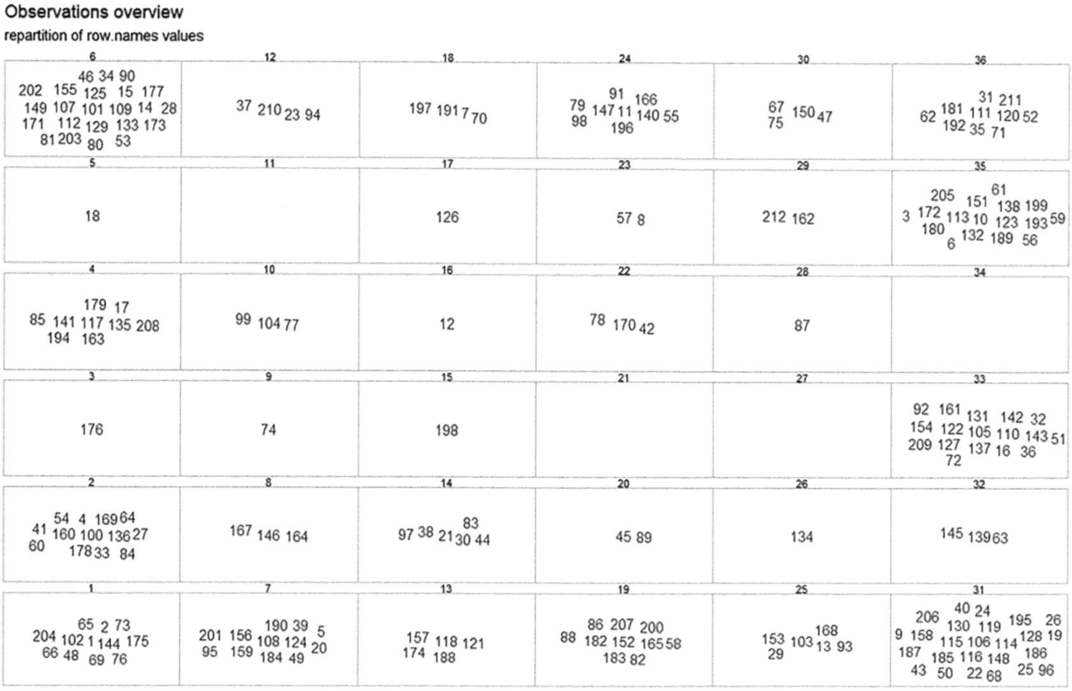

Fig. 5.2 Cluster from the analysis of Amazonian municipalities. Each municipality is represented by a number, which is aggregated into a group based on similar characteristics

5.3.3 Analysis of Clusters

The municipalities that integrate the pink group have in common a median Firjan Municipal Development Index (less than 0.65) and a high rate of hospitalization due to diseases of the respiratory system. Regarding the demand for water, the demand for irrigation is generally low, with a maximum demand of 75%, which is the proportion of cubic meters of water consumed for animal grazing. Municipalities in the green group present the Firjan Municipal Development Index ranging from 0.45 to 0.75, which is above the average of the municipalities studied. The values for the Firjan Municipal Development Index of education are also considered the highest compared to the other municipalities, ranging from 0.45 to 0.9. In terms of health, the municipalities have a high rate of hospitalization due to respiratory diseases. A particularity of municipalities in this group is highly dependent on water for animal watering (up to 90% of the water flow for watering) and on the flow of water consumed for irrigation (85% of the flow for irrigation).

The municipalities in the purple group and the blue group have as a common characteristic the Firjan Municipal Development Index lower than 0.65, with an average below the other groups. It is worth mentioning the low values for the Firjan Municipal Development Index for the education sector (max 0.6). Both groups also have high values for the admission rate for acute diarrhea. The difference between the purple and blue groups is in the demand for the water flow consumed for animal watering and the percentage of the water flow consumed for irrigation. While the purple group consumes up to 90% of water for irrigation, the blue group has almost zero demand.

Cross-analysis of the distinct characteristics of groups of municipalities with climate change scenarios suggests that the situation for these two groups of municipalities can become particularly problematic. In coastal regions and its adjacent municipalities in the states of Amapá, Pará, and

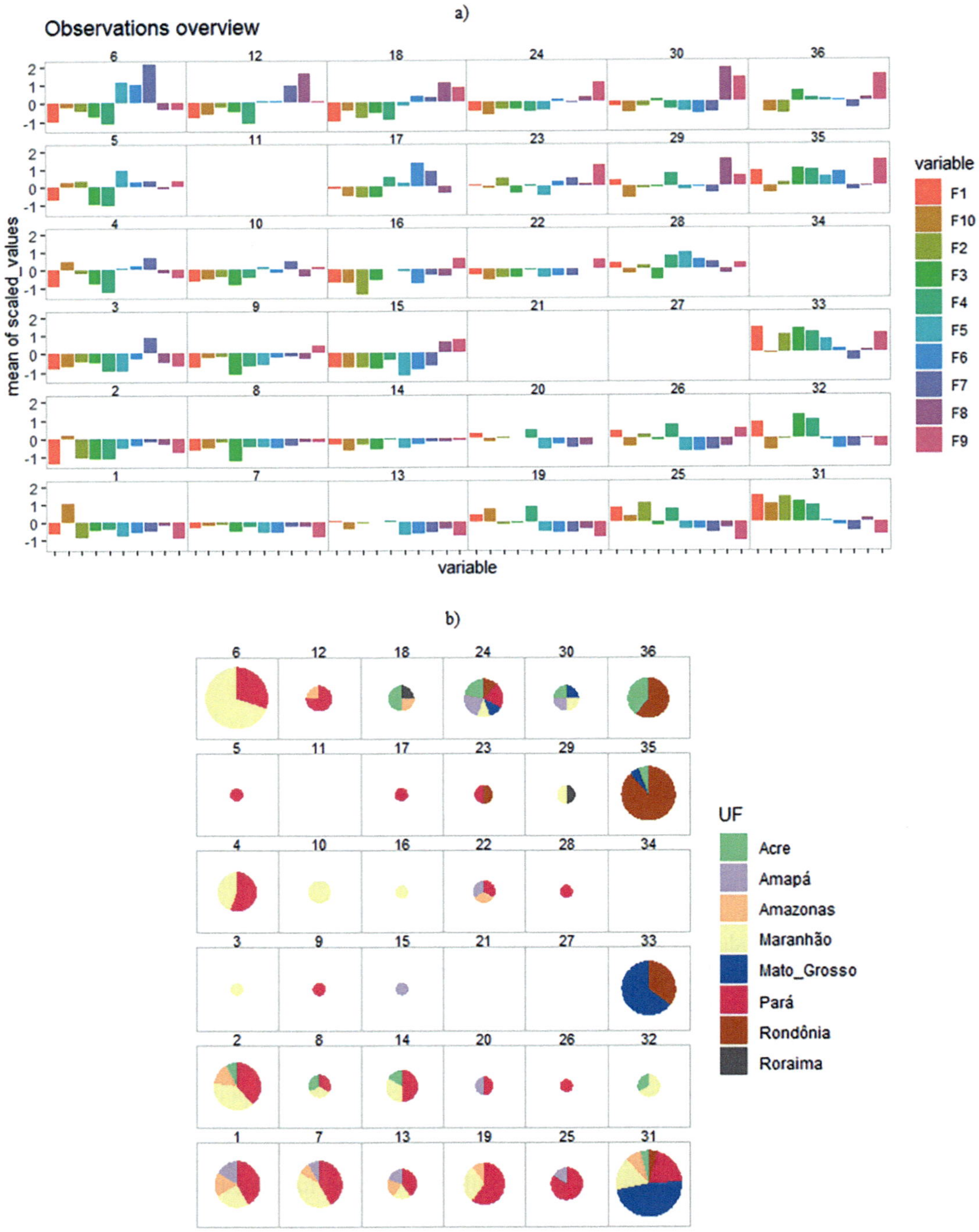

Fig. 5.3 (**a**) Pattern of cluster of municipalities by variables. (**b**) Composition of each cluster based on the participation of Amazonian federal units

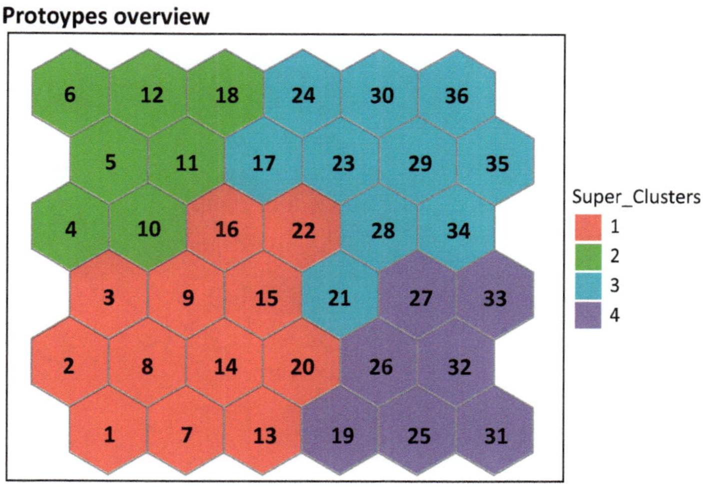

Fig. 5.4 Supercluster of the municipalities from the SOM analysis

Maranhão, where an increase in the number of days of drought and increased temperature is expected, 27 municipalities have a high demand for water for irrigation and animal watering and may suffer from the scarcity of this resource, compromising local agricultural production (livestock and crop) with indirect effects on the municipal economy and the food security of citizens. In addition, dry weather and increased temperature can increase fire rates, thus impacting the increase in hospitalizations for respiratory problems, which is already high in the region.

For the western Amazon region, three have associated potentially disastrous results. Although they are part of the groups with low demand for water for animal watering and irrigation, they are municipalities that have low values in all sectors of the Firjan Index and high rates of hospitalization for acute diarrhea, which indicates that they are municipalities that should have a low sanitation index basic and capacity to deal with climatic extremes as predicted for the region. In these cases, the increase in the number of consecutive days of rain, which can result in flooding problems, can bring serious problems of access to water and waste treatment.

5.4 Discussion

The Amazon is an environmentally and socially complex region. The diversity of its ecosystems associated with human activities conducted in different subregions gives rise to a mosaic of diverse and, at the same time, complementary socio-spatial dynamics. Similarly, due to the extension of the region, the climate scenarios for the region point to different dynamics in different zones, with the extension of dry days, increased temperature, and increase in the number of consecutive days of rain being remarkable (Marengo et al., 2012). The municipalities selected for this study have the highest projected climate anomaly values, which means that they are municipalities that urgently need actions to adapt to climate change.

The results of the clusters point out that the water demand and the values for the development indices are crucial characteristics in the differentiation between the classes of clusters. An interesting result is the distribution of different classes in the study area. The groups of municipalities are distributed throughout the region, without following an apparent pattern (Fig. 5.5); however, this may be related to locational issues such as urban infrastructure, access to services, and availability of water resources (especially the proximity to rivers), which can directly affect the adopted variables in this work. The concretization of climate scenarios has a great potential to cause a ripple effect capable of affecting all sectors of human life, being particularly serious for vulnerable populations, that is, those with low income, who live in unhealthy conditions and without access to health systems.

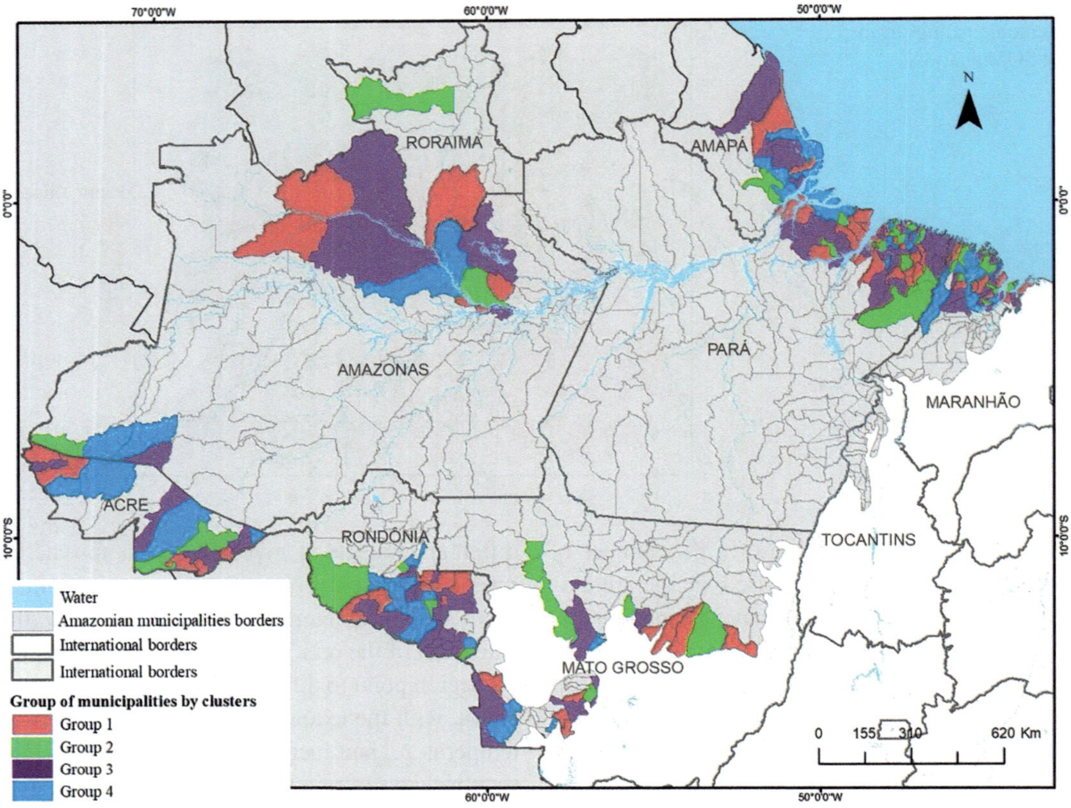

Fig. 5.5 Spatialization of the studied municipalities by cluster group

According to an official document from the Ministry of Science, Technology, and Innovations (2021) in the fourth national communication, the number of days with high temperatures in Brazil increased 30%, and regional rainfall patterns have changed in the last 40 years (MCTI, 2021). Temperature increases are forecast for regions with intense agricultural activity, such as the center of the state of Mato Grosso and Rondônia. Based on climate analysis conducted for the period 1980 to 2018 by the MCTI, agriculture in central Brazil is one of the most vulnerable sectors to climate change, due to the increase in temperature and the number of dry days.

The analysis of the impact of rainfall decrease in Southern Brazilian Amazon (Leite-Filho et al., 2021) shows that there is a high correlation between deforestation and reduced rainfall in the region, which has already caused large economic losses in soy and livestock production, which can add up to US$ 1 billion annually by 2050. In addition, there is an aggravating potential regarding the regional response to climate change, which can be enhanced by the absence of forests and forest degradation (Brando et al., 2020; Fonseca et al., 2019; Tasker & Arima, 2016).

Temperature increases are also expected for the whole metropolitan region of Manaus. These events could be aggravated by urban infrastructure with low rates of tree cover, favoring the phenomenon of heat islands and increasing demand for water and energy (Lapola et al., 2018; MCTI, 2021). This represents not only the need to think about strategies to continue agricultural production in scenarios of temperature changes but also the need to prepare urban spaces for a new climate reality, where heat can affect labor productivity and overload systems of health, thus increasing the population's vulnerability (Brondizio, 2016).

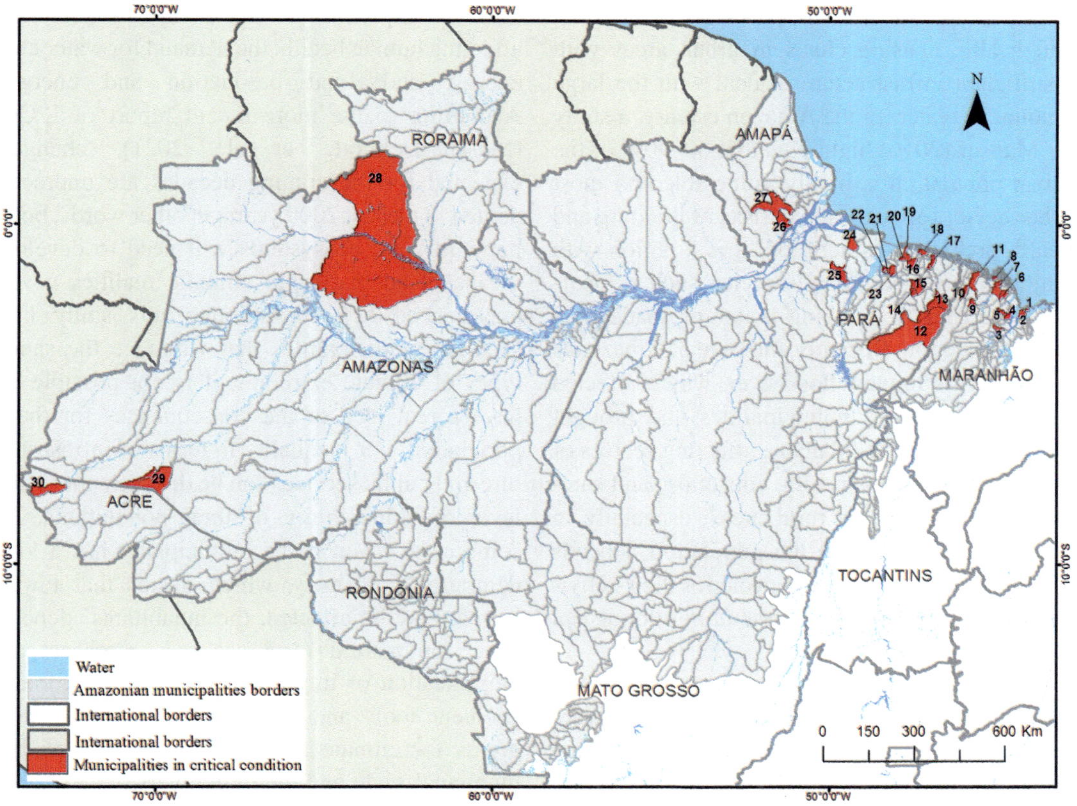

1 Paço do Limiar	11 Cândido Mendes	21 Benevides
2 São Luís	12 Paragominas	22 Ananindeua
3 Matinha	13 Nova Esperança do Piriá	23 Santa Izabel do Pará
4 Bacurituba	14 Mãe do Rio	24 Santa Cruz do Arari
5 Peri Mirim	15 Irituia	25 São Sebastião da Boa Vista
6 Mirinmizal	16 Igarapé - Açu	26 Santana
7 Serrano do Maranhão	17 Tracuateua	27 Porto Grande
8 Apicum Açu	18 São João de Pirabas	28 Barcelos
9 Presidente Médice	19 Marapanim	29 Envira
10 Maracaçumé	20 Terra Alta	30 Rodrigues Alves

Fig. 5.6 Municipalities that may reach critical condition due to climate change and their socioeconomic and water demand indices

At the same time, increases in the number of dry days in Amapá, Pará, and Maranhão have the potential to affect agricultural production zones and urban areas with high population density. The absence of rain can compromise agriculture, especially in municipalities that have a high dependence on water for agricultural activities, such as those highlighted in Fig. 5.5. It is noteworthy that these are municipalities with high Firjan indexes, especially in the education sector. However, in the context of climate change, their main economic activities can be impacted, leaving these municipalities in a particular situation of vulnerability (Brondizio, 2016). Furthermore, the prolonged deficiency of rain can affect the production of hydroelectric energy, increasing its cost in the region, increasing water and energy demand and the occurrence of diseases such as dengue fever (Horta et al., 2014).

Another important aspect identified concerns the scenario of an increase in the number of consecutive days of rain, which have the potential to cause major problems for populations in both urban and rural areas in the Amazon. Consecutive days of rain can generate problems of flooding and health, causing chaos in urban areas with insufficient infrastructure to deal with the large amount of water. In the Amazon estuary, a study by Mansur (2016) highlights that 60–90% of the urban population is highly vulnerable and most urban sectors are exposed to risks of flooding and health problems. In addition, it is a region with major basic structural deficiencies, such as insufficient access to drinking water or inadequate waste collection. Recently, the state of Amazonas registered its biggest flood ever, which affected 99% of the state's municipalities and brought chaos to the capital, Manaus, affecting sectors of commerce, services, health, education, and sanitation (G1, 2021). In rural areas, especially in floodplain areas, where the population depends on the natural hydrological cycle for their activities, such as hunting and agriculture, floods can have disastrous effects.

Considering, also, a deficient and precarious environmental sanitation and health infrastructure for many of the municipalities in the Legal Amazon, the climatic variability of rainfall and temperature can also aggravate the health condition of residents in this group of municipalities analyzed. Previous studies show the effect of changes in temperature and rainfall on the behavior of diarrheal infectious diseases (Aguiar et al., 2020; Duarte et al., 2019; Horn et al., 2018; Franca, 2015) and cardiovascular diseases (Casas et al., 2016). Duarte et al. (2019), for example, found a positive association between hospitalizations for infectious diarrheal diseases and the increase in the level of the Acre River, which is an important river for the region. The study of Ellwanger et al. (2020) shows the association between anthropogenic action in the Amazon rainforest, climate change, and alterations in vector dynamics, human migration, genetic changes in pathogens, and the poor social and environmental conditions can give rise to the "perfect storm" for the emergence of infectious diseases in Amazonian municipalities.

Climate change poses a serious challenge to Amazon cities because regardless of the type of extreme climate that municipalities will experience, they produce a chain effect capable of affecting human health, the demand for water and energy, and food production and energy. According to the more recent report of IPCC (Masson-Delmotte et al., 2021), changes expected for the coming decades are unprecedented in the last 2000 years; in other words, both humanity and ecosystems will need to develop means of adaptation to climatic realities never experienced before. Although this study has grouped municipalities that will face the same types of climate extremes, it is not possible to have a real idea of the consequences for their population. For inequality in terms of infrastructure and public services can be determinant in the level of vulnerability of their populations. In some cases, even if the municipality has a low demographic density, which means that fewer people may be affected, the inhabitants' dependence on natural resources may represent an amplification of impacts. For this reason, management tools and regional and national programs for climate adaptation need to consider municipal scale and the high diversity and complexity realities in Amazon municipalities.

In this context, the WEF nexus approach can be integrated in municipal and state management tools, contributing so that the different sectors to be affected by climate change are managed in an integrated and inseparable manner and trying to encompass the complexity of the impacts that climate change can bring. The urban scenario represents a challenge and an opportunity to understand and direct resources to create more resilient cities – in particular, by mobilizing action and understanding that the WEF nexus in cities can provide benefits by addressing some of the most complex global problems, including climate change, as well as helping to achieve sustainable development goals (Benites-Lazaro & Giatti, 2021). Thus, urban planning with a nexus focus can provide the lens and basis to achieve urban resilience in an integrated way in different but interlinked sectors considering both ecological and built infrastructures (Nhamo et al., 2021).

However, although the interrelationship between the three axes of the nexus may seem noticeably clear, in practice, the public administration still has a long way to go in the development of policies and action plans that integrate the water, food, and energy sectors. At the same time, different methodologies have been developed within the scope of the nexus approach that allows designs that are easy to apply in government management. Based on the indicators adopted in this work, the great discrepancy in social and economic indices between Amazonian municipalities is evident, which may require different actions. In these cases, inducing the adoption of integrated actions based on the nexus approach can be induced by a higher level of public management, strengthening the relationship between various levels of the government. Furthermore, strengthening participatory governance and involving different stakeholder groups is one of the key elements of the nexus approach. Democratic management and integrated, systematic sectoral planning, with a clear notion of the cause and effect, are fundamental for the design of multisectoral actions that are more likely to be applied.

Another important aspect is related to the resolution of social problems, which are fundamental for reducing vulnerability to climate change. Improving basic sanitation indices and access to health and education are crucial for the Amazon population to have a better chance of facing the effects of climate change. In this sense, although the nexus approach is strongly based on the water, food, and energy sectors, the demands for better governance in these sectors demand its inclusion in other sectors of society.

The results of this work can help the management of municipalities as follows: the municipalities highlighted in this work can receive priority attention in adaptation strategies to climate change, and the sectors that involve the production and demand of water, food, and energy can be planned together considering both conscious demand and the autonomy in production, in this case, prioritizing alternative sources of energy and solving problems of basic sanitation and access to water. Due to the environmental importance of the Amazon region and the forecasts of increased temperature and drought, managers could prioritize nature-based solutions with investment in green areas that can lower temperature and increase humidity in urban areas, reducing the effects of heat islands. Cities also need to think about intelligent drainage strategies, reducing impermeable surfaces and guaranteeing the fluidity of rivers. One of the characteristics of the nexus approach is to consider social equality as one of the main measures to guarantee the sustainability of the system as a whole. In this context, governments must prioritize the reduction of social inequality and increase the quality of life of its inhabitants, consequently, reducing the social and climate vulnerability of its population.

5.5 Conclusion

Increases in temperature, number of dry days, and consecutive rainy days are the main effects of climate change estimated for 244 municipalities in the Brazilian Amazon. The analysis of climate data with social variables indicates that some municipalities can be highly impacted due to their water demand in regions where floods are likely to occur, or their inability to deal with floods in regions where the number of consecutive rainy days may increase, or even the increase of up to 3 °C in large regional capitals, which can generate intense heat islands. Our results indicate that none of the studied municipalities is prepared for the predicted climate future, and, therefore, impact mitigation and adaptation strategies are urgent and essential. In this context, the water-food-energy nexus integrated approach can be considered a crucial tool for local and regional managers to reduce the vulnerability of their cities without compromising the quality of life of their population and economic growth.

Acknowledgments Lira Luz Benites Lazaro acknowledges financial support received from São Paulo Research Foundation (FAPESP), Grant: 2017/17796-3 and 2019-24479-0.

References

Aguiar, K. C. G., Cohen, S. C., Maciel, E. M. G. S., & Kligerman, D. C. (2020). Risk factors for diarrhea occurrence in children living in Guaratiba Island (RJ). *Saúde Debate, 44*, 205–2019.

Alencar, A. S., do Nascimento, L. B. T., de Lima Garcia, M., et al. (2020). The impact of fires in amazon rainforest on public health. Amadeus International Multidisciplinary Journal, 4(8), 1–6.

Amaral, M. H., Benites-Lazaro, L. L., Antonio de Almeida Sinisgalli, P., et al. (2021). Environmental injustices on green and blue infrastructure: Urban nexus in a macrometropolitan territory. *Journal of Cleaner Production, 289*, 125829.

Ardila, J., Arieria, J., Bauch, S. C., et al. (2021). Latin American and Caribbean forests in the 2020s: Trends, challenges, and opportunities.

Baccini, A., Walker, W., Carvalho, L., et al. (2017). Tropical forests are a net carbon source based on aboveground measurements of gain and loss. *Science, 358*, 230–234.

Benites-Lazaro, L. L., & Giatti, L. (2021). O nexo água-energia-alimentos – uma abordagem para cidades sustentáveis e o desenvolvimento sustentável. In P. Jacobi & L. Giatti (Eds.), *Inovação para governança da macrometrópole paulista face à emergência climática*.

Benites-Lazaro, L. L., Giatti, L. L., Sousa Junior, W. C., & Giarolla, A. (2020). Land-water-food nexus of biofuels: Discourse and policy debates in Brazil. *Environment and Development, 33*, 100491.

Boelaert, J., Bendhaiba, L., Olteanu, M., & Villa-Vialaneix, N. (2014). SOMbrero: An R package for numeric and non-numeric self-organizing maps. In *Advances in intelligent systems and computing*. Springer.

Bouwer, L. M. (2019). Observed and projected impacts from extreme weather events: Implications for loss and damage, 63–82.

Brando, P., Macedo, M., Silvério, D., et al. (2020). Amazon wildfires: Scenes from a foreseeable disaster. *Morphology, Distribution, Functional Ecology of Plants, 268*, 151609.

Brondizio, E. (2016). The elephant in the room: Amazonian cities deserve more attention in climate change and sustainability discussions*. *ClimaCom Cultura Científica, 5*.

Brondízio, E. S., de Lima, A. C. B., Schramski, S., & Adams, C. (2016). Social and health dimensions of climate change in the Amazon. *Annals of Human Biology, 43*, 405–414.

Casas, A. L. F., Santos, G. M. D., Chiocheti, N. B., & de Andrade, M. (2016). Effects of temperature variation on the human cardiovascular system: A systematic review. In W. Leal Filho, U. Azeiteiro, & F. Alves (Eds.), *Climate change and health. Climate change management*. Springer. https://doi.org/10.1007/978-3-319-24660-4_5

Coe, M. T., Brando, P. M., Deegan, L. A., et al. (2017). The forests of the Amazon and Cerrado moderate regional climate and are the key to the future. *Tropical Conservation Science, 10*, 194008291772067.

Duarte, J. L., Diaz-Quijano, F. A., Batista, A. C., et al. (2019). Variabilidade climática e internações por doenças diarreicas infecciosas em um município da Amazônia Ocidental brasileira. *Ciência & Saúde Coletiva, 24*, 2959–2970.

Ellwanger, J. H., et al. (2020). Beyond diversity loss and climate change: Impacts of Amazon deforestation on infectious diseases and public health. *Anais da Academia Brasileira de Ciências, 92*.

Van Der Ent, R. J., Savenije, H. H. G., Schaefli, B., & Steele-Dunne, S. C. (2010). Origin and fate of atmospheric moisture over continents. *Water Resources Research, 46*, 1–12.

Esquivel-Muelbert, A., Baker, T. R., Dexter, K. G., et al. (2017). Seasonal drought limits tree species across the Neotropics. *Ecography (Cop), 40*, 618–629.

Feldmeyer, D., Wilden, D., Kind, C., Kaiser, T., Goldschmidt, R., Diller, C., & Birkmann, J. (2019). Indicators for monitoring urban climate change resilience and adaptation. *Sustainability, 11*(10), 2931.

Figueiredo, L., Honiden, T., & Schumann, A. (2018). Indicators for resilient cities.

Franca, R. R. da. (2015). Eventos pluviais extremos na Amazônia meridional: riscos e impactos em Rondônia.

Fonseca, M. G., Alves, L. M., Aguiar, A. P. D., et al. (2019). Effects of climate and land-use change scenarios on fire probability during the 21st century in the Brazilian Amazon. *Global Change Biology, 25*, 2931–2946.

Gatti, L. V., Basso, L. S., Miller, J. B., et al. (2021). Amazonia as a carbon source linked to deforestation and climate change. Nature, 595(7867), 388–393.

G1. (2021). Cheia do Rio Negro alcança recorde histórico em Manaus; veja o fenômeno em FOTOS | Amazonas | G1.

Gephart, J. A., Golden, C. D., Asche, F., et al. (2020). Scenarios for global aquaculture and its role in human nutrition. *Reviews in Fisheries Science & Aquaculture, 29*, 122–138. https://doi.org/10.1080/23308249202017823342

Giannini, T. C., Costa, W. F., Borges, R. C., et al. (2020). Climate change in the eastern Amazon: Crop-pollinator and occurrence-restricted bees are potentially more affected. *Regional Environmental Change, 20*, 9.

Gomes, V. H. F., Vieira, I. C. G., Salomão, R. P., & ter Steege, H. (2019). Amazonian tree species threatened by deforestation and climate change. *Nature Climate Change, 9*, 547–553.

Horn, L. M., Hajat, A., Sheppard, L., Quinn, C., Colborn, J., Zermoglio, M. F., Gudo, E. S., Marrufo, T., & Ebi, K. L. (2018). Association between precipitation and diarrheal disease in Mozambique. *International Journal of Environmental Research and Public Health, 15*, 1–10.

Horta, M., Bruniera, R., & F K, et al. (2014). Temporal relationship between environmental factors and the occurrence of dengue fever. *International Journal of Environmental Health Research, 24*, 471–481.

IBGE. (2015). *Pesquisa Nacional por Amostra de Domicílios – PNAD | IBG*. https://www.ibge.gov.br/estatisticas/sociais/educacao/9127-pesquisa-nacional-por-amostra-de-domicilios.html?=&t=o-que-e. Viewed 31 Oct 2020.

IPEA – Instituto de Pesquisa Econômica Aplicada. (2017). Atlas da Vulnerabilidade Social. Available in: http://ivs.ipea.gov.br

IPCC. (2013). Climate Change 2013: The Physical Science Basis. Contribution of Working Group I to the Fifth Assessment Report of the Intergovernmental Panel on Climate Change [Stocker, T.F., D. Qin, G.-K. Plattner, M. Tignor, S.K. Allen, J. Boschung, A. Nauels, Y. Xia, V. Bex and P.M. Midgley (eds.)]. Cambridge University Press, Cambridge, United Kingdom and New York, NY, USA, 1535 pp, https://doi.org/10.1017/CBO9781107415324

Lapola, D. M., Pinho, P., Quesada, C. A., et al. (2018). Limiting the high impacts of Amazon Forest dieback with no-regrets science and policy action. *Proceedings of the National Academy of Sciences of the United States of America, 115*, 11671–11679.

Lazaro, L. L. B., Giatti, L. L., Bermann, C., Giarolla, A., & Ometto, J. (2021). Policy and governance dynamics in the water-energy-food-land nexus of biofuels: Proposing a qualitative analysis model. *Renewable and Sustainable Energy Reviews, 149*, 111384.

Leite-Filho, A. T., Soares-Filho, B. S., Davis, J. L., et al. (2021). Deforestation reduces rainfall and agricultural revenues in the Brazilian Amazon. *Nature Communications 2021 121, 12*, 1–7.

Leite-Filho, A. T., Sousa Pontes, V. Y., & Costa, M. H. (2019). Effects of deforestation on the onset of the rainy season and the duration of dry spells in southern Amazonia. *Journal of Geophysical Research – Atmospheres, 124*, 2018JD029537.

Longo, M., Saatchi, S., Keller, M., et al. (2020). Impacts of degradation on water, energy, and carbon cycling of the Amazon tropical forests. *Journal of Geophysical Research Biogeosciences, 125*.

Machado-Silva, F., Libonati, R., de Lima, T. F. M., et al. (2020). Drought and fires influence the respiratory diseases hospitalizations in the Amazon. Ecological Indicators, 109, 105817.

Mansur, A. V., Brondízio, E. S., Roy, S., et al. (2016). An assessment of urban vulnerability in the Amazon Delta and Estuary: a multicriterion index of flood exposure, socio-economic conditions and infrastructure. Sustainability Science, 11(4), 625–643.

Masson-Delmotte V., Zhai, P., Pirani, A., Climate Change (2021). The Physical Science Basis. Contribution of Working Group I to the Sixth Assessment Report of the Intergovernmental Panel on Climate Change. Cambridge University Press, Cambridge, UK2021.

Marengo, J. A., & Souza, C. (2018). *Climate change: Impacts and scenarios for the Amazon*. São Paulo.

Marengo, J. A., Tomasella, J., Soares, W. R., et al. (2012). Extreme climatic events in the Amazon basin. *Theoretical and Applied Climatology, 107*, 73–85.

Marengo, J. A., Camarinha, P. I., Alves, L. M., Diniz, F., & Betts, R. A. (2021). Extreme rainfall and hydro-geo-meteorological disaster risk in 1.5, 2.0, and 4.0° C global warming scenarios: an analysis for brazil. Frontiers in Climate, 3, 13.

MCTI – Ministério da Ciência Tecnologia e Inovação. (2021). Quarta comunicação nacional do brasil à convenção-quadro das nações unidas sobre mudança do clima. Brasília, DF.

Müller, H., Griffiths, P., & Hostert, P. (2015). Long-term deforestation dynamics in the Brazilian Amazon—Uncovering historic frontier development along the Cuiabá–Santarém highway. *International Journal of Applied Earth Observation and Geoinformation, 44*, 61–69.

Nhamo, L., Rwizi, L., Mpandeli, S., Botai, J., Magidi, J., Tazvinga, H., Sobratee, N., Liphadzi, S., Naidoo, D., Modi, A. T., & Slotow, R. (2021). Urban nexus and transformative pathways towards a resilient Gauteng City-Region, South Africa. *Cities, 116*, 103266.

Nobre, A., Sampaio, G., Borma, L., et al. (2016). Land-use and climate change risks in the Amazon and the need of a novel sustainable development paradigm. *Proceedings of the National Academy of Sciences of the United States of America, 113*, 10759–10768.

Nunes, K. V. R., Ignotti, E., & Hacon, S. D. S. (2013). Circulatory disease mortality rates in the elderly and exposure to PM2. 5 generated by biomass burning in the Brazilian Amazon in 2005. Cadernos de saude publica, 29(3), 589–598.

OECD-FAO. (2019). *OECD-FAO Agricultural Outlook 2019–2028*. Food and Agriculture Organization of the United Nations/OECD.

Prado-Lima, M., & Val, A. L. (2016). Transcriptomic characterization of Tambaqui (Colossoma macropomum, Cuvier, 1818) exposed to three climate change scenarios (M Portolés, Ed). *PLoS One, 11*, e0152366.

Raju, E., Boyd, E., & Otto, F. (2022). Stop blaming the climate for disasters. *Communications Earth & Environment, 3*(1), 1–2.

Rasul, G., & Sharma, B. (2016). The nexus approach to water–energy–food security: An option for adaptation to climate change. *Climate Policy, 16*, 682–702.

Scott, M., Lennon, M., Tubridy, D., et al. (2020). Climate disruption and planning: Resistance or retreat? *Planning Theory & Practice, 21*, 125–154. https://doi.org/10.1080/14649357202017041 30

Spracklen, D. V., Baker, J. C. A., Garcia-Carreras, L., & Marsham, J. H. (2018). The effects of tropical vegetation on rainfall. *Annual Review of Environment and Resources, 43*, 193–218.

Tasker, K. A., & Arima, E. Y. (2016). Fire regimes in Amazonia: The relative roles of policy and precipitation. *Anthropocene, 14*, 46–57.

United Nations, Department of Economic and Social Affairs PD. (2017). *World Population Prospects: The*

2017 Revision, Methodology of the United Nations Population Estimates and Projections. New York.
Vialaneix, N., Elise, M., Jerome, M., et al. (2020). Using self-organizing maps with SOMbrero to cluster a numeric dataset • SOMbrero viewed.
Wisner, B., Gaillard, J. C., & Kelman, I. (2015). Framing disaster: Theories and stories seeking to understand hazards, vulnerability and risk, handbook of hazards and disaster risk reduction. In B. Wisner, J. C Gaillard, I. Kelman (Eds.), 2012. *Disaster Prevention* (pp. 58–76). Routledge.
Yalew, S. G., van Vliet, M. T. H., Gernaat, D. E. H. J., et al. (2020). Impacts of climate change on energy systems in global and regional scenarios. *Nature Energy 2020 510, 5*, 794–802.
Zscheischler, J., Martius, O., Westra, S., et al. (2020). A typology of compound weather and climate events. *Nature Reviews Earth & Environment 2020 17, 1*, 333–347.

Part II

Innovation and Transformative Change

6 The Water-Energy-Food Nexus and the Micro-politics of Everyday: A View from Institutional Bricolage

Thainara Granero de Melo

Abstract

Brazil's current conjuncture exacerbates an inescapable task to shed light on the UN 2030 Agenda for Sustainable Development Goals (SDGs) and recent experiences of sustainable rural development policies across the country. While Brazilian scholars from a range of disciplines have long documented the policy achievements and the persistence of siloed responses, others have argued for the need to transform the knowledge about these experiences into new policy strategies integrated into the SDG targets, particularly in the case of rural populations. However, neither the calls for integration nor the well-designed institutional solutions are self-evident or unproblematic assumptions. This chapter explores the everyday experiences of rural communities with integrated resource management and sustainable development interventions in Brazil. I argue that to better understand these interrelationships in their dynamic, uneven, and ambiguous manifestations, a critical and politicized interpretation of integration and institutions is needed. Drawing on the concepts of "water-energy-food nexus" and "institutional bricolage," I analyze an empirical case study of small-scale farmers and their local organizations while participating in an intervention to support the integrated management of land, water and ecosystems in São Paulo state. I suggest that a dynamic bricolage for managing the local nexus can be observed, but with contradictions that suggest its limitations. I contend that we will need more than just designed arrangements or creative bricolage to integrate policies and natural resources to support transformative social justice, within and beyond the SDGs.

Keywords

Sustainable rural development · Water-energy-food nexus · Institutional bricolage · Micro-politics · Brazil

6.1 Introduction

In Brazil's current conjuncture, the task to shed light on the UN 2030 Agenda for Sustainable Development Goals (SDGs) is inescapable. The effects of austerity measures and the dismantling of existing policies, aggravated by the criminal government of Jair Bolsonaro, evoke profound skepticism toward social and environmental commitments (Cernov & Pietricovsky, 2020). The rapid deterioration of institutions and partici-

T. G. de Melo (✉)
Pontifical Catholic University of Paraná (PUCPR), Curitiba, Brazil
e-mail: thainara.granero@pucpr.br

L. L. B. Lazaro et al. (eds.), *Water-Energy-Food Nexus and Climate Change in Cities*, Sustainable Development Goals Series, https://doi.org/10.1007/978-3-031-05472-3_6

patory modes of resource governance, especially those related to climate change, food security, and biodiversity conservation, marks significant backtracking for the advancements of recent decades (David, 2020; Sabourin et al., 2020). While hunger levels moved 20 years backward during the Covid-19 pandemic (IBGE, 2020), structural adjustments challenge the rural poor and small-scale farming sector that has been a crucial part of Brazil's food production. Increasingly, environmental deregulations – from walking back pesticide bans to contemporary forms of land grabbing – act in corporate interests (Sauer et al., 2020). Combined, these effects on the erosion of biodiversity, drought, and energy crisis are evident across the country, as well as their implications for inequalities and injustices (Fernandes et al., 2020).

At the same time, this shift demands a critical reconsideration of the recent experience of Brazil with sustainable rural development policies, which had unforeseen political consequences. In the past two decades, the creation of policies and programs integrating territories to poverty reduction and natural resource management was promising. With significant impacts on food security and rural poverty reduction, these arrangements became globally acknowledged as a success model (Niederle et al., 2019), bringing together civil society groups, social movements, scholars, and practitioners in the making of interventions (Grisa et al., 2018). However, the promised articulation did not have the same advances, revealing limits to consolidate outcomes at the local level (Buainain et al., 2018; Wojciechowksi et al., 2020). As such, the prevalence of techno-managerial rationality in some programs resonated with a conception of local territories as passive recipients of interventions rather than active elements in social and political life. Participatory arrangements for decision-making then failed to reflect local concerns, diversity, and heterogeneity in terms of resource use and livelihoods (Lotta & Favareto, 2016). Furthermore, the lack of synchronicity between infrastructure (provision of water, energy, and technical assistance) and food programs (financial loans and public food procurement) hindered the inclusion of rural families, though these programs have changed their vulnerable conditions substantially (Favareto, 2018). Another fundamental contradiction of these policies relied upon the attempts to accommodate sustainable rural development within the global large-scale export agenda (Favareto, 2019) and its "commodities consensus" (Svampa, 2015), which is characterized by neoextractivism, processes of dispossession, and enclosure of natural resources across Latin America.

Reflecting upon these experiences, recent studies call for a transformation of the existing knowledge into designed arrangements for policy integration (Favareto, 2019; Valencia-Perafán et al., 2020). By making visible the interdependencies and trade-offs between resources, urban-rural interactions, and livelihoods, Favareto (2018) argues, for instance, that researchers could move an agenda focused on the SDGs toward a tangible policy change. But the calls for integration and its designed arrangements do not necessarily add self-evident or unproblematic assumptions. A growing body of critical scholarship has identified problematic tendencies in mainstream approaches of integration (Williams et al., 2019). As Allouche, Middleton, and Gyawali (2019) demonstrate, though discursively constituted as technical and apolitical tools for resources management, such constructions often obfuscate the lived experiences of communities and unequal power relations that disproportionally affect poor and vulnerable communities. Other critiques suggest that the growing engagement of Brazilian scholars with integrated frameworks has contributed significantly to decision-making recommendations, yet empirical evidence and context-specific issues are rarely made explicit in this literature (Dalla Fontana et al., 2020). Furthermore, a common problem among social scientists studying sustainable development issues is the need for bringing more "on-the-ground" realities to inform realistic interventions, as well as to offer more nuanced lenses for policy solutions.

To address this problem, this chapter explores the everyday experiences of rural communities with integrated resource management, using a sustainable development project in Brazil as a case study. I aim to understand the ways in which resources and institutions are accessed, integrated, and organized in the daily life of policies and programs. More specifically, I seek to understand how the issue of "integrated resource management" is differently framed by key actors involved in interventions and designed arrangements, how the local arrangement for managing varied resources is formed, and what outcomes they produce. Following the work of recent critical scholarship, I draw on the concepts of "water-energy-food nexus" (Allouche et al., 2019) and "institutional bricolage" (Cleaver, 2012) to understand these relationships as dynamic, uneven, and ambiguous domains intersecting people, resources, and the everyday policy/politics. By discussing an empirical case study of small-scale farmers and their local organizations while participating in an intervention for supporting the integrated management of land, water, and ecosystems in São Paulo state, I suggest that people act themselves as "bricoleurs," shaping both formal and informal arrangements for resources management, while designed institutions alone do not generate the expected outcomes of integration. On the other hand, I argue that the adapted institutions are not sufficient to challenge persistent inequitable patterns of access to environmental resources and disconnection from significant institutions and infrastructures.

Contributions from this chapter can be envisioned in two directions. First, by combining both the nexus and institutional bricolage, the chapter brings a conceptual framework that locates and analyzes the linkages between natural resources and interventions as socially constructed relationships, thereby contributing to the methodological and practical challenges to identify institutional arrangements less visible in the everyday (Whaley & Cleaver, 2018). Second, by bringing to the fore the everyday experiences of an emblematic case from Brazil, this chapter seeks to contribute, both theoretically and empirically, to the efforts of critical scholarship aimed at pluralizing the nexus research through "bottom-up" analysis and re-politicizing the SDG discussion from the perspective of emancipatory projects and social justice (Allouche et al., 2019).

In what follows, I begin by unpacking some of the key aspects of the nexus and institutional bricolage framework. I elaborate how they contribute to understand better the complex, nuanced domains of resource management. This is followed by a description of the method and case study context. I justify why this case is illustrative to explore the interrelationships between sustainable development interventions, different arrangements for resources management, and the everyday of an agrarian reform area. I then move to discuss the case study in three parts. I start by situating the *Microbacias* arrangement and how the actors and institutions frame the integrated resource management in the Sepé Tiaraju context. The subsequent section understands the local arrangements and how bricolage is limited by structural problems, social relationships, and missing integration to relevant institutions and infrastructure, (re)producing unequal relationships. Finally, I discuss the ambiguous and unexpected outcomes of these bricolage processes, suggesting the creation of an agroforestry multi-use platform that mediates the local nexus. In the conclusions, I contend that we will need more than just designed arrangements or creative bricolage to integrate policies and practices to support transformative social justice within and beyond the SDGs.

6.2 Conceptual Framework: Unpacking the Nexus and Institutional Bricolage

6.2.1 The Water-Energy-Food Nexus

The last decades have seen the emergence of approaches to improve sustainability by adopting integrated solutions in response to climate change (Urbinatti et al., 2020). A recent development of these ideas is reflected in the notion of water-energy-food (WEF) nexus. The nexus is often defined as the idea that water, energy, and food, along with the natural resources that underpin them, are inextricably linked and articulate one another. Presented by the literature as a concept, an analytical approach, a system thinking, or a governance framework (Harwood, 2018), the nexus is often used to refer to the pressures, trade-offs, conflicts, resilience, and potential synergies between these sectors (Endo et al., 2015). It is also argued as a potential basis to guide security and efficiency use of resources across sectors and the coordination of global policymaking and the future of natural resource governance (Bazilian et al., 2011; Hoff, 2011a, b). Although its origins date back to the World Bank's preliminary formulations on water and food security linkages (Allan, 2003), the nexus gained traction in the wake of 2008s financial crisis and Hoff's seminal background paper in preparation for the Rio + 20 summit. Hoff presents the "ecosystem services," "creating more with less," and "accelerating access, integrating the poorest" as the fundamental principles guiding the nexus approach and the transition toward a green economy (Hoff, 2011a, b, p. 13). Since then, the nexus has been widely applied by the international community of researchers, experts, and practitioners on the front lines of policymaking, including the UN support for the nexus to promote the implementation of the SDGs (Hoff, 2011a, b; Schmidt & Matthews, 2018).

Scholars from critical social sciences have examined how the nexus emerged as a "buzzword" (Cairns & Krzywoszynska, 2016) and apolitical concept (Allouche et al., 2019; Leese & Meisch, 2015; Moss & Hüesker, 2019). Because of its compelling arguments for integration, the nexus has become a highly discussed frame, coming to functioning as an umbrella term, encompassing a diversity of definitions, methods, lenses, and policy recommendations. Indeed, the nexus' imprecise meanings and slippages are getting captured in different ways. However, it has also been gradually established as a consensual idea regarding concerns, research, and methodological tools. As Allouche et al. (2019) argue, much of the nexus literature adopt a technical rather than a political stance, proposing modeling techniques to quantify interactions, eliminate tensions, and explore win-win scenarios among sectors. These perspectives constantly draw on simplistic understandings of resource access and availability and naturalized discourses of scarcity, which overlook how other interfaces mediate natural resources at the local level, such as institutional arrangements and the power asymmetries and inequalities embedded in them. While democratic decision-making is claimed, the commonly held assumption is that the integration between sectors is merely a management and reengineering issue. Arguing more forcefully, Williams et al. (2019) note in these techno-managerial dispositive an alignment of the nexus with neoliberal logics, providing markets more space for the ongoing accumulation and enclosure of resources. Similarly, Schmidt and Matthews (2018) see the economic and political use of the nexus, expressed in ideas like the "green economy," as critical for the expansion of global financial networks.

One might then ask why the nexus concept is coherent for this chapter. First, Allouche et al. (2019) argue for reconceptualizing the nexus within politics, power, and justice. Second, building on these landmarks, the nexus must be cognizant of the ways that "distributive inequalities, resource grabbing, and enclosure of commons, or systemic sources of social injustice" are entangled across a range of levels and scales (Allouche et al., 2019, p. 131). Such an alternative view entails trade-offs and policy integration across sectors as fundamentally sociopolitical processes. Consequently, a shift to politics can make more visible the works of power "amongst different actors with distinct perceptions, interests, and practices" (Allouche et al., 2019, p. 9), as well as potentially transformative solutions. These alternative framings are emerging in Brazilian literature, though techno-managerial perspectives still dominate (Dalla Fontana et al., 2020). Critical scholarships have approached the nexus in terms of an engagement with social and environmental justice, in which questions about inequalities, power, youth, citizenship engagement, agency, and plurality are rendering visible the complexity and depth of the nexus in Brazil (Börner, 2021; Carvalho, 2021; Jacobi & Giati, 2017; Kraftl et al., 2018; Urbinatti et al., 2021). Thus, I turn to these critical, alternative approaches on the nexus. As I have argued elsewhere (Melo and Scopinho, 2022), this perspective is consistent with the ontological commitment to think with the communities' livelihoods and, through them, understand the nexus as transformative rather than apolitical knowledge. It calls for transdisciplinarity and spaces that acknowledge nonacademic modes of thinking, including plural perspectives, cultures, and contexts of asymmetrical powers (Dalla Fontana et al., 2020).

I argue that these ideas share a further strength with the concept of "institutional bricolage" (Cleaver, 2012). In addition to dialogue with a politicized concept of integration, bricolage goes beyond tracking practices and social relationships concerning natural resources. This also involves interrogating the complex entanglements of "meanings, world views, forms of legitimization and authority," which may be less visible in decision-making and formal arrangements for resource management (Cleaver & de Koning, 2015, p. 9). Institutional bricolage then provides an additional conceptual tool to a critical nexus perspective by making the interrelations between the local nexus and broader governance structures more legible. The next subsection shows how institutional bricolage draws on assumptions like complexity, unevenness, and ambiguity, converging in the "dynamic thinking" proposed by Leach, Scoones, and Stirling (2010) for an alternative nexus perspective. Moreover, the emphasis of bricolage on the everyday experiences with interventions can sharpen what Allouche et al. (2019) have called "plural pathways," to proceed a bottom-up nexus approach more attentive to the ways in which local systems of resources work.

6.2.2 Institutional Bricolage

Before unpacking the concept of institutional bricolage, it is worth reflecting on critical institutionalism. Embracing the complexity of natural and social worlds, this school of thought centers on how "institutions dynamically mediate relationships between people, natural resources and society" (Cleaver & de Koning, 2015, p. 1). According to Cleaver, critical institutionalism asserts its key premises from social theory, focusing on the interplay between nature and human action, the centrality of power, and concerns with social justice. Crucially, it frames institutions in contrast to mainstream institutionalism and the influential ideas of Douglass North and Elinor Ostrom to new institutional economics, criticized for its optimistic and often apolitical proposals about shaping local institutions for natural resources at the community level. As Cleaver argues, this approach gained traction among international development agencies by bridging neoliberal ideas of resource governance and decentralized local management. Conversely, critical institutionalism suggests that local negotiations over resources are more likely to be a dynamic, ambiguous, and uneven combination between global and local factors, formal structures, and everyday relationships, and thus partially amenable to designed arrangements. In recognizing the role of different agents in specific contexts, the local history, politics, power asymmetries, social identities, and intersecting privileges, critical institutionalism sees these aspects as influential over the livelihood choices on resource use (Cleaver, 2012). Nonetheless, Cleaver acknowledges that critical institutionalism often falls into romanticized notions of agency as empowerment or contestation when external interventions come to the fore.

Along with these critiques, Cleaver proposes "institutional bricolage" to understand the formation of institutions for resources management at the local level, how they really work in everyday life, and the outcomes they produce. Institutional bricolage is conceptualized as the process in which people "consciously and non-consciously, assemble or reshape institutional arrangements, drawing on whatever materials and resources are available, regardless of their original purpose" (Cleaver & de Koning, 2015, p. 4). This "patchwork of institutions" (Cleaver, 2012) that results from bricolage should be understood not as a simple combination but a necessary improvisation to deal with everyday challenges and changing situations. This process involves borrowing arrangements enmeshed in social relationships (e.g., meaningful experiences, reinvention of traditions, taken-for-granted thinking, local identities) that seem familiar to make institutions fit with local livelihoods. Thus, the adapted institutions are unbounded by bureaucratic institutions or sectoral delimitations and can be used for various purposes.

Drawing on social theory and extending the works of Mary Douglas, Pierre Bourdieu, and gendered theories, Cleaver (2012) builds this concept supported by structure-agency theories and the mediation of institutions. She suggests, for example, that people creatively innovate through bricolage, but these adaptations are both exercised and constrained in multiple ways. Key aspects of the everyday, such as the physical environment and material structures, power asymmetries, legitimized roles and authority, routinized practices taken-for-granted thinking, social identities, embodiment, and livelihoods are all crucial for bricolage. Taken together, these complex interactions shape different possibilities for people to access, use, control, make decisions, and negotiate solutions over multiple resources.

The sense of unevenness and ambiguity of bricolage then reflects its possibilities for reproducing inequities and transformative change. Thus, one important challenge when analyzing the dynamics of local resource management is tracking the functioning of institutional arrangements and the outcomes of bricolage. Studies in institutional bricolage have identified that its practices were critical for positive outcomes of interventions to solve production and resource management problems. Improvised arrangements by local people along with intervention rules can form a dynamic process that enables innovation. The authors conclude that the more open the interventions are for bricolage practices, the more they can trigger changing processes (de Koning & Cleaver, 2012; Merrey & Cook, 2012; van Mierlo & Totin, 2014). However, most of these studies address areas where people have lived for generations. These cases suggest that social cohesion is a critical element that enabled communities to manage their shared interests in bricolage. In turn, in recent areas like some rural settlements the outcomes of designed interventions might be even more unpredictable.

What this framework highlights is that the different uses, access, and ways of organizing integrated natural resources are intrinsically connected to the ordinary moments of structure and agency. However, as Cleaver (2012) remembers, these dynamics are neither limited to local-level practices, nor to simple definitions (Allouche et al., 2019). As these authors suggest, we can make a better sense of these "wicked problems" when we approach the everyday interaction between people, natural resources, and institutions within wider frames, scales, and levels of governance. Thus, my concern is not to demonstrate how water, energy, and food systems are objectively interconnected. Instead, in the sense of Stirling's argument, I seek to explore the "different constellations" that relate resources, institutions, and people, which are very much a subjective process (Stirling, 2015, p. 9). To do so, I choose to foreground the farmers' perspective and their micro-politics as a field of action, but also as a process built on structural forms of exclusion. The following section describes how I approached these everyday dynamics through the case of Sepé Tiaraju rural settlement and its participation in a sustainable rural development policy.

6.3 Method and Case Study Area

This chapter draws on a qualitative case study completed between 2015 and 2019. The original study analyzed the interrelationship between sustainable rural development policies, organizing processes and subjectivities, turning empirically to the Sepé Tiaraju rural settlement, an agrarian reform community that participated in the Sustainable Rural Development Project Micro Basins II (*Projeto de Desenvolvimento Rural Sustentável Microbacias* II – hereafter *Microbacias*), between 2014 and 2017. Designed as a partnership between the World Bank and the São Paulo State, this project composed the state's environmental policies. *Microbacias* financed productive activities of low-income farming organizations (e.g., associations, cooperatives, or environmental NGOs) aiming to integrate ecosystem goods and services to conserve biodiversity, natural resources, and climate change adaptation. In Sepé Tiaraju, 3 local organizations (one cooperative as the lead proponent and the other two as co-partners) and 35 families received funds from *Microbacias* to implement integrated land-water use management through agroforestry systems (SAF). Organizations and farmers, in turn, were made responsible for collectively managing the project, including the SAFs, the improvements in local food processing infrastructures and goods acquired.

The participation of Sepé Tiaraju in *Microbacias* is instructive on many scales and levels. Primarily, rural settlements materialize the connections between the national policy of agrarian reform and the landless' political struggle for land and livable lives (Fernandes & Welch, 2019). Moreover, rural settlements exacerbate the sociopolitical, contradictory nature of the local nexus. While they promote significant social and environmental reconstruction, regulatory controversies and persistent inequalities remain around resources (Melo et al., 2020). Located at the heart of Brazil's largest sugarcane producer – the macroregion of Ribeirão Preto – and crossing the Pardo river basin, the Sepé Tiaraju settlement is a Sustainable Development Project, a model of agrarian reform area created by the federal government to grant collective land tenure and regulate the sustainable use of natural resources (Brasil, 2000). The settlement was formalized in 2004, after 4 years of informal occupation organized by the Landless Workers' Movement (MST – *Movimento dos Trabalhadores Rurais Sem Terra*) to reclaim the rights over a degraded sugarcane plantation by the ethanol industry. Under an agreement with the federal government, 80 families were settled in the area. They became responsible for restoring the land and ecosystem through agroforestry systems (Ramos Filho, 2013) and collectively managing the natural and productive resources through formal organizations (Scopinho, 2012). These conditions were vital to defining the Sepé Tiaraju's eligibility to *Microbacias*.

While these issues around the resources management discussion are worthy of attention, the key reason to choose this case study is my own involvement with agrarian reform areas as researcher in social sciences. Since 2013, I have been working with sustainable development projects in rural settlements, following food-related local organizations and communities. These activities gave me the opportunity to understand context-specific challenges of environmental transformation and how people portray their own experiences and perspectives of those changes within an institutional setting. At the same time, I was curious about their everyday agency and its limits, particularly on how the interaction between various institutions, practitioners, and communities can serve as a locus of struggle or constraint around resources. The case study, then, could make more evident empirical findings that are instructive to socially nuanced perspectives of resource integration and policy interventions.

Drawing on ethnographic perspectives and triangulation of qualitative methods (Minayo, 2012), I started the fieldwork in 2015 by closely following the *Microbacias* phases of implementation for 2 years. Methodological strategies included participant observations of the project's decision-making forum – the Management Committee and its routine: the project's meetings; the *mutirões* (schemes of cooperative working in which every member participate in alternating tasks) to implement agroforestry systems; preparatory courses in agroforestry; food markets and food handling. My personal contact during observations at the settlement allowed me to invite all the participants for interviews. I conducted semi-structured, face-to-face interviews with the 35 residents that participated in the *Microbacias* (9 women, 26 men, ranging in age from 30 to 70) to explore their perspectives about the project, such as their involvement with decision-making and implementing actions, preliminary evaluations and expectations about agroforestry systems and food processes facilities. Additionally, other 35 interviews took place with nonparticipant residents (3 women, 32 men) to understand people's motivations for not participating in the project. Interviews were fully audio-recorded and transcribed. Alongside the fieldwork, I examined official documents and academic publications related to the project.

My involvement in *Microbacias* project also allowed me to invite people who wanted to continue to participate in timeline workshops. These workshops were adapted from the Reflexive Monitoring in Action's (RMA) tools for monitoring and evaluating sustainable development projects (van Mierlo et al., 2010). Along with the group, I aimed to retrace the experiences and outcomes in retrospect to earlier experiences of collective management of resources and relationships with other institutions and development policies implemented in the area, year by year. I organized seven timeline workshops with the settlement's residents, participants, and nonparticipants of the project, from October 2018 to March 2019. First, I invited all residents for the first cycle of workshops; invitations included in-person talks, phone calls, and messaging apps. The first cycle was completed in 12 hours of activity divided into four meetings, with nearly eight participants. The second cycle of workshops took place to generate a greater diversity of participants and opinions. This cycle was completed in 9 hours and three meetings, with nearly five participants. Coincidentally, all groups participated in *Microbacias* project. I invited the participants to remember the main events, ebb and flows of *Microbacias* and former projects during the meetings. While the participants were remembering and discussing, I registered the facts and keywords into post-its and placed them on a large timeline poster to make more visible. Each event was fixed in line with the corresponding year. In the final workshop, both groups compared their timelines to discuss a broad picture of the events and outcomes from these experiences.

Here, I do not try to provide a methodological protocol that can be replicated. Other researchers will be able to judge whether the strategies used here make sense to be reproduced in similar problems and scenarios. But a crucial point is that the internal triangulation ensured some reliability. By combining different sources of information and comparing the results, I could look at the contradictory accounts that emerged from empirical data from different angles, ensuring the diversity of perspectives expressed by the interlocutors rather than a single truth. Additionally, the adaptation of reflexive monitoring techniques was another fundamental part of reliability. The dialogue with the local actors and other research peers during the timeline workshops functioned as a member checking, allowing clarification and validation of the preliminary findings after 3 years of fieldwork.

To prepare this chapter, I further analyzed primary data gathered from the specific moment of the Sepé Tiaraju participation in *Microbacias*. After ordering the empirical material and secondary data, I approached the material by paying attention to the relevant institutional structures presented by the community members and the internal logic of the groups to access, integrate, and organize different resources, which resulted in the empirical units discussed in the following sections. I start the analysis by positioning the implementation of *Microbacias* within broad governance frames: from the designed arrangement by the World Bank to the formation of the local Management Committee in Sepé Tiaraju. Though I do not exactly focus my analysis on official discourses of the World Bank for financing sustainable rural development interventions (see also Gameiro & Martins, 2018) or the political uses of mainstream perspectives of the nexus and natural resources management, I briefly discuss how the issue of "integrated resources management" is differently framed by key agents involved with the *Microbacias* and the proposed arrangements.

6.4 The *Microbacias* Project: Framing the "Integrated Resource Management" in Sepé Tiaraju

The *Microbacias* was initially designed in 2000, around a two-term partnership between the World Bank and São Paulo state. In 2010, the project was redesigned in its second phase – the *Microbacias* II. This new version shifted the focus from environmental actions to the competitiveness of small-scale farmers and local organizations (Neves Neto & Clemente, 2014). Back to that year, the World Bank released the report "Development and Climate Change," making claims for integrated natural resources management in the face of climate change: "Only integrated management and flexible, long-term, large-scale planning can meet the growing demands on natural resources for food, bioenergy, hydropower and ecosystem services, while conserving biodiversity and maintaining carbon stocks in soil and forests" (World Bank, 2010, p. 136). The document does not explicitly address the nexus thinking but follows with several recommendations to build resilience and synergies across sectors in every level of governance. For rural communities, for instance, it recommends actions under the jargon "helping people to help themselves" (World Bank, 2010, p. 87), ranging from adaptation of agri-systems for water protection to reduce vulnerability, efficient use of energy, pilot projects for environmental services in micro basin areas, increased productive capacity, and participatory implementation to community-based management and design of robust rules and institutions. The latter was a reference to the "polycentric model."

Some of these recommendations got translated into the loan agreement of the *Microbacias* II, especially regarding the environmental subprojects (the group of interventions in which Sepé Tiaraju participated). For example, such a translation is noted when the World Bank describes "environmental subprojects" as those promoting "improved environmental practices in rural production systems to ensure the sustainability of the productive resource-base (land and water)," "pilot subproject at farm-level," "implementation of the borrower's environmental compliance system in rural areas combined with environmental awareness-raising," and "a new program of payments for environmental services to rural areas" (World Bank, 2010, p. 7). Additionally, the Bank defined the state's obligations to adopt three institutional arrangements – a Project Management Unit, a Steering Council, and a Consultative Forum – formed by bureaucratic "competent staff" to carry out the projects. Other obligations determined to conduct the projects with "due diligence and efficiency and following sound technical, agricultural, economic, financial, managerial, environmental, and social standards and practices satisfactory to the bank." Moreover, the obligation to "ensure that the Project Management Unit not to assign, amend, abrogate, or waive any Grant Agreement" did not resemble the claimed flexibility in the bank's report (World Bank, 2010, p. 10).

At the state level, the São Paulo's Secretariat for the Environment (SMA) and the Biodiversity and Natural Resources Coordination (CBRN) issued a call for proposals, which explicitly linked the terms for the environmental subprojects with a funding opportunity for local organizations to implement projects on agroforestry systems (SAF). Here, the SAF was defined as a land use system and occupation that manages high diversity of plants and crops. According to official documents, the SAF integrates the farmers' productive activities to comply with the existing environmental legislation (Coordenadoria de Biodiversidade e Recursos Naturais, 2014). Nonetheless, agroforestry was never mentioned by the World Bank as one of the sustainable practices of resource management recommended in the agreement or even in its reports. In Brazil, agroecology and agroforestry systems have been at the forefront of civil society groups and social movements, such as the MST. They advocate sovereignty over land and natural resources through agroecology, particularly against the agribusiness hegemony. From this perspective, the SAF not only suggests integrated and sustainable use of resources by providing "many ecological services to farmers, such as soil fertility, pest and disease regulation, and pollination" but a political means to "enhancing autonomy, resilience, and food sovereignty" (Altieri & Nichols, 2020, p. 205).

These first examples suggest that both arrangements prescribed by the World Bank, SMA, and CBRN framed the issue of integration differently: one to be resolved by strict acceptance of managerial rules, and the other to be resolved by legal procedures. When analyzing the hybrid governance of the nexus in Nepal, Allouche et al.

(2019) explain that the more the nexus is approached as a hierarchic regulatory issue, the more it moves from the local communities to the national and international agencies. Consequently, the more entrenched silofication takes place, though the discourses point to the contrary direction. On the other hand, the nexus finds more linkages in the closer social organizing, such as communities, given other dynamic social relationships, such as solidarity. Indeed, if we look at how these formal arrangements worked in Sepé Tiaraju, we will see that much of the local nexus was mediated by a combination between formal and informal arrangements, particularly by those meaningful social relationships that "recur over time and space" (Cleaver, 2012).

For example, when the *Microbacias* started in Sepé Tiaraju, these prescribed rules resulted in creating a local Management Committee, an official arrangement composed of representatives of the three local organizations, farmers, researchers from public institutions, and SMA/CBRN staff. The committee was responsible for making decisions and executing everyday actions, such as implementing the SAFs, upgrading food facilities, and training and rendering accounts. In practice, the committee configuration, its sub-commissions, meeting routines, and participatory methodologies used by the researchers were like the earlier arrangements of collective decision-making organized by the MST at the beginning of the settlement. These researchers had a deep relationship with the settlement, either in earlier interventions or by supporting the landless struggle. For example, during an activity to evaluate the project, a group of farmers discussed how the proposed discussion should be done: "that is the same way we used to do in the movement (MST)."

The fact that people found a way to survive in a land occupation organized by social movements in precarious circumstances is crucial for this dynamic configuration. The social and political role that the MST had in organizing the means to secure vital access to other natural resources shaped the primary basis to reconfigure the local nexus. These first experiences of organizing the struggles for water, energy access, and conditions to produce food established ways of thinking and using resources with significant meanings to the people. Therefore, these meaningful relationships can leak from one old arrangement to another, along with their formal similarities. In institutional bricolage, the leakage of meaning from a symbolic reference of the past helps to give social applicability to the new institutions (Cleaver, 2012). But other sources of authority shift over time, such as the "modern" stakeholders from the World Bank and SMA. Furthermore, people also turn to these authorities and their projects to reach more space to access other resources.

According to Jones (2015), this is a crucial area where external agents can use their own experiences at the local level to find more room for maneuver and adapt more practical interventions. In acknowledging and supporting practices of bricolage, these partners can work with the existing community arrangements by drawing on both tradition and new ways of organizing. On the other hand, there are also limits in these bricolage processes, as this way of working cannot be easily accommodated within the interventions' rules and timelines, budget restrictions, or other social relationships. Moreover, these adaptations do not necessarily mean equitable outcomes for everyone, as I discuss next.

6.5 Local Arrangements, Nexus Disconnections, and Uneven Relations

In this chapter, I have argued that mainstream perspectives of the nexus and institutionalism often lose that people do not passively incorporate new rules and arrangements – neither have unlimited capacity to exercise agency to contest the formal rules. Moreover, interventions often assume that every rural community will cooperate with homogeneous rules and interests (Cleaver, 2012). Rural settlements like Sepé Tiaraju indeed have a substantial background of cooperation and experiences with former development interventions. However, it does not necessarily mean that these relationships are free from constraints or necessarily positive. This subsection discusses how these constraints in bricolage and structural disconnections of the nexus produced unequal outcomes.

Some of these uneven effects became explicit during the functioning of *Microbacias*, specifically with the adaptations made by the Management Committee in the informal arrangements, like *mutirões* to implement the SAFs in 35 household lots. While these traditional practices are routinized in the livelihoods of many rural communities, Sepé Tiaraju has a troubled history with *mutirões*. These arrangements were vital in steering the local environmental change, particularly in the first years of the settlement when land degradation, drought, lack of income, credit for inputs, and infrastructure were unbearable. Nevertheless, *mutirões* also meant to intensify the work in individual and collective reforestation tasks, fetch water, participate in courses and meetings, and take care of family. As Cleaver (2012) reminds, all these physical activities are experienced under different conditions of embodiment, which can enable or constrain how people take part in the collective work. Considering other sociocultural conditions, most families were migrants from different regions, with various cultural backgrounds, precarious work experiences in urban areas, and fragmented memories about agriculture. The Management Committee was aware of these experiences. However, *mutirões* were the strategy to compensate for the disconnections between the bureaucratic rules for buying seedlings and the time to design the SAF layouts and implement them during the wet season. Moreover, these resources were under the risk of being "de-nexused" (Allouche et al., 2019) by the bureaucratic timelines, as a farmer observed:

> The farmer is not concerned with layouts; he wants to put the plant in the soil as soon as possible because of the rainy period. They put the layout in our hands, but we do not understand it. It is very slow, and the farmer needs to work.

Mutirões then contributed to saving physical and financial resources, while the Committee fulfilled the rules defined by the higher arrangements. However, the outcomes of this bricolage were not equally appropriated among the farmers. For example, during the first round of *mutirões*, I joined a working group with very few residents. According to a woman, people were skeptical about this scheme, as it usually benefited the lots scheduled in the early days. She referred to the past experiences of joint efforts in reforestation and house construction at the very beginning of the settlement. When Scopinho (2012) analyzed these experiences in Sepé Tiaraju years ago, she noted that *mutirões* and collective schemes were dissolved to the extent that people ceased to take turns, motivated by conflicts and divergent styles of working. Instead, people drew on these taken-for-granted thinking and affective motivations to informally reject the rule in the everyday. These examples confirm how the interplay between natural resources and formal and informal arrangements are uneven. Even when bricolage shapes arrangements to fit the local reality better, facilitating the nexus activities, it can produce unequal outcomes, especially for those most dependent on the cooperative efforts. On the other hand, it can evolve unexpectedly, producing new avenues to reshape the local nexus, as I will discuss in the following subsection.

Another example that illustrates these complex interrelationships between interventions, bricolage, and the local nexus is found in the formal recommendations for the SAFs. Some farmers were instructed to use in vitro seedlings of banana as a solution more resilient to pests and the lack of irrigation and, therefore, more productive and profitable. However, an interviewee explains why many farmers preferred their own in vivo seedlings:

> What makes better for families in doing so? If we plant the way they say is correct, you have a banana tree with 450 bunches. If you cut and sell everything, the woman wants to buy a sofa, the son wants a brand-new cell phone, the husband wants to go to the bar. At the end of the day, there is no money anymore. But the way we plant is simpler. It is not much money, but you can go to the bar here, the son gets 50 reais, the wife, too.

Affective aspects, social relationships, and powers intersect the complex, often contradictory, formal rules and how people manage natural resources (Cleaver, 2012). The quotation above suggests a rationality that accounts for the material and affective motivations to manage the resources, not for the innovative solutions brought by the experts. This necessary improvisation meant for this worker a fairer distribution of economic resources and leisure possibilities for all family members, though resulting in less scalable or resilient food production. Moreover, it illustrates what Kraftl et al. (2018) describe as a caring sense of people when dealing with these "trade-offs" in the everyday experiences of the nexus.

On the other hand, recommendations for in vitro seedlings were made due to the lack of water access to irrigation. The uneven distribution of water in the area is a chronic infrastructure problem across Brazilian rural settlements (Simonato et al., 2019). The temporary supply system network implemented by the federal government often cracks, leaks, and clogs, increasing the pressure over the productive activities. These constraints also imposed limits on the practices of bricolage and the possibilities that people found in creatively managing the resources to ensure the food production integrated with water and ecosystem protection. In this absence of integration with infrastructure programs, some farmers considered the rules of sustainable management too unjust to be realistically institutionalized in the everyday.

6.6 Unexpected Outcomes

As discussed above, development interventions and participatory arrangements bring to the fore a diversity of framings about resource management and integration. These framings are informed by a normative vision that idealizes how a particular problem must be resolved. However, these perspectives are partial because they assume roles that simplify the complexity of local identities. For example, the World Bank publications often stressed the aim to integrate resource management to the small farmers' organizations and their entrepreneurial potential. Alongside, the SMA focused on the farmers' potential in consolidating their environmental obligations. There is extensive literature in Brazil about the social categories in dispute and the presupposed identities to rural communities by development policies. Whether low-income, smallholders or family farmers (Garner & Campos, 2014; Neves, 2007), new peasants (Robles, 2001; van der Ploeg, 2008), or agroecologists (Altieri, 2018), the implications of such denominations are controversial as they imply antagonistic logics of development, production systems, and ways of organizing resources and livelihoods (Fernandes, 2008). Though this chapter does not address this discussion, it is worth mentioning that these ambiguous social identities necessarily come into play in the relationships between people, resources, and interventions.

As Allouche et al. (2019) suggest, local identities matter to the ways in which people negotiate or contest the external rules for resource use. Though these complex identities are often tenacious, they also are amenable to change by bricolage, intersecting various networks of relationships (Cleaver, 2012). I mentioned before that the recent formation of Sepé Tiaraju was marked by a cultural "cauldron" of people, making evident the diversity and blurred identities in the area (Scopinho, 2009). These identities were also expressed in how people framed their roles and expectations regarding the *Microbacias*. For example, some participants that identified themselves as "agroecologists" often talk about the *Microbacias* as an opportunity to have proper resources to make "mixed-and-match" agriculture and improve the environmental change of the area. For the organizations' leaders, the project would showcase their entrepreneurial capacity and, thereby, a means to establish new connections with other institutions and markets. For others, despite their lack of identification with agroforestry, the new SAFs became meaningful as they were reinterpreted by childhood memories of conventional agriculture practices.

The mechanisms through which the local arrangement of *Microbacias* evolved suggest the reinvention of an "agroecological" identity. For Cleaver (2012), people can reconfigure these ambiguous identities as an instrumental strategy for managing resources, based on the understanding of wider cultural frames and shared past histories. The reinvented *mutirões* were crucial for this process. At the beginning of this section, I discussed that *mutirões* did not engage families in work together, considering the past experiences of collective work. On the other hand, this reinvented arrangement also brought together students, interns, and practitioners in agroecology to work and learn with the farmers. The research institutes that composed the committee envisioned in *mutirões* an opportunity to promote knowledge exchange and facilitate the lived experiences of students and practitioners with the SAFs. Farmers, on one side, welcomed students as they added a necessary workforce to the SAFs. Students, on the other, contributed to farmers by paying the accommodation costs. These experiences were described as spaces of mutual learning and co-creation of knowledge.

As the *mutirões* took place during the *Microbacias*, researchers and a group of farmers engaged with the SAFs gave new applicability to these activities. Gradually, *mutirões* were designed and institutionalized to fit into courses and field days. According to an interviewee, these activities also included cultural celebrations based on the meaningful lived experiences since the occupation period: "This is something new. But we are going back to what we used to do." Taken together, these outcomes of bricolage suggest what de Koning and Cleaver (2012) define as practices of "aggregation." Aggregation combines local sociocultural elements with formal institutions and authoritative roles, giving them new meanings and purposes that shape a multiuse platform. These platforms are not limited to a single function and become influential in making decisions, accessing other resources, and shaping a new identity for the group. Nonetheless, the outcomes of such a creative encounter between different types of knowledge involve thinking carefully about the ambiguities and invisible works of power in bricolage. Cleaver (2012) argues that processes of design and bricolage are not immune from reproducing inequalities, marking how these can set apart dissent voices (or identities) and institutionalize dominant ideas. Moreover, as Jones (2015) observes in his study, arrangements developed through bricolage can change from relative flexibility to an ambiguous expectation about who shares the responsibilities and costs regarding resource management.

Another unexpected outcome of this multiuse platform started with *Microbacias* is the creation of new connections to urban markets. Since the end of the project, this platform has evolved toward a network of agroecology, integrating the agroecological production to crowdfunding, consumer groups, and an agroforestry network with another rural settlement, researchers, supporters, and organizations. Recently, these arrangements have reached other vulnerable families affected by Covid-19 in urban areas by donating fresh food from the settlements. Finally, the platform also suggests a possible reconfiguration of the local nexus, articulating land, water, and biodiversity regeneration to collective capacities and knowledge exchange (Levidow et al., 2021).

To what extent do these outcomes suggest mechanisms to deal with the nexus precarity or a transformative basis for a local nexus system toward social justice and environmental change? As Cleaver reminds us, bricolage is subject to the necessary everyday agency but also to the wider structural constraints that reproduce inequalities. When *Microbacias* was implemented, the national political conjuncture anticipated the backtracks that occurred since 2016, with significant pressures and impacts over the reproduction of families, especially on the food production of local organizations and market channels. Austerity measures also affected the budget of several public institutions, including those working with Sepé Tiaraju during the *Microbacias*. As such, these dynamic, uneven, and ambiguous outcomes of bricolage can be channeled into compensating for the injustices from different sources while preserving broad logics of unequal resource distribution. Still, these experiences also illuminate that the everyday construction of alternative and integrated avenues toward social and environmental justice is a matter of politics (Allouche et al., 2019).

6.7 Conclusions

At the beginning of this chapter, I questioned the normative claims for integration emerging in the Brazilian literature of public policies engaged with alternative pathways for sustainable development. In many ways, the ideas of integrations have been sitting comfortably with the growing apolitical, techno-managerial solutions to integrate resource management while perpetuating historical inequalities. I took this contradiction as the starting point of this chapter along with the SDGs recent challenges as an opportunity to make a better sense of how these calls for integration can take place in the everyday of rural communities and their experiences with development interventions. Critically and politically, alternative perspectives of water-energy-food nexus and critical institutionalism (institutional bricolage) highlight the dynamic, uneven, and ambiguous features of the interrelationships between people, institutions, and the mediations over natural resources. These aspects imply that integration and institutions are necessarily a political process, far from being a simple matter of technical solution or cross-sectoral governance. As such, structure-agency dynamics, power, politics, cultural, and social relationships are some of the aspects that need to be centered in any analysis concerned with the interrelationships and multiple dimensions of poverty (SDG 1–5), infrastructure (SDG 6–9), environmental justice (SDG 13–15), and institutions (SDG 16–17).

I then focused on the everyday experiences of a rural community while participating in a public project aiming at integrated management of land, water, and ecosystems to discuss how these dynamism, unevenness, and ambiguities of integrated resources produce different outcomes concerning the proposed interventions. The analysis of Sepé Tiaraju rural settlement and the *Microbacias* suggest that a dynamic reconfiguration of the nexus can be observed but with contradictions that suggest its limitations. Through bricolage processes, people shaped both formal and informal arrangements by drawing on past arrangements, meaningful experiences, different rationalities, and reinvented traditions. These combinations helped to give social applicability to the agroforestry systems. Moreover, bricolage also resulted in an unexpected multiuse platform around agroforestry, suggesting that other reconfigurations of the local nexus are to be made. However, these practices were also contested in the face of structural inequalities and past relationships, making it evident that bricolage alone cannot resolve the missing integration to other significant institutions and infrastructures, which preserve broad logics of unequal resources distribution and social injustices.

There are several implications of the lived experiences of communities with interventions that evade this case and its partial analysis, including how we should position it in broad frames of governance and the future challenges for public policy. Either way, the case highlights the importance of rethinking integration to problematize the easy association of these discourses with the oversimplistic or over-romantic view of rural communities. Through bricolage, rural settlements and local supporters demonstrate their painstaking capacity to forge new alliances and meaningful experiences, remaking the micro-political space to find solutions under uncertain circumstances, and contesting injustices through a fundamental transformation of the local nexus materialized in the landscape. Nonetheless, bricolage is not just a response to get by with the few resources at hand. It is also socially produced by more profound socioeconomic and environmental inequalities, by the missing integration to fundamental infrastructures constraining the use and access to natural resources. Too often, however, we fall into a trap to naturalize these creative solutions of rural settlements and other vulnerable populations as manifestations of "resilience" and resistance, while the concentration of power and the persistent inequalities became normative. Therefore, structural inequalities and environmental issues that have long been present will not disappear due to well-designed arrangements or institutional bricolage alone. Ensuring fair and stable use and access to land, water, energy, food, ecosystems, sources of income, and livelihoods of rural communities must be our fundamental commitment to social justice within and beyond the SDGs.

Acknowledgments The author acknowledges the financial support provided by Fapesp-São Paulo Research Foundation, Brazil. Process n. 2014/25042-0.

References

Allan, J. A. (2003). Virtual water – the water, food, and trade nexus. Useful concept or misleading metaphor? *Water International, 28*, 1.

Allouche, J., Middleton, C., & Gyawali, D. (2019). *The water-food-energy nexus: Power, politics, and justice*. Routledge.

Altieri, M. (2018). *Agroecology: The science of sustainable agriculture*. CRC Press.

Altieri, M. A., & Nichols, C. (2020). Agroecology: Challenges and opportunities for farming in the Anthropocene. *The International Journal of Agriculture and Natural Resources, 47*(3), 204–215.

Bazilian, M., Rogner, H., Howells, M., et al. (2011). Considering the energy, water and food nexus: Towards an integrated modelling approach. *Energy Policy, 39*(12), 7896–7906.

Börner, S. (2021). Vozes silenciosas sobre o nexo e as mudanças climáticas: Diálogos cotidianos na periferia urbana de São Paulo. *Diálogos socioambientais na macrometrópole paulista, 4*(10), 12–15.

Brasil. (2000). Ministério do Desenvolvimento Agrário. In *Projeto de Desenvolvimento Sustentável – PDS*. MDA.

Buainain, A. M., Favareto, A., Garcia, J. R., & Vieira Junior, P. A. (2018). Challenges for a sustainable agriculture: An approach based on the Brazilian biomes. *Desenvolvimento em Debate, 6*(2), 71–97.

Cairns, R., & Krzywoszynska, A. (2016). Anatomy of a buzzword: The emergence of 'the water-energy-food nexus' in UK natural resource debates. *Environmental Science & Policy, 64*, 164–170.

Carvalho, C. M. (2021). A contribuição das abordagens participativas para o nexo urbano na região macrometropolitana de São Paulo. *Diálogos socioambientais na macrometrópole paulista, 4*(10), 9–11.

Cernov, A., & Pietricovsky, I. (2020). *Brazil at the crossroads: The SDGs, the crises and the impact of COVID-19*. https://www.socialwatch.org/node/18506. Accessed 10 Jul 2021.

Cleaver, F. (2012). *Development through bricolage. Rethinking institutions for natural resource management*. Routledge.

Cleaver, F., & de Koning, J. (2015). Furthering critical institutionalism. *International Journal of the Commons, 9*(1), 1–18.

Coordenadoria de Biodiversidade e Recursos Naturais. (2014). *Projeto de Desenvolvimento Rural Sustentável Microbacias II – Acesso ao Mercado. Edital – Chamada PDRS/SA 02*. Secretaria do Meio Ambiente.

Dalla Fontana, M., Moreira, F., Di Giulio, G., & Malheiros. (2020). The water-energy-food nexus research in the Brazilian context: What are we missing? *Environmental Science & Policy, 112*, 172–180.

David, G.C. (2020). The unreality of promoting the SDGS without a sufficient budget. *Social Watch*. https://www.socialwatch.org/node/18072 Accessed 10 Jul 2021.

de Koning, J., Cleaver, F. (2012). Institutional bricolage in community forestry: an agenda for future research. In B. Arts, S. van Bommel, M. Ros-Tonen, & G. Verschoor, G. (Eds.), Forest-people interfaces. Wageningen Academic Publishers.

Endo, A., Tsuritab, I., Burnett, K., & Oencio, P. M. (2015). A review of the current state of research on the water, energy, and food nexus. *Journal of Hydrology: Regional Studies, 11*, 20–30.

Favareto, A. (2018). O combate à pobreza rural na América Latina e os Objetivos do Desenvolvimento Sustentável – a necessidade de um enfoque relacional. *Revista Grifos, 27*(45), 13–52.

Favareto, A. (2019). Transição para a sustentabilidade no Brasil e o desenvolvimento territorial nos marcos da Agenda 2030 e dos Objetivos de Desenvolvimento Sustentável. *Parc. Estrat., 24*(49), 49–72.

Fernandes, B. M. (2008). Questão agrária: Conflitualidade e desenvolvimento territorial. In A. M. Buainain (Ed.), *Luta pela terra, reforma agrária e gestão de conflitos no Brasil* (pp. 173–224). Editora da Unicamp.

Fernandes, B. M., & Welch, C. (2019). Contested landscapes: Territorial conflicts and the production of different ruralities in Brazil. *Landscape Research, 44*, 1–16.

Fernandes, B. M., Cleps Junior, J., Sobreiro Filho, J., et al. (2020). A questão agrária no Governo Bolsonaro: pós-fascismo e resistência. *Caderno Prudentino de Geografia, 42*(4), 333–362.

Gameiro, M. P., & Martins, R. C. (2018). O desenvolvimento rural sob regime de verdade: o discurso do Banco Mundial. *Sociedade e Estado, 33*(1), 13–37.

Garner, E., & Campos, A. P. O. (2014). Identifying the "family farm": An informal discussion of the concepts and definitions. In *ESA Working Paper No. 14–10*. FAO.

Grisa, C., Sabourin, E., & Le Coq, J. (2018). Políticas públicas para a agricultura familiar na América Latina e Caribe. *Raízes: Revista de Ciências Sociais e Econômicas, 38*(1), 7–21.

Harwood, S. A. (2018). In search of a (WEF) nexus approach. *Environmental Science and Policy, 83*, 79–85.

Hoff, H. (2011a). *Understanding the Nexus. Background Paper for the Bonn 2011 Nexus conference*. SEI.

Hoff, H. (2011b). Managing the water-land-energy nexus for sustainable development. UN Chronicle. https://www.un.org/en/chronicle/article/managing-water-land-energy-nexus-sustainable-development Acessed 2 Sep 2021.

IBGE. (2020). *POF 2017–2018: proporção de domicílios com segurança alimentar fica abaixo do resultado de 2004*. https://agenciadenoticias.ibge.gov.br/agencia-sala-de-imprensa/2013-agencia-de-noticias/

releases/28896-pof-2017-2018-proporcao-de-domicilios-com-seguranca-alimentar-fica-abaixo-do-resultado-de-2004 Accessed 12 Jul 2021.

Jacobi, P. R., & Giati, L. L. (2017). Nexos para a sustentabilidade: a busca por uma nova racionalidade. *Ambiente & Sociedade, 20*(2), 1–5.

Jones, S. D. (2015). Bridging political economy analysis and critical institutionalism: An approach to help analyse institutional change for rural water services. *International Journal of the Commons, 9*(1), 65–86.

Kraftl, P., Balastieri, J. A. P., Campos, A. E. M., et al. (2018). (Re)Thinking (Re)Connection: Young People, "natures" and the Water–energy–food Nexus in São Paulo State, Brazil. *Transactions of the Institute of British Geographers*, 1–16.

Leach, M., Stirling, A. C., & Scoones, I. (2010). *Dynamic sustainabilities: Technology, environment, social justice*. Routledge.

Leese, M., & Meisch, S. (2015). Securitising sustainability? Questioning the 'water, energy and food-security nexus'. *Water Alternatives, 8*(1), 695–709.

Levidow, L., Sansolo, D., & Schiavinatto, M. (2021). Agroecological innovation constructing socionatural order for social transformation: Two case studies in Brazil. *Tapuya: Latin American Science, Technology and Society, 4*, 1.

Lotta, G., & Favareto, A. (2016). Desafios da integração nos novos arranjos institucionais de políticas públicas no Brasil. *Revista de Sociologia e Política, 24*(57), 49–65.

Melo, T. G., & Scopinho, R. A. (2022). Situating the subject on the WEF nexus research: Insights from a critical perspective. In F. A. Moreira, M. Dalla Fontana, T. F. Malheiros, & G. M. Di Giulio (Eds.), *The Water-energy-food nexus: What the Brazilian research has to say*. (pp. 113-128). Faculdade de Saúde Pública da USP.

Melo, T. G., Souza, B. L., & Scopinho, R. A. (2020). Peri-urban territories and WEF nexus: The challenges of Brazilian agrarian reform areas for social justice. *Journal of Integrative Environmental Sciences, 47*, 45–67.

Merrey, D., & Cook, S. (2012). Fostering institutional creativity at multiple levels: Towards facilitated institutional bricolage. *Water Alternatives, 5*(1), 1–19.

Minayo, M. C. S. (2012). Qualitative analysis: Theory, steps and reliability. *Ciência & Saúde Coletiva, 17*(3), 621–626.

Moss, T., & Hüesker, F. (2019). Politicised nexus thinking in practice: Integrating urban wastewater utilities into regional energy markets. *Urban Studies, 56*(11), 2225–2241.

Neves, D. P. (2007). Agricultura familiar: Quantos ancoradouros! In B. M. Fernandes, M. I. M. Marques, & J. C. Suzuki (Eds.), *Geografia Agrária: teoria e poder*. Expressão Popular.

Neves Neto, C., & Clemente, E. (2014). Análise da Atuação do Programa Estadual de Microbacias Em São Paulo em suas duas fases (I E II). *Revista Do Departamento De Geografia, 27*, 147–169.

Niederle, P., Grisa, C., Picolotto, E., & Soldera, D. (2019). Narrative disputes over family-farming public policies in Brazil: conservative attacks and restricted countermovements. *Latin American Research Review, 54*(3), 707–720.

Ramos Filho, L.O. (2013). *Reforma agraria y transición agroecológica en una zona de grandes monocultivos de caña de azúcar: el caso del Asentamiento Sepé Tiaraju, región de Ribeirão Preto*, Brasil. 381p. Tese de Doutorado, Pós Graduação em Agroecologia, Universidade de Córdoba, Córdoba.

Robles, W. (2001). The landless rural workers movement (MST) in Brazil. *The Journal of Peasant Studies, 28*(2), 146–161.

Sabourin, E., Grisa, C., Niederle, P., et al. (2020). Le démantèlement des politiques publiques rurales et environnementales au Brésil. *Cah. Agric., 29*(31), 1–8.

Sauer, S., Leite, A. Z., & Tubino, N. L. G. (2020). Agenda política da terra no governo Bolsonaro. *Revista da ANPEGE, 16*(29), 285–318.

Schmidt, J. J., & Matthews, N. (2018). From state to system: Financialization and the water-energy-food-climate nexus. *Geoforum, 91*, 151–159.

Scopinho, R. A. (2009). Em busca de "elos perdidos": projeto de assentamento e modos de identificação entre trabalhadores rurais assentados. *Cadernos de Psicologia Social do Trabalho, 12*(2), 257–270.

Scopinho, R. A. (2012). *Processo organizativo em assentamentos rurais: Trabalho, condições de vida e subjetividades*. Anablume.

Simonato, D. C., Figueiredo, R. A., Dornfeld, C., et al. (2019). Saneamento rural e percepção ambiental em um assentamento rural – São Paulo – Brasil. *Retratos de Assentamentos, 22*(2), 264–280.

Stirling, A. (2015). *Developing 'nexus capabilities': Towards transdisciplinary methodologies*. The Nexus Network.

Svampa, M. (2015). Commodities consensus: Neoextractivism and enclosure of the commons in Latin America. *South Atlantic Quarterly, 114*(1), 65–82.

Urbinatti, A. M., Benites-Lazaro, L. L., Carvalho, C. M., & Giatti, L. L. (2020). The conceptual basis of water-energy-food nexus governance: Systematic literature review using network and discourse analysis. *Journal of Integrative Environmental Sciences, 17*(2), 21–43.

Urbinatti, A. M., Dalla Fontana, M., Stirling, A., & Giatti, L. G. (2021). Opening up' the governance of water-energy-food nexus: Towards a science-policy-society interface based on hybridity and humility. *Science of the Total Environment, 744*, 140945.

Valencia-Perafán, M. E., Le Coq, J., Favareto, A., et al. (2020). Políticas públicas para el desarrollo territorial rural en América Latina: Balance y perspectivas. *Eutopía, 17*, 25–40.

van der Ploeg, J. D. (2008). *The new peasantries: Struggles for autonomy and sustainability in an era of empire and globalization*. Earthscan.

van Mierlo, B., & Totin, E. (2014). Between script and improvisation: Institutional conditions and their local operation. *Outlook on Agriculture, 43*(3), 157–163.

van Mierlo, B., et al. (2010). *Reflexive monitoring in action. A guide for monitoring system innovation projects*. Communication and Innovation Studies, WUR.

Whaley, L., & Cleaver, F. (2018). Understanding process, power, and meaning in adaptive governance: A critical institutional reading. *Ecology and Society, 23*(2), 49.

Williams, J. G., Bouzarivski, S., & Swyngedouw, E. (2019). The urban resource nexus: On the politics of relationality, water-energy infrastructure, and the fallacy of integration. *Environment and Planning C: Politics and Space, 37*(4), 652–669.

Wojciechowksi, M. J., Favareto, A., Vega, S. M. R., et al. (2020). Uma leitura territorial e escalar dos processos inovadores da transição agroecológica em dois municípios do Vale do Paraíba e do Litoral Norte de São Paulo, Brasil Redes. *Revista do Desenvolvimento Regional, 25*(1), 59–83.

World Bank. (2010). *Desenvolvimento e Mudança Climática*. São Paulo.

7 WEF Nexus Innovations: The Institutional Agenda for Sustainability

Maria Ester Soares Dal Poz, Alessandro Luis Piolli, Erika Cristina Francisco, and Marcos Rehder Batista

Abstract

Cities must embrace the transition towards sustainability, since they are intricately linked to the context of peri-urban agriculture – farming that provides food and competes for resources such as water and energy and is therefore a critical process in intensifying or mitigating climate change. Additionally, the city-rural borders are blurred and not easy to define.

The governance of natural common-pool resources (the *commons*) is central to this urban transition to sustainability. To make this change happen, technological, organizational, social, financial and human behaviour innovations are needed.

This chapter presents the context in which the Water-Energy-Food Nexus (WEF Nexus) concept can be considered a mission-oriented *institution* regarding the governance of the sustainability transition, within the evolutionary process to create urban systems. The WEF Nexus is used as a prime mover to prompt a transition in the innovation agenda – organized in a three-dimensional matrix composed of an ecosystemic services approach to the future sustainable *commons* ensembles, as follows: *physical and material conditions* of production systems, *attributes of community*, a sustainable perspective of desired economic behaviours concerning the use of *commons* and *rules in use*, the agreements that constrain the exploration and removal of *commons*.

Keywords

Water · Energy · Food nexus · Urban sustainability · Transition agenda · Institutional economics · Innovation

7.1 Introduction

The so-called transition to sustainable systems, concerning climate change challenges, is possibly the greatest shift facing modern civilizations. The governance of common-pool resources, natural or technological, is the central point of this challenge.

Sustainability is a new milestone of civilization, an urgent collective transition goal, given the severity of climate change. A great deal of uncertainty surrounds the new constitutions and

M. E. Soares Dal Poz (✉) · A. Luis Piolli · E. C. Francisco
School of Applied Sciences & Institute of Economics, University of Campinas, Campinas, Brazil
e-mail: esterdal@unicamp.br; apiolli@unicamp.br; erikafra@unicamp.br

M. R. Batista
Institute of Economics, University of Campinas, Campinas, Brazil
e-mail: marcosrehder@gmail.com

L. L. B. Lazaro et al. (eds.), *Water-Energy-Food Nexus and Climate Change in Cities*, Sustainable Development Goals Series, https://doi.org/10.1007/978-3-031-05472-3_7

commitments that should be taken into account, considering a new pattern of economic growth and market development, from the perspective of maintaining a sustainable global human system.

Climate change – the prime mover of efforts towards the sustainability challenge – is a political economics issue: the management of this social and planetary dilemma of the risk of scarcity, which attempts to include social, political, and economic factors. The sustainable transition demands, in this context, *institutions that foster an institutional transition*.

This chapter presents the analysis of an institutional agenda for the transition to sustainability, based on the Water-Energy-Food Nexus (WEF Nexus) conceptual tool. The transition agenda involves an exploratory analysis of a set of constitutional factors that must be considered, so that such an innovative transition process can be planned and implemented, reaching an expected level of sustainability governance for which it is designed.

The WEF Nexus was conceived as an analytical promoter of and guide for actions to enable the many innovative processes involved in the transition processes. Plainly speaking, it is an *institution for institutional change*, assuming that institutions are made up of a broad set of human behaviours, culture and modes of common-pool natural and technological resource uses.

"Institutions" are systems of established and prevalent social rules and norms that provide structure to social interactions. The transition, in this sense, is an innovative shift of institutions which represents a huge shift in the cognitive rules by which humans live. A sustainable world requires massive changes at the structural level of social organizations and how they relate to resource systems. Beyond this, the transition process involves dealing with a complex system of intentionally – and innovative – targeted changes in human factors towards more sustainable living. These changes are addressed by technological, organizational, social, financial, educational, behavioural, legislative and regulatory innovations related to the sustainable use of commonly owned resources of the WEF Nexus. The nexus involves productive, distributive and consumption behaviours and national and global commitments that foster a higher level of environmental resilience combined with human development. Rules and regulatory bodies of common-pool resources, such as national or international level laws, also foster the development and adoption of clean energy as well as a multilevel set of agreements concerning the use of common-pool resources. Another group of factors is market and non-market incentives for innovative funding and monetary devices for more sustainable productive arenas, national and international policies to foster sustainable energy and sustainable environmental conditions or, in a word, innovation and innovative institutions.

According to the World Economic Forum (2011), "any strategy that focuses on one part of the water-food-energy nexus without considering its interconnections risks serious unintended consequences". The "nexus-oriented approach" therefore acts as a tool for planning and decision-support frameworks to manage the transition. The WEF Nexus is constantly evaluating such analyses, since its primary intention is to point out, address and lead the concept that the transition is an interconnected complex system, which is founded on the idea that there are nexuses between common property natural resources: water, energy and food.

What kind of tool, backed by what kind of collective, shared and commonly accepted concept and comprehension of the transition phenomenon, provides common ground for humanity to meet this challenge? How should the different levels of cognitive rules assumed by a huge diversity of human beings in their political, social, cultural and economic contexts be changed, considering the so-called climate change problems that impact efforts towards transition?

The WEF Nexus is a concept that, even if unable to provide these ex ante conditions, is a productive space for experimentation and a collective definition of desired institutional values that can, in the future, legitimate and support the interests of diverse stakeholders concerning the climate change issue. It can be viewed as a procedural tool for future learning about the transition process, and not as an established and unchangeable instrument.

Let us analyse the relations among urban innovations and the WEF Nexus, as complementary human constructs useful to drive the urban transition to sustainability challenge. The concept of innovation is intertwined with the climate change problem as a political, technological, social, organizational and financial prime mover. The Latin term *innovate* – from which the word *innovation* originates – can be understood as innovative forms of governance. In turn, the WEF Nexus is a conceptual tool for change, that is, for achieving new sustainable forms of life on the planet.

Thus, this chapter sees the WEF Nexus as a mission-oriented institution for institutional transition (from now on, SMOI). This idea is guided by the fact that the design of a sustainable future is actually operated by a variety of actions to change cognitive rules by which productive and commercialization flows use common-pool natural resources. There is one type of institution, at the level of countries' political and social actions or born from international initiatives, that may be involved in creating the rules and the instruments of economic, environmental and social policy. Such rules must be implemented and scaled in different ways, such as rules, behaviours, markets dynamics and citizenship rights. The goal of changing the cumulative hazardous effects of current modes of production and consumption is probably highly dependent on new and systemic forms of the governance of common-pool resources (Ostrom et al., 1994). Going beyond this thinking, a complex systems approach is required, given that common-pool resources are spread across the globe, with weak borders for their extraction and no borders for their dispersion nor for solving the problems resulting from their extensive use: i.e. the WEF Nexus.

As will be explored in this chapter, it seems essential to view the climate change problem as a complex system and acknowledge the concept and the WEF Nexus tools as a mandatory rationality to face the problem. In this system, cities are intricately linked to peri-urban agriculture, either because it provides food or because it competes for resources such as water and energy and is therefore a critical factor in intensifying or mitigating climate change. For this reason, it is not possible to think about the WEF Nexus without considering these complex relations between rural and urban areas, in terms of food production, transport, distribution and consumption dynamics.

In order to frame and induce the creation of innovation roles in the sustainability transition processes, the agenda follows an Institutional Analysis and Development (IAD) framework (Ostrom, 2009), assuming three main dimensions of the ecosystemic services approach and socio-ecological systems, as pointed out by Ostrom (2011). Each of these dimensions reveals a set of integrated complexity factors related to the dynamics of the ecosystemic services approach: (a) the physical and material conditions used to explore common-pool natural resources for human life; (b) the attributes of community, human involvement in and perceptions of the use of common resources; and (c) rules in use collective commitments – such as laws and regulatory devices – that humans agree among themselves, concerning the use of common-pool resources.

According to Ostrom (1990), the IAD framework is:

> …an evolving method to identifying and analysing interactions between the physical environment and sociocultural and institutional realms. The framework links the characteristics of the physical world – as forests, with those of the cultural settings – the villages (or, in this more specific case, the production systems that use common-pool resources of the WEF nexus).

This framework is designed to address the expected evolution of the governance of common property resources around and within cities, as given by the WEF Nexus. In addition, the IAD framework (Ostrom, 1998) is designed to promote and foster "situation arenas" of discussions and decision-making processes, involving relevant stakeholders and policy-makers – and the state – as collective learning, accountability-based, iterative and interactive evolutionary processes.

Finally, it proved essential to detail additional levels of the analytical structure of the innovative urban transition agenda. The matrix of institutional factors that can support action situation arenas – or the transition phenomenon in itself – is presented according to five factors (Teisman & Edelenbos, 2011): strategy (expected actions to foster sustainability), object (the structured planning to achieve the transition goals), form (the expected institutional evolution of human systems that could turn the transition into a reality), method (tools and ways to pursue the transition process) and logic (actor engagement to manage the transition).

The chapter is organized as follows:

- Section 7.1 presents the context in which the WEF Nexus will be considered as an institution for sustainability governance, considering the institutional economics approach.
- Section 7.2 explores the role of the WEF Nexus in the evolution of a new rationality for sustainable urban systems.
- Section 7.3 uses Institutional Analysis and Development to prompt the WEF Nexus transition innovation agenda regarding urban sustainability.
- Final considerations about the WEF Nexus agenda challenge.

In this context, the WEF Nexus agenda presented in this chapter helps understand the nonlinear dynamics and evolutionary character of the forthcoming sustainable transition in which innovation is a type of pivoting shaft.

7.2 The WEF Nexus as an Institution for Urban Sustainability Governance

Following the idea that one of the major challenges surrounding the study of institutions is their identification and explanation (Ostrom, 2005; Märker et al., 2018), in addition to the complex institutional factors related to the dynamics of a relatively new and unique type of institution, those that should promote the transition to more sustainable human systems are the SMOIs.

The WEF Nexus – as a conceptual path that allows for the analysis of a complex system's transition – considers the dynamics of climate change, which cannot be dealt with using basic, isolated variables. Despite being an aggregate macro-level approach, the WEF Nexus is able to embrace climate change target synergies, at the same time as providing a framework for its dynamics. Thus, it is necessary to explore an institutional framework for the sustainability challenge, considering the complex systems approach to deal with a broad multidimensional problem, which is not sensitive to segmented or sectorial treatments.

Criticism of the WEF Nexus concept, its diffusion and use as a systemic approach tool for climate change and transition analysis is, of course, expected as part of the human phenomenon of analysing how to deal with the global sustainability challenge. The institutional change demanded by this transition process is however undeniable, since the transition itself constitutes the changes in a set of rules, behaviours and social commitments underlining this sustainable condition. This is why a systemic concept such as the WEF Nexus is required, linked with an institutional transition agenda, as suggested in this chapter.

The transition represents an innovative and urgent journey that requires technological, social, financial and environmental innovations focused on sustainable growth and a resilient productive environment. The arrangements humans use to govern and access resources and their actions – including natural innovative resource use, such as WEF production systems, distribution, consumption and problem-solving actions to manage these resources – involves a new set of innovative rules, behaviours, conventions, cognitive and learning patterns of choice and social interactions. Thus, the sustainability transition is understood here as an innovation-based dynamic process. In turn, innovative transition management – to address, at least in the short term, changing human motivations and rules governing the access to resources – manifests itself as the management of institutional change.

This urgent situation creates the "transition-institution" duality concept, in which the evolutionary process of change can be identified by the major challenges regarding the management and governance of the resources system. This duality, on the other hand, brings us back to the "tragedy of the *commons*" (Hardin, 1968) issue versus current efforts – as this chapter does – to deny that such a tragedy is in fact destiny. On the contrary, following the logic to implement systems for "governing the *commons*" (Ostrom, 1990) – orchestrating the use and exploitation of common-pool natural properties and resources – this chapter remains in line with the ongoing rationale behind the entangled challenge surrounding a sustainable urban future.

The present analysis is based on the idea of improving institutional arrangements to utilize common resources (the *commons*) in such a way that does not exhaust them. Referring to Dietz et al. (2002), the *commons* concept – resources such as ecosystems, water supplies and the atmosphere – can be investigated by understanding institutions dynamically.

From the eco-economic common-pool resources perspective, this chapter argues that WEF Nexus is an innovative institution in itself, a prime mover for urban food production related to the transition of water and energy resources; it discusses two main issues regarding the WEF Nexus governance challenge, using (a) the profile of institutions – rules, agreements, behaviours, incentives and laws promoting transition – that seek to change "less sustainable contexts and ensembles" to create a new ruling over the future of the management of urban common resources and (b) as an extension of the first topic, a new pattern of human "cognitive rules", understanding that any change depends on cognition and beliefs (North, 2005). The common thread between both approaches is the set of institutions and factors that must be systematically analysed to create the agenda for the transition. The political science model for this transition proposal is given by the collective choice and public economics theories and the idea that neither markets nor states (Ostrom, 1990) can carry out such a change; polycentric governance systems may be required to manage the natural common-pool resources.

From a macro-level approach, the urban transition process – whereby new "more sustainable" ensembles could arise – depends largely on the adaptive capacity of this evolutionary (urgent) path: society's effectiveness to create institutions that are, considering the global demand for sustainability, productive, stable, fair, broadly accepted and, importantly, flexible enough to be changed or replaced in response to political and economic feedback (North, 2005).

The WEF Nexus approach, from a middle-range perspective, is strongly related to the diffusion of sustainability-oriented innovation (Geels, 2007). The evolutionary economics concept of governance sees economic processes as dynamic and constantly changing; technological, social, organizational or financial innovations leverage competition and evolution, epitomizing the goal of the WEF Nexus approach. This is why the so-called innovation-driven efforts towards new urban sustainability standards make complex demands on an evolving agenda, in which a highly diverse plethora of actors including stakeholders, shareholders, policy-makers and public and private representatives must be involved.

A huge variety and number of initiatives and efforts are currently in place worldwide to tackle the future sustainability challenge, such as national ministries, national and supranational commissions, national and supranational rules embedded in trade agreements, social policy and national and transnational policies for environmental protection, among others. Sustainability is undoubtedly viewed as a permanent, cross-sectional issue.

The constitutionality and functions of the WEF Nexus – as SMOI entities – are to carry out actions, such as public policies, regulatory contexts, laws, rules, routines, productive practices and education for sustainable consumption, essentially all the "innovation-driven institutions for sustainability". The core mission of these entities is to address a set of new human and environmental ensembles that increase sustainability – whether local, regional or global.

From the evolutionary economics perspective, the terms "transition," "change" and "evolution" are highly polysemic. Complementary to this is the difference between institutional change and changes derived from institutions' performance, including the level of rationality, intentionality and collective choice that led to these changes. Clarifying this issue is the starting point for further discussions on the so-called institutions for institutional transitions, whose mission is to promote a higher level of sustainability for human systems, such as urban ensembles. Thus, it refers to a type of institution that purposely aims to promote transformations in human systems. Institutions, themselves, evolve; in this case, they face the challenge of purposely designing institutions that can manage coordination systems for sustainability.

This is why the institution-transition duality is the unit of analysis of this institutional exploratory framework. The objective concerns deciphering which "governance core competencies (Penrose, 1971) and capabilities (Teece and Pisano, 2003, *apud* 1997) and profile of such transition-oriented institution, or SMOI, could strengthen the role of institutions for urban sustainability". As Teece (1997) points out, there is an ongoing endeavour to specify the nature and micro-foundations of the capabilities necessary to sustain superior enterprise performance in an open economy with rapid innovation and globally dispersed sources of invention, innovation and manufacturing capability.

In our case, this enterprise is the WEF Nexus tool, designed to guide the complex, collective and evolutionary process of change.

Cities are embedded in diverse local, regional, national and international variables, institutional factors and policy and political contexts in which each kind of transition outlines the transitional spectrum challenge – on a global level – to analyse the role and forms of urban sustainability governance. Concerning all the factors explored by the institutional approach and literature in previous decades, what are the main institutional capabilities of such push-and-pull institutions, and what is the main objective of generating and diffusing a new global rationale regarding natural and human resources? The process and the drivers of urban institutional evolution of sustainability involve a collective rationality for transition, such as that which the WEF Nexus concept addresses.

The institutional diversity (Ostrom, 2005) of the transition-driven efforts for a more sustainable world is focused on the representative case of the reach and capacity of the WEF Nexus as an SMOI entity; thus, the huge and multifaceted challenge surrounding the institution-transition duality is analysed, in terms of the expected evolution of a "new" rationality for sustainable cities.

The WEF Nexus conceptual tool has been broadly used for investigation and policy-making on climate change and common-pool resources. A great deal of literature has been generated to better understand the complex interactions among multiple and common resource systems.

According to Cairns and Krywoszynska (2016), the WEF has become increasingly prominent in international science policy and natural resource governance circles; it is also a buzzword in that it combines ambiguity of meaning, with strong normative resonance, and is being strategically appropriated into powerful managerial discourses, making it an open "matter of concern" for social sciences.

Thus, using a Nexus approach to manage common-pool resources is not supposed to be a noncritical activity: this conceptual tool, as mentioned in the introduction, is adopted as a productive space for experimentation and collective definition of desired institutional values that can, in the future, legitimate and support the interests of a diverse set of stakeholders concerning the climate change issue. It is important to highlight that the WEF Nexus was only created and is being used because the problem regarding the use of common resources already appears to have crossed the limits of future planet survival.

Governance – of the sustainable economic exploitation of common-pool resources – positions itself in political sciences and political economy discussions. This line of thinking came from Machiavelli and gained a predominantly economic guise from Adam Smith and Thomas Malthus. Scarcity conditions, used as a starting point by neoclassical economics, are at the heart of price theory given the balance between supply and demand.

In contrast to the neoclassical dimension, this chapter's theoretical framework follows Elinor

Ostrom's (1990) analysis of the need for a powerful authority, anchored in public or private ownership of resources; Ostrom studies both the most conservative and progressive views of economic management of the *commons*. The author establishes – as a main factor of change – the efficiency of institutional processes of collective learning and diffusion, providing the scope for the development of a new paradigm of innovation economy.

The problem of the disorderly and unlimited use of resources shared by various economic agents, which can range from pastures to consumer markets, was emblematically resumed in Garret Hardin (1968), as previously highlighted; the author argues that no agent will exhaust resources if an authority (public or private) that appropriates the good imposes rules of use (Hardin, 1968). This authority would also be fundamental to the "free rider" problem, exposed by Mancur Olson (1965) when addressing the problems of collective action, where some agents take advantage of the efforts of others and reap the rewards, discouraging others to cooperate (Olson, 1965). In the face of these assumptions, Ostrom (1990) asserts that if we consider that authority is never omniscient, it is unable to exercise this control. Thus, collective monitoring, with rules and a governance system that respects certain conditions (the rules in use), would exercise more efficient control and would instil a long-term learning culture.

That said, the governance of shared resources would be focused on the collective regulation of two basic aspects of resources: the difficulty of "exclusion" of beneficiaries and the "subtractability" of resources. The "exclusion" deals with the limitations of use, particularly in selecting who and how many will have access to the resource, thus regulating the offer; the "subtraction" refers to the flow, of how much the use of a unit decreases the quantity available for new uses.

There are a multitude of shared resources related to the matrix of subtractability of use versus the difficulty of excluding potential beneficiaries (Ostrom, 2009). The common-pool resources are positioned at the intersection between high levels of these two factors. This approach precisely follows the chapter's main heuristic: the transition towards sustainability is an evolutionary, iterative and interactive process continuously performed by a huge range of actors, at different levels of policy demands and actions, and involving a sort of geographical sphere and level of integration. This option places the WEF Nexus concept in the category of institutions that carry out sustainability governance, or, more specifically, those that are mission oriented.

Institutions (North, 2005) are systems of established and prevalent social rules that provide structure to social interactions; they enable ordered thought, expectation and action by imposing form and consistency on human activities through conventions and collective rules. These rules include behavioural norms and social conventions as well as legal rules. Members of a relevant community share tacit or explicit knowledge of these rules: in the case of climate change, we must address human engagement as a whole, since the challenge lies in creating a global context and sharing knowledge about how to make sure common-pool natural resources are available for future generations.

In this chapter, which aims to explore the role of the WEF Nexus as a sustainability mission-oriented institution, with a specific and urgent mission to develop more sustainable cities, it is essential to review such institutionality.

Sustainability is, in itself, an institutionally cross-cutting theme: it would be unthinkable to have only sustainability-oriented institutions, as this "entity" should be implemented and pursued through different initiatives in a systemic, uninterrupted and comprehensive way. It must be for this reason that in the last three or four decades, and in parallel with the sustainable thinking process, the number of institutions directly or indirectly dedicated to the theme has increased exponentially. The sustainability challenge has also been viewed as a transdisciplinary agenda, which has led to a wide range of institutional transformations and the creation and promotion of new kinds of sustainability-oriented ones.

It is difficult to characterize and categorize the role of sustainability challenges in the restructuring of objectives and considering institutional dynamics. When analysing institutional diversity, the national environmental ministries, environmental protection institutes, sustainable production departments and NGOs for forest preservation and sustainable production, among others, exist at the international level, as:

1. Political and scientific-political organizations, whose main objectives are to monitor, synthetize and disseminate knowledge on climate change and related actions with global reach, highlighting its causes, effects and risks to humanity and the environment, such as the IPCC7, FAO8 and Belmont Forum, which suggest and design ways to combat problems
2. Funding and policy-making institutions, such as the World Bank
3. NGO global initiatives, such as the Worldwide Wildlife Fund (WWF10) and Greenpeace – organizations which were initially more focused on preservation issues and now include sustainable development on their agenda

There is a clear overlapping effect of these institutional functions and efforts, in a *continuum* of combined objectives and with different levels of institutional mandates.

According to Ménard (2014), institutional economics has at least two main branches: the "institutional environment" (Davis & North, 1971), concerning the societal institutions that refer to the general rules that frame and constrain the behaviour and domain of economic entities' actions; this type of institution is much more focused on maintaining and conserving status quo. The second type of institutions, again quoting Davies and North, are those called "institutional arrangements", or "organizational arrangements", such as firms, NGOs, strategic alliances and so forth. Ménard uses this differentiation to introduce the question about player's structures and their transaction dynamics. He calls these arrangements meso-institutions; the WEF Nexus can be characterized in this way.

In the face of the analytical and practical demands on institutions involved in the intense promotion of sustainability transitions – considering the emergence of new economic and societal dynamics – and their supposed and desired evolution, a third type of institution seems to be emerging: the institutions for institutional change, such as the WEF Nexus. These institutions are cross-structural institutions and are key prime movers of change. The WEF Nexus is among the main institutions presented in this category given its scope on the governance of complex systems.

This institutional role of the WEF Nexus constitutes an agenda for change, given:

1. The size, complexity and the long-term character of such a challenge, considering the expected assertive evolution; this process would be the result of interactive and iterative processes that drive the transition, resulting in new, more sustainable contexts – related to natural resources scarcity – that constrain the future behaviour and domain of institutions.
2. The potential structural change in the two previous types of institutions as a result of this challenge – i.e. a new collective rationale for new rules, codes, behaviours and conventions, in the form of a new pattern of established human relations.

These transitions may be seen as part of a planned evolution – even more so in the Lamarckist form of biological evolution theories – as clear and systemic efforts are made to make it happen.

As an SMOI, the WEF Nexus permeates the majority of transition experiments – such as Belmont Forum's Urban Living Labs (ULLs) – which assumes that the WEF Nexus is a fundamental concept in understanding the system's complexity. The experimental variations of this kind of initiative involve the understanding and definition of an ecosystem for action, a theory of urban development, a tool for pilot services and an approach to experimental partnerships between researchers, citizens, companies and local governments (Lerhmann et al., 2015; Voytenko et al., 2016; Keith & Headlam, 2017; Scozzi et al., 2017). Interestingly, according to Lehrman et al. (2015), sectors involved in these initiatives tend to have different conceptions of it. Government institutions generally perceive ULL-like experiments as a regional innovation platform for sustainable production, while the private sector considers it more as a method of innovation. The WEF Nexus concept is however the cornerstone of sustainable systems, despite the diversity of initiatives and experiments. Moreover, such initiatives present multiple functional logic, given their institutional settings, the time horizons and research functions (Keith & Headlam, 2017). Some of these experiments can be established by local public administration, looking for cost-effective solutions for problems at the city level. This could involve a neighbourhood, be part of a global initiative or appear in the form of a partnership between research institutions and a municipal government as well as a network of city stakeholders and industry.

A central point in harvesting the concept of the WEF Nexus-based experiments is to find common ground; in other words, urban living labs will typically combine field experimentation and innovation methods that allow for the translation of long-term visions into more concrete and operational services and therefore provide a more suitable environment for behaviour change in users (Voytenko et al., 2016; Blühdorn, 2011, 2013). The employment of a service that is understood can shed light on how new behaviours and new ideas are supported by existing institutions. In environmental analysis, for example, institutions and adaptive capacity are deeply linked. According to Wise et al. (2013) the prevalence of changes and responses that cross spatial scales, sectors and jurisdictional boundaries can lead to threshold effects. Urban "borders", in this context, are difficult to define. This is why cooperative interdisciplinarity in the form of transition experiments mediated by the WEF Nexus concept is such an important tool in tackling the transition to sustainable systems. Despite the diversity of experiments and their partial results in terms of understanding transition processes, they must not be seen as isolated cases of institutions that carry out institutional transition. Their structure and dynamics can be evaluated considering the conceptual evolutionary capability of the WEF Nexus, in order to prompt the transition processes.

7.3 The Role of the WEF Nexus in the Evolution of a New Rationality for Sustainable Urban Systems

The unit of analysis of this institutional framework analysis is the institution-transition duality. This duality is based on the assumption that humanity must proceed to a more sustainable future. But who establishes and governs the patterns and rules concerning what is "sustainable"? Institutional values do not exist in a vacuum but result from sets of social relations that legitimate and support the interests of groups with a stake in existing constraints and that have the resources to make such constraints dominant (Matthews & Sydneysmith, 2010; North, 1990; Blühdorn, 2011, 2013).

What is the WEF Nexus's role in the institutional transition known as the "sustainable world"? Who are the prime movers and agents of such a transition process? Which kind of governance is expected for a more sustainable world – since global sustainable scenarios are collective – that goes beyond the national development perspective that is currently deeply asymmetric?

Consider that the problem is dependent on the assumption that climate change is a complex system, and then assume the concept and the WEF Nexus tools are a new rationality seemingly essential to face the problem. It is necessary to end current consumption tendencies to resolve this issue; the "collective irrational choice" concerning the uses and exploration of actual natural resources by humanity may be a starting point for this discussion. The institutional approach has struggled with the bounded or unbounded rationality; in this case, the issue appears to be the unbounded irrationality concerning the global modes of production and consumption. It is, of course, a common-pool resources issue (Ostrom et al., 1994), but may be the most complex case civilization will face for some time.

Adaptation governance (Bisaro & Hinkel, 2016) or, in this paper's approach, the governance of transition, is in this sense, a pillar of the discussion on why institutions emerge and how they enable or constrain adaptation. This branch of literature assumes the term "adaptation" as a semantic approach present in the technical literature on climate change, as the IPCC reports.

The institutional approach has been using the term "transition" as a correlate of the "adaptation process" that permeates the technical literature, such as the IPCC. The latest report (IPCC AR5) highlights the governance concept 36 times; in 21 of them, the relations between governance and institutionality are expressed in the following 5 groups of related constructs:

1. Common enabling factors and constraints for adaptation responses: institutions, governance and innovations.
2. Building adaptative capacity: complex challenges that demand governance structures, new institutions and new institutional arrangements.
3. Adaptation and mitigation are constrained by the inertia of global and regional trends in economic development: institutions as well as improving governance coordination and cooperation can help overcome constraints.
4. Policy coordination and the ability of actors to adapt depend on institutional arrangements.
5. International *forums* have been focused on addressing climate change with nearly universal participation: a growing number of institutions are organized at different levels of governance, which has resulted in the diversification of international cooperation on climate change.

In order to understand the potential mechanisms of transition and its governance challenge, it is fundamental to explore the ways in which attitudes and norms, as well as individual behaviours, express the outcomes of social dilemmas (Borjesson, 2009). Social dilemmas occur when society or group members find themselves in conflict over creating or using shared public goods. Although each individual would be better off fulfilling their own private interest and denying socially cooperative choice, it is a fact that all individuals, as a group, are better off if all of them cooperate (Dawes, 1980; Gifford, 2006). This is a basic question of rational and irrational decision-making. Putting it another way, we could ask, as in Blühdorn (2013), why, if the ecological discourse and urgency of culture change and structural transformation has never been so widely accepted, have efforts of many activist movements and academics, as well as some government agencies, from a local to global level, still not achieved the change in the development trajectory of modern societies? One reason may be due to contemporary societies' "(...) ever expanding needs in terms of mobility, technology or shopping opportunities having become essentially non-negotiable" (Blühdorn, 2011, 2013). This is particularly representative of urban living scenarios.

Assuming that well-being and quality of life must meet these expanding needs, how is it so that at the same time, society is being flooded by awareness campaigns on the crucial role of sustainability? One pathway may be understanding that such consumer habits should not be seen as obstacles to progress but as new information. References to paradigm shifts, old and new theories and methods have always coexisted (Lara, 2015; Borjesson, 2009). There is therefore an opportunity to develop further knowledge on societal rationality. People tend to disregard things as "old", and this allure towards all things new is "(...) one driving force in the mismanagement of our resources and not least in the creation of waste" (Borjesson, 2009).

A serious analysis must take into account the reconfiguration of modern societies concerning principles of individual freedom, choice and self-determined self-realization. These principles appear to conflict with the sustainability discourse as they are often "(...) based on accelerated consumption; highly complex, flexible and open to internal contradiction in ways that are incompatible with any notion of ecological virtues or an ethics of ecological duty or responsibility" (Blühdorn, 2013). In essence, this is the social dilemma, the conflict shown by Hardin's "tragedy of the *commons*", in which he stated that each person in society was locked into a system that compels him to increase his win without limit, in a world that is limited (Dawes, 1980). However, to acknowledge and manage this constant influence of non-rationality is fundamental to developing transition methodologies.

According to the founders of the concept of the "wicked problem", that is, one with a very complex solution (Rittel & Webber, 1973), the most burdensome problem is defining what a specific problem is (to distinguish an observed condition from a desired condition) and locating where, within a complex network, the problem really lies. Hence, the challenge lies in how to address rationality within the sustainability debate. Unlike rational choice theory, which assumes that the individual has unlimited analytical capabilities and perfect information at hand, Elinor Ostrom states that human beings have bounded rationality, that is, it is limited and adaptive (Lara, 2015). In line with this statement, policy approaches, directly determined by citizens and looking to respond to their demands, may well imply even less sustainability (Blühdorn, 2013). The social dilemma research acknowledges that decision-makers do not all make the same choices because choices arise from the individual's experience of interacting with the world and patterns of influence, as well as shared culture and preferences, or shared mental models (Lara, 2015; Gifford, 2006; Gargia-Parpet, 2013). Hence, given the presence of incomplete information combined with the limited and selective analytical ability to identify problems, and all the possible solutions and consequences, individuals will sometimes make irrational decisions (Lara, 2015).

Irrational problems are not "wicked", but they require novel approaches that acknowledge irrationality, according to Blühdorn (2013), by addressing the mechanisms that enable unsustainability to be sustained and their social and ecological implications managed. Again, considering "complex systems governance", or "wicked systems" of the collective irrationality that permeates the recent modes of production and consumption, and "adaptative" or "evolutionary" overly generic terms used to understand the transition challenge through sustainability, the scope and success of action for institutional change seems to be strongly linked to the learning concept (Hodgson, 2017), particularly collective learning processes.

7.4 Understanding WEF Nexus Dimensions Through Institutional Development Analysis for Future Sustainable Systems

The management of the social and planetary dilemma of the risk of resource scarcity – the iceberg being climate change – is treated here as a matter of political economy. The WEF Nexus is a conceptual tool to deal with this context.

This section offers an institutional framework for the WEF Nexus as sustainability mission-oriented institutions (SMOI), based on the following assumptions that they:

1. Are essentially innovation-driven institutions, in which performance is shown through social, technological, organizational and institutional innovation (Geels & Schot, 2008; Kemp et al., 1998); policy-making dynamics is a key issue in the diffusion of sustainable ensembles.
2. Are supposed to be structured as learning adaptative institutions, demanding iterative and interactive permanent behaviours such as networking, learning-by-doing and learning-by-interacting cooperative and interorganizational processes (Cohen & Levinthal, 1989, 1990; Sabel, 1993), presenting strong absorptive capacity to understand and react to new sustainable demands (Mowery & Rosenberg, 1989; Arora & Alfonso, 1990, 1994).
3. Make demands on the huge plethora of institutional constructs: rules, incentives, laws, behaviours and expectations (Hodgson, 2017).

Integrated with these assumptions, the present work is based on the analytical structure of socio-ecological systems, as pointed out by Ostrom (2011), in order to operationalize the prioritization rule using the IAD conceptual framework.

The *commons* issue must be revisited, in order to qualify the WEF Nexus as a conceptual tool, in the context of the political economy of natural resources. There is a distinction (Ostrom, 1990) between the characteristics of "provision", or the regulation of the conditions to supply resources, and "flow", the regulation of modes of use in terms of the demand for resources (their "subtractability"). In IAD, each rule is classified according to the impact it has on the analysed "action situation", and patterns resulting from interactions can be defined and evaluated.

Three dimensions of the *commons* exploration by humankind frame the IAD tool, as follows:

1. The physical and material conditions, concerning the world's physical (and biological) attributes, such as water or land, and the kind of human exploration processes that will be applied, such as capital
2. The attributes of community: the actors involved and their level of cohesion, in terms of compliance with rules that can lead to the governance of the ecosystems
3. The rules in use: the set of human agreements, laws, regulatory devices and incentives, as decisive in defining the rules of interactions between the participants of the commons' exploration and uses

The IAD approach was chosen due to its intrinsic rationale and relevance to the SMOI problem, namely:

- It provides a systematic approach to analysing the institutions that govern actions and outcomes in collective arrangements (Ostrom, 2009), just as the SMOI case represents, in particular due to its nature – the rational planned transition goal and demand for collective decision-making.
- SMOI goals are highly dependent on interactive and iterative processes, and, from the IAD perspective, institutions are defined as a set of prescriptions that people use to organize all forms of structured and repetitive interactions, such as rules and norms (Ostrom, 2005).
- The spectrum of the IAD approach – in which analysis of the problem can take place at multiple levels – operational, where individuals make daily choices and collective decisions, focused on operational rules, or at the constitutional level, where rules, incentives or any other collective constraints represent collective choice.
- The analytical focus of the IAD is the action arena, composed of actors who take part in action situations, involving choices of social or individual order, from which patterns of interaction derive and a new pattern of interaction, concerning the sustainable future, is expected.

The action situation, as the *social space in which people interact and make trade*, can be used to describe, analyse and explain behaviours within an institutional arrangement since cities are, by definition, civilian spaces, and actors within the action situation may make assumptions regarding variables that are directly related to the issue of urban sustainability governance: (i) the resources actors bring to the action situation; (ii) the channels actors use to obtain and maintain information on the problem they are involved in; (iii) the state of the art and the state of the world and actions actors bring to the situation and (iv) the path that the action actors choose.

Since the institution-transition construct presents revolutionary and evolutionary characters, a multi-criteria approach is designed, considering the IAD's three dimensions – physical and material conditions, attributes of community and rules in use with five institutional governance factors:

1. "Strategy" represents the action-oriented system through which the sustainability transition presents an objective to be achieved.
2. "Object" deals with the formal structure of the transition actions, subject to human behaviour.
3. "Form" concerns the structure and evolution of the institutions that can evolve gradually or in leaps, in order to achieve the transition.
4. "Method" considers the way in which change is promoted and which prime movers are involved, the desired future planning and how the transition is achieved.
5. "Logic" is the idea that the transition is supposed to be carried out by actual involved actors – who are capable of taking the transition processes forward, in a managed and complete way.

The following tables present the WEF Nexus Innovation Urban Governance Agenda,[1] from the IAD dimensions, to qualify the actions and competences needed to achieve the urban sustainable transition (Penrose, 1971) (Tables 7.1, 7.2, 7.3, 7.4, and 7.5).

[1]For greater immersion in the topic, an expanded and annotated version of the agenda is presented in the supplementary material in this chapter (Annex 7.1).

Table 7.1 Strategies of WEF Nexus governance: From the IAD dimensions

Institutional and Development Analysis dimensions/factors	Attributes of community	Rules in use	Physical and material conditions
Strategy: action orientation to achieve sustainability	Promote collective values, since the main attribute of a community to achieve the transition is awareness of the issue (Geels & Schot, 2008). Motivating local and regional education, skills and capabilities for sustainable practices and markets (Wieczorek et al., 2015). Foster micro-adaptive capacity and resilience capabilities exhibited by small farmers and native communities, given the strategic importance in facing climatic change. Foster meso-adaptative capacity and resilience capabilities according to the industrial scaling up of agribusiness and any other industrial activities. Foster network and producer cooperatives and network effects (Katz & Shapiro, 1994), since the low diversity of small holders' production may have little value alone, but generate higher value when combined with others, in a structured business (Ferrari et al., 2019). Promote iterative experiments of sustainable production and consumption – at the city or district level – to foster new arrangements of rules (Madsen & Hansen, 2019; Coutard & Rutherford, 2010; McCormick et al., 2013; Williamson, 1993). Create integration capabilities, since the mission to monitor the change depends on the ability to integrate new actors, groups and demands, in addition to new institutions arising during the process (Giddens, 1984; Nelson, 2008). Monitor science's methods of communication and how it relates to what may or may not be perceived as truth (Servaes & Lie, 2014).	Apply permanent efforts to understanding the transition process of cognitive rules (Greif & Mokyr, 2017), since the orientation of the action, regarding the new rules for sustainable ensembles, depends on a framework of informal rules that are internalized in the community (Greif & Mokyr, 2017; Ostrom, 2005). Perform permanent and effective actions surrounding sustainable policy analysis, policy decision-making and policy framework proposals (Geels, 2011; Lindblom, 1979; Polski & Ostrom, 1998; Ostrom, 1999). Adopt and address the public choice approach regarding public policy demands on the circular economy (Geissdoerfer et al., 2017; Meira, 2014; Dagnino, 2012; Ostrom et al., 1961, 1994; Ostrom, 1990).	Remain up to date with policy-making and implementation based on collective constraints and mechanisms for the diffusion of sustainable innovation (Kemp et al., 1998; Kemp, 2010; Ostrom et al., 1994; Kemp & Rotmans, 2009). Provide incentive systems for sustainable innovations (Scotchmer, 2004) through property rights and patent policies, regulation, social norms in production chains and consumption systems, reduction of transaction costs for sustainable system value chains and tax incentives for sustainable innovation (Pereira & Pereira, 2019; Silva et al., 2021; Lliso et al., 2021; Aza et al., 2021). Conceptual use and adaptation of the national innovation system, its elements and how it interacts in the production, diffusion and use of new knowledge related to a sustainable future (Freeman, 1987; Dosi & Orsenigo, 1988; Lundvall, 1992; Nelson & Rosenberg, 1993; Dal Poz, 2006). Perform policy-making to foster and maintain prevalence of complex and diversified modes of production, such as traditional knowledge-based cropping systems and alternative uses of biodiversity (Esposti, 2012). Policy-making for environmental sustainability – such as natural biomass and biodiversity preservation and their sustainable use, combined with social and economic development.

Table 7.2 Objects of WEF Nexus governance: From the IAD dimensions

Institutional and development analysis dimensions/factors	Attributes of community	Rules in use	Physical and material conditions
Object of the transition goals	Adopt a transition-oriented long-term perspective (Ostrom, 2011; Halliday & Glaser, 2011; Martínez-Fernández et al., 2021; Felicetti et al., 2015). Monitor the evolution of the concept of a transition towards sustainability, regarding its policy and political dimensions (Kemp & Rotmans, 2005; Kemp et al., 2007; Geels, 2011). Promote education, skills and communication for transition: based on the linear economy of the "take-make-use-dispose" model of production, distribution and consumption (Andrews, 2015), as a planned obsolescence market rational (Coase, 1972; Waldman, 1993) to a regenerative approach of resource use (Ellen MacArthur Foundation, 2016; Fracalanza, 2013).	Make effective sustainable policy prescriptions (North, 1990; Ostrom, 2005), to ensure the rules for the desired transition remain transparent. Define and implement rational planned transition goals (Nill & Kemp, 2009; Elzen et al., 2005; El Bilali, 2020). Promote the rule of law from a sustainable future perspective, at the national and international levels, according to the United Nations Development Goals, upward and downward accountability in institutions, enhanced participation, freedoms and capabilities, in order to improve the baseline for developing countries and poorer global populations (Gupta & Vegelin, 2016; Gupta & Baud, 2015).	Follow sustainable governance indicators, data and reports on national and international climate change investigation initiatives and their impacts in countries, sustainability efforts or counterfactual efforts (Schmidt-Traub et al., 2017; Sustainable Development Solutions Network, 2015; IAEG-SDGs, 2016; OECD, 2013).

Table 7.3 Form of the WEF Nexus governance: From the IAD dimensions

Institutional and Development Analysis dimensions/factors	Attributes of community	Rules in use	Physical and material conditions
Form: the structure and evolution of mission-oriented institutions	Promote the creation of transition-oriented institution networks, as meso-level initiatives – from global to country dimensions. Perform permanent effective efforts to model complex systems: mapping the nexus among resource uses in order to reduce its dysfunctionalities (Bazilian et al., 2011; Zhang et al., 2018; Gupta & Vegelin, 2016).	Adopt a model to understand and follow the diffusion processes of clean or eco-innovation (Kemp & Volpi, 2008). Analyse transnational coalitions and coordinated efforts through sustainability in the form of international agreements, as pool resources and coordinate state policies, that provide cooperative gains for sustainability global patterns (Böhmelt e Butkutė, 2018; Roggero et al., 2019); this procedure can guide the emergence and evolution of new SMOI forms (Bennich et al., 2021; Zihare et al., 2021).	Foster technological, social, environmental and organizational sustainable innovation solutions under an ecological economics perspective (Romeiro, 2012) and eco-innovation perspective (Kemp, 2010; Rennings, 2000; Boulding, 1966; Kemp, 2010)

Table 7.4 WEF Nexus Governance methods: From the IAD dimensions

Institutional and Development Analysis dimension/factors	Attributes of community	Rules in use	Physical and material conditions
Methods: the prime movers to promote transitions	Carry out educational initiatives with sustainability as the theme (Vare & Scott, 2007; McKeown, 2002) at local, regional, national and international levels.	Remain up to date with public regulation and self-regulation arrangements through certifications, codes of conduct and standards (Fuchs, 2007; Benites-Lazaro et al., 2018; Ostrom, 1998). Creation of markets to upscale new green technologies (Sengers et al., 2019; Gaganis et al., 2021; Gómez-Baggethun et al., 2010). Implementation of tax exemptions in order to create economic incentives to move towards more sustainable production (Madsen & Hansen, 2019; CEU, 2021; Hatta & Haltiwanger, 1986; Danish Energy Agency, 2015; Berlingske Business, 2016). Deregulation of high negative environmental impact utilities industries, exposing this sector to new pressures, leading to the emergence of more sustainable alternative technologies, more sustainable organizational and resource innovative modes of production and natural resource uses (Berkhout et al., 2009; Magat & Viscusi, 1990; Laplante & Rilstone, 1996; Dasgupta et al., 2001; Foulon et al., 2002; Kemp & Rotmans; 2001; Joskow, 1998). Closely monitor "operational rules": the regulatory dimension of the management methods and the level of diffusion of shared sustainable protocols and practices (Kemp et al., 1998). Test and validate instruments related to understanding the transition process, since the goal to end poverty and build shared prosperity is sustainability (World Bank Agenda 2030, 2021; Gupta & Vegelin, 2016).	Decision-making methods based on environmental monitoring techniques, such as satellite and earth monitoring systems for climate change and scientific production analysis (Martínez-Fernández et al., 2021; Belucio et al., 2021; Feroz et al., 2021; Tsaples & Papathanasiou, 2021; Zihare et al., 2021). Use the existing tools and methods to create new ones with technological, social and organizational foresight, in order to understand the evolution of the transition of complex systems towards sustainability (Martin, 1995, 2010; Magruk, 2011).

Table 7.5 Logic of the WEF Nexus governance: From the IAD dimensions

	Attributes of community	Rules in use	Physical and material conditions
Logic: management of actors involved in sustainable policy processes	Analyse the multiple narratives, epistemes and paradigms of sustainability from a political perspective: governance, rational choice, agency and institutional approaches (Bevir & Rhodes, 2001; Thaler, 2000; Etzioni, 1996; Ostrom, 2000).	Analyse collective governance processes (Ostrom, 1990).	Foster regulatory, legal and tax advantages for sustainable production and environmental services (Gómez-Baggethun et al., 2010). Include "social technology" and "solidary economy" concepts (Dagnino, 2012; Meira, 2014, Rodrigues & Barbieri, 2008), as channels to strengthen small user-oriented innovations (Rodrigues & Barbieri, 2008; Benini et al., 2012).

7.5 Final Considerations

The analytical study of the future sustainable systems agenda presented in this chapter is not intended to be exhaustive or complete. It represents an ongoing effort, which can be used as a starting point in order to move forward and increase the understanding of and efforts devoted to collective action aimed at transition. It is also necessary to consider that the sustainability agenda is a dynamic object, a horizon that moves as some degree of transition is achieved or fails to evolve as initially thought. This is particularly important since agendas such as this are defined as ex ante by agents whose character and demands are also changeable and whose implementation potential and conditions can be easily changed.

In this context, an effort like this, to prompt the transition phenomenon challenge – from the perspective of the WEF Nexus concept – can show, at the very least, the multiplicity, complexity and tangled character of collective thinking and actions aimed at achieving the transition to sustainability. This tentative agenda can be seen as a skyline, a moving horizon that, despite incomplete, captures the multifaceted nature of this human challenge.

The political, social, economic and environmental aspects involved in this agenda are limitless, in that it is configured as a multi- and meta-disciplinary ensemble cast viewed as an open collective agenda. Add to this the fact that managing an innovative set of sustainable scenarios is a complex system of factors that must be considered in the diffusion of innovation as an uncertain and highly risky set of phenomena.

The WEF Nexus concept can be adopted so that it can continue to apply to new patterns of organization demanded by the transition processes towards sustainability. It can be understood as a collective new mindset that guides sustainability-driven (a) policies, and their related policy dynamics; (b) markets, and their entry barriers, competitive struggles and necessary financial innovation, remembering the creative destruction of "old" markets; and (c) capacity building for new, sustainable production and distribution processes, new consumption behaviours and new rules in use, in an accountability-based, iterative and interactive evolutionary process that must be faced together, given the current demands surrounding the exploration of common resources.

Acknowledgments The authors would like to thank FAPESP, the Sao Paulo Science and Technology Research Foundation, Brazil, for the Sao Paulo in Natura Lab grant.

Supplementary Material

Annex 7.1 The WEF Nexus agenda: expanded and annotated version

Institutional and Development Analysis Dimensions/Factors	Attributes of Community	Rules in Use	Physical and Material Conditions
Strategy: Action orientation to achieve sustainability	**• Promote collective values, since the main attribute of a community to achieve the transition is the awareness of the issue (Geels and Schot, 2008).** One strategy to maintain sustainability as a transition process is to address and sustain a relationship with the most important collective values in a community, even if only to question them. This depends on the dissemination of knowledge in a way that the population understands the importance of	**• Apply permanent efforts to understanding the cognitive rules applied to the transition process (Greif and Mokyr, 2016), since the action, regarding the new rules for sustainable ensembles, depends on a**	**• Remain up to date with policy making and implementation based on collective constraints and mechanisms for the diffusion of sustainable innovation (Kemp et al., 1998; Kemp, 2010; Ostrom et al., 1994).** The diffusion of clean technologies through new

Annex 7.1 continued

Institutional and Development Analysis Dimensions/Factors	Attributes of Community	Rules in Use	Physical and Material Conditions
	change, since, as a public policy, institutions – and their actions - committed to promoting change towards more sustainable standards must be legitimate. This is the case of the Multi-level Perspective for Strategic Niche Management (Geels and Shot, 2008), surrounding the diffusion of new clean technologies which are less likely to reach markets (such as solar technologies), due to the prevailing technologies (such as fuel-based energy technologies). Thus, the authors created a third dimension, which goes beyond the management of niches and market regimes, avoiding the trap of market power in order to maintain the "old" technology with no competition. They call it "Landscape", which is the socio-cultural and political environment capable of strengthening the process of adopting new clean technologies (Geels and Schot, 2008), for instance, when consumers do not buy food in non-renewable plastic packaging. • **Motivating local and regional education, skills and capabilities for sustainable practices and markets (Wieczorek et al., 2015).** Scaling up agroecological or any other sustainable approaches for production and distribution chains is essential to making the transition a reality. Wieczorek (2015) argues for a move away from a predominantly national approach to transition studies to embrace a multi-scalar understanding of transition processes. For instance, The Brazilian R&D Agroecological Enterprise (EMBRAPA) "Sisteminha Project" ("Little System"), has been successfully developed, both in environmental and in food security terms. It consists of a fish tank, chicken coop, earthworm farm, hydroponics and a compost shelter, in addition to a peripheral vegetable patch. The entire system reutilizes water from the fish tank, which reduces production costs and increases food supply, since fish faeces are a powerful fertilizer for plants and is wasted in conventional fish farming. The system was developed to guarantee food security in small subsistence properties, or small production units for sale. However, the idea and central scheme of the almost complete reuse of resources, the reuse of fish farming water and fenced farm animals for fertilization, autonomy in local fertilizer production, could be incorporated into medium and large productions in more sustainable models of production. Nevertheless, for this to take place, the development of innovations and outputs with international and local impact is fundamental. • **Foster micro-adaptive capacity and resilience capabilities exhibited by small farmers and native communities, given the strategic importance in facing climatic change.**	**framework of informal rules that are internalized in the community (Greif and Mokyr, 2016; Ostrom, 2005).** The implementation of innovations (sustainable) involves rational market games. In this interactive dynamic, informal rules are shared between agents that may or not be formalized over time. These rules emerge in a latent manner from the practical movements of society. These tacit patterns are known as "cognitive rules"; they emerge as adaptations of new realities resulting from the economic disputes that occur along the diffusion processes of new technologies. Understanding how these standards emerge and their impacts are fundamental to monitoring the change processes. • **Perform permanent and effective actions regarding sustainable policy analysis, policy decision-making and policy framework proposals (Geels, 2011; Lindblom, 1978; Polski and Ostrom, 1998).** Sustainable policy analysis does, of course, fall under the complex umbrella of policy making theories. The main issues regarding policy making, despite the intertwined theoretical and conceptual approaches, are policy affordability, access and accountability factors, which also apply to the challenge of the sustainability policies analysis. There is a multiplicity of approaches – within a complex set of variations that takes relative factors into account – such as the	markets presents strong entry barriers, due to the competitive advantages of market incumbents' mature technologies. Sectors tend to be taken-over by sustainability-oriented change to create market competition strategies that prevent transition, such as oligopoly effects and market lock-in, thus holding on to mature technological innovation markets. These collective control mechanisms based on market structures prevent eco-innovation from gaining scale. Hence the need for a policy approach to sustainable innovations, which goes beyond a mere extension of sustainability policy. The systemic approach to eco-innovation therefore needs to be on a multi-scale perspective (micro, meso and macroeconomics), and interdisciplinary (sociological and political science). Such policies focus on socio-political demands directed towards the development and diffusion of sustainable technologies. The literature on this transitioning policy issue (Kemp and Rotmans, 2009) discusses the so-called windows of opportunity for those technologies from a new integrated policy ensemble: a) effective governmental policy integration regarding sustainable technologies, b) long-term thinking for short-term action – in order to maintain sustainable project experiments of search and selection as ex-post processes, c) keeping multiple options of substitution technological candidates open, d) learning-by-doing during technological selection and learning-by-doing processes to foster the diffusion of sustainable technologies. • **Incentive systems for sustainable innovations (Scotchmer, 2004)**

Annex 7.1 continued

Institutional and Development Analysis Dimensions/Factors	Attributes of Community	Rules in Use	Physical and Material Conditions
	Indigenous communities and small farmers who live close to or within preserved natural areas are examples of harmonious coexistence with the environment, through less aggressive forms of production. These groups demand certain specific innovations to guarantee subsistence and food security as well as to maintain their sustainable practices. Many organic farming techniques such as biological control, companion plants and agroforestry systems, are traditionally used by indigenous communities, riverside dwellers, and other traditional communities. These techniques can be improved to enhance their economic use and environmental performance in traditional communities themselves or implemented into processes of transition to sustainability in other contexts, such as a scalable bioeconomy. There are many experiments on agrarian reform settlements and small farmers that have successfully used these traditional techniques, with scientific and technological advancements, to produce certified organic foods. This combination of traditional ancestral agriculture and innovations in agricultural and organizational processes means the difference between the poverty line and the sustainable insertion into the market. Thus, operating through cooperative solidarity economy ventures combined with technical support and the development of social technologies is what has made the difference. • **Foster meso-adaptative capacity and resilience capabilities by the industrial scaling-up of agribusiness and other industrial activities.** Some of the techniques used in traditional agriculture can be used or adapted for more resilient scalable production. But, just as for smallholders and traditional peoples, demand-oriented technologies, whether organizational, processes or products, are essential to increasing the degree of sustainability in agribusiness large-scale operations. The no-till system, the use of biological control, the use of fertilizers of animal and vegetable origin are some of the examples of the use of traditional techniques in the agro-industry. However, agro-industry developments accompany the latest technologies. The transition to agricultural sustainability on an industrial scale therefore requires adapting sustainable technologies from other agricultural or industrial contexts, as well as developing technologies specific to the context. Different types of innovations, at the different stages of the production process must constantly be adapted: including new storage techniques and structures, less pollution and less expensive industrial processes, transport and production with lower greenhouse gas emissions and water consumption.	rationality pattern of the decision makers and the role and extension of information in the decision-making process. The main polarization occurs between incrementalism – based on the bounded rationality concept of the decision-making process, and the rational model - in which problems are understood **as "technical" (and not "political"), the** environment is consensual, and the process is permanently under control. Carrying on the issue of policy analysis for the case of climate change and the sustainability challenge, and given the complexity of the topic, at least two analytical streams can illustrate the socio-technical reinforcement of the roles and dynamics of actors in decision-making processes, concerning transition as system innovation, and not a partial redesign or change in the optimization system. The first – which addresses the idea of demands for collective decisions - was analysed by Geels (2011), considering that: sustainability transitions are goal-oriented or **'purposive', in the sense** that they address persistent environmental problems, whereas many historical transitions were **'emergent' (e.g.,** entrepreneurs exploring commercial opportunities related to new technologies). The second is driven by the institutional economics strand (Ostrom, 1999), in which sustainable ecosystemic services depend on collective future design and implementation. These	**through property rights and patent policies, regulation, social norms in production chains and consumption systems, reduction of transaction costs for sustainable systems chains, and tax incentives for sustainable innovation.** Governmental policy and integration of global efforts are essential to promoting innovative sustainable systems, like resource substitution incentives, penalizing polluting resources and supporting non-polluting ones (Silva et al., 2021). However, these subsidies can face opposition depending on the origin of the revenues used to finance them (Pereira and Pereira, 2019). In the case of The Clean Development Mechanism (CDM) and Renewable Energy Sources (RES), which are considered independent alternative subsidies, it is possible to follow the logic of an Environmental Tax Reform (ETR) in which the use of polluting resources is taxed, and the revenues are used to finance the use of renewables (Silva et al., 2021). Considering the Payments for Ecosystem Services (PES) that represent voluntary agreements meant to compensate individuals or communities for supplying socially valuable ecosystem services, (Lliso et al., 2021), studies show that these projects do not always produce the best results due to inefficiencies in sustainable development policies. According to Aza et al., (2021) it is necessary to implement PES systems through the integrated approach, aiming to increase benefits and offer solutions to some of their problems. Additionally, the integration should provide an optimal combination of both short and long-term ecosystem services, ensuring resilience to

Annex 7.1 continued

Institutional and Development Analysis Dimensions/Factors	Attributes of Community	Rules in Use	Physical and Material Conditions
	• Foster the formation of network and producers' cooperatives and network effects (Katz and Shapiro, 1993), since the low diversity of small holders' production may have little value in isolation, but generate higher value when combined with others, in a structured business. Regarding the WEF Nexus context, the effects of cooperative production on reducing production costs and increasing market advantages, such as collective purchase power – from neoclassical economics – are still valid. The concept of network effects has been used to understand technology-based markets (Katz and Shapiro, 1993) and innovation-based markets, characterized by rapid innovation, such as market-places global platforms or IT sectors. This does not mean that sectors such as agriculture or cattle raising, in the context of the WEF Nexus – that are not traditionally seen as innovation-based - cannot be assessed in terms of the benefits of network effects. Agriculture, energy or other resource-based sectors can be (positively) affected by these effects. This is particularly true since innovations have been increasingly present in the competitive dynamics of these sectors. Network effects, in terms of action-oriented strategies to deal with food production losses and waste must be taken into account: we must only remember the effects on agricultural competitiveness of the evolution of markets based on technologies of genetically modified organisms (Ferrari et al., 2019) and, more recently, the editing of genes by methodologies like CRISP Ca. In the economy in general these effects are characterized as lock-in generators, since they induce a monopoly; but, in the case of eco-innovation market entry barriers, suffering from the effects of competitiveness against mature and monopolistic markets. Network effects come from the fact that the benefits given by the user to a given product or service depend on the number of other users who are on the same network. So, if the intention is to foster sustainable food production, conventional market barriers given by the conventional produced food markets must be eliminated. This would be an interesting strategy for agroecological food markets, for instance, to drop entry barriers, since network effects make sustainable food production more valuable as the number of users increases. **• Promote iterative experiments of sustainable production and consumption – at the city or district level - to foster new arrangements of rules.** The transition challenge is not an *ex-ante* process nor one that can be individually conceived, since there is a plethora of themes and options of "more sustainable" contexts. It concerns innovative modes of production and	entities are known as "situation arenas", or polycentric systems of evolutionary decision cycles performed by a huge range of actors. **• Adopt and address a public choice approach for public policy demands regarding the circular economy (Geissdoerfer et al., 2017; Meira, 2014; Dagnino, 2012).** The public choice approach is defined as "the application of economic reasoning to "collective," "political" or "social" decision making" (Ostrom el al., 1961). It assumes that neither the State nor the market are able to effectively regulate environmental problems arising from the unrestricted use of natural common pool resources (Ostrom, 1994). A higher level of sustainability to food production activities may be reached through situation arenas (Ostrom, 1990) that involve several stakeholders in the formulation of public policies for sustainable innovations. In this context, the "economic man" could be replaced by a new "decision-making man"; the challenges of sustainability permeates the circular economy (Geissdoerfer et al., 2017) diffusion. Concerning social diversity and modes of production and consumption chains, the solidary economy and social technology (known as social innovation) approaches are two key concepts when thinking about the circular economy within precarious economic and social conditions (Meira, 2014; Dagnino, 2012). Firstly,	environmental and social change. **• Conceptual use and adaptation of the national innovation system, its elements and relationships that interact in the production, diffusion and use of new useful knowledge for a sustainable future.** The concept of National Innovation Systems (NIS) (Freeman, 1987; Dosi et al., 1988; Lundvall, 1992; Nelson and Rosenberg, 1993) has been developed to explain the dynamics of economic subsystems in countries where various organizations and institutions interact and influence each other, throughout the innovation process. It focuses not only on the introduction of new technologies and organizational forms in a country, but also on the research and development efforts carried out by companies and the public sector, as well as the determinants of innovation, such as the learning processes and mechanisms to encourage innovation (Dal Poz, 2006). In this context, it proves to be useful and pertinent in the case of the generation and diffusion of sustainable innovations, since it refers to the processes of organizational and institutional change that can emerge from the interactive learning systems demanded by the substitution of mature and less sustainable technologies. **• Policy making to foster and preserve the prevalence of complex and diversified modes of production, such as traditional knowledge-based cropping systems and alternative uses of biodiversity.** Traditional ecological knowledge systems capable of adapting to both external changes and internal pressures have been a

Annex 7.1 continued

Institutional and Development Analysis Dimensions/Factors	Attributes of Community	Rules in Use	Physical and Material Conditions
	consumption, finance, funding, and new market structures as a result of the diffusion of clean technologies, new business models for technological and social innovation, among others. Interactive and iterative cycles of experimentation should be implemented, in order to understand the optimal collective scenarios. This requires stakeholder engagement, generating trust and legitimacy in the implemented actions (Madsen and Hansen, 2019; Coutard and Rutherford, 2010; McCormick et al., 2013); the evolution of these arrangements is essential to understanding the possibilities of: a) the collective learning processes to make sustainable markets viable; and b) transaction cost reduction and the best governance for resilient sustainable market structures (Williamson, 1993). Complementary to this, behavioural economics is a promising knowledge area to develop this issue, since it analyses the interactive dynamics concerning human behaviour and the division of goods. • **Create integration capabilities, since monitoring change depends on the ability to integrate new actors, groups and demands, in addition to new institutions arising during the process (Giddens, 1984; Nelson, 2008).** The transition process demands that innovations addressed by actors that often do not communicate with each other directly are adapted and diffused. Complex demands on technology substitution, for instance, may conflict in this context. This is the case of the dispute between cars powered by biofuels and electricity, which compete for the same design space, something that may have been solved by the regulation and development of hybrid vehicles. This kind of institutional innovation proposes solutions to the dilemmas that are generated by innovations. Understanding the transition as a process of creative destruction (in which new technologies can replace old ones), it is possible to bring together and agglutinate actors in interest groups; for instance, food waste can be reduced via mechanisms of Community Supported Agriculture (CSA), in which direct sales from producers to consumers leads to a fall in waste and losses; the diffusion of CSA mechanisms is able to change the dynamics of food commercialization, with a shorter production and consumption chain. • **Pay attention to scientific communication methods and its links with what may or may not be perceived as the truth.** In addition to political-economic approaches, socio-cultural approaches can help build the resilience levels of public belief. Building resilient communities must be a priority in the field of sustainable development communication (Servaesa & Lieb 2014). The	considering the poverty threshold, it is necessary to recognise the need for socioeconomic improvements before demanding ecological improvements. From a general point of view, social technologies are those that apply to social inclusion: they must be adapted to small sized enterprises (physical and financial) and not tied to employer-employee relations; in summary, self-employed and small business enterprises, such as cooperatives, small farm holders common initiatives and small producers' networks, must be able to carry out economically viable ventures.	mainstay of human ecology for some time (Berkes et al. 2000). According to Esposti (2005), the Agricultural Knowledge and Innovation Systems (AKIS), in an evolutionary process, is represented as: "the set of agricultural organisations and/or persons, and the links and interactions between them, engaged in the generation, transformation, transmission, storage, retrieval, integration, diffusion and utilisation of knowledge and information, with the purpose of working synergistically to support decision making, problem solving and innovation in agriculture". These systems can provide solutions for the adaptive response to new environmental, social, or economic conditions in the transition to sustainability. This is particularly so as they are able to maintain genetic stocks of cultivars or varieties of livestock species, given the dynamic of a sustainable exploitation of the *Commons*. Regenerative agroforestry is the most illustrative example of a complex diversified system based on traditional knowledge. It has the potential to increase livestock productivity (by at least three times per area), in addition to being able to totally neutralize the carbon balance of final products. There are several other intermediate forms, such as crop-livestock forest integration, consortia of different species and rotation of crops in the same area, and, of course, agroecological food production. Public policies and R&D of this type of initiative can be a powerful starting point for the sustainable transition. • **Perform policy-making for environmental sustainability – such as natural biomass and biodiversity preservation**

Annex 7.1 continued

Institutional and Development Analysis Dimensions/Factors	Attributes of Community	Rules in Use	Physical and Material Conditions
	main challenges in communication (World Communication for Development Congress, 2006) also concern the transition to sustainability, as they presuppose scientific dissemination and good governance practices, as follows: • Good governance and accountability, • The challenging character and complexity of the communication process; • The different levels at which participation processes can occur - decision-making, benefits, evaluation and implementation; • To strengthen independent and pluralistic media to promote good governance and transparency; • To make the most of radio's potential, which in some regions may be the most effective tool (despite the multitude of medias); • To guarantee permanent adjustments and adequate policies and resources for communication processes; • A legal and supportive framework implemented by national governments that favours the right to freedom of expression. And the emergence of free and pluralistic information; • Provision of mechanisms that give a voice to the poorest, so that they are engaged in politics and decision-making on sustainable development.		**and their sustainable use, combined with social and economic development.** Promoting policies on the preservation and sustainable use of resources is a central point to the transition to sustainability. Richly biodiverse economic activities must be regulated – either to protect the genetic patrimony associated with biodiversity or because it is a water-producing area - to guarantee the livelihood of local residents, without compromising the renewal of the critical resource. Without minimum conditions for survival of the poor, or profit perspective for agribusiness, it becomes almost unviable for local governments to promote the preservation of all these areas. There are several possibilities for sustainable systems that combine obtaining income with preserving biodiversity, including: ecotourism in areas of more restricted use, regenerative agroforestry systems in degraded areas, planned extraction in preserved areas, crop-livestock-forest integration to reduce the impact of livestock on the environment, among others. In addition to the development of the legal framework on the delimitation, use and management of priority preservation areas, it is necessary to develop public policies to encourage innovations, financing, technical assistance, education and scientific dissemination in order to increase awareness of the possibilities and advantages of the transition toward sustainability.

Annex 7.1 continued

Institutional and Development Analysis Dimensions/Factors	Attributes of Community	Rules in Use	Physical and Material Conditions
. Object: of the transition goals.	• **Adopt a transition-oriented long-term perspective (Ostrom, 2011).** According to Ostrom (2011), it is necessary to adopt a long-term perspective to transition. Additionally, the transition correlated with socio-ecological systems (SES) is highly complex, since such systems are complex by nature, meaning conventional scientific approaches are insufficient to understand them and guide decision-making processes (Halliday and Glaser, 2011). Martínez-Fernández et al. (2021) mention that the integration is a keyword in all these approaches, regarding the transition from complex systems, both about what to integrate as well as who is involved in such integration, consequently, integrating actors who create knowledge with decision makers, through the deliberative participation of stakeholders and social actors in the decision-making process using the knowledge available (Felicetti et al., 2015). • **Monitoring the evolution of the transition to sustainability concept, concerning policy and political dimensions (Kemp and Rotmans, 2005; Kemp et al., 2007; Geels, 2011).** Monitoring the transition toward sustainable change has, in addition to the technological and economic aspects, a political dimension. Political aspects – such as how they related to public perception and opinion about new technologies, institutional innovations (such as the WEF Nexus) that provide sustainable public policies, and even those related to the pattern of negotiation between political agents - need to be monitored by SMOIs, in order to maintain a collective shared perception of the transition challenge. The basis for a widespread public conception of sustainability demands depends on innovative and technological actions – such as the adoption of green markets. It is however a particularly political decision, both in the sense of defining	• **Strong performance on sustainable policy prescriptions (North, 1990; Ostrom, 2005), so that the rules for the desired transition remain transparent.** Technological and economic demands for sustainability transition involves new collective agreements, rules and the implementation of commitment processes. The purpose of the rules is to define the way that the game is played (North, 1990), and transition demands many new games. These processes must be monitored and evaluated in terms of the strength of the performance of these new rules in promoting change. Some of **the "objects" of this agenda matrix, such as** economic enforcement mechanisms including tax exemptions, can be monitored by extensive and recursive experiments. The monitoring of compliance with the rules and constant evaluation of their *ex-cursus* impact on the performance of the implementation of the transition processes enable the new rules themselves to become more effective. • **Define and implement rational planned transition goals (Nill and Kemp, 2009).** The radical change in the production and consumption of sustainable transition requires new, viable technologies so, rational planned efforts must be taken into account. The governance of change of this process **presupposes "direction" and "coordination"** of the interaction systems of the factors that lead the change. Technologies, regulatory aspects, user practices and behaviours, culture, markets, supplier networks, and infrastructures must change, which forms an innovation system. Innovation policies – in order to foster the diffusion of sustainable technologies, are crucial to the sustainable transition. Improvements in environmental efficiency present at least three levels of evolutionary character, from an incremental level to a new system design (Elzen et al., 2004): a) system optimization – as an incremental process, is not comparable with the level of change necessary to consider such a phenomenon as a transition; b) system redesign, a partial process of transformation, and c) system innovation. The focus of the transition theory, in this context, is the dynamics of change and its drivers (Elzen et al., 2004): technologies that go beyond mere input–output relations, taking into account trajectories and different characteristics of innovation; they are also able to describe circumstances under which established technologies might persist even when they are to some extent inferior to their new competitors (lock-in). **Six main approaches in** studies on transition policies (El Bilali, 2018) illustrate the multi-character of this issue: a) Technological Innovation System (TIS); b) Social Practice Approach (SPA); c) Strategic Niche Management (SNM); d) Multi-level Perspective (MLP); e) Transition	• **Follow sustainable governance indicators, data and reports such as national and international climate change investigation initiatives and their impacts on countries, sustainability efforts or counterfactual efforts (Schmidt-Traub et al., 2017).** The United Nation's (UN) Sustainable Development Goals (SDGs) adopted in 2015 chart out a universal, holistic set of objectives to help set the world on a path towards sustainable development, by addressing all three dimensions of economic development, social inclusion, and environmental sustainability to be achieved by 2030 (Sustainable Development Solutions Network, 2015). Reaching these goals will require transformations in every country, as well as close monitoring and measuring of progress through indicators (Schmidt-Traub et al., 2017). The UN Statistical Commission recommended an initial set of 230 global indicators to measure the SDGs achievements. The indicators are considered the backbone of monitoring progress towards the SDGs at local, national, regional, and global levels. An indicator framework should be strong enough to turn the SDGs and their targets into a management tool to help countries and the global community develop implementation strategies and allocate resources accordingly (Sustainable Development Solutions Network, 2015). However, some suggested indicators lack comprehensive, cross-country data and some even lack agreed upon statistical definitions (IAEG-SDGs, 2016). According to Schmidt-Traub et al. (2017), the indicators must be robust and capable of application in a broad range of countries,

Annex 7.1 continued

Institutional and Development Analysis Dimensions/Factors	Attributes of Community	Rules in Use	Physical and Material Conditions
	politics and in the political games that make the execution of these projects feasible; an illustrative case concerns the Amazon rainforest preservation policies in South America, from the 1990s to the 2010's, which due to intense governmental communication on the efforts to control forest burning and illegal land occupation, have been successful. This is known as the "eco-innovation" niche, that can foster public perception and address positive policies for environmental protection. • **Promote education, skills and communication for transition: from the linear economy of the "take-make-use-dispose" model of production, distribution and consumption (Andrews, 2015), as a planned obsolescence market rational (Coase, 1972; Waldman, 1993) to a regenerative approach of resource use (Ellen MacArthur Foundation, 2016).** According to Andrews (2015), the origins of the linear economy date back to the Industrial Revolution and the global economy developed around this model. Current social, economic and environmental factors – such as climate change - mean that the linear model is no longer sustainable, since it is based on the planned obsolescence of products and encourages waste. The Circular Economy, as a radical new model, is being advocated but as yet is not widely practiced. The transition to sustainability requires cultural change, which promotes a change in habits and behaviours. In some cases, organizational innovation is more important and necessary than product or production process innovation. One of the obstacles related to the education and scientific communication for the transition to sustainability is the notion of "infinite resources" associated with the linear economy. Although more laborious, decisions made in more regenerative governance	Management (TM) and f) Time Strategies (TS). • **Promote the rule of law from a sustainable future perspective, at the national and international levels, according to the United Nations Development Goals, upward and downward accountability in institutions, enhanced participation, freedoms and capabilities, in order to improve baselines in developing countries and poorer global populations (Gupta and Vegelin, 2016).** It can be said that the transition to sustainability, composed of ecological, social and economic aspects, presents a major challenge in terms of optimizing all three aspects so that present and future generations achieve desirable sustainable development (Gupta and Baud 2015), as highlighted in the introduction of this chapter: sustainability is a recent milestone of civilization, an urgent collective transition goal, since the issue of climate change must be dealt with. Aiming to create a better world by 2030, in 2015, world leaders agreed on 17 Global Goals (officially known as the Sustainable Development Goals or SDGs). The goals aim to create a better world by ending poverty, fighting inequality and addressing the urgency of climate change. The objectives have become a guide for governments, companies, civil society and the general public in building a better future for all.	followed by better and more frequent data collection and dissemination, particularly in underdeveloped countries, to allow for the reliable estimation of trends over time. The metrics used to improve and monitor indicators require collaboration between statistical offices, policymakers and academic communities, a sector that has dedicated great efforts to developing indicators used in monitoring the sustainability of complex systems. Dialogue could be promoted by international organizations aimed at producing handbooks on new measurements, such as the OECD Guidelines on Measuring Subjective Well-being (OECD, 2013).

Annex 7.1 continued

Institutional and Development Analysis Dimensions/Factors	Attributes of Community	Rules in Use	Physical and Material Conditions
	systems tend to be more socially and ecologically robust, since they foresee a more limited use of natural resources, such as water (Fracalanza, 2013).		

	Attributes of Community	Rules in Use	Physical and Material Conditions
Form: the structure and evolution of the mission-oriented institution	• **Promote the creation of transition-orientated institution networks, such as meso-level initiatives – from a global to country level.** Expanding actor networks and mechanisms to ensure lessons are mobile and accessible to other places and cause knowledge spillovers (Wieczorek et al., 2015; Dawes, 1980; Gifford, 2006) is crucial to achieving a structured transition system. Meso-level transition-oriented institutions have spread exponentially in recent years (see Section 1 of this chapter). The meso-level character of these institutions can be determined by their cross-national reach: they may be based in countries or in economic blocks, but their objectives have common and cross-border goals. As with other types of transition-oriented institutions, the meso-level ones have short-term and long-term goals and strategies. The challenges they face lie in establishing bridges between a large set of production, cultural and economic gaps embodied in the change process, such as eco-innovation and technological transition, consumption behaviour, and market barriers to sustainable industrial production. The Belmont Forum initiative, for instance, is one of these institutions. • **Carry out effective and permanent efforts on modelling complex systems: mapping the nexus among resource uses in order to reduce its dysfunctionalities (Bazilian et al., 2011).** The Nexus approach has gained increasing global attention, including scientific studies on calculations of flows and dependencies between different	• **Adopt a model to understand and follow the diffusion processes of clean or eco-innovation.** The diffusion of clean and sustainable technologies, as with the diffusion of any innovation, is governed by endogenous mechanisms, such as epidemic learning and learning economies and by exogenous mechanisms (Kemp and Volpi, 2008). The market's incumbent power, and the costly and uncertain processes linked to technological replacement are considered the main market barriers. These processes are governed by endogenous mechanisms, like epidemic learning and learning economies; and by exogenous mechanisms (Kemp and Volpi, 2008). To ensure sustainability is a widespread objective, institutions and public policies must hold common values geared towards transition in order to adopt mechanisms that allow communities to understand the reasons for and importance of adopting eco-innovations. The mobilization of endogenous mechanisms so that communities take ownership of these innovations diffuses positive values that encourage the advancement of transition. This is the case of agricultural and aquaponics production, which integrates two "old" technologies – production of fish and greens – and is innovative in character, since environmental impacts are reduced and families are able to obtain higher incomes. The diffusion of innovation depends on: a) the decision-making processes – from the micro-institutional level, such as entrepreneurship, to policy and economic incentives; it is possible to have a "niche" of interests, i.e., an exchange of technologies that has greater positive collective rationality and can be more easily diffused (as, for instance, in the case of the production of renewable energy from biomass); b) the planning of the exchange process, which is uncertain –again due to market barriers and the fact that it is an evolutionary and adaptative process – such as the impact on the diffusion of electric car technology by the oil and gas markets dominant technological trajectories; the adoption of complementary communication mechanisms for global warming consumption behaviours – such as strategies to strengthen the diffusion of clean technologies. • **Follow transnational coalitions and coordinated efforts through sustainability - in the form of**	• **Foster technological, social, environmental and organizational sustainable innovation solutions from an ecological economics perspective (Romeiro, 2012) and an eco-innovation perspective (Kemp, 2010; Rennings, 2000).** Considering ecological economics, the environment, as one of its subsystems, represents an absolute limit to the expansion of the economy. However, if, by definition, a subsystem cannot be larger than the system that contains it, its size in relation to the whole is not limited by the system. The subsystem's carrying capacity is defined by thresholds of ecosystem resilience. This is one of the fundamental premises of ecological economics, that is rooted in the work of Kenneth E. Boulding (Boulding, 1966)" (Romeiro, 2012). Overcoming this issue depends on solutions generated by science and the design of clean technologies, concerning both scalable production industries and socio-technical systems that promote a more rational (and resilient) use of common pool resources. This is the proposal of ecological economics, and transition theories based on eco-innovation efforts. Concerning eco-innovation perspectives of the diffusion of sustainable technologies, according to Kemp (2010), the diffusion of clean technology works in the same way as the diffusion of any other innovation. These processes are governed by endogenous mechanisms and are ruled by many of the

Annex 7.1 continued

Institutional and Development Analysis Dimensions/Factors	**Attributes of Community**	**Rules in Use**	**Physical and Material Conditions**
	resources; this approach can be performed via the complex systems modelling approach, assessments of technology and policy applications, in order to understand the sustainable gains and the necessary interlinkages to develop sustainable management possibilities (Zhang et al., 2018). This is due to the fact that a global society is required to solve a set of complex, interrelated problems characterized as a fundamental threat to human civilization (Bazilian et al., 2011). Closely follow the global processes of capacity-building to enable the effective participation of a range of stakeholders (Gupta and Vegelin, 2016).	**international agreements, such as pooling resources and coordinating state policies, that provide cooperative gains for sustainability global patterns (Böhmelt e Butkutė, 2018; Roggero et al., 2019); this procedure can guide the emergence and evolution of new SMOI's forms.** The implementation of sustainability measures presents a challenge to policymakers, while also dealing with parallel and sometimes conflicting policy agendas, long-term policy impacts, and contested interpretations of sustainability. These aspects show the difficulty in simultaneously making progress with goals belonging to different sustainability agendas (Bennich et al., 2021). However, global transition towards sustainable development has been one of the major objectives in recent years, which has led to national or regional development resulting in strategies, instruments and tools including the expansion of the bio-based economy (Zihare et al., 2021) and the implementation of Environmental-Social-Governance (ESG) concepts in companies.	traditional economic entry barriers, such as monopoly effects, mature technological trajectories, etc. Policy making, of course, is a central issue in technology diffusion: the characteristics of clean technology, absorptive capacities of potential adopters and the age structure of capital are important factors to take into account. According to Kemp (2010) "it is often overlooked that companies have a choice: they can choose between an end-of-pipe solution, a process change (adaptation) and a change of process (substitution). This means that the diffusion and evolution of one clean technology will be at the expense of the diffusion of another clean technology, something overlooked in studies on clean technology diffusion. Further research is needed on the influence of public policy on clean technology choice, expectations (about learning economies and prices), adjustment costs, network externalities and complementary innovations on clean technology adoption choices".

	Attributes of Community	**Rules in Use**	**Physical and Material Conditions**
Methods: the prime movers to promote transitions	**• Carry out educational initiatives with sustainability as the theme (Vare & Scott, 2007; McKeown, 2002) at local, regional, national and international levels.** Educational systems need to dovetail their sustainability efforts on sustainable development. Like most factors linked to the evolution of sustainable systems, education involves actions that must be	**• Remain up to date with public regulation and self-regulatory arrangements, through certifications, codes of conduct and standards (Fuchs, 2007; Benites-Lazaro et al., 2018).** This factor is possibly the most complicated and intangible planned method for the evolution of sustainable systems. It involves individual and collective behaviours concerning the consumption of goods and services from different technological trajectories – some more sustainable under one set of criteria, some under another set of criteria. Choosing the most sustainable type of energy implies selecting bioethanol fuels or wind energy, which requires a complex capital and resource-based industry organization. The regulatory efforts of the market on each of the energy matrices, as well as each of these industries. Many authors concerned with the common-pool resources problem (Ostrom 1998) argue that it is neither government nor	**• Decision-making methods based on environmental monitoring techniques, such as satellite and earth monitoring systems for climate change and scientific production analysis.** This item is related to the WEF Nexus conceptual tool, since it involves different techniques used to monitor climate change. Sustainability must be analysed using a holistic and integrated approach that facilitates the analysis of the social, economic, institutional and environmental factors and

Annex 7.1 continued

Institutional and Development Analysis Dimensions/Factors	Attributes of Community	Rules in Use	Physical and Material Conditions
	carried out at different and integrated levels, from local, regional, national, international and global. When it comes to policies and theory in education, any approach can be seen as reductionist. This issue involves a challenge typical of education itself: to understand what we are communicating and teaching, how we are doing it and, most importantly, why we are doing it. Thus, the only assumption that can be made is that the science and educational initiatives present an extra challenge when it comes to the transition to sustainability. For a comprehensive understanding of this challenge, it is possible to consider **McKeown's (2002)** broad analysis about the adoption of new concepts, theories and assumptions necessary to develop educational initiatives related to sustainable development.	markets, but communities and stakeholders themselves, that are capable of managing environmental public goods with a set of rules and regulations framed for that purpose. The self-regulation issue involves participation and empowerment of the involved partners, or stakeholders, in any programme designed to facilitate a sustainable future, not just for one generation, but for all future generations. Its human geographical dimension is local, regional and global. This approach reveals that the self-regulation processes can be more efficient under collective commitment processes. It is necessary also to consider the role of the effectiveness of enforcement, governance and legal compliance mechanisms implemented by the States, such as environmental fines, legislation on land use and forest preservation. • **Creation of markets to upscale new green technologies (Sengers, 2019).** Until recently, climate change was mostly viewed as a reputational risk that could be addressed through the environmental, social and governance (ESG) agenda; it now seems that this perception is found more in the financial field (despite the perception of reputational risks also continuing to be valid). Many firms have attempted to integrate financial risks with sustainability management frameworks. Aiming to supply the market demand to upscale new green technologies, recent initiatives such as the 2019 recommendations from the Network of Central Banks and Supervisors for Greening the Financial System (NGFS) and the 2018 EU Action Plan, that promote the inclusion of climate risks into financial institutions, appear as risk management policies and the potential integration of climate risk into prudential regulations and capital requirements (Gaganis et al., 2021). The growing number of ecosystem services that are reproducing the market logic to tackle environmental problems has contributed to attracting political support to tackle environmental problems (Gómez-Baggethun et al., 2010). • **Implementation of tax exemptions to create economic incentives to move towards more sustainable production (Madsen and Hansen, 2019).** Strategies and policies aimed at the transition to sustainable systems and based on economic mechanisms are under development around the world. Aiming at achieving sustainability, the actions are focused on several areas such as clean energy, sustainable industry, construction and renovation, sustainable mobility, biodiversity, elimination of pollution, climate actions and food security (CEU, 2021). Economic incentives and enforcement mechanisms to foster clean production and consumption systems can be considered as both strategies and methods to promote transition, in the context of market mechanisms related to limit a firm's pollution levels. Economic instruments to foster the sustainability process range from government	their interactions, that characterize complex socio-ecological systems (SES) (Martínez-Fernández et al., 2021). The WEF Nexus is believed to present this character. Several technical tools can be used to analyse the sustainability of environmental and social sectors considering a complex system approach: software and methods for System Dynamics analysis and Agent-Based Modelling, validated models for simulations of environmental impacts (soil, water, air, biodiversity), Data Envelopment Analysis (DEA) method, Environmental life-cycle assessment (LCA), digital technologies - artificial intelligence (AI), big data analytics, mobile technologies, Internet of Things - and social platforms generate positive improvements for society and industry assessment. Of course, different methodological approaches and techniques are welcome, since science is the new knowledge generation that must be fed by recursive cycles of experiments; a weakness, however, in the integrated use of tools is identified (Belucio et al., 2021; Feroz et al., 2021; Tsaples and Papathanasiou, 2021). Study groups have been developing multicriteria decision analysis (MCDA) methods, in an attempt to supply this demand. These tools have been used to form a broader view of decision-making results by creating consolidated results between different methods (Zihare et al., 2021); this approach is capable of adding political and social dimensions to technical factors, making monitoring systems more extensive and useful for defining comparative climate change scenarios. • **Use existing tools and methods to create new ones with technological, social and organizational**

Annex 7.1 continued

Institutional and Development Analysis Dimensions/Factors	Attributes of Community	Rules in Use	Physical and Material Conditions
		policies to financial innovations derived by markets – such as cap and trade on greenhouse gas emissions and carbon credit, driving limits on pollution or buying and selling allowances to emit certain amounts of greenhouse gases. In the case of tax exemption mechanisms, the need for many comparative experiments appears necessary, in order to understand different scenarios of the welfare gain of each mechanism, depending on the kind of market or sector it is related to. The trade-off between capital allocation – from fuels to wind technology markets, for instance, and the tax exemption impacts on greenhouse emissions must be better understood. In the case of this energy matrix allocation case, tax exemptions contribute to lower greenhouse gas (GHG) emissions without distorting consumer choices, since consumers have a low individual capacity to choose which kind of energy to use. Unlike the case of electric car mechanisms replacing fuel cars, the taxing effects in the case of close substitutes - with different rates - in relation to the generation of substantial welfare costs is a useful example of this trade off. According to Hatta and Haltiwanger, (1986), two strong substitutes have different ad valorem tax rates, so a consumer can avoid substantial tax payments by substituting the good with the lower rate for the other. Thus, tax wedges are important differential aspects to be taken into account, since they can distort consumers' choices between two substitute goods. Some European experiments on the effects of economic incentives for pollution reduction on urban climate change suggest that projects are influenced by external circumstances, particularly national legislation. This seems to point to a strong relation between the rules in use and the power of law dimension on environmental control. Some of these experiments have proved that tax exemptions are effective as economic incentives aimed at a transition to sustainable systems (Madsen and Hansen, 2019). National taxation on fuel influences the infrastructural developments carried out by local suppliers, biomass-based production, for instance (Danish Energy Agency, 2015). However, the politically unstable environment can contribute negatively to large infrastructural investments, such as when the Danish government announced intentions to reopen the energy agreement and withdraw economic resources from the wind industry (Berlingske Business, 2016). These experiments are necessary, as this short discussion shows, to understand the extent to which tax exemptions or other sustainable economic incentives are value-added mechanisms for sustainability transitions. • **Deregulation of high negative environmental impact utilities industries, exposing this sector to new pressures, leading to the emergence of more sustainable alternative technologies, more sustainable organizational and innovative-resource modes**	**foresight, in order to understand the evolution of the transition of complex systems through sustainability (Martin, 1995).** According to Martin (2010), "Foresight is the process involved in systematically attempting to look into the long-term future of science, technology, economy and society with the aim of identifying the areas of strategic research and the emerging generic technologies likely to yield the greatest economic and social benefits". This concept strongly justifies the use of foresight methods in the case of the sustainable transition. Complementary, foresight is not foreseeing future, but creating it, which fits with the climate change challenge. A comprehensive analysis of foresight methods can be found in Magruk (2011), such as multi-criteria and decision analysis, cross analysis or bibliometric approaches. These studies and experimental processes, evaluation and social learning, built upon stakeholder engagement and participation are central to the design of a sustainable future. They are a set of tools that can be used to understand the agent's perception of a desired future, prompting the path to attain it. The climate change challenge requires comprehending future scenarios of sustainable life: it demands transformations in the socio-technological structure of human society to address these challenges. The diffusion of sustainable innovation is a knowledge area for future studies. It demands ex ante planning and the engagement of relevant stakeholders, shareholders and policy makers. Interactions among these actors must create the desired collective transition in the future.

Annex 7.1 continued

Institutional and Development Analysis Dimensions/Factors	Attributes of Community	Rules in Use	Physical and Material Conditions
		of production and uses of natural resources (Berkhout et al., 2009). The traditional "fines and penalties" enforcement strategies are the most highly diffused pollution control mechanisms by governmental institutions around the world and that are used to achieve sustainability in the industrial sector. Although this mechanism induces significant reductions in the pollution levels of the different environmental compartments (Magat and Viscusi, 1990; Laplante and Rilstone,1996; Dasgupta et al., 2001), studies highlight the need for the integration of different mechanisms to improve the environmental performance of polluters. Foulon et al. (2002) suggest that the complementation or supplementation of traditional enforcement actions with structured information programs (or public disclosure programs), by which the environmental performance of polluters is revealed, could urge industries to undertake environmental protection actions. The most efficient action to address this problem may be the emergence of new, more resource-efficient socio-technical systems as the basis of more sustainable development pathways (Berkout et al.,2009); this means the application of an emerging 'systems innovation'. These systems address the interaction between domestic and globalized markets, knowledge flows and governance. Both technological and organizational innovation challenges must be taken into account: not only production or distribution and logistics innovation, but new organizational patterns for markets must be redesigned. The most shocking aspect is that of food loss and waste in sustainable systems. New agronomic technologies must compose the same complex system as the productive organizations, the supply chains, food and nutritional policies and organizational aspects of the WEF Nexus. The negative impacts of food loss due to productive dysfunctionalities can be added to those derived from the failed governance of food demands and prices, contracts or transaction costs. Food loss on small farms negatively impacts environmental indicators, since water, energy and supplies are wasted if the commercialization of food is not properly managed. This case reveals the importance of socio-technical regimes and socio-technical landscapes (Kemp and Rotmans (2001), that are situated in broad political, economic and institutional contexts. Landscapes can be considered selection environments for regimes and niches related to emerging clean and environmentally positive technologies. One example is the deregulation of utility industries in some countries, exposing electricity supply technologies to new challenges, resulting in the possibility of alternative technologies emerging rapidly from niches in which they had previously been operating (Joskow, 1998). • **Closely monitor "operational rules", the regulatory dimension of the management methods.**	

Annex 7.1 continued

Institutional and Development Analysis Dimensions/Factors	Attributes of Community	Rules in Use	Physical and Material Conditions
		Agents are the main promoters of sustainable transition, and usually operate on a system of participatory governance, with collective decisions and empowerment of stakeholders (Kemp et al., 1998). At the same time, transition causes huge changes in social values and the agent's roles become central to the co-creation and development of innovations that will drive these changes. The main question becomes: how can agents -who coordinate the current economy – carry out processes whose rules are based on a new sustainable rationale? Dealing with these new rules in iterative processes has a direct impact on the participants in decision-making processes and other stakeholders, and thus generates new conjunctures of agents. This process affects participants both in terms of the constraints they must accept (the sustainable rules of economic games) and in empowerment gains – so they are motivated to remain committed to change. • **Test and validate instruments related to understanding the transition process, since goals to end poverty and build shared prosperity rely on sustainability (World Bank Agenda, 2030; Gupta and Vegelin, 2016).** The concept of *social technology* assumes that not all technologies can be used in any context, such as those based on scale or capital-intensive production. Special tailored technological development platforms must be developed, often to facilitate the small farmers' activity. Thus, developing an environmentally advantageous technology that, at the same time, can contribute to poverty reduction, must also consider the user's access and its economic viability for rural (including but not restricted to small producers) producers. Otherwise, it will not help solve environmental problems. Social technologies, in this case, are those that can support family farming, even with traditional technologies for plant resistance. Resistant plant varieties, when available, have to be produced and distributed to family land holders. This implies that the technological issue becomes a public policy in agriculture and regional sustainable development.	

	Attributes of Community	Rules in Use	Physical and Material Conditions
Logic: actor management for sustainable policy processes.	• **Analyse the multiple narratives, epistemes, and paradigms of sustainability from the political perspective: governance, rational choice, agency and institutionalism approaches (Bevir and Rhodes, 2001; Thaler, 2000; Etzoni, 1996; Ostrom, 2000).** In order to monitor the adoption of a new process of sustainable change, considering the most optimal evolutionary framework, it is	• **Analyse collective governance processes (Ostrom, 1990).** The transition to sustainability has largely been considered legitimate by society. Collective decision-making, however, and the integration of agents and their demands – which will make the transition real - depend on clear, inclusive and fair	• **Foster regulatory, legal and tax advantages for sustainable production and environmental services (Gómez-Baggethun et al., 2010).** Environmental issues have no boundaries nor strata; at the global level, regulation takes place through international agreements with future goals – such as the Kyoto Protocol and the Paris Agreement. The compliance logic involves long-term cycles of countries' negotiations and trade enforcement tools that are strongly shaped

Annex 7.1 continued

Institutional and Development Analysis Dimensions/Factors	Attributes of Community	Rules in Use	Physical and Material Conditions
	necessary to understand the different concepts of sustainability and how they are opposed in the political spectrum. Different views of sustainability (such as conflicting market interests) form pressure groups that sometimes compete and sometimes cooperate in political and social dilemmas. For example, some advocate replacing gasoline with bioethanol, others with electricity. Multiple narratives must be contemplated, understood, and implemented – despite the internal conflicts that can arise - as they represent the views of different actors who are involved in disputes and cooperation during the transition process. Different logics arise from the different economic, political, technological and social demands of the community that generate other demands for sustainable innovations. The implementation of a project that involves several stakeholders needs to be focused on shared objectives – in this case, the pursuit of sustainability. For instance, the substitution of traditional plastics for renewable alternatives can be carried out through cooperative research, involving governance between two originally conflicting epistemes. The development of projects carried out by heterogeneous groups depends on finding and even spreading common values to be pursued, such as sustainability. For stakeholders in the traditional plastics industry, the new technological trajectory can be perceived as increased market share and therefore present a competitive advantage; while for environmentalists, the potential to benefit from ecological economics is the motive for a partnership.	governance rules, or institutions. At the same time, these rules must be adopted by multiple groups and innovations that are most efficient in disseminating and improving the degree of sustainability of technological systems and their political, social and economic results. That is, achieving the sustainability tripod.	by political contexts. Within the scope of national legislations, regulatory-oriented, command-controlled instruments (standards, environmental impact and evaluation studies, environmental licensing for exploration and zoning) are defined as those that establish the norms, rules, procedures, and standards determined for economic activities in order to ensure that the objectives of the sustainability policy are complied with. These rules and standards are, at the micro-institutional level, national in nature, since they make up the sovereign legislations of the countries. They can include tax incentives or payments for ecological services, or as fines, for example. Environmental defence mechanisms – such as the granting of water resources and reporting negative environmental impacts – also make up these regulatory systems. In the case of national regulation mechanisms, the enforcement devices must generate socially and environmentally desirable behaviour; they are subject to transaction costs involved in the application of sanctions such as the collection of fines and permanent inspection of environmental damages. • **Include "social technology" and "solidary economy" concepts (Dagnino, 2012; Meira, 2014, Rodrigues and Barbieri, 2008) as channels to strengthen small user-oriented innovations.** The concept of social technology has been used by several authors (Dagnino, 2012; Dagnino et. al., 2004; Meira, 2014, Rodrigues and Barbieri, 2008) to describe the technological platform of solidarity economy ventures. Social technology presupposes the development of technology in close relationship with the end user and not only in the top-down model. In agriculture, social technology is closely related to the acquisition of extra economic gains due to certified organic production, such as in enterprises in the solidarity economy model in agrarian reform settlements (Benini, et. al. 2012).

References

Andrews, D. (2015). The circular economy, design thinking and education for sustainability. *Local Economy, 30*(3), 305–315.

Arora, A., & Alfonso, G. (1990). Evaluating technological information and utilizing it: Scientific knowledge, technological capability, and external linkages in biotechnology. *Journal of Economic Behavior and Organization, 24*, 91–114.

Arora, A., & Alfonso, G. (1994). Complementarity and external linkages: The strategies of large firms in biotechnology. *The Journal of Industrial Economics, 38*, 361–379.

Aza, A., Riccioli, F., & Di Iacovo, F. (2021). Optimising payment for environmental services schemes by integrating strategies: The case of the Atlantic Forest, Brazil. *Forest Policy and Economics, 125*, 102410.

Bazilian, M., Hogner, H., Howells, M., et al. (2011). Considering the energy, water and food nexus: Towards an integrated modelling approach. *Energy Policy, 39*(12), 7896–7906.

Belucio, M., Rodrigues, C., Antunes, C. H., et al. (2021). Eco-efficiency in early design decisions: A multimethodology approach. *Journal of Cleaner Production, 283*, 124630.

Benini, É. A., Faria, M. S. D., Novaes, H. T., et al. (2012). *Gestão pública e sociedade: Fundamentos e políticas públicas de Economia Solidária*. Outras Expressões.

Benites-Lazaro, L. L., Giatti, L., & Giarolla, A. (2018). Sustainability and governance of sugarcane ethanol companies in Brazil: Topic modeling analysis of CSR reporting. *Journal of Cleaner Production, 197*, 583–591.

Bennich, T., Belyazid, S., Stjernquist, I., et al. (2021). The bio-based economy, 2030 Agenda, and strong sustainability – A regional-scale assessment of sustainability goal interactions. *Journal of Cleaner Production, 283*, 125174.

Berkes, F., Colding, J., & Folke, C. (2000). Rediscovery of traditional ecological knowledge as adaptive management. *Ecological Applications, 10*, 1251–1262.

Berkhout, F., Angel, D., & Wieczorek, A. J. (2009). Asian development and sustainable sociotechnical regimes. *Technological Forecasting and Social Change, 76*(2), 218–228.

Berlingske Business. (2016). *Politisk uro skræmmer konsortie fra vindmølleprojekt*. http://www.business.dk/energi/politisk-uro-skraemmer-konsortie-fra-vindmoelleprojekt. Accessed 27 May 2021.

Bevir, M., & Rhodes, R. A. W. (2001, March 9). *A decentered theory of governance: Rational choice, institutionalism and interpretation*. Paper presented to a colloquium at the Institute of Governmental Studies, University of California, Berkeley, CA. Retrieved February 13, 2003, from http://www.igs.berkeley.edu:8880/events/governance/colloqMark.html

Bisaro, A., & Hinkel, J. (2016). Governance of social dilemmas in climate change adaptation. *Nature Climate Change, 6*(4), 354–359.

Blühdorn, I. (2011). The politics of unsustainability: COP15, post-ecologism, and the ecological paradox. *Organization and Environment, 24*(1), 34–53.

Blühdorn, I. (2013). The governance of unsustainability: Ecology and democracy after the post-democratic turn. *Environmental Politics, 22*(1), 16–36.

Böhmelt e Butkutė. (2018). The self-selection of democracies into treaty design: insights from international environmental agreements. International Environmental Agreements: Politics, Law and Economics, Springer, vol. 18(3), pages 351-367.

Borjesson, K. (2009). Affective Sustainability. Is this what timelessness really means? In *Undisciplined!* Paper presented at the Design Research Society Conference 2008, Sheffield Hallam University, Sheffield, 2009.

Boulding, K. E. (1966). The economics of the coming spaceship earth. In H. Janet (Ed.), *Environmental quality in a growing economy*. Johns Hopkins University Press.

Cairns, R., & Krzywoszynska, A. (2016). Anatomy of a buzzword: The emergence of 'the water-energy-food nexus' in UK natural resource debates. *Environmental Science & Policy, 64*, 164–170.

CEU. (2021). *Council of the European Union*. https://ec.europa.eu/info/energy-climate-change-environment_pt. Accessed 3 May 2021.

Coase, R. H. (1972). Durability and monopoly. *Journal of Law and Economics, 1*(1), 143–149.

Cohen, W., & Levinthal, D. (1989). Innovation and learning: The two faces of R & D. *The Econometrics Journal, 99*, 569–596.

Cohen, W., & Levinthal, D. (1990). Absorptive capacity: A new perspective on learning and innovation. *Administrative Science Quarterly, 35*, 128–152.

Coutard, O., & Rutherford, J. (2010). Energy transition and city–region planning: Understanding the spatial politics of systemic change. *Technology Analysis & Strategic Management, 22*(6), 711–727.

Dagnino, R. (2012). Why science and technology capacity building for social development? *Science and Public Policy, 39*(5), 548–556.

Dagnino, R., Brandao, F. C., & Novaes, H. T. (2004). Sobre o marco analítico-conceitual da tecnologia social. In *Tecnologia social: Uma estratégia para o desenvolvimento* (pp. 65–81). Fundação Banco do Brasil.

Dal Poz, M. E. S. (2006). *Redes de inovação em biotecnologia: Genômica e direitos de propriedade industrial*. Thesis, University of Campinas

Danish Energy Agency. (2015). *Regulation and planning of district heating in Denmark*. Danish Energy Agency.

Dasgupta, S., Laplante, B., Mamingi, N., et al. (2001). Inspections, pollution prices, and environmental performance: Evidence from China. *Ecological Economics, 36*, 487–498.

Davis, L. E., & North, D. C. (1971). *Institutional change and American economic growth*. Cambridge University Press.

Dawes, R. M. (1980). Social Dilemmas. *Annual Review of Psychology, 31*, 169–193.

Dietz, T., Dolšak, N., Ostrom, E., et al. (2002). The drama of the *commons*. In E. Ostrom, T. Dietz, N. Dolšak, et al. (Eds.), *Committee on the human dimensions of global change, division of behavioral and social sciences and education, the drama of the Commons* (pp. 3–35). National Academy Press.

Dosi, G., & Orsenigo, L. (1988). Industrial structure and technical change. In A. Heertje (Ed.), *Innovation, Technology and Finance*. Basil Press.

El Bilali, H. (2020). Transition heuristic frameworks in research on agro-food sustainability transitions. *Environment, Development and Sustainability, 22*, 1693–1728.

Ellen MacArthur Foundation (2016) - Towards the circular economy - Economic and business rationale for the accelerated transition. Ellen MacArthur Foundation. Available via https://ellenmacarthurfoundation.org/towards-thecircular-economy-vol-1-an-economic-and-business-rationale-for-an. Accessed 27 Mar 2021.

Elzen, B., Geels, F. W., & Green, K. (2005). *System innovation and the transition to sustainability: Theory, evidence and policy*. Edward Elgar, University of Manchester.

Esposti, R. (2012). Knowledge, technology and innovations for a bio-based economy: Lessons from the past, challenges for the future. *Bio-based and Applied Economics, 1*(3), 231–264.

Etzioni, A. (1996). *The new golden rule: Community and morality in a democratic society*. Basic Books.

Felicetti, A., Niemeyer, S., & Curato, N. (2015). Improving deliberative participation: Connecting mini-public to deliberative systems. *European Political Science Review, 8*(3), 427–448.

Feroz, A. K., Zo, H., & Chiravuri, A. (2021). Digital transformation and environmental sustainability: A review and research Agenda. *Sustainability, 13*, 1530.

Ferrari, V., Silveira, J. M. F. J., & Dal Poz, M. E. (2019). Patent network analysis in agriculture: A case study of the development and protection of biotechnologies. *Economics of Innovation and New Technology, 30*(2), 111–133.

Foulon, J., Lanoie, P., & Laplante, B. (2002). Incentives for pollution control: Regulation or information? *Journal of Environmental Economics and Management, 44*, 169–187.

Freeman, C. (1987). *Technology and economic performance: Lessons from Japan*. Pinter.

Fuchs, D. (2007). *Business power in global governance*. Lynne Rienner Publishers.

Gaganis, C., Pasiouras, F., Tasiou, M., & Zopounidis, C. (2021). CISEF: A composite index of social, environmental and financial performance. *European Journal of Operational Research, 291*, 394–409.

Gargia-Parpet, M. F. (2013). A sociologia da economia de Pierre Bourdieu. *Sociologia & Antropologia, 3*(5), 91–117.

Geels, F. W. (2011). The multi-level perspective on sustainability transitions: Responses to seven criticisms. *Environmental Innovation and Societal Transitions, 1*, 24–40.

Geels, F. W., & Schot, J. (2008). Strategic niche management and sustainable innovation journeys: Theory, findings, research agenda, and policy. *Technology Analysis & Strategic Management, 20*(5), 537–554.

Geissdoerfer, M., Paulo, S., Bocken, N. M. P., et al. (2017). The circular economy – A new sustainability paradigm? *Journal of Cleaner Production, 143*, 757–768.

Giddens, A. (1984). *The constitution of society. Outline of the theory of structuration*. Polity.

Gifford, R. A. (2006). General model of social Dilemmas. *International Journal of Ecological Economics and Statistics, 4*(6).

Gómez-Baggethun, E., de Groot, R., Lomas, P. L., et al. (2010). The history of ecosystem services in economic theory and practice: From early notions to markets and payment schemes. *Ecological Economics, 69*(6), 1209–1218.

Greif, A., & Mokyr, J. (2017). Cognitive rules, institutions, and economic growth: Douglass North and beyond. *Journal of Institutional Economics, 13*(1), 25–52.

Gupta, J., & Baud, I. S. A. (2015). Sustainable development. In P. Pattberg & F. Zelli (Eds.), *Encyclopedia of global environmental politics and Governance* (pp. 61–72). Edward Elgar.

Gupta, J., & Vegelin, C. (2016). Sustainable development goals and inclusive development. *International Environmental Agreements: Politics, Law and Economics, 16*(3), 433–448.

Halliday, A., & Glaser, M. (2011). A management perspective on social ecological systems: A generic system model and its application to a case study from Peru. *Human Ecology Review, 18*(1), 1–18.

Hardin, G. (1968). The tragedy of the *commons*. *Science, 162*, 1243–1248.

Hatta, T. & Haltiwanger, J. (1986). Tax Reform and Strong Substitutes. International Economic Review, 27(2) 303-315.

Hodgson, G. M. (2017). Institutional economics. In L. Fisher, J. Hasell, P. Christopher, et al. (Eds.), *Rethinking economics – An introduction to Pluralist economics*. Francis & Taylor.

IAEG-SDGs (2016) *Inter-agency and expert group on sustainable development goal indicators. Update on the work to finalize the proposals for the global indicators for the sustainable development goals* (pp. 1–7).

Joskow, P. L. (1998). Restructuring, competition and regulatory reform in the U.S. electricity sector. In H. P. Chao (Ed.), *Designing competitive electricity markets* (pp. 11–33). Kluwer.

Katz, M. L., & Shapiro, C. (1994). Systems competition and network effects. *The Journal of Economic Perspectives, 8*(2), 93–115.

Keith, M., & Headlam, N. (2017). Comparative international exemplar urban and living labs. In *The urban living global challenge: A prospectus*. Urban

Transformations. Centre on Migration, Policy and Society (COMPAS), University of Oxford.

Kemp, R. (2010). Eco-innovation: Definition, measurement and open research issues. *Economia Política, 27*(3), 397–420.

Kemp, R., Rotmans, J. (2001). *The management of the co-evolution of technical, environmental and social systems*. Paper presented at the international conference towards environmental innovation systems, Garmisch-Partenkirchen.

Kemp, R., & Rotmans, J. (2005). The management of the co-evolution of technical, environmental and social systems. In M. Weber & J. Hemmelskamp (Eds.), *Towards environmental innovation systems* (pp. 33–55). Springer.

Kemp, R., & Rotmans, J. (2009). Transitioning policy: Co-production of a new strategic framework for energy innovation policy in the Netherlands. *Policy Sciences, 42*, 303–322.

Kemp, R., & Volpi, M. (2008). The diffusion of clean technologies: A review with suggestions for future diffusion analysis. *Journal of Cleaner Production, 16*(1), 14–21.

Kemp, R., Shot, J., & Hoogma, R. (1998). Regime shifts to sustainability through processes of niche formation: The approach of strategic niche management. *Technology Analysis & Strategic Management, 10*(2), 175–198.

Kemp, R., Loorbach, D., & Rotmans, J. (2007). Transition management as a model for managing processes of co-evolution towards sustainable development. *International Journal of Sustainable Development & World Ecology, 14*, 78–91.

Laplante, B., & Rilstone, P. (1996). Environmental inspections and emissions of the pulp and paper industry in Quebec. *Journal of Environmental Economics and Management, 31*, 19–36.

Lara, A. (2015). Rationality and complexity in the work of Elinor Ostrom. *International Journal of the Commons, 9*(2), 573–594.

Lerhmann, V., Frangioni, M., & Dubé, P. (2015). Living Lab as knowledge system: An actual approach for managing urban service projects? *Journal of Knowledge Management, 19*(5), 1087–1107.

Lindblom, C. E. (1979). Still muddling, not yet through. *Public Administration Review, 39*, 517–526.

Lliso, B., Pascual, U., & Engel, S. (2021). On the role of social equity in payments for ecosystem services in Latin America: A practitioner perspective. *Ecological Economics, 182*, 106928.

Lundvall, B. A. (1992). *National systems of innovation: Toward a theory of innovation and interactive learning*. Pinter.

Madsen, S. H. J., & Hansen, T. (2019). Cities and climate change – Examining advantages and challenges of urban climate change experiments. *European Planning Studies, 27*(2), 282–299.

Magat, W. A., & Viscusi, W. K. (1990). Effectiveness of the EPA's regulatory enforcement: The case of industrial effluent standards. *Journal of Law and Economics, 33*, 331–360.

Magruk, A. (2011). Innovative classification of technology foresight methods. *Technological and Economic Development of Economy, 17*(4), 700–715.

Märker, C., Venghaus, S., & Hake, J. F. (2018). Integrated governance for the food–energy–water nexus – The scope of action for institutional change. *Renewable and Sustainable Energy Reviews, 97*, 290–300.

Martin, B. R. (1995). Foresight in science and technology. *Technology Analysis & Strategic Management, 7*(2), 139–168.

Martin, B. R. (2010). The origins of the concept of 'foresight' in science and technology: An insider's perspective. *Technological Forecasting and Social Change, 77*, 1438–1447.

Martínez-Fernández, J., Banos-González, I., & Esteve-Selma, M. A. (2021). An integral approach to address socio-ecological systems sustainability and their uncertainties. *Science of the Total Environment, 762*, 144457.

Matthews, R., & Sydneysmith, R. (2010). Adaptive capacity as a dynamic institutional process: Conceptual perspectives and their application. In *Adaptive capacity and environmental governance* (p. 307). Springer.

McCormick, K., Anderberg, S., Coenen, L., et al. (2013). Advancing sustainable urban transformation. *Journal of Cleaner Production, 50*, 1–11.

McKeown, R. (2002). *Education for sustainable development toolkit*. University of Tennessee. https://www.ramsar.org/sites/default/files/documents/library/esd_toolkit_version_2.pdf. Accessed 03 May 2021

Meira, F. B. (2014). Liminal organization: Organizational emergence within solidary economy in Brazil. *Organization, 21*(5), 713–729.

Ménard, C. (2014). Embedding organizational arrangements: Towards a general model. *Journal of Institutional Economics, 10*, 567–589.

Mowery, D. C., & Rosenberg, N. (1989). *Technology and the Pursuit of economic growth*. Cambridge University Press.

Nelson, R. R. (2008). What enables rapid economic progress? What are the needed institutions? *Research Policy, 37*(1), 1–11.

Nelson, R. R., & Rosenberg, N. (1993). Technical innovation and national systems. In R. R. Nelson (Ed.), *National Innovation Systems – A Comparative Analysis*. Oxford University Press.

Nill, J., & Kemp, R. (2009). Evolutionary approaches for sustainable innovation policies: From niche to paradigm? *Research Policy, 38*, 668–680.

North, D. C. (1990). *Institutions, institutional change, and economic performance*. Cambridge University Press.

North, D. C. (2005). *Understanding the process of economic change*. Princeton University Press.

OECD. (2013). *Organisation for economic co-operation and development. Guidelines on measuring subjective well-being*. OECD Publishing.

Olson, M. (1965). *The logic of collective action: Public goods and the theory of groups* (Revised ed). Harvard University Press.

Ostrom, E. (1990). *Governing the commons: The evolution of institutions for collective action*. Cambridge University Press.

Ostrom, E. (1998). A behavioral approach to the rational choice theory of collective action: presidential address. *American Political Science Review, 92*, 1–22.

Ostrom, E. (1999). Coping with Tragedies of the Commons. Workshop in Political Theory and Policy Analysis; Center for the Study of Institutions, Population, and Environmental Change, Indiana University, Bloomington, Indiana 47408-3895.

Ostrom, E. (2000). Collective action and the evolution of social norms. *The Journal of Economic Perspectives, 14*, 137–158.

Ostrom, E. (2005). Doing institutional analysis: Digging deeper than markets and Hierarquies. In C. Ménard & M. M. Shirley (Eds.), *Handbook off new institutional economics*. Springer.

Ostrom, E. (2009). Beyond markets and states: Polycentric governance of complex economic systems. *The American Economic Review, 100*(3), 641–672.

Ostrom, E. (2011). Background on the institutional analysis and development framework. Policy Stud. J. 2011, 39, 7–27.

Ostrom, V., Tiebout, C. M., & Warren, R. (1961). The organization of Government in metropolitan areas: A theoretical inquiry. *American Political Science Review, 55*, 831–842. Reprinted in McGINNIS, Michael D (ed) 1999.

Ostrom, E., Gardner, R., & Walker, J. (1994). *Rules, Games, and Common-Pool Resources*. University of Michigan Press.

Penrose, E. T. (1971). *The growth of firms, Middle East oil and other essays*. Frank Cass.

Pereira, R., & Pereira, A. (2019). Financing a renewable energy feed-in tariff with a tax on carbon dioxide emissions: A dynamic multi-sector general equilibrium analysis for Portugal. *Green Finance, 1*(3), 279–296.

Polski, M. M., & Ostrom, E. (1998). *An institutional framework for policy analysis and design*. Paper presented at the workshop in political theory and policy analysis. Indiana University, Bloomington.

Rennings, K. (2000). Redefining innovation – Eco-innovation research and the contribution from ecological economics. *Ecological Economics, 32*(2), 319–332.

Rittel, H. W. J., & Webber, M. M. (1973). Dilemmas in a general theory of planning. *Policy Sciences, 4*(2), 155–169.

Rodrigues, I., & Barbieri, J. C. (2008). A emergência da tecnologia social: revisitando o movimento da tecnologia apropriada como estratégia de desenvolvimento sustentável. *Revista de Administração Pública, 42*(6), 1069–1094.

Roggero, M., Käler, L., & Hagen, A. (2019). Strategic cooperation for transnational adaptation: Lessons from the economics of climate change mitigation. *International Environmental Agreements, 19*, 395–410.

Romeiro, A. R. (2012). Sustainable development: An ecological economics perspective. *Estudos Avançados, 26*(74), 65–92.

Sabel, C. (1993). Learning by monitoring: The institutions of economic development. In N. J. Smelser & R. Swedberg (Eds.), *The handbook of economic sociology* (pp. 137–165). Princeton University Press.

Schmidt-Traub, G., Kroll, C., Teksoz, K., et al. (2017). National baselines for the sustainable development goals assessed in the SDG Index and Dashboards. *Nature Geoscience, 10*, 547–556.

Scotchmer, S. (2004). *Innovation and incentives*. MIT Press.

Scozzi, B., Bellantuono, N., & Pontrandolfo, P. (2017). Managing open innovation in urban labs. *Group Decision and Negotiation, 26*(5), 857–874.

SDSN. (2015). *Indicators and a monitoring framework for sustainable development goals: Launching a data revolution for the SDGs*. Sustainable Development Solutions Network.

Sengers, F., Wieczorek, A. J., & Raven, R. (2019). Experimenting for sustainability transitions: A systematic literature review. *Technological Forecasting and Social Change, 145*, 153–164.

Servaes, J., & Lie, R. (2014). New challenges for communication for sustainable development and social change: a review essay. *Journal of Multicultural Discourses, 10*(1), 124–148. https://doi.org/10.1080/17447143.2014.982655

Silva, S., Soares, I., & Afonso, O. (2021). Assessing the double dividend of a third-generation environmental tax reform with resource substitution. *Environment, Development and Sustainability*. https://doi.org/10.1007/s10668-021-01290-7

Teece, D. J., Pisano, G., & Shuen, A. (1997). Dynamic capabilities and strategic management. *Strategic Management Journal, 18*(7), 509–533.

Teisman, G. R., & Edelenbos, J. (2011). Towards a perspective of system synchronization in water governance: A synthesis of empirical lessons and complexity theories. *International Review of Administrative Sciences, 77*(1), 101–118.

Thaler, R. (2000). From Homo Economicus to Homo Sapiens. *The Journal of Economic Perspectives, 14*(l), 133–114.

Tsaples, G., & Papathanasiou, J. (2021). Data envelopment analysis and the concept of sustainability: A review and analysis of the literature. *Renewable and Sustainable Energy Reviews, 138*, 110664.

Vare, P., & Scott, W. (2007). Learning for a change: Exploring the relationship between education and sustainable development. *Journal of Education for Sustainable Development, 1*, 191–198.

Voytenko, Y., McCormick, K., Evans, J., et al. (2016). Urban living labs for sustainability and low carbon cities in Europe: Towards a research Agenda. *Journal of Cleaner Production, 123*, 45–54.

Waldman, M. (1993). A new perspective on planned obsolescence. *Quarterly Journal of Economics, 108*(1), 273–283.

Wieczorek, A. J., Raven, R., & Berkhout, F. (2015). Transnational linkages in sustainability experiments. The case of solar PV in India. *Environmental Innovation and Societal Transitions, 17*, 149–165.

Williamson, O. E. (1993). Calculativeness, trust, and economic organization. *Journal of Law and Economics, 34*, 453–502.

Wise, R., Fazey, Y., Smith, M. S., et al. (2013). Reconceptualising adaptation to climate change as part of pathways of change and response. *Global Environmental Change, 28*, 325–336.

World Bank Agenda 2030. (2021). *US Washsington.* https://www.worldbank.org/en/programs/sdgs-2030-agenda. Accessed 27 Mar 2021.

World Economic Forum. (2011). *Committed to improving the State of The World.* http://www3.weforum.org/docs/WEF_WI_WaterSecurity_WaterFoodEnergyClimateNexus_2011.pdf. Accessed 16 Mar 2021.

Zhang, C., Chen, X., Li, Y., Ding, W., & Fu, G. (2018). Water-energy-food nexus: Concepts, questions and methodologies. *Journal of Cleaner Production, 195*, 625–639.

Zihare, L., Kubule, A., Vamza, I., Muizniece, I., & Blumberga, D. (2021). Bioeconomy triple factor nexus through indicator analysis. *New Biotechnology, 61*, 57–68.

8 Innovations Towards "the Nexus" in the Science-Politics-Society Interface: What Transformations Do We Really Seek?

Alberto Matenhauer Urbinatti

Abstract

The debate on the nexus between water, energy, and food (WEF) has generated expectations on the global stage about possible innovative transformations towards sustainability. On the one hand, "the nexus" has been an important umbrella concept for innovative research on intersectoral and multilevel governance of WEF systems, responsible management of resources, and integrated methodologies. On the other hand, the argument is that there is little discussion about the communicative processes in which knowledge is bargained to pursue innovation. In this context, the questions that arise here are as follows: What kind of transformations do we really seek from this approach? How can this debate be related to the Brazilian context? This chapter aims at discussing these questions in the light of two projects friendly to the nexus approach in Brazil. For this purpose, the Sustainable Cities Innovation Observatory (OICS) database will be used. Finally, this chapter provides insights on how to move towards more responsible nexus innovations.

Keywords

Water-energy-food nexus · Science-politics interface · Knowledge production · Innovation · Sustainability transformations

8.1 Introduction

The interconnection between water, energy, and food that the nexus approach proposes is already well known to many: the production of energy involves water; the distribution of water, in turn, involves energy; and water and energy are necessary conditions for food production. Not surprisingly, resources have had a close interdependence to supply the needs of different civilizations. In the last decades, this overlapping of resource flows has generated, on the one hand, expectations for sustainable development but, on the other, several uncertainties related to the security of resources in a climate change scenario.

While seeking to reduce trade-offs between resources, the nexus approach requires different synergies. According to Hoff (2011, p. 5), "a nexus approach can support a transition to sustainability, by reducing trade-offs and generating additional benefits that outweigh the transaction costs associated with stronger integration across sectors. Such gains should appeal to national interest and encourage governments, the private sector and civil society to engage." Therefore, the

A. M. Urbinatti (✉)
Public Health from the University of São Paulo, São Paulo, Brazil

L. L. B. Lazaro et al. (eds.), *Water-Energy-Food Nexus and Climate Change in Cities*, Sustainable Development Goals Series, https://doi.org/10.1007/978-3-031-05472-3_8

approach was designed to encourage collaboration between different sectors.

Shared knowledge is the reference point for this collaboration. For example, when scientists engage in politics, they are likely to be emphasizing the validity of their claims to knowledge. On the other hand, when politicians engage in scientific debates, they are likely to claim that they make decisions based on the best available knowledge. "This does not mean, however, that knowledge claims are translated directly into political decisions or that scientists would be the ultimate power holders" (Grundmann, 2007, p. 416). In this case, it is vital to understand how this knowledge is produced, the disagreements, and the acceptability of evidence in pluralistic societies (Jasanoff, 1994).

In this context, this chapter assumes that it is crucial to grasp the nexus debate as a negotiated knowledge at the science-politics-society interface. What is under negotiation is the innovative dimension of sustainability for public and private governance of resources. Firstly, this chapter aims to discuss theoretical aspects of knowledge at a science-politics-society interface to analyze some practical projects in the Brazilian context that can help us answer the following starting question: What sustainable transformations do we really seek?

8.2 Methodology

The methodology employed in this chapter is divided into two parts. First, a literature review was carried out in order to investigate how the interrelation between science and politics was approached throughout the last century, with greater focus on its second half. Overall arguments were sought in the field of sociology, philosophy, and planning. The aim is to conduct a review of the main authors who proposed to think about the place of knowledge at the science-policy interface. From this, the chapter goes closer to the relationship between knowledge and society, delving into the literature of Science and Technology Studies (STS). Second, the debate turns to the nexus topic to discuss the intertwining of knowledge and innovation. To this end, authors who address the subject were also reviewed. Finally, the chapter brings practical examples of projects that seek sustainable innovation, from the pillars of water, energy, and food in Brazil. This chapter is oriented by a constructivist approach to the science-politics-society interface in the nexus literature. A constructivist approach is a way to collectively understand controversies, to address questions to be investigated, and to reestablish the basis for scientific autonomy within a democratic scope (Jasanoff, 1996).

8.3 Knowledge at the Science-Politics Interface

The role that knowledge plays in scientific production or policy-making is an object of dispute. Science and politics have been treated in normative ways throughout recent history, suggesting different modes of governance (Habermas & Shapiro, 1971; Raman, 2014).

One of them is the decisionist model, based on Weber's work in the final of the nineteenth century. Millstone (2007) shows that this author has argued that modern industrial societies work on bureaucratic forms of governance. The decisionist model was defined in that context, considering deliberations of bureaucrats as secondary to political decisions defined from previous objectives. Confronting "politics" and "science," this model considers the first as an initial point to the second: "In the decisionist model, politics dominate science. Politicians use the results of research selectively and ad hoc in pursuit of their own interests, while scientific discoveries and their political consequences hardly gain a hearing in public and in the political decision-making process" (Sager, 2007, p. 5). The problem is the strong anchorage in rationalizing choices through calculated strategies and automatic decision procedures (Habermas, 1994). In other words, in this model, one may find political, tactical, interactive, and dialogical associations of science utilization instrumentally by policy-makers (Edwards, 1999).

In contrast, at the end of World War II, the progress of science became attractive to governments, particularly when scientists influenced governance agendas and political parties (Sager, 2007). Thus, inspired by the ideas of Saint-Simon and Auguste Comte, the so-called technocratic model emerged (Millstone, 2007). Roughly speaking, if in the decisionist model, politics dominates science, in the technocratic model, science dominates politics. The problem is that this model does not consider the existence of empirical uncertainty, pluralism of perspectives, and the social construction of science itself (Edwards, 1999). Habermas (1994) states that the dilemma is the imminent need for technical progress that would appear to be an independent and self-regulating process based on social interests operating in the model. In addition, the assumption that there is a continuous rational process in dealing with technical and practical problems does not necessarily happen.

In this context, Habermas suggested pragmatism. The opposition between the expert and the politician is replaced by a critical interaction. In this interaction, the exercise of power with ideological and non-legitimized bias would be eliminated to make the "scientifically informed discussions" accessible, in Habermas' words (1994, p. 67). For the author, this is the only model that necessarily presupposes democracy. Sager (2007) says that Habermas defined this model as a nonhierarchical relationship between science and politics. The public would mediate this relationship for Habermas as a political institution. It indicates a communicative process in which there is rooting in a particular social lifeworld with social interests and values (Habermas, 1994).

Years later, Habermas dealt with the theme of democracy from the concept of "deliberative politics" (Habermas, 1996). The author showed that the idea is linked to communicative presuppositions insofar as it includes the cultural and social context of the political public sphere. Habermas understands that democracy is discursive to the extent that the participants involved should have the capacity to deliberate, from an epistemic and political path to mutual understanding and consensus, and discursive ethics would provide regulatory principles (Durant, 2011; Habermas, 1996).

Some similar discussions took place in the field of planning. Considering planning as "the application of scientific methods to policy-making processes" (Faludi, 1973, p. 1), this discussion also fits here. The tradition that was considered dominant in planning theory was exactly the rationalist one. To some extent, it considers that the decision precedes the action (Friedmann & Hudson, 1974). David Collingridge called a "justificationist model," in which "only those decisions which can be fully justified are seen as rational" (Genus & Stirling, 2018, p. 66). Lindblom (1959) called it a "synoptic model." As Collingridge and Douglas show (1984, p. 344): "On this view, a policymaker ought to achieve a synoptic view of his problem before making decisions." For Stirling (1994), all justificationist approaches tend to measure the performance of a set of options that meet some implicit or explicit weighted appraisal criteria. Two examples are cost-benefit and risk analysis, which tend to reduce the possibility of recognition for uncertainties, ambiguities, and ignorance (Genus & Stirling, 2018).

Charles Lindblom (1959) criticized the literature of decision-making and planning by focusing on strategic and mechanical choices. That is, they were interested in choosing the best means to achieve certain ends. He called rational-comprehensive methods or the "root" method. In opposition to this, Lindblom proposed the successive limited comparisons method or the "branch" method. The second approach is appropriated to complex problems since the first would not tackle them. The branch method put together values, goals, and empirical analysis to show that they are intertwined. In this way, means and ends are not distinct. Dahl and Lindblom were propellers of what came to be called "incrementalism," in which they questioned how incremental strategies at the individual level might fit within broad processes of social control (Atkinson, 2011). The core of incrementalism can be summarized as the action of decision-makers when trying to solve problems in the absence of certainty "will usually engage in a local search for options," and consequently, "this results in small adjustments from the status quo premised on what is practical and what is possible" (Atkinson, 2011, p. 10).

Twenty years after this publication, Lindblom (1979) divided the general idea into two approaches: disjointed incrementalism as an analytical method and political incrementalism as a political pattern, incorporating changes in small steps. In this way, incrementalism became an integral part of what Lindblom called a partisan mutual adjustment. The main question for him was the rational adaptation of one act to the existing policies and the anticipated actions. To some extent, incrementalists consider a path dependency. Lindblom stressed the idea that self-interested partisans practicing incrementalism could achieve policy coordination and play an essential role in intelligent policies without the need for a synoptic model (Atkinson, 2011).

Smith and May (1980) understood this separation between rationalists and incrementalists as artificial for two reasons. The first is a confusing relationship between what "is" and what "ought to be," somehow determining that incrementalist models have explanatory tendencies while rationalist models have prescriptive ones. The second is that the debate does not consider the need to act on any set of decision-making rules and, precisely because of this, neglects how policymakers may use "decision-making" as a glossary for different practices. It is from this context that the mixed-scanning model appears. This model includes elements from both rationalist and incrementalist models. That is, it looks at both the richness of detail strategically gathered for a decision and similar patterns of development (Etzioni, 1967). "Etzioni accepts Lindblom's criticism of the synoptic ideal, but rejects disjointed incrementalism on the grounds that it gives too great a weight to the powerful in policymaking, that many decisions are not incremental, and that it encourages bureaucratic inertia" (Collingridge & Douglas, 1984, p. 362).

In the mid-1980s, planners incorporated into the planning theory ideas recently published by Jürgen Habermas, which can be called a "communicative turn" in planning theory (Healey, 1996). Incrementalism was an important bedrock of this debate. John Forester (1980) was one of the pioneers, discussing the communicative character of planning actions through structures such as the organizational and political contexts of planning practice as provocateurs of constant and systematic distorted communications. The question in this way was: "Does the planner speak in a way that people can understand, or are they mystified?" (Forester, 1980, p. 276). He inquired about the political and selective orientation of decision-making agendas and the unequal participation of citizens. Sager (1995) re-signified the debate that opposed synoptic planners to the incrementalists from Habermas' communicative theory, inserting the norm of dialogue into disjointed incrementalism. For him, knowledge is constructed in discourse. In this way, a consensus among informed people discussing in an "undistorted communication" way creates true knowledge. This claim can be considered a controversy, but he insisted that it is a starting point for formulating both the synoptic and the incrementalist planning models (Sager, 1995).

This theoretical background was an overview of how the interfaces between science and politics in decision-making were approached differently. In the next section, I will bring up elements that emphasize dialogue at the interfaces of science and politics with society.

8.4 The Production of Knowledge and the Interface with Society

Authors like Sheila Jasanoff and Brian Wynne, scholars in the Science and Technology Studies (STS) field, have something to share with Habermas' so-called discursive ethics (Durant, 2011). They are not Habermasian; however, they share, through different views, the concern with the deliberative process that presupposes democracy and public participation. In the pursuit of this attempt, STS has made significant contributions to broadening out more robust understandings of the science-politics interface. STS is a crucial field in which scientific knowledge and policy-making become part of the same process because it is a distinctly hybrid object between science and politics (Raman, 2014). This means that efforts by scientists and politicians to separate areas of expertise have created hybrid processes that make political decisions more difficult (Gieryn, 1997; Jasanoff, 1994; St Clair, 2006).

In this way, this field has produced efforts to democratize science, making technoscientific research closer to the ideals of inclusive opinion-making and accountable decision-making (de Vries, 2007; Marres, 2007). Lemos and Morehouse (2005) summarized that the science produced for politics generated a dichotomy between "applied" and "decision-driven," and, secondly, science turned only to research was called "basic" and "knowledge-driven." The third way for these authors happens when the division between science and politics is dissolved and "usable" knowledge is co-produced in the context of everyday interaction between scientists, policy-makers, and the public. Here, as in STS literature more widely, what is at stake is the degree and quality of participation as much in the making of science as policy. Public participation in science was often seen as non-science (see Kuhn, 1977; Merton, 1973). Some models proposed by STS authors were unfairly called "populists" (Raman, 2014). To some extent, this also indicates an indirect critique of knowledge production within the policy formulation process, or in other words, a critique of policy appraisal.

Scientific knowledge has undergone significant changes in the last century. To some extent, experts have had to deal with the limitations of their outcomes. This is not to say that they have found reliable previsions of the unpredictable, but they probably have learned from past mistakes (Stirling, 2011). The idea of mistakes here is associated with the historical tendency to specialize scientific procedures making whatever it takes. However, these attempts may have failed in many circumstances since complex problems often require complex solutions (Funtowicz & Ravetz, 1997). If we agree that contemporary phenomena are endowed with complexities, any narrow worldview is prone to error if outlined in a very specialized way to find out solutions.

As Weber showed at the beginning of the twentieth century, the "Western rationalism" was born of a culture of experts, which would have made cognitive, aesthetic-expressive, and moral-practical elements specialize (Habermas, 1994). This rationalism was shaped throughout the Industrial Revolution and may be considered a fundamental process in the constitution of the modern world. At least, this new world was inaugurated by the premises of a capitalist model that expanded in Europe in the nineteenth century. This process placed new paradigms for economic, political, social, and scientific relations. Consequently, just as the changes themselves brought new uncertainties, the old uncertainties about the "unknown" were reinforced.

Undoubtedly, environmental issues are compelling examples of this. The future of environmental governance is permeated by constant "insecureness" at different scales. Firstly, this probably happens because of the relationship between uncertainty and insecurity: Uncertainty here may represent feelings like fear or helplessness, showing a link to risk. Secondly, assertive scientific responses are very complicated and can be reduced in probability scenarios. Wynne (1992) differentiated risk assessment and uncertainty by showing that the first is a scientific way to analyze risk and safety problems related to technology. According to this author, this kind of assessment reduces uncertainty artificially because it treats "all uncertainties" as if they were determinate by a cause-effect system.

Collingridge (1980) shows that if one of the outcome options cannot be identified, it can be said to be at risk, and the Bayesian rule can be applied, enabling the decision-maker to optimize the expected value. The same rule has been applied for those decisions called "under uncertainty," in which it is impossible to determine the objective probability distribution. To differentiate between those decisions that can receive Bayesian optimization and those that cannot, the author calls the latter "decisions under ignorance." Technology control, for example, would fall into this second category. Collingridge bases his analysis primarily on Popper's philosophy of fallibilism, showing that errors are inherent parts of humans and contrasting with the still-dominant synoptically rationalized decisions (Genus & Stirling, 2018).

Stirling provides a sophisticated picture of these decisions, which he called "contrasting states of incomplete knowledge" (2007) or "uncertainty matrix" (2010). For him, there are at least four categories: risk, ambiguity, uncertainty, and ignorance. Certain scientific methods tend more to one of these categories, but it is problematic. So he says that "[a] definitive science-based decisions are not just potentially misleading — they are a fundamental contradiction in terms" (Stirling, 2010, p. 1030). Even when experts try to address uncertainty, they tend to reduce it to measurable risk (Stirling, 2011). The acceptance of unknowns in traditional scientific methods is unusual. This is why experts are always trying to block off them in some specific approach. From this, Stirling suggests the old term "incertitude" because it may represent better all kinds of unruliness in knowledge production, avoiding merely simplifications of uncertainty as a probability (Stirling, 2019). Notwithstanding, incertitude is not a simple condition but an endemic status of science, policy, and finance.

Utmost presumably to avoid the obscure, the production of knowledge has crystallized into methods that have not been extended to the appraisal of society. Knowledge, therefore, would be more prone to "closing down," i.e., treated in a prescriptive and instrumental way to justify a choice, than to "opening up," elucidating broad, plural, and democratic decision-making processes (Stirling, 2007, 2015). "Knowledge is routinely represented as a discrete quantifiable medium – to be appropriated, accumulated, measured and distributed according to universal templates. This is a convenient construct" (Stirling, 2009, p. 34). These failures in participatory and precautionary constructions of knowledge are reflected in policy responses and policy thinking (Stirling & Scoones, 2009).

As far as much nexus-oriented science is concerned, there is no explicit recognition of this status of incertitude. This might be reproducing some errors in treating the concept under normative aspects that, instead of "open up" knowledge production, tend to "close it down" in specific scientific consensus groups (Urbinatti et al., 2020). In the next section, the innovative dimension of the nexus will be addressed.

8.5 Nexus Innovations

The literature review conducted by Wiegleb and Bruns (2018) concludes that a prominent view on the "WEF nexus" is the technoscientific approach, often based on the theories of "ecological modernization." This involves an essentially positive and optimistic understanding of the potential for the claimed "win-win" dynamics around sustainable technologies or green innovations, such as rainwater harvesting, desalination based on renewable energy, and photovoltaic water pumps (Hoff, 2011). Wiegleb and Bruns (2018) highlight the non-neutrality of these discourses and how wider sociopolitical dimensions of nexus discourses are poorly addressed in the resulting "solutionism." Greater weighting is thereby implicitly given to concepts and bodies of evidence structured in terms of narrow economic cost-effectiveness and resource efficiency as seen from technical-administrative perspectives. The point is not to be at all pessimistic but simply to admit that overly intense preoccupations with technology can minimize the acknowledgment of broader social, economic, and political drivers and solutions.

An overly uncritical obsession with innovation in the debates over the WEF nexus can be seen as part of the problem rather than contributing to the solution. Approaches founded too exclusively on *technical* definitions of problems and solutions can hide crucial *political* decisions around technological pursuits (Allouche et al., 2015). Here, the hybrid political-technical nature of the nexus becomes crucial. Since the diversity of social forces is complicit in shaping the forms that the nexus takes, it is more accountable to be explicit about co-production processes. With particular efforts to resist the barriers to the least powerful groups, diverse social actors should be involved in recognizing and defining viable infrastructural and institutional options to address the nexus challenges in particular settings, weigh up their respective pros and cons, and explore uncertainties and context-dependencies. Thus, access to water, energy, and food for vulnerable populations can be promoted.

The key issue is not the designed responses to the nexus problems but the frequent necessity of technological interventions to assume that these responses can be sufficient. Somehow, it is necessary to recognize the hubris behind scientific and technological achievements throughout history and deal with the hybrid implications that have arisen (Hård & Jamison, 2013). The most critical question about technical contributions to nexus problems is not in the binary resolution of better or worse solution within private interests. Rather, it is the queries about "what are the alternatives," "who says," and "why" about the public good. These are more challenging questions for participation and democracy than for technical expertise.

Over and above the dubiousness about the technological contributions to addressing "the nexus," greater appreciation is demanded for the roles of transparency, accountability, and participation in determining the appropriate *directions* for technological change. It is also crucial to respect the potential of social technologies, grassroots innovations, and deeper innovation democracies, thereby recognizing particular directions and diversity. The benefits of innovations around recycling, reuse, waste reduction, and renewable resources, for instance, lie in the speed with which any given technology is pursued in any particular area, and which technology is chosen, and in what form. To address such dilemmas robustly requires effective innovation along with ongoing participatory deliberation and democratic debate.

The responsible research and innovation (RRI) concept is getting traction to address the resulting challenges. RRI deals directly with the challenges of anticipation and accountability around innovation (Grinbaum & Groves, 2013). It highlights the problem famously recognized by Collingridge (1980); the period that there is the most significant potential for influencing technological change is typically at the earliest stages when understandings of the consequences are least developed and distributed. Moves towards "responsible innovation" can therefore span the divide between caricature pessimism and optimism, such as recognizing the importance of deliberate societal steering of which particular technological pathways get to be pursued and those get set aside (Genus & Stirling, 2018). Technologies are getting perpetually inseparably entwined into the structure of society, and as funders push for quicker development and innovation, this process requires comprehending the audience, perceptions, and values related to those technologies (Monteiro, 2018).

Hoolohan et al. (2019) suggest that the potential of technological innovations throughout the WEF nexus depends on recognizing the unintended consequences of policy mechanisms to support technological adaptation and restructuring of incentives for the achievement of multisectoral benefits, for example, energy production processes via anaerobic digestion, concluding from environmental, economic, and social aspects that it is a positive innovation towards "the nexus." According to the authors, it is vital to ensure transparency about both harm and public benefits, as well as ethical accountability in how a decision can affect people's lives.

This leads to qualities of learning that are central to the "nexus of humility" framework proposed by Urbinatti et al. (2020). This dimension is based on plurality and permeates all the others. The scientific knowledge is essential but not in itself sufficient. To recognize and materially enact this in existing governance processes (rather than merely rehearse it rhetorically) is a significant undertaking for learning by policy and scientific actors in particular. "To be dialogical, even the humbler social actors must be involved in the process, having the opportunity, through a self-reflection on their context, to evaluate the risk and to consider which kind of alternatives can be better to them" (Giatti, 2019, pp. 31–32). Furthermore, with imperatives operating in multiple directions, social learning can become robust only if it is constituted in plural processes – not merely one-way science/technical communication nor even just "two-way" dialogue between experts and society. Let us look at some examples to discuss these interactions.

8.6 Exploring the Sustainable Cities Innovation Observatory

The Sustainable Cities Innovation Observatory (Observatório de Inovação para Cidades Sustentáveis/OICS, in Portuguese) is part of CITinova, a multilateral project conducted by the Ministry of Science, Technology, and Innovations (MCTI) to promote sustainability in Brazilian cities through innovative technologies and integrated urban planning. The project is funded by the Global Environment Facility (GEF), implemented by the United Nations Environment Programme (UNEP), and run in partnership with Recife Agency for Innovation and Strategy (Agência Recife para Inovação e Estratégia/ARIES) and Porto Digital, Center for Management and Strategic Studies (Centro de Gestão e Estudos Estratégicos/CGEE), Sustainable Cities Program (Programa Cidades Sustentáveis/PCS), and Secretariat of Environment of Federal District (Secretaria do Meio Ambiente/SEMA/DF).

A practical example of how CTinova is within nexus thinking is a project involving magnetized water in agriculture. It is a pilot project being implemented in Chácara Colina, Brazlândia, in the Federal District, by the Secretariat of Environment (Sema-GDF). According to the researchers involved in the project, water exposed to static magnetic induction (SMI) enables greater hydration, altering cellular metabolism.[1] This makes more energy available for plant growth and reduces the consumption of water and pesticides. High-flow magnetizers aim for higher magnetic intensity and efficiency, enabling increased productivity in conventional irrigation systems. We can easily recognize here the synergies between water, energy, and food systems. For sure, it is a great advance for the country; however, CTinova contributes even more in the way the generation and application of knowledge has been encouraged.

The OICS has defined typologies of city regions that are organized to address specific urban challenges, considering characteristics of a territory based on geophysical data and indicators aligned with the Sustainable Development Goals (SDGs). The observatory also carries out a compilation of solutions that involve big data analysis, co-production processes, and consultation with experts. The proposal, therefore, is to articulate public managers, civil society, companies, and academia in favor of the urban agenda, co-producing alternatives for the transition of cities towards sustainability. The choice to look at this project makes us ground the debate on nexus innovations in the Brazilian territory from real examples. This is because the OICS project categories include the main nexus-related themes: low-carbon mobility and access to and in the city; affordable and rationally used clean water and decentralized and efficient sanitation; decentralized and efficient renewable energy; solid waste, circular economy, and efficient treatment; nature-based solutions and green and blue infrastructure for greater resilience to climate change; participatory vision and integrated long-term planning; low-carbon and socially beneficial built environment; and innovation, policies, and vocational strategies for regional development and the strengthening of value chains.

The OICS platform is interactive, and we can filter the case studies based on their connections to the SDGs. I will look at two projects on the platform related to SDG 6 (clean water and sanitation) and SDG 7 (renewable and affordable energy). The intention here is to show two different scales of innovative nexus-friendly projects.

The first project that draws attention is Aquapolo, created in 2009, which uses sophisticated water treatment processes for reuse and distribution to the industry. After going through the treatment process performed by the Basic Sanitation Company of the State of São Paulo (Companhia de Saneamento Básico do Estado de São Paulo/SABESP), the water goes to Aquapolo,

[1] See https://citinova.mctic.gov.br/projeto-piloto-sobre-o-uso-de-agua-magnetizada-na-agricultura-tem-bons-resultados/ (accessed May 5, 2021).

which collects 650 liters per second in an automated way and sends the water back to the company for treatment. In this process, the organic matter present in the water is oxidized with carbon dioxide, and the ammonia resulting from the oxidation is transformed into nitrate. This, in turn, is removed by an anoxic reactor. In the next step, the water passes through an aerobic reactor to remove the remaining organic matter and is directed to the membrane bioreactor (MBR) filter. In the first module, the storage tank receives and distributes water to industries that do not need demineralized water. In the second module, the reverse osmosis process extracts all impurities from the water for companies that require demineralized or clarified water. In this process, only part of the water is used; however, the water that is not used returns to the first module where it is distributed. Thus, by taking reused water, petrochemical industries in the metropolitan region of São Paulo can save around 1.7 billion liters of drinking water in their facilities.

This project directly involves science through the advanced technique of water treatment. However, there is no direct involvement of universities. It involves SABESP and the many companies in the ABC Petrochemical Complex that are at the cutting edge of the process and benefit from saving drinking water for less noble purposes, such as cooling towers and boilers. There is an evident nexus here between the water and energy sectors (Fig. 8.1).

The second project is part of the Médio Juruá Territory Program, which receives funding from the United States Agency for International Development (USAID), Coca-Cola Brazil, and Natura, through a cooperation agreement signed with SITAWI Finanças do Bem and the Médio Juruá Territory Forum. The project was carried out in 2018 and sought to illuminate 57 riverside communities of the Médio Juruá, in the municipality of Carauari, state of Amazonas, with the production and distribution of 620 solar energy lanterns, being 600 lanterns financed by the Association of Rural Producers of Carauari (ASPROC), 15 lanterns donated by Campus Brasil, and 5 lanterns offered by the organization Litro de Luz. The action was carried out collectively with the local population, creating access to new knowledge that fostered community development and increased the social impact.

Of course, there is technoscientific knowledge at stake for the construction of the lanterns based on sunlight. However, no direct relationship with universities is specified, only private partners and foundations. Like the first one, this project involved large companies in the actions, in this case in a remote region of Brazil, in the middle of the Amazon forest. The big difference is that the installation was carried out together with the residents of the region. Throughout the process, 20 residents were trained to be ambassadors, undergoing training related to basic electrical concepts and maintenance of photovoltaic solutions. They also had the opportunity to play leadership and engagement roles with other local residents.

The platform does not provide information about the costs of both projects. This shows that responsible and accountable innovation needs to improve in Brazil. This data would be important for a quantitative understanding of the involvement of companies and associations and possible financial returns from the actions. Nevertheless, it is worth noting that the OICS provides a "how to replicate" tab, giving details about the methodologies used in each action (Fig. 8.2).

In the last section, I will discuss how these examples help us reflect on what transformations we want from the nexus approach, taking up specific aspects of the theoretical discussion about knowledge at the science-policy-society interface.

8.7 Concluding Remarks: What Transformations Do We Really Seek?

I sought to show here that innovation is a process of knowledge sharing within science, policy, companies, and society. The literature review on the interface between these domains elucidates a complex relationship at play, framed by decisionist, technocratic, or pragmatic models and by synoptic, incrementalist, communicative, and mixed-scanning perspectives. Regrettably, these

Source: GS Inima Industrial Brasil website.

Fig. 8.1 Aquapolo project in the São Paulo metropolitan region. (Source: GS Inima Industrial Brasil website)

Source: Litro de Luz website.

Fig. 8.2 Implementation of solar energy lanterns in riverside communities in the Amazon region, Brazil. (Source: Litro de Luz website)

dynamics have been under-scrutinized due to tendencies (which they themselves foster) to treat relations between science, politics, and society in rather simplistic normative ways, such as instrumental use of knowledge (Owens et al., 2006). The hard truth is that it is not easy to identify each of these theoretical frameworks in real cases nor to say that there are better analysis conditions from one or the other. It is more tangible to find traces and tendencies that are more or less present in each case. Thus, innovations friendly to the nexus approach will also present these characteristics.

According to Al-Said and Elagib (2017), it is possible to find positive attitudes inserted in the idea of innovation that the nexus approach brings, enhanced by Western narratives of good policies, which sometimes fail during translation or due to lack of context. This is an essential point because it leads us to the still little explored cultural dimension in the nexus literature. This cultural dimension can be strongly embedded in contextual learning processes that respond differently to the global problem of access to water, energy, and food resources. It is critical to allow space for a more multidirectional plurality of learning processes from the humility of experts and policymakers themselves in the discursive arenas in which they participate (Urbinatti et al., 2020). These responsibilities underlie the recognition of the need for a more active "governance of knowledge-for-governance" (Raman, 2014) around the nexus. Nevertheless, constant challenges are posed to the conceptions of reality of the actors involved in the governance process (Alasuutari & Qadir, 2014).

Considering the two projects of the OICS platform, one can understand two possibilities of innovation towards the nexus. The rationality of efficiency exists in both cases, but the way they are constructed is different. What changes is the way society is involved in the projects. In the Aquapolo project, society benefits indirectly, mediated by technical and economic interests of political decisions, in a kind of business-to-business (B2B) relationship. In the lantern installation project in vulnerable communities, even though there is economic interest from the financial sponsors, society is immersed in learning about energy sustainability.

It is true that we can find more decisionist and rationalist characteristics in the first project and pragmatists and incrementalists in the second. However, it would be inconvenient to define them only through this lens. Each trait may work better for a large- or small-scale project. Nonetheless, I emphasize here that one must first recognize that there are power relations behind any decision and, only then, encourage communicative processes of co-production of knowledge about sustainability.

There is no ideal scale of innovation. The fact is that both solutions were designed to solve problems. In this case, the RRI approach can be a means to ensure that, independently of the actors involved, there needs to be precaution and social responsibility ultimately. This approach fosters the engagement of public, scientific, political, and private actors to produce ethically acceptable, sustainable, and socially desirable research and innovation outcomes.[2]

Another example from the OICS platform describes this kind of thinking in practice. This is the Urban Living Laboratory for Intelligent Building Solutions (Laboratório Urbano Vivo Soluções Construtivas Inteligentes/LCI), led by Coppe/Federal University of Rio de Janeiro. The objective of this laboratory is to place the user at the center of an ecosystem that covers a given territory (such as a neighborhood, city, or region) and integrates simultaneous research, development, and innovation processes and promotes collaboration among players in academia, government, public and private institutions, and society. Through a process of co-production, the LCI is based on three pillars: (1) the development of innovative and intelligent constructive solutions; (2) the intelligent occupational use of buildings by considering aspects such as comfort, health, and air quality; and (3) the rapid insertion of these solutions into society.

Thus, we return to the question posed at the beginning of this chapter about what transformations we want from the concept of nexus. To answer it, I echo the distinction made by Stirling (2014) between transition, which brings more emphasis on "control," and transformation for sustainability, which aims to bring "care" as a democratic decision-making tool. In his words (p. 22), "In this 'caring' mode, the knowing and doing of transformation are not separate, but intimately interlinked. Neither alone is sufficient." It is in this sense that the nexus approach tends to become more robust by considering its innovation potential as strictly linked to social progress. Notwithstanding, it can facilitate and transform the realities of injustice in accessing water, energy, and food around the world through more inclusive decisions and knowledge sharing.

[2]Definition used by the RRi-Practice project. See https://www.rri-practice.eu/about-rri-practice/what-is-rri/ (accessed May 6, 2021).

References

Alasuutari, P., & Qadir, A. (2014). Epistemic governance: An approach to the politics of policy-making. *European Journal of Cultural and Political Sociology, 1*, 67–84. https://doi.org/10.1080/23254823.2014.887986

Allouche, J., Middleton, C., & Gyawali, D. (2015). Technical veil, hidden politics: Interrogating the power linkages behind the Nexus. *Water Alternatives, 8*, 17.

Al-Saidi, M., & Elagib, N. A. (2017). Towards understanding the integrative approach of the water, energy and food nexus. *Science of the Total Environment, 574*, 1131–1139. https://doi.org/10.1016/j.scitotenv.2016.09.046

Atkinson, M. M. (2011). Lindblom's lament: Incrementalism and the persistent pull of the status quo. *Policy and Society, 30*, 9–18. https://doi.org/10.1016/j.polsoc.2010.12.002

Collingridge, D. (1980). *The social control of technology*. Frances Pinter.

Collingridge, D., & Douglas, J. (1984). Three models of policymaking: Expert advice in the control of environmental Lead. *Social Studies of Science, 14*, 343–370. https://doi.org/10.1177/030631284014003002

de Vries, G. (2007). What is political in sub-politics?: How Aristotle might help STS. *Social Studies of Science, 37*, 781–809. https://doi.org/10.1177/0306312706070749

Durant, D. (2011). Models of democracy in social studies of science. *Social Studies of Science, 41*, 691–714. https://doi.org/10.1177/0306312711414759

Edwards, A. (1999). Scientific expertise and policy-making: The intermediary role of the public sphere. *Science and Public Policy, 26*, 163–170. https://doi.org/10.3152/147154399781782473

Etzioni, A. (1967). Mixed-scanning: A "third" approach to decision-making. *Public Administration Review, 27*, 385–392. https://doi.org/10.2307/973394

Faludi, A. (1973). *Planning theory* (Vol. 7). Pergamon Press, Urban and regional planning series. ISSN 0305-5582.

Forester, J. (1980). Critical theory and planning practice. *Journal of the American Planning Association, 46*, 275–286. https://doi.org/10.1080/01944368008977043

Friedmann, J., & Hudson, B. (1974). Knowledge and action: A guide to planning theory. *Journal of the American Institute of Planners, 40*, 2–16. https://doi.org/10.1080/01944367408977442

Funtowicz, S., & Ravetz, J. (1997). *Ciência pós-normal e comunidades ampliadas de pares face aos desafios ambientais*. https://doi.org/10.1590/s0104-59701997000200002

Genus, A., & Stirling, A. (2018). Collingridge and the dilemma of control: Towards responsible and accountable innovation. *Research Policy, 47*, 61–69. https://doi.org/10.1016/j.respol.2017.09.012

Giatti, L. L. (2019). *Participatory research in the post-normal age: Unsustainability and uncertainties to rethink Paulo Freire's pedagogy of the oppressed*. Springer International Publishing. https://doi.org/10.1007/978-3-030-27924-0

Gieryn, T. F. (1997). Boundaries of science. In A. I. Tauber (Ed.), *Science and the quest for reality* (pp. 293–332). Main Trends of the Modern World. Palgrave Macmillan UK. https://doi.org/10.1007/978-1-349-25249-7_12

Grinbaum, A., & Groves, C. (2013). *What is "responsible" about responsible innovation? Understanding the ethical issues, responsible innovation*. Managing the Responsible Emergence of Science and Innovation in Society. p. Forthcoming. https://doi.org/10.1002/9781118551424.ch7

Grundmann, R. (2007). Climate change and knowledge politics. *Environmental Politics, 16*, 414–432. https://doi.org/10.1080/09644010701251656

Habermas, J. (1994). Three normative models of democracy. *Jurgen Habermas, 1*, 10.

Habermas, J. (1996). *Between facts and norms: Contributions to a discourse theory of law and democracy*. Polity.

Habermas, J., & Shapiro, J. J. (1971). Toward a rational society: Student protest, science and politics. *Science and Society, 35*, 373–375.

Hård, M., & Jamison, A. (2013). *Hubris and hybrids: A cultural history of technology and science. Hubris and hybrids: A cultural history of technology and science* (pp. 1–335). Routledge. https://doi.org/10.4324/9781315024400

Healey, P. (1996). The communicative turn in planning theory and its implications for spatial strategy formation. *Environment and Planning. B, Planning & Design, 23*, 217–234. https://doi.org/10.1068/b230217

Hoff, H. (2011). Understanding the nexus. Background Paper for the Bonn2011 Conference: The Water, Energy and Food Security Nexus. Stockholm Environment Institute, Stockholm

Hoolohan, C., Soutar, I., Suckling, J., Druckman, A., Larkin, A., & McLachlan, C. (2019). Stepping-up innovations in the water–energy–food nexus: A case study of anaerobic digestion in the UK. *The Geographical Journal*, 391–405. https://doi.org/10.1111/geoj.12259@10.1111/(ISSN)1475-4959.Geography_and_the_Water-Energy-Food_Nexus

Jasanoff, S. (1994). *The fifth branch: Science advisers as policymakers*. Harvard Univ Press.

Jasanoff, S. (1996). Is science socially constructed—And can it still inform public policy? *Science and Engineering Ethics, 2*, 263–276. https://doi.org/10.1007/BF02583913

Kuhn, T. S. (1977). Objectivity, value judgment, and theory choice. In *The essential tension: Selected studies in scientific tradition and change* (pp. 320–339). University of Chicago Press.

Lemos, M. C., & Morehouse, B. J. (2005). The co-production of science and policy in integrated climate assessments. *Global Environmental Change, 15*, 57–68. https://doi.org/10.1016/j.gloenvcha.2004.09.004

Lindblom, C. (1959). The science of 'muddling through. *Public Administration Review, 19*, 79–88. https://doi.org/10.2307/973677

Lindblom, C. E. (1979). Still muddling, not yet through. *Public Administration Review, 39*, 517–526. https://doi.org/10.2307/976178

Marres, N. (2007). The issues deserve more credit: Pragmatist contributions to the study of public involvement in controversy. *Social Studies of Science, 37*, 759–780. https://doi.org/10.1177/0306312706077367

Merton, R. K. (1973). *The sociology of science: Theoretical and empirical investigations*. University of Chicago Press.

Millstone, E. (2007). Can food safety policy-making be both scientifically and democratically legitimated? If so, how? *Journal of Agricultural and Environmental Ethics, 20*, 483–508. https://doi.org/10.1007/s10806-007-9045-x

Monteiro, M. (2018). Ethnography and interdisciplinary work: Experiences from the US and Brazil. *Tapuya: Latin American Science, Technology and Society, 1*, 153–169. https://doi.org/10.1080/25729861.2018.1521091

Owens, S., Petts, J., & Bulkeley, H. (2006). Boundary work: Knowledge, policy, and the urban environment. *Environment and Planning. C, Government & Policy, 24*, 633–643. https://doi.org/10.1068/c0606j

Raman, S. (2014). Science, uncertainty and the normative question of epistemic governance in policymaking [WWW Document]. *Knowledge, Technology and Law*. https://doi.org/10.4324/9780203797600-9

Sager, T. (1995). Dialogical incrementalism: Competing rationalities and the cultivation of the synoptic/incremental distinction. *Journal of Architectural and Planning Research, 12*, 121–140.

Sager, F. (2007). Habermas' models of decisionism, technocracy and pragmatism in times of governance: The relationship of public administration, politics and science in the alcohol prevention policies of the Swiss Member States. *Public Administration, 85*, 429–447. https://doi.org/10.1111/j.1467-9299.2007.00646.x

Smith, G., & May, D. (1980). The artificial debate between rationalist and incrementalist models of decision making. *Policy & Politics, 8*, 147–161. https://doi.org/10.1332/030557380782629005

St Clair, A. L. (2006). Global poverty: The co-production of knowledge and politics. *Global Social Policy, 6*, 57–77. https://doi.org/10.1177/1468018106061392

Stirling, A. (1994). *Power technology choice: Putting the money where the mouth is?* https://doi.org/10.13140/2.1.5123.4084

Stirling, A. (2007). "Opening up" and "closing down". *Science, Technology & Human Values, 33*, 262–294. https://doi.org/10.1177/0162243907311265

Stirling, A. (2009). Risk, uncertainty and power. *Seminar Magazine, 597*, 33–39.

Stirling, A. (2010). Keep it complex. *Nature, 468*, 1029–1031. https://doi.org/10.1038/4681029a

Stirling, A. (2011). Pluralising progress: From integrative transitions to transformative diversity. https://doi.org/10.1016/j.eist.2011.03.005.

Stirling, A. (2014). *Emancipating transformations: From controlling 'the transition' to culturing plural radical progress*. STEPS Centre Working Papers, University of Sussex.

Stirling, A. (2015). Developing 'Nexus Capabilities': Towards transdisciplinary methodologies. https://doi.org/10.13140/RG.2.1.2834.9920

Stirling, A. (2019). Politics in the language of uncertainty [WWW Document]. *STEPS Centre*. https://steps-centre.org/blog/politics-in-the-language-of-uncertainty/. Accessed 15 Jan 2020

Stirling, A., & Scoones, I. (2009). From risk assessment to knowledge mapping: Science, precaution, and participation in disease ecology. *Ecology and Society, 14*. https://doi.org/10.5751/ES-02980-140214

Urbinatti, A. M., Dalla Fontana, M., Stirling, A., & Giatti, L. L. (2020). 'Opening up' the governance of water-energy-food nexus: Towards a science-policy-society interface based on hybridity and humility. *Science of the Total Environment, 744*, 140945. https://doi.org/10.1016/j.scitotenv.2020.140945

Wiegleb, V., & Bruns, A. (2018). What is driving the water-energy-food nexus? Discourses, knowledge, and politics of an emerging resource governance concept. *Frontiers in Environmental Science, 6*. https://doi.org/10.3389/fenvs.2018.00128

Wynne, B. (1992). Uncertainty and environmental learning: Reconceiving science and policy in the preventive paradigm. *Global Environmental Change, 2*, 111–127. https://doi.org/10.1016/0959-3780(92)90017-2

9 The Water-Energy-Food Nexus in Latin America and the Caribbean: Priority Interconnections

Antonio Embid and Liber Martin

Abstract

This chapter analyzes implementing the water-energy-food (WEF) nexus in Latin America and the Caribbean (LAC), focusing on the identification of obstacles, opportunities, and priority interconnections for the region. Based on a review of the theoretical background to the nexus concept and its current global configuration, we consider the main aspects of the WEF nexus to establish the debate in the LAC region. The chapter discusses different nexus features in LAC while identifying the main challenges for implementation, harmonization with the human rights legal framework, and national legal priorities for water use. By briefly describing the interconnections (water-energy, water-food, water-food-energy), we identify and discuss those which may prove critical for nexus implementation. The interconnections that should be prioritized include agriculture and food; irrigation modernization and aquifer overexploitation; hydropower, oil, and mining; and water and sanitation services and biofuels. With one of the highest urban population percentages worldwide and many megacities, the urban nexus become also critical for the region. After summarizing the priority interconnections, we conclude that scant consideration has been given to the nexus approach in a region where it could help tackle climate change and transform the current development model of intensive, yet unsustainable, use of natural resources.

Keywords

Water-energy-food nexus · Planning · Latin America and the Caribbean · Nexus priorities · Sustainability

Abbreviations

ECLAC	Economic Commission for Latin America and the Caribbean
LAC	Latin America and the Caribbean
WEF	Water-energy-food nexus

9.1 Introduction

Water-energy-food (WEF) nexus is not a new concept, but literature on WEF nexus does not stop growing since its arrival on the international agenda (WEF, 2011) in Bonn 2011 Conference "The Water, Energy and Food

A. Embid (✉)
University of Zaragoza, Zaragoza, Spain

L. Martin
CONICET/University of Cuyo, Mendoza, Argentina

L. L. B. Lazaro et al. (eds.), *Water-Energy-Food Nexus and Climate Change in Cities*, Sustainable Development Goals Series, https://doi.org/10.1007/978-3-031-05472-3_9

Security Nexus" – Solutions for the Green Economy (Martin-Nagle et al., 2012; Tashtoush et al., 2019; De Andrade Guerra et al., 2020). Nevertheless, very little of this literature focus on the LAC region, where the approach has been driven by some relevant international organizations, including the pioneering works of ECLAC (Embid & Martín, 2017, 2018), then also reproduced by others (Mahlknecht et al., 2020).

Despite this literature, scant consideration has so far been given to the water-energy-food nexus in Latin American and Caribbean countries, where the nexus has not been yet incorporated into the design of policies, plans, or management of natural resources (Naranjo & Willaarts, 2020; Urquiza, & Billi, 2020; Willaarts et al., 2021). The relative abundance of natural resources coupled with their own restrictions on their underdeveloped status and a focus on pressing short-term needs has led to the nexus not being considered in intensive development models dependent on the exploitation of natural resources that have shown clear signs of environmental unsustainability and social inequality. Consequently, some of the region's characteristics, such as the relative wealth of these resources with huge potential for future development, weak governance, and hardly any capacity to establish and implement public policies, alongside high levels of inequality and dissatisfaction in guaranteeing and realizing human rights, show the potential for appropriate consideration of the nexus for future improvement.

9.1.1 Materials and Methods

This chapter presents the conclusions of a qualitative analysis of the key variables for implementing the nexus approach in LAC. Following a nonexperimental observational methodological design, the chapter describes the interconnections seeking to identify those that can be considered priority or critical for nexus implementation in LAC. This definition is based on the following variables: (i) considering water as a predominant nexus element, (ii) impacts on the other nexus elements, (iii) focus on sectors of economic relevance with development potential and increasing complexity, and (iv) realizing and respecting human rights, in particular economic, social, and cultural rights, as an obligation and priority goal for the region's sustainable development.[1] For an interconnection to be a priority or highly relevant, it must be a critical activity for the region or some of its countries and present an opportunity to improve its performance from implementing a nexus viewpoint.

9.2 Interrelations Between Water and Energy

The use of water to produce energy is a traditional approach that refers to hydropower above all others. Water use for energy production represents the critical element of the water-energy nexus. This is proved by 15% of worldwide water abstraction being allocated to energy production (IRENA, 2015). Such a significant figure obviously merits clarification, since it is far less in desert or arid areas where water is scarce, but, on the contrary, far more where water abounds, which is the general case in many countries in the region where hydropower predominates and its appreciable potential still needs to be tapped into.[2]

The generation of hydroelectricity is currently by far the main source of power in the region, but it is distributed very unevenly among the countries. The Southern Cone produces 68% of its electricity by hydroelectric means, while in the Andean Community of countries, this percentage

[1] These premises with an emphasis on institutional consolidation and guaranteeing economic, social, and cultural rights are consistent with risk assessment (WEF, 2016). Unlike other regions that seem more threatened by natural catastrophes, water crises, or abrupt variations in energy prices, the concerns of Latin America and the Caribbean are firstly governance failures and secondly profound social instability associated with other economic factors, which come in third place.

[2] In keeping with its mandate, IRENA (2015) recommends decreasing water use for hydropower production and replacing it with renewable energy (mainly wind and solar).

rises to 71%. In Central America, Mexico produces 15% of its national electricity production by hydroelectric means; however, all the other countries produce a higher proportion of their electricity with this source, led by Costa Rica with 78%. The Caribbean countries are the exception since they do not depend on hydropower as they do not have much surface water (Escobar et al., 2011).

South America is the third region worldwide that added the most hydropower capacity in 2015 (IHA, 2016). At the same time, hydroelectric projects come second, after mining and oil, in causing social unrest due to the use of water resources (Martín & Justo, 2015). These data suffice to demonstrate that hydroelectricity production is and will be one of the key nexus interactions in the region. With this potential and growth rate, this sector provides an opportunity to apply a nexus focus combining the multipurpose nature of reservoirs (although the number of multipurpose reservoirs built since 2000 has dropped) more efficiently with optimization of their economic, social, and environmental impact.

Regional hydroelectric potential must be re-evaluated from a nexus and trinomial planning focus, as must hydropower projects in the design, construction, and development phases, based on future scenarios of flow rates, precipitation, and water use, since more installed capacity does not necessarily mean more generation. Instead of static plans and designs, hydroelectric infrastructure should be planned within the ranges of uncertainty imposed by climate and precipitation trends (Santos Da Silva et al., 2018). Consequently, the recommendation is for plans to contain provisions for specific operations depending on the climate and demand for water, based on a flexible infrastructure, and pumped-storage plants and smaller hydroelectric systems must be considered as options to overcome this climatic variability (Escobar et al., 2011).

Hydroelectricity production gives the system reliability, but it must be complemented by other forms of renewable energy, even if they are intermittent, to reduce the risks from depending excessively and exclusively on the hydroelectric source.[3] Permanent or seasonal changes or decreases in generation due to drought or reduced flow rates prove that the situation is worsening in some regions, countries, or catchment areas with the prediction of future scenarios (Vallejo, 2013; Flavin et al., 2014; Recalde, 2016). For example, one pessimistic scenario for the Chixoy plant in Guatemala and Cerrón Grande in El Salvador predicts generation reductions above 20% by 2020, more than 40% for 2050 and over 70% by the end of the century. A less pessimistic scenario predicts an increase of 4–6% by 2020, after which a decrease of 26% is estimated at Chixoy and 17% at Cerrón Grande by 2100 (CEPAL et al., 2015).

The effects of El Niño phenomenon on the region are drought (Caribbean, Colombia, northeast Brazil, and Venezuela) and flooding (Peru, south Brazil, Paraguay, Uruguay, Ecuador, Bolivia, and Argentina) with huge impacts on the energy sector. Unlike Peru and Ecuador, where the impacts of this phenomenon mainly affect infrastructure, in several cases in Colombia and Venezuela, they have led to rationing water and power, and production losses and higher costs have been associated with the increase in thermoelectric power to compensate for the drop in hydroelectricity generation. Andean countries are especially vulnerable to this phenomenon, which means a normal to extraordinary Niño could deduct 0.6–1.7% GDP from these countries (CAF, 2016; Martín, 2016).

Analyses of the region from a nexus point of view include one conducted in Bolivia using the United Nations Food and Agriculture Organization method (Flammini et al., 2014), which, in relation to the multipurpose San Jacinto hydroelectric project, found that

> more pressure is exercised on the food/land aspect given that crop productivity is below average. The

[3] This search should not be limited to the classic options, wind or photovoltaic, whose establishment is increasing in the region, but also others that are still in an experimental phase, such as wave power and tidal power. Chile's potential wave power is estimated to be unique in the world (Hassan, 2009). Exploiting just 10% of this potential would double installed capacity of the entire Central Interconnected System in Chile.

> above could be due to the quality of the vine for later vinicultural processing being championed more than gross sales in the area ... Furthermore, although the construction of the reservoir has limited the flow of fish downstream, the type and number of species have increased in the reservoir itself, in some cases without proper control, leading to some species becoming the predators of other preexisting species... the project... besides being beneficial for the region, also has trade-offs. The most significant is the impact on water quality in the reservoir due to agricultural returns containing pesticides. (Rojas & Heiland, 2015)

The obvious priority production of hydrocarbons and mining can also fall within the interaction between water and energy, as it uses both, due to reasons that include the extreme dependence of many countries in the region on these activities and proven reserves (Altomonte & Sánchez, 2016). This is the case of the Andean countries (Chile, Bolivia, Peru, and Ecuador) and others, such as Argentina, Brazil, Colombia, Mexico, and Venezuela. Water consumption in mining in Argentina, Brazil, Chile, Colombia, Costa Rica, Mexico, and Peru can account for up to 6% of all water use (Willaarts et al., 2014). Although detailed and reliable information is lacking on the impacts these activities cause on nexus elements, the implications of both hydrocarbon exploitation and (formal and informal) metal and nonmetal mining in nexus terms are relevant to the extent that they unlawfully occupy the first level in terms of potential for social conflict (Martín & Justo, 2015).

The region has very significant proven reserves of oil and natural gas (20% and 4%) compared with other reserves worldwide (Altomonte & Sánchez, 2016). One example within this group of recently developed activities that should be nexus focused is the exploitation of nonconventional hydrocarbons. The investigation and exploitation involved in the hydraulic fracturing process have far higher energy, water, and environmental impacts than the methods used for conventional resources. The region has more than a quarter of total worldwide technically recoverable natural shale gas resources, mostly located in Argentina, Mexico, and Brazil, which are in a full exploitation phase (Arroyo & Perdriel, 2015). The main risk of exploiting shale affects water, the key nexus element, while its essential role for the region is explained by (i) possible self-supply of energy in countries that have this resource; (ii) possible economic impact of lower power prices, less volatility in these prices, more economic growth, fewer inequality levels, generation of employment sources, and so on; and (iii) possible strengthening of energy integration in updated regional geopolitics.

The same occurs when water and energy are considered as supplies for metal and nonmetal mining, an activity that can involve an intensive use of both resources and which several countries in the region depend greatly on as potential for growth, especially gold (Peru), copper and silver (Chile, Mexico, and Peru), and iron (Brazil). Mining is also one of the fastest-growing sectors in the region, which is a main destination for mining investments planned worldwide (Altomonte & Sánchez, 2016).

Mining can impact both water quantity and quality. Impacts on quantity are more noticeable in areas where this resource is scarce. This is the situation in semiarid areas, such as central Mexico, north Colombia, and northeast Brazil, or directly arid areas, such as north Chile, northwest Argentina, east Bolivia, and south Peru. Exploiting nonrenewable groundwater or groundwater with a very low renewal rate can severely affect people's water security, as occurred in north Chile (Willaarts et al., 2014). Where water resources are abundant, the problem is avoiding polluting them.

Similarly, mining can involve intensive energy use of both hydrocarbons and electricity, thus competing with other sectors with the possibility of significant impacts. In some cases, the high profitability of mining can meet the energy cost of desalination in arid or semiarid areas.

9.2.1 Interrelations Between Energy and Water

Concerning the relationship between energy and water, the first point worth mentioning is the essential role energy plays in the more "traditional" techniques of seawater desalination, debrining of brackish water, pumping of groundwater, and irrigation modernization,[4] making it possible for these processes to take place. In addition, the complete urban water cycle is an example of high energy use to address the various sector processes this cycle requires, from supply and conveyance to treatment, if we consider only water and no other type of waste. Although desalination is not generally a priority in the region, it is beginning to become an increasingly attractive option, especially where there are no alternatives, as in the Caribbean (Hoff, 2011) or in arid locations, to perform high-profitability activities, such as copper mining in Chile or urban use in Mexico.

A recent study on Chile using a particular conceptual focus of the nexus in four regions in the country with a variety of problems (Antofagasta, Copiapó, Maipo, and Maule) confirms several of the interactions identified as priority in this study for the region – mining, aquifer overexploitation, urban expansion, hydroelectricity, and agriculture – and concludes that reuse, recycling, and desalination are becoming attractive management options in some arid and highly vulnerable areas (Meza et al., 2015). But its impacts and negative externalities must be well explained and understood by decision-makers, private managers, and society to coordinate and define adaptation policies and strategies aimed at consolidating the nexus.

Highlighted in this interconnected scenario are drinking water and sanitation services in regard to the expansion of their coverage and, especially, the improvement of their quality and the increase in municipal wastewater treatment for agriculture involving electro-intensive activities that are often inefficient. It is estimated that a significant proportion of the energy consumption of water supply and sanitation services occurs in the conveyance and distribution or collection phases, particularly the pumping of fluids in the drinking water supply. In contrast, in the sanitation system, electricity is most consumed in wastewater treatment. Sludge treatment and disposal represents significant consumption, although it can also generate energy (Ferro & Lentini, 2015). The electricity costs of drinking water and sanitation system providers represent 5–30% of total operating costs, and it is estimated that the total energy cost can be reduced by 5–15%.

Long-distance interbasin transfers require the use of extremely high amounts of energy for the conveyance (including lifting). In some places in the region (for example, Mexico City, San Pablo in Brazil), this is important even when these lifts can often be compensated by drops in the conveyed water and the possibility of installing hydropower plants in these sections. The above relationship is very important due to constant growth in energy consumption in the abovementioned processes. This is related to resorting to desalination once technical advances have led to a substantial drop in the cost of desalted water, which, in any event, is still far more than the cost of water collected by other means. The same can be said of the increased use of groundwater in agriculture, especially when the price of energy is subsidized in several countries. In view of high energy consumption, the main proposals comprise more use of renewable energy[5] and an increase in energy efficiency (Ferro & Lentini, 2015), and, strictly from an energy-water nexus perspective, including the lack of water in the energy decision-making process is insisted upon.

[4]"Irrigation modernization" usually refers to the process of transforming the irrigation technique from gravity (flooding) to localized or sprinkle irrigation. In theory, this saves water, but it also involves an increase in energy consumption inherent in any pressurized irrigation system. Saving water is not usually substantive given that a consequence of irrigation modernization is the increase in the number of crops and often also the expansion of irrigated land, so the supposed savings disappear. In any event, the inevitable increase in energy consumption and, therefore, energy costs, even with subsidized electricity prices, must always be considered.

[5]According to IRENA (2015), "With renewable energy technology that decreases costs, technological advances and an increase in economies of scale, renewable desalination will probably become significantly more cost-effective and will represent a key solution to mitigate the increasing development of risks posed by resource restrictions. The competitiveness of renewable desalination will be further improved if the volatile costs of fossil fuels are considered."

Finally, we must consider agricultural aspects that more clearly involve energy. The agrifood chain is estimated to consume around 30% of worldwide energy (FAO, 2011). The central place agriculture represents in the region is obvious: it uses around 80% of the water and occupies 25% of the total surface area. At the same time, there is considerable potential for expansion and possibilities to increase productivity associated, above all, with the implementation and modernization of irrigation, which, in most cases, is linked to an increase in energy consumption. Irrigation modernization and expansion and the increasing use of and dependence on groundwater form a relevant variable in some of the region's countries and could become critical for the nexus in many others in the future.

Based on the above considerations, we can conclude that irrigation expansion and modernization and aquifer overexploitation are the priority interconnection in this area. A more intensive or efficient use of water in agriculture involves an increase in energy consumption, which must be previously addressed using a trinomial focus. Modernizing irrigation without appropriate consideration of the hydrological, environmental, or energy impacts in regions with an energy deficit, supply insecurity, or price volatility can further producer vulnerability and increase their costs with the subsequent impact on agriculture and livestock production and its profitability and the future demand for subsidies from the public sector. These are factors that must be considered in the design of a nexus-focused policy for a sector that offers major implementation and expansion opportunities.

The overexploitation of some aquifers involves a series of problems and inefficiencies that could be modulated or prevented using a nexus focus. With low energy prices that are often subsidized to promote agricultural development, the overexploitation of aquifers can lead not only to a non-sustainable management of aquifers in hydrological and environmental terms with significant impacts on quality and quantity but also energy inefficiency and, in some cases, social injustice. Inefficiency stems from the higher energy cost of pumping, both on a global and individual scale, which increases as water-table levels drop. Also worth considering is the injustice for the losers in the re-entry race, higher costs of extraction, or use of electrical subsidies designed by policies that do not generally segment beneficiaries and end up favoring the major players, even to the detriment of priority uses, such as a municipal water supply as opposed to agriculture or mining. The unbridled race for re-entry drilling in overexploited aquifers often complicates municipal supply and human consumption due to the decrease in levels or quality, emphasizing its high individual and social cost on the one hand and no or weak controls in the land characterizing the region on the other. Examples of this situation can be found in Copiapó and Antofagasta (Chile), Valle de Ica (Peru), Guanajuato, Sonora, and Mexico City.

This interaction has received attention based on the water-energy-climate nexus focus in Mexico, for example, where implementing an energy price policy has been proposed to discourage pumping to reduce overexploitation of aquifers and improve their sustainable use (Scott & Shah, 2004). The conclusion reached is the need to consolidate nexus-based policies, such as an increase in the agricultural electricity tariff, the elimination of the low night tariff, the application of regulations conditioning groundwater extraction on energy use, and the establishment of differential energy prices based on a drop in aquifer levels or the limitation of new electrical connections for groundwater extraction (Scott, 2011). This type of specific proposals for overexploited aquifers should not overlook the fundamental issue that both energy and water have historically been – and still are – a means to make small-scale irrigation possible or make an agricultural sector characterized by direct and indirect subsidies of all types competitive.

9.3 Interrelations Between Water and Food

The relationship between water and food can be considered in the rural environment, where it refers to agriculture either with or without irrigation, and in the urban environment, where it is mainly linked to the provision of public drinking water and sanitation services, the supply of food to cities, and waste management (Covarrubias, 2019). Firstly, connections are established using water for two coincident purposes in food production: irrigation and aquaculture, with far more importance placed on the former than the latter. Water use percentages in this area compared with other sectors can vary depending on the country in question, but they are around 70–80% and perhaps even 80–90% (Hoff, 2011). This is the most important use and consumption since returns to the cycle do not usually exceed 20–30% in plot irrigation, even when much higher losses occur in the conveyance to the plot when the channels are not lined, which is usually the case. Similarly, this water use enjoys a privileged position in countries whose legislation is based on a hierarchy of uses (irrigation, as the same cannot be said of aquaculture), normally only exceeded by the municipal water supply.

A first conclusion of this situation is the need for the expansion of the "agricultural frontier" to be subject to rigorous controls due to the implications of water and energy consumption. That is why developed countries have reformist policies focusing on reducing agricultural water use so that this "excess" water can be earmarked for other uses that can create more added value. These policies aimed at reducing agricultural water use seek to relaunch some industrial uses without affecting food production. Normally these countries are either self-sufficient, produce a surplus, or, because of their economic capacity, can acquire the food in international markets or establish policies (for irrigation modernization) to save water and achieve the same level of food production.

We should also be aware that this usually causes conflicts between the organized structures representing farmers and those representing other uses, even with their respective governments. It is partly for this reason that appropriate use of the nexus focus could bring about a decrease in the conflict between different water uses (between its owners and also considering its informal users, who are important in many parts of the region). This is primarily due to the predictability of impacts on the various nexus components arising from policies designed with knowledge of their effects and, therefore, with the ability to correct them at source.

In developing countries, however, policies to intensify irrigated land and a desire to achieve food security or, in other words, food sovereignty are also common. One example is the *Ley de la Década del Riego* (Irrigation Decade Act) 2015–2025, No. 745 dated 5 October 2015, in Bolivia. Its objective is to provide one million hectares with irrigation for those 10 years "to promote agricultural and livestock production through investments made by the State and autonomous regional entities for irrigation development."

Above all, the prime aim is to achieve maximum effectiveness from existing irrigated land. In general, irrigation efficiency in the region is not very good, an issue often primarily determined by water conveyance infrastructures, which are not always lined to prevent leaks. This type of action requires a sufficient quantity and quality of water to be available, the financial means for the investments, and then, an aspect that is often overlooked, appropriate organizational structures (both within the irrigation organization and government-level administration) to effectively and productively manage this increase in irrigated agricultural land. Some of the conditions also required are an administration with appropriate technical levels to attribute with legal certainty the necessary water flow rates (by means of concessions or other systems) to address this new irrigated surface area and a registry of public water that provides an exact picture of the legal water distribution and use situation. This is extremely hard to find and often compromises the success of public policies.

The land grabbing process that is taking place above all in Africa is part and parcel of the same negative dynamics. Some private companies or even countries are buying land in bulk in developing countries (Borras et al., 2011; Mehta et al., 2012) to produce food whose future destination is their own countries of origin. Even when food production is wielded as the primary justification for these purchases, the underlying reason for them is, above all, the desire to seize the water linked to this land, especially in countries with weak management, tax, and control systems.

To prevent some of the risks involved with irrigation, emphasis is sometimes placed on more capacity to improve food production with "green water," which comes from rain, is in the ground, and is linked to nonirrigated crops, in contrast to "blue water," which is artificially conveyed by various channel types and leads to irrigated crops. The argument is that where there is not enough water, improvements can only be made by increasing the productivity of existing water and limiting population growth, which will lead to food security (Falkenmark & Rockström, 2011). The relationship can also improve with an increase in municipal wastewater treatment reaching quality levels that enable its use for irrigation. This is feasible near large inhabited population centers where the costs of wastewater treatment and regeneration facilities can be defrayed. Rainfall harvesting, in other words collecting precipitation, has a role to play in this relationship.

Finally, in the water-food relationship, we need to consider the theoretical construction of what is known as "virtual water." This is the water transported (embodied) with the food a country imports (or exports). A focus of this kind can help justify public policies based on the country's position (area, basin, or region) as an exporter or importer of virtual water in relation to some food. Evaluations of this situation together with a water situation that is generally surplus or deficient can make it advisable to intensify some production policies of some food types or, in contrast, lead to importing them, which entails considering the economic and social conditions of each case.

The major issue raised by the concept of virtual water trade is whether this will result in water savings. Hoekstra and Chapagain (2008) assessed the global water savings that can be gleaned from agricultural trade at 352,000 million cubic meters per year. Given that worldwide trends point towards an increase in consumption and reduction in the availability of water resources, these focuses can obviously help countries, especially those with water shortages, manage their water and agricultural policies better based on improved information. Research into virtual water trade in LAC has shown that the region is a net exporter of virtual water and that the cases of Argentina and Brazil are quantitatively important (Willaarts et al., 2014).

No exploration of the water-food relationship for the purposes of the nexus can ignore the possibilities virtual water trade offers. From a nexus perspective, ceasing the production of a food product with a large water footprint and replacing it for consumption purposes by the same imported product may make it possible to free up the water for other uses, such as producing energy or caring for the environment. This country's specific situation of water and energy requirements can rely on a powerful instrument to guide it along a path that augers positive possibilities, at least theoretically.

However, despite an extraordinary potential in water and energy resources and food production capacity, in most of the countries, the region still owes its own population a huge debt where food is concerned (Martínez & Palma, 2016). In other words, the potential implementation of a nexus focus cannot sidestep that it is not the food, the water resources, the energy sources, or the agricultural development that is lacking in the region but rather a minimal parity in sustainable development models and their distribution, as posited by some nexus focuses (Biggs et al., 2015). It follows, therefore, that agriculture is not directly equivalent to food.

This paradox highlights the need to complement and connect the new nexus concern and water-food-energy security with the immediate scope of minimum parity levels and satisfaction of human rights required by the legal system and closely linked to the management of each of the nexus elements, in particular, the right to food being indiscernible from the right to water. If the situation is examined from a security point of view, the paradox can be seen in the key role the region has been given in ensuring global food security (Bellfield, 2015) while it seems incapable of guaranteeing minimum acceptable security levels in the three elements for itself.

One aspect that has better links to food security, generating employment and reducing poverty, and more potential to improve them – observed cross-sectionally in all the countries of the region and which characterizes as a production method – is family agriculture. This form of economic organization coexists in the region with the medium-sized and large enterprises, but it has been undervalued by governments and society.

9.4 Interrelations Between Water, Energy, and Food

The urban nexus become critical in the region for several reasons, including the following: (i) 80% of the population is urban with forecasts for more increases; (ii) current urban expansion has hardly been planned; (iii) the existence of several megacities or large urban conglomerates as critical spaces for urban supplies and increasing energy costs; (iv) the lack of sustainability, vulnerability, and spatial segregation; and (v) the deficient and low coverage of public services, in particularly sanitation and treatment systems (CEPAL, 2014). This phenomenon, which clearly characterizes the region as having one of the highest urban population percentages worldwide, represents a cross-sectional problem as it involves the three nexus elements with specific interactions in urban contexts in developing countries in general (Covarrubias, 2019; Babette, 2016) and Latin American megacities in particular.

Interactions in the production and management of the three nexus elements are essential; however, interactions are also important in the transport and consumption method of food and waste production, treatment, and disposal predominantly in urban contexts. This dependence and reciprocal influence of the city on the locality producing goods and services, food, water, and energy has firstly led to reassessing the scale of land-use planning that has usually focused on its urban aspects (Artioli et al., 2017; Heard et al., 2017). Some urban nexus priority interconnections should be included in the main urban problems in the region: no or deficient public services, pollution and environmental degradation, transport, waste, poverty, social segmentation, and insecurity. This leads us to posit that land-use planning and nexus interactions in urban contexts, especially referring to the provision of public services in the region, are priority or critical areas.

Based on the above considerations, we conclude that drinking water and sanitation services are priority interconnections. The urban water cycle includes conveyance from the supply point (normally located outside the urban area and sometimes even outside the municipality), drinking water treatment, conveyance of treated drinking water to the various connection points, collection of wastewater, conduits to the associated wastewater treatment plants, wastewater treatment, sludge treatment, and, ideally, the regeneration of urban wastewater so that it can be reused with suitable quality levels for a variety of purposes including agriculture.

However, the best example of the trilateral relationship[6] might be biomass production using plant products considered as a source of energy. The growth of this phenomenon has been substantial in recent years in some geographical areas, especially in the region (Saulino, 2011; Hoff, 2011). However, the positive elements of a new energy source classified as "renewable" need to be clarified given that this production of plants to generate energy can involve a decrease in food production (resulting from the associated reduction in agricultural land for that purpose) and also remove some of the water linked to irrigated agriculture.[7] If the biomass is forest waste, the ecological function of the forests can be affected. Another consequence can be an increase in food prices, although the cause–effect relationship between biomass production (worldwide) and

[6] There is an FAO method to measure the impacts of water and energy decisions on food (Flammini et al., 2014).

[7] In this respect, the European Commission (2011) stated that: "Intensive agriculture whose purpose is to produce more food and biomass, could intensify freshwater demand for irrigation and, therefore, pressure on water resources. Water solutions need to be established quickly to increase irrigation efficiency, reduce water consumption and manage and preserve aquifers sustainably."

agricultural prices is not unanimously accepted (Martín Mateo, 2008).

These risks can be prevented with appropriate public intervention, which can operate on several levels. The first is the authorization of energy production facilities with this origin. Another is the intervention that can take place on agricultural or forest land that has been earmarked for these purposes. Normally both intervention types are necessary.[8] At the 2011 Bonn conference, a preventive approach was outlined to the effect that "developing countries considering bioenergy must weigh up every factor and consequence, such as land suitability, water availability, competitiveness, socio-economic costs and benefits, food security, economic growth and poverty alleviation" (Hoff, 2011).

Another important aspect that needs to be considered is that one-third of food produced is thrown away or lost, which is also a source of concern about the nexus (Martin-Nagle et al., 2012). Globally this food loss represents 15% of the total energy demand, energy consumed in vain to produce unused food. Even more significant is the impact of this food loss on water use, since non-used food is calculated to represent 20–30% of total water use, and also on land use (Hoff, 2011). This is an important aspect that should be considered when establishing policies (food storage in refrigerated systems, safer transport, education for small-scale farmers and their families, and so on) to prevent this outcome.

Finally, concerning the interrelations between agriculture and climate change, the agricultural activity globally generates 10–12% of greenhouse gases and 47% of methane emissions (IRTA, 2016). Similarly, electricity and heat production contributes 27% to the total greenhouse gases worldwide (Martin-Nagle et al., 2012). Environmental considerations are very important in the nexus: perspectives that are only sectoral (from the energy and food production angle), purely productivist, involve the risk of resulting in maximum water use with the subsequent degradation of the resource, which would eventually affect energy and food production, whose maximization was precisely the cause of the degradation of the water resource. Based on the above considerations, we can conclude that biofuels are the priority interconnection in LAC in this area.

The region is a net exporter of raw materials and food with a huge potential still for more development. In particular, Brazil and Argentina are among the major biofuel producers, although other countries, such as Colombia, Paraguay, and Peru, are producers to a lesser extent. Nevertheless, we should not overlook countries moving in opposite directions. An example is Bolivia, where the *Ley Marco de la Madre Tierra y Desarrollo Integral para Vivir Bien* (Framework Act for Mother Earth and Comprehensive Development to Live Well) No. 300 dated 15 October 2012, prohibits "the production of biofuels and the selling of agricultural products to produce biofuels since it is the priority of the Plurinational State of Bolivia to safeguard sovereignty with food security."

In the cases of Brazil and Argentina, the production of these biofuels has greatly increased, as has their consumption, with a relationship between production and consumption level and the price of oil. The result of this production activity is a huge increase in water demand, an increase that is less in Brazil than in Argentina, since sugarcane is not irrigated in Brazil but in Tucumán (Argentina), where the most production is concentrated. In both cases, however, water pollution has risen due to the use of fertilizers and pesticides in the production process of plants, and similarly, pollution has increased in the water arising from the biofuel production process (Saulino, 2011). Out of the food-energy interactions, biofuel production is especially relevant for the region (Mirzabaev et al., 2015). From this perspective, the nexus focus involves the need to revise the conditions in which some countries' commitment to producing biofuels has occurred (Saulino, 2011; Scott et al., 2015).

[8]In Spain, for example, electrical facilities for this energy production are subject to authorization. In addition to this is intervention, above all with an environmental purpose, on forest land for biomass production. Intervention on agricultural land is mainly guided by a water perspective, since concessions must expressly refer to irrigated land and crops needing the water.

9.5 Priority Interrelations for LAC Region

Unlike previous considerations, the nexus focus examines the multiple interactions of the three elements in the climate change context to define priorities and avoid harmful or undesired effects, with environmental protection always in the background. Out of the activities involving two or more nexus elements, considering the following is obviously a priority for the region:

(a) Water-energy: Hydropower, hydrocarbons, and mining/water abstraction, use, and desalination.

Most energy production methods require water, although hydroelectricity is the most important method in the region as the main source of power offering most future growth in the majority of South and Central American countries. Large-scale hydropower demonstrates multiple interconnections, and excessive dependence on this source, considering climate change and variability, simultaneously threatens water, energy, and food security in countries such as Colombia, Venezuela, and some in the Caribbean.

The exploitation of hydrocarbons and mining demands variable quantities of water and energy, which can seriously affect the environment and the quality of water resources. This interconnection is highly relevant throughout almost the entire region, but very especially in the Andean countries, Brazil, Mexico, Venezuela, and some in Central America. The relationship could become especially intense when hydraulic fracturing techniques are used. The use of water for energy does not compare with water use for agriculture in quantity (except in arid or semiarid regions), but it is the use that causes the most social unrest due to the displacement of people, its associated consequences, and the impact on the sources' quality.

On the other side, the largest energy cost in relation to water in the region occurs in the abstraction stages of groundwater, conveyance, and use, which include irrigation. This interconnection must especially consider the subsidy level for extraction, aquifer overexploitation, and the inefficiency of irrigation systems and pumping equipment. The relevance of groundwater and growing dependence on it are shared throughout the region with an emphasis on Central America and Mexico, where it amounts to 65% of consumed water and in the desert or semidesert areas of Argentina, Brazil, Chile, Bolivia, Mexico, and Peru. The increasing overexploitation of aquifers presents interrelations with the three nexus elements by impacting water quantity and quality, removing land from production and increasing the energy costs of extracting it. Currently, energy consumption is not significant in water treatment or seawater desalination, which are confined to isolated areas for highly profitable activities (mainly in Chile, Mexico, Peru, and some Caribbean countries).

(b) Water-food: Agriculture.

The importance of agriculture must be understood in relation to regional peculiarities where large-scale practice and expansion, mainly for exporting, have a direct relationship with deforestation, single crop growing, the subsequent increase in the risk of diffuse pollution, sedimentation, erosion and flooding, the displacement of the local population, and the impact of family or subsistence agriculture, essential for food security in the region. Its relevance is key for the region in terms of water consumption, participation in the gross domestic product, and employment forecasts, and it affects almost all the countries in the region.

(c) Water-energy-food: Drinking water and sanitation services, biofuels, and irrigation modernization.

Drinking water and sanitation services are one of the high priority interconnections in urban contexts since the region has one of the highest urban population percentages worldwide, with deficient and low coverage of the sanitation and treatment system. All these features along with other specific interconnections represent major cross-sectional challenges for LAC developing countries and in non-planning growing megacities such as Buenos Aires, Sao Pablo, Rio de Janeiro, Lima, México, Santiago, and Bogotá.

The three nexus elements take part in biofuel production given that they usually consume water, are used to produce energy, and can affect food production due to removal. Agriculture for energy production or biofuels not only share the impacts of large-scale agriculture referred to above but also have a considerable effect on food availability and price. The development of biofuels is particularly relevant in Argentina, Brazil, Paraguay, and, to a lesser extent, Peru, Colombia, and Central American countries, such as Costa Rica. The relationship between the three elements can also be observed in the agricultural sector when irrigation modernization is involved (which leads to more energy use, more water consumption, and could increase food production) or policies are established to promote electricity, fostering aquifer overexploitation, based on discounted tariffs, given that water abstraction increases.

9.6 Conclusions

Although the previous summary is not exhaustive, it synthesizes the priority nexus interactions that are crucial for most of the region's countries. When considered in public policy planning and adoption, they can help prevent proven harmful effects and instead extract all the advantages and potential the focus entails.

The countries of the region have not yet incorporated the nexus approach for designing their public policies, planning, and management of natural resources. In addition, most of the countries present serious deficiencies in governance systems and lack of reliable information that makes planning, the cornerstone for the implementation of the nexus approach, very difficult.

In order to prioritize interconnections and avoid conflicts, this planning must be adaptive, multiscalar, pluri-temporal, intersectoral, built in a participative process, and based on a broad consensus of all political forces, guaranteeing both its immediate feasibility and its permanence over time. A coordinated planning of the three nexus elements can constitute a new impulse to reinforce, integrate, and give greater consistency to sectoral processes of territorial, water, energy and environmental planning, in different scales – regional, integrated, national electrical system, local, and basin. However, hydrological basin planning will be a preferred instrument for these purposes.

The nexus approach must include the entire satisfaction of the economic, social, and cultural rights associated with the elements of the nexus as the first objective of nexus policies. This is not only because these human rights are expressly recognized in regional instruments but also because there cannot be a development, an increase in synergies and efficiency or sustainability, which do not have as an immediate objective the urgent satisfaction of these minimum vital needs, still pendent in many countries of the region. This statement connects directly with the idea of water, food, and energy security underlying the nexus approach.

In the current development phase of the region, faced with the challenge of a costly energy transition in a threatening climate change context, the nexus approach can make significant contributions to the adjustment, diversification, adaptation, or transformation of development patterns models that are absolutely natural resource use intensive but clearly unsustainable and inequitable from the environmental and the social points of view, respectively.

References

Altomonte, H., & Sánchez, R. (2016). *Hacia una nueva gobernanza de los recursos naturales en América Latina y el Caribe* (LC/G.2679-P). Comisión Económica para América Latina y el Caribe (CEPAL).

Arroyo, A., & Perdriel, A. (2015). *Gobernanza del gas natural no convencional para el desarrollo sostenible de América Latina y el Caribe: Experiencias generales y tendencias en la Argentina, el Brasil, Colombia y México* (LC/L.3948). Comisión Económica para América Latina y el Caribe (CEPAL).

Artioli, F., Acuto, M., & McArthur, J. (2017). The water-energy-food nexus: An integration agenda and implications for urban governance. *Political Geography, 61*(2017), 215–223. https://doi.org/10.1016/j.polgeo.2017.08.009

Babette, N. (2016). *Wastewater systems and energy saving in urban India: Governing the water-energy-food nexus series* (Discussion paper, 12/2016). Deutsches Institut für Entwicklungspolitik.

Bellfield, H. (2015). *Water, energy and food security nexus in Latin America and the Caribbean. Trade-offs, strategic priorities and entry points*. Global Canopy Programme.

Biggs, E., Bruce, E., Boruff, B., Duncan, J., Horsley, J., Pauli, N., McNeill, K., Neef, A., Van Ogtrop, F., Curnow, J., Haworth, B., Duce, S., & Imanari, Y. (2015). Sustainable development and the water-energy-food nexus: A perspective on livelihoods. *Environmental Science & Policy, 54*, 389.

Borras S.M. (jr.), Franco J.C., Kay C., Spoor M. (2011), El acaparamiento de tierras en América Latina y el Caribe visto desde una perspectiva internacional más amplia, FAO.

CAF (Corporación Andina de Fomento). (2016). *Informe El Niño en América Latina: ¿Cómo mitigar sus efectos en el sector eléctrico*. CAF.

CEPAL. (2014). *Panorama Social de América Latina 2014* (LC/G.2635-P).

CEPAL (Comisión Económica para América Latina y el Caribe) (2015), Anuario Estadístico de América Latina y el Caribe 2015, LC/G.2656-P, Santiago de Chile.

Comisión Europea (2011), Recomendación de la Comisión, de 27 de octubre de 2011, relativa a la iniciativa de programación conjunta de investigación "El desafío del agua para un mundo en mutación", C 317/1.

Covarrubias, M. (2019). The nexus between water, energy and food in cities: Towards conceptualizing socio-material interconnections. *Sustainability Science, 14*, 277–287.

De Andrade Guerra, J. B. S. O., Berchin, I. I., Garcia, J., et al. (2020). A literature-based study on the water–energy–food nexus for sustainable development. *Stochastic Environmental Research and Risk Assessment, 35*, 95–116. https://doi.org/10.1007/s00477-020-01772-6

Embid, A., & Martín, L. (2017). *El nexo entre el agua, la energía y la alimentación en América Latina y el Caribe. Planificación, marco normativo e identificación de interconexiones prioritarias. SRNI* (71 pp). CEPAL.

Embid, A., & Martín, L. (2018). *Lineamientos de políticas públicas. Un mejor manejo de las interrelaciones del Nexo entre el agua, la energía y la alimentación, SRNI* (83 pp). CEPAL.

Escobar, M., López, F. F., & Clark, V. (2011). *Planificación de Energía, Agua y Clima para el Desarrollo sin Carbono en Latinoamérica y el Caribe*. Stockholm Environment Institute (SEI).

Falkenmark, M., & Rockström, J. (2011). Back to basis on water as constraint for global food production: Opportunities and limitations. In A. Garrido & H. Ingram (Eds.), *Water for food in a changing world*. Routledge.

FAO (Organización de las Naciones Unidas para la Alimentación y la Agricultura) (2011), Estado de los recursos de tierras y aguas del mundo para la alimentación y la agricultura. Gestión de sistemas en peligro, Roma.

Ferro, G., & Lentini, E. (2015). *Eficiencia energética y regulación económica en los servicios de agua potable y alcantarillado, Comisión Económica para América Latina y el Caribe (CEPAL)* (LC/L.3949)

Flammini, Alessandro; Manas Puri; Lucie Pluschke y Olivier Dubois (2014), Walking the Nexus Talk: Assessing the Water-Energy-Food Nexus in the Context of the Sustainable Energy for All Initiative, Organización de las Naciones Unidas para la Alimentación y la Agricultura (FAO), Roma.

Flavin, C., González, M., Majano, A. M., Ochs, A., da Rocha, M., & Tagwerker, P. (2014). *Study on the development of the renewable energy market in Latin America and the Caribbean* (OVE/WP-02/14). Banco Interamericano de Desarrollo (IDB).

Hassan, Garrad (2009), Preliminary site selection - Chilean Marine Energy Resources, Garrad Hassan and Partners Limited.

Heard, B. R., Miller Shelie, A., Liang, S., & Xu, M. (2017). Emerging challenges and opportunities for the food–energy–water nexus in urban systems. *Current Opinion in Chemical Engineering, 17*, 48–53. https://doi.org/10.1016/j.coche.2017.06.006

Hoekstra, A. y Chapagain, A.K. (2008), Globalization of water; Sharing the planet's freshwater resources. Oxford, Blackwell Publishing.

Hoff, H. (2011). *Understanding the nexus. Background paper for the Bonn 2011 conference: The water, energy and food security nexus*. Stockholm Environment Institute.

IHA (International Hydropower Association). (2016). *Briefing. 2016 key trends in hydropower*.

IRENA (International Renewable Energy Agency). (2015). *Renewable energy in the water, energy & food nexus*.

IRTA (Instituto de Investigación y Tecnología Agroalimentarias) (2016), http://www.irta.cat, 20 de febrero.

Mahlknecht, J. G.-B., Loge, R., & Frank, J. (2020). Water-energy-food security: A nexus perspective of the current situation in Latin America and the Caribbean. *Energy, 194*, 116824.

Mehta L., Veldwisch G.J., Franco J. (2012), "Introduction to the special issue: Water Grabbing? Focus on the (Re)appropriation of Finite Water Resources", Water Alternatives, pp. 193–207.

Martín, L., & Justo, J. (2015). *Análisis, prevención y resolución de conflictos por el agua en América Latina y el Caribe* (LC/L.3991). Comisión Económica para América Latina y el Caribe (CEPAL).

Martin Mateo, R. (2008). *La verde energía de la biomasa*. Civitas.

Martín, Lucía (2016), "¡Es Niño! Impacto económico en la Región Andina", Nota Técnica, N⁰ IDB-TN-951, Banco Interamericano de Desarrollo (IDB).

Martínez, R., & Palma, A. (2016). *Plan para la seguridad alimentaria, nutrición y erradicación del hambre de la CELAC 2025. Una propuesta para seguimiento y análisis* (LC/L.4136). Comisión Económica para América Latina y el Caribe (CEPAL).

Martin-Nagle, Renee; Elizabeth Howard; Alyssa Wiltse David Duncan (2012), Bonn 2011 conference "The water, energy and food security nexus" – Solutions for the green economy, 16–18 November 2011. Conference synopsis. Federal Ministry for the Environment, Nature Conservation and Nuclear Safety (BMU), Federal Ministry for Economic Cooperation and Development (BMZ), OOSKAnews, Inc.

Meza, F., Vicuña, S., Gironás, J., Poblete, D., Suárez, F., & Oertel, M. (2015). Water-food-energy nexus in Chile: The challenges due to global change in different regional contexts. *Water International, 40*(N° 5–6).

Mirzabaev, A., Guta, D., Goedecke, J., Gaur, V., Börner, J., Virchow, D., Denich, M., & von Braun, J. (2015). Bioenergy, food security and poverty reduction: Trade-offs and synergies along the water-energy-food security nexus. *Water International, 40*(N° 5–6).

Naranjo, L., & Willaarts, B. A. (2020). *Guía metodológica: diseño de acciones con enfoque del Nexo entre agua, energía y alimentación para países de América Latina y el Caribe* (LC/TS.2020/117) (83 pp). Cepal.

Recalde, M. Y. (2016). The different paths for renewable energies in Latin American countries: The relevance of the enabling frameworks and the design of instruments. *Wiley Interdisciplinary Reviews: Energy and Environment, 5*(N° 3).

Rojas, Franz O. & Heiland, Michael, Informe Final Primera evaluación en torno al Nexo Agua -Energía-Seguridad Alimentaria. Estado Prurinacional de Bolivia. MMAyA/GIZ, 2015.

Santos Da Silva, S. R., et al. (2018). *Energy-water-land nexus in Latin America and the Caribbean: A perspective from the Paris agreement climate mitigation pledges* (IDB working paper series, no. IDB-WP-901). Inter-American Development Bank (IDB). https://doi.org/10.18235/0001179

Saulino, F. (2011). *Implicaciones del desarrollo de los biocombustibles para la gestión y el aprovechamiento del agua* (LC/W.445). Comisión Económica para América Latina y el Caribe (CEPAL).

Scott, C. & Shah, T, (2004), "Groundwater Overdraft Reduction through Agricultural Energy Policy: Insights from India and Mexico", Water Resources Development, Vol. 20, No. 2, 149–164, June 2004.

Scott, C., Kurian, M., & Wescoat, J. (2015). The water-energy-food nexus: Enhancing adaptive capacity to complex global challenges. In M. Kurian & R. Ardakanian (Eds.), *Governing the nexus: Water, soil and waste resources considering global change.* Springer.

Scott, C. A. (2011), The water-energy-climate nexus: Resources and policy outlook for aquifers in Mexico, Water Resources., 47, W00L04, https://doi.org/10.1029/2011WR010805.

Tashtoush, F. M., Al-Zubari, W. K., & Shah, A. (2019). A review of the water–energy–food nexus measurement and management approach. *Int J Energ Water Res, 3*, 361–374. https://doi.org/10.1007/s42108-019-00042-8

Urquiza, A., & Billi, M. (2020). *Seguridad hídrica y energética en América Latina y el Caribe: Definición y aproximación territorial para el análisis de brechas y riesgos de la población* (Documentos de Proyectos (LC/TS.2020/138)). ECLAC.

Vallejo, M. C. (2013). Seguridad Energética y Diversificación en América Latina: el Caso de la Hidroenergía. *Revista Retos, 6*.

WEF (World Economic Forum) (2011), Water security: The water-food-energy-climate nexus: The world economic forum water initiative.

WEF (World Economic Forum) (2016), The Global Risks Report 2016. 11th Edition, Ginebra.

Willaarts, B., Garrido, A., & Llamas, R. (Eds.). (2014). *Water for food security and Well-being in Latin American and the Caribbean.* Earthscan Studies in Water Resource Management.

Willaarts, B., et al. (2021). *Análisis comparativo de acciones con enfoque del Nexo Agua-Energía-Alimentación: lecciones aprendidas para los países de América Latina y el Caribe* (serie Recursos Naturales y Desarrollo, N° 204 (LC/TS.2021/18)). ECLAC.

10 Methods for Evaluating Food-Energy-Water Nexus: Data Envelopment Analysis and Network Equilibrium Model Approaches

Paulo Nocera Alves Junior,
José Eduardo Holler Branco,
Daniela Bacchi Bartholomeu,
José Cesar Cruz Junior,
and José Vicente Caixeta Filho

Abstract

Understanding the food, energy, and water nexus (FEW Nexus) and more broadly the environment and socioeconomic impact of the food supply chains has become an important topic on policy-makers' agendas all over the world. Particularly in Brazil, this analysis is of great value because of the country's importance in world food production, and it's expected that Brazil will be among the main players in supplying the additional food demand that will be requested because of the world population expansion. In this chapter, our objective is to present a model and describe modeling frameworks that can be used to measure the FEW Nexus performance in production and transportation, among other systems, and present cases of the utilization of this kind of tool in Brazil. We present a dual-step procedure that combines the Data Envelopment Analysis (DEA) method with the Network Equilibrium Model (NEM). This approach is used to evaluate the cost and energy from transportation and its expansion, at the same time promote food production and productivity growth, considering the occurrence of proper water balance promoted by natural rainfall, for a sustainable agricultural frontier expansion in Brazil under the FEW Nexus context. We also include the mitigation of CO_2emissions in our model.

P. N. A. Junior (✉)
Escuela de Ingeniería de Coquimbo (EIC), Universidad Católica del Norte (UCN), Coquimbo, Chile
e-mail: paulo.alves@ce.ucn.cl

J. E. H. Branco
Group of Research and Extension in Agro-industrial Logistics at the Luiz de Queiroz College of Agriculture (ESALQ-LOG) University of São Paulo (USP), Piracicaba, SP, Brazil
e-mail: jehollerbranco@gmail.com

D. B. Bartholomeu
Luiz de Queiroz College of Agriculture (USP/ESALQ), University of São Paulo (USP), Piracicaba, SP, Brazil
e-mail: daniela.bartholomeu@alumni.usp.br

J. C. C. Junior
Department of Economics, Federal University of São Carlos (UFSCAR), São Carlos, SP, Brazil
e-mail: cesarcruzjr@gmail.com

J. V. C. Filho
Department of Economics, Management and Sociology of the Luiz de Queiroz College of Agriculture (USP/ESALQ), University of São Paulo (USP), Piracicaba, SP, Brazil
e-mail: jose.caixeta@usp.br

L. L. B. Lazaro et al. (eds.), *Water-Energy-Food Nexus and Climate Change in Cities*, Sustainable Development Goals Series, https://doi.org/10.1007/978-3-031-05472-3_10

Keywords

Water-energy-food nexus · Network Equilibrium Model (NEM) · FEW Nexus · Data Envelopment Analysis (DEA)

10.1 Introduction

Agriculture in the twenty-first century faces multiple challenges (FAO, 2009, 2017). The population is estimated to reach 9.7 billion people in 2050 (United Nations, 2019), while urbanization is expected to increase, accounting for 70 percent of the world population. Simultaneously, the world will still be managing the issue of economic deprivation and malnutrition of significant parts of the population (FAO, 2009). Demand for cereals – for both food and animal feed uses – is projected to reach about 3 billion tonnes by 2050 (European Comission, 2019). These trends mean that food security will continue to be a key driver of sociopolitical priorities at the global, regional, and national levels (European Comission, 2019).

Together, these movements in the world food market imply many challenges, such as increasing total food availability; sustainably improving agricultural productivity; satisfying the increasing diversification of consumers' basket; meeting quality, safety, environment, welfare, and ethical standards; ending hunger and malnutrition; addressing climate change; and keeping food affordable (European Comission, 2019; FAO, 2017).

Some studies analyze how production would respond to these trends and challenges. The pressure on natural resources, such as arable land and water use, will necessarily increase. FAO (2009) and Alexandratos and Bruinsma (2012) point out that the problem is that these natural resources are very unevenly distributed, with an increasing number of countries or regions reaching alarming levels of land and water scarcity (FAO, 2009).

In general, the FAO (2009) remarks that 80 percent of the growth in crop production in developing countries is expected to come from higher yields and increased cropping intensity, with the remainder coming from land expansion. Crop yields would continue to grow but at a.

slower rate than in the past (FAO, 2009). Resource constraints for agricultural production have become relatively more stringent than in the past, while the growth of yields is slowing down (Alexandratos & Bruinsma, 2012).

The expansion in food production will occur mainly in developing countries, especially in sub-Saharan Africa and Latin America (FAO, 2009). Brazil figures as an important world food supplier, being the world leader in the exports of soybeans, beef meat, poultry, sugar, orange juice, and coffee (USDA, 2018; apud Calil et al., 2019).

Soybeans and corn are the main grains produced in Brazil. Their production was about 210 million tonnes in 2019 and is expected to reach 266 million tonnes in 2029. To address this expansion, the cropland area used for soybean and corn production should increase from 23 million ha in 2019 to 34 million ha in the next 10 years (MAPA, 2019).

More resources, such as land, energy, water, and fertilizer, will be requested for increasing food production in Brazil. Moreover, distances between production areas and consumers (urban centers or ports, in the case of exports) have been increasing due to the expansion of the agricultural cropland frontier to more remote areas in the last decades. This expansion is advancing without a good planning for improving the sustainability performance of the supply chains. Providing integrated planning including production land use, transportation, processing, and distribution until the final consumer, aiming at minimizing the use of natural and energy resources, can make the food system more efficient and sustainable.

The best regions for expanding the soybean and corn production in Brazil depend on the availability of suitable cropland area without legally protected natural forests and the occurrence of a good water balance promoted by natural rainfall and will be affected by the configuration of the transportation system. There is an important trade-off that must be considered: there are regions that have good agriculture eco-performance but present higher distances and higher transportation costs and emit higher levels of greenhouse gases (GHG). In view of those characteristics regarding land use and transportation, it is desirable that the planning of the Brazilian grain production expansion goal at balancing this trade-off.

In this chapter, our overall objective is to present modeling frameworks that could be applied to evaluate and improve the FEW Nexus performance of various agriculture systems, including the Brazilian food supply chains.

Based on the modeling results, we intend to present an analysis of the most efficient regions to produce soybeans and corn in Brazil, considering criteria such as rainfall availability, yield, energy consumed, and CO_2 emissions in the transport.

This chapter will present a dual-step procedure using a Data Envelopment Analysis (DEA) and a Network Equilibrium Model (NEM). While DEA classifies suitable regions for a production growing according to their agriculture eco-performance, NEM finds the optimal spatial distribution of the soybean and corn production expected in the future and the optimal interregional transportation flows of cargo between the supply and demand regions.

We show two case studies to illustrate the application of this procedure for planning future corn and soybean production in Brazil.

Our approach can be used to guide strategies aiming at the expansion of the food production. The results obtained with our model reveal an efficient set of policies that can be used to reach the expected future food demand while improving supply chain sustainability. The model formulation avoids that production expansion occurs in areas covered with natural forests, reduces the GHG emission by land use and transportation, minimizes the energy from transportation, and reduces the need of water via supplemental irrigation, benefiting lower and competitive costs.

10.2 Data Envelopment Analysis (DEA) and Network Equilibrium Model (NEM): Some Insights About the Models and Applications

10.2.1 Data Envelopment Analysis (DEA)

According to Lampe and Hilgers (2015) and Chen et al. (2015), usually stochastic frontier analysis (SFA) is applied to the economic field, while usually DEA is applied to the operations research field. And comparing these methods, generally, common SFA models consider one output and multiple inputs and it needs a functional form, while DEA always allows multiple outputs and multiple inputs, besides not requiring a specific functional form.

According to Farrell (1957), one of the types of efficiency is technical efficiency. T echnical efficiency reflects obtaining the maximum outputs with the minimum inputs, i.e., obtaining the optimal efficiency (best weights) for each DMU. Data Envelopment Analysis (DEA) was developed by Charnes et al. (1978) for efficiency analysis based on the concept of technical efficiency from Farrell (1957).

Data Envelopment Analysis (DEA) models are widely applied in various areas for multicriteria efficiency level assessment, comparison, and ranking across decision-making units (DMUs). According to Stewart (1996), DEA is a satisfactory method for grading systems or DMUs' performance, by comparing how efficiently these DMUs convert inputs to outputs. DEA is a non-parametric method, based on mathematical programming, which makes it possible to minimize or maximize functions with or without restrictions.

The main advantage of this method is that it allows measuring the performance of systems with multiple inputs and outputs, without requiring a production function specification nor the prior definition of inputs and outputs weights, making the weighting process less subjective. When compared to other methods, such as econometric models that depend on the estimation and tests of several parameters, the flexibility given by DEA models is often preferable. However, the formulation of DEA models results in a separate linear program for each DMU, and their solution may require large computational capacity when the problem comprehends a large number of DMUs (Raju & Kumar, 2006).

Several studies have used DEA to evaluate performance in agriculture, focusing mainly on the analysis of efficiency, often measured by the total factor productivity (TFP). According to Fare et al. (1994), TFP is an index widely used to measure the economic efficiency of agricultural production. The index embodies the average productivity of all inputs with market value, such as land, labor, and capital (tractors, machinery, fertilizers, livestock, etc.), measured in terms of market value. The main objective of efficiency analysis is to understand whether a production unit is delivering the maximum yield from a given set of inputs (Kalirajan et al., 1996).

DEA method is also applied to measure the eco-efficiency of agri-food supply chains. This type of analysis has become more popular given the increasing concerns about the possible damage to the environment caused by agriculture (Masuda, 2016; DeSimone & Popoff, 2000; Mwambo et al., 2020). The assessment of irrigation efficiency is another popular application of DEA in agriculture and has been studied by Kibirige et al. (2019), Yilmaz et al. (2009), Raju and Kumar (2006), and Rodrigues-Díaz et al. (2004), for example.

There is a lack of studies applying DEA to evaluate FEW Nexus systems. Li et al. (2016) and Zhang and Xu (2019) developed DEA-Malmquist models to compare the Nexus efficiency for Chinese cities or regions. While Ibrahim et al. (2019) evaluated the efficiency of Organization for Economic Co-operation and Development (OECD) countries in terms of land-FEW Nexus at a transnational level. Results showed that the outcome obtained from the land-FEW Nexus efficiency scores is more related to an adequate use of resources than the scores obtained using the minimum land-FEW Nexus resources.

10.2.2 Network Equilibrium Model (NEM)

According to Alves Junior et al. (2021), NEM applications in green and sustainable cargo transportation are increasing, but they are still rare. Since the mid-nineteenth century, the researchers and urban and transportation planners recognized the importance of developing tools for modeling the interaction of land use and transportation, with a special contribution of Carey (1858), who proposed a macroeconomic model to predict people and commodity spatial flows (Nijkamp, 2007).

According to Holguín-Veras et al. (2001), the conditions of the transport supply and costs influence the business and production location decisions. At the same time, there are other aspects influencing the land-use configuration and thus impacting indirectly the transportation system. For this reason, scientific researchers perceived the necessity for developing strategic planning models that take into account the mutual relation between land use and transportation. More recently, other aspects, such as environmental impacts and energy consumption on transportation, are being frequently considered in this kind of modeling.

The freight transportation models have the as main purpose to reproduce the transportation system consciously, including the main components of the system and their interrelationships, and it is desirable they take into account spatial variations of the supply and demand levels, allowing the planners to assess the impacts of policies, infrastructure improvements, and management actions upon the current and future transportation performance (Holguín-Veras et al., 2001).

The modeling of spatial production and transportation models are analytical tools of freight movement patterns and economic interaction over geographic space. The methods and models developed for evaluating these phenomena have been improved since the second half of the nineteenth century in order to support the transportation and land-use planners. The initial foundations of multiregional goods interchange modeling are based on the gravitational force theory consolidating a category named gravity models (Batten & Boyce, 2007). The gravity models are widely applied in the distribution phase of the traditional four-step method used in transportation planning: (i) trip generation, (ii) distribution, (iii) mode split, and (iv) traffic assignment.

Holguín-Veras et al. (2001) based on an extensive review of regional freight models (RFM) classified them into three main families: Input-Output models, Spatial Interaction models, and Origin-Destination synthesis formulations.

The Input-Output (I-O) models are analytical formulations representing the interrelation among economic sectors, based on functions that describe the number of inputs required by a sector to produce a given economic output. The I-O models are derived from the initial concepts proposed by Leontief (1936), and generally, they predict the intersectoral flows of the economic production when the equilibrium between total supply and total demand occurs, considering that the products are homogeneous and assuming the average technology of the production sectors. The single-region I-O model does not allow assessing the commodity or monetary regional flows, as a consequence of its natural structure. However, improved methods like the Multiregional Input-Output Models (MRIO) overcome this limitation. The literature indicates that I-O models can be well applied for interregional freight transportation forecasting. One negative aspect of this method is that it requires a significant amount of data, and the estimation of the different technical coefficients can be difficult.

Spatial Interaction models are a family of models that try to estimate the commodity interregional flows as a function of the interactions among supply and demand regions in space, and this model's category includes gravity, direct demand, and equilibrium models.

In the case of the direct demand models, the transportation flows and the mode split are determined using an econometric model that predict the interregional freight flows based on a set of regions and transportation network aspects, such as population, production, income, travel time, transportation costs, and others (Holguín-Veras et al., 2001).

The equilibrium models are based on "Wardrop's principle," assuming that all cargo shippers are identical, non-cooperative, and rational and that they select the shortest (or lowest cost) route for delivering their cargo. Considering that all shippers select the routes according to this principle, the flows through the transportation network reach an equilibrium. Generally, the equilibrium models are classified into two main categories: the spatial price equilibrium (SPE) model and the network equilibrium model (NEM).

Since the Harvard Model (Kresge & Roberts, 1971) proposed a method for finding the optimal distribution of freight flows from suppliers to consumers that minimize total transportation cost over a simplified transportation network, the world's researchers have developed other equilibrium models that permit a more detailed simulation of logistics operations and transportation network aspects.

As an example of these advances, specialized models are presented in the literature proposing – among others – the use of nonlinear transport cost functions sensitive to economies of scale, a multimodal transportation network with capacity constraints, delay functions reproducing the effect of congestions, and the traffic of empty or specialized vehicles, allowing a more detailed and realistic representation of transportation systems (Branco et al., 2020; Branco et al., 2019; Caixeta-Filho & Macaulay, 1989; Crainic et al., 1990; Crainic & Laporte, 1997; De La Cruz et al., 2010; Friesz et al., 1983; Friesz and Harker, 1983; Gédéon et al., 1993; Guélat et al., 1990; Labys & Yang, 1991).

According to Holguín-Veras et al. (2001), it is expected that a transportation planning tool provides a good estimative of commodity interregional flows and freight traffic, supporting the analysis of the impacts caused by capacity enhancement or new multimodal transport infrastructures upon the freight transportation system as a whole. In view of that, the authors highlight two important aspects of freight transportation models:

- It should take into account the interrelationship between transportation activity and the economy as a whole because the typical use of a regional freight model is on the analyses of how the level of economic activity could impact the transportation demand and vice versa.
- It should simulate the shippers' choice about the different transportation modes, being able to model as real as possible the complex freight transportation systems. So, it could be applied to the analyses of multimodal projects.

10.2.3 Data Envelopment Analysis (DEA) and Network Equilibrium Model (NEM)

The traditional interregional freight models allow finding the optimal spatial distribution of the production and the optimal interregional flows through the multimodal transportation network, resultant of the supply and demand levels in each region, that minimize the total cost. However, it does not consider other important aspects that can affect the production spatial distribution, like the agriculture performance of the production regions. In other words, normally this kind of modeling did not take into account the productivity of the supply regions when choosing the best regions that could increase their production for supplying the expected future demand.

In view of that, over time, transportation models have become progressively fused with models that allow describing and predicting economic production behavior (Batten & Boyce, 2007).

In this context, we used a dual-step procedure for determining the optimal future grain production and its interregional transportation flows into the multimode network. First, we applied a Data Envelopment Analysis (DEA) aiming to classify suitable regions for soybean growing according to their agriculture yield. In the second step, we applied a Network Equilibrium Model (NEM) to find the optimal spatial distribution of the soybean and corn production expected by 2050 and the optimal interregional transportation flows of cargo between the supply and demand regions. The application is detailed in the next sections.

10.3 A Dual-Step Procedure Using Data Envelopment Analysis (DEA) and Network Equilibrium Model (NEM): An Application to Few Nexus Performance Evaluation

The Brazilian freight transportation system is highly dependent on trucks, even for large distance routes. Around 60% of the Brazilian corn and soybean production is transported by road, 30% by rail, and 10% by waterway (BRASIL, 2018). However, the total public investment in freight transport infrastructure fell to 37% between 2010 and 2017, almost freezing the expansion of waterways, railways, and roads in the country (BRASIL & EPL, 2019). During the same period, Brazilian corn and soybean production increased 61% (IBGE, 2020), occupying new cropland located even farther from the main grain exporting ports terminals. Consequently, the eco-efficiency of freight transportation in Brazil weakened, increasing costs, diesel consumption, and GHG emissions. The Brazilian transport sector is responsible for about 35% of the total fossil fuel consumption and for over 48% of the GHG emissions in the country (BRASIL and EPL, 2019).

In 2020, the Brazilian government scheduled the next steps of the railway concession plan, aiming to attract private companies to build and operate the following railway projects: Ferrogrão railway (FG), West-East Integration Railway (FIOL), and Center-West Integration Railway (BRASIL, 2020; VALEC, 2020).

Due to the importance of Brazilian agriculture, especially the production and exports of soybean, the case goals to determine the efficient locations where production should be expanded to reach the estimated oilseed production in 2050. The analysis prioritizes production in suitable areas located in the most efficient producing regions and only allowing the use of pasture areas, without additional deforestation. Therefore, a holistic and integrated perspective is applied, considering the interaction between the transport system and the land use.

The developed model consists of a two-step process, in which we describe the interaction between transportation and land use. Other models that integrate those two variables often consider the association between production location and transport. With that said, the level of transport accessibility affects the decision regarding the production location, and each production location generates different travel demands. As the changes in the transport system can cause long-term effects on the spatial distribution of production and transportation demands, the connection between land use and transport (and vice versa) is a pillar to the transportation modeling (Mackett, 1985).

In this section, we describe the application of a Data Envelopment Analysis (DEA) and Network Equilibrium Model (NEM) models used to evaluate the FEW Nexus performance in various regions and to direct cropland area expansion in Brazil. This model is used to find the best transportation flows, considering the trade-off between environmental and economic gains. In our FEW Nexus approach, we use the average soybean yield and soybean yield risk to represent "food," the average temperature to represent "energy," and the reduction of potential soybean yield due to water deficit to represent "water." In addition to the FEW variables, we include the energy consumed in soybean transportation in a Network Equilibrium Model (NEM).

After applying the DEA model to determine the most appropriate spatial distribution for the current production, we use the Network Equilibrium Model to find the optimal spatial distribution for future production. We considered as total cost the sum of the following individual costs: (i) the loss of productivity due to water deficit in each region, (ii) the cost of corn and soybean production, and (iii) the transportation cost from the producing regions to the exporting ports.

10.3.1 DEA Model Formulation

DEA model was used to determine soybean production efficiency in Brazilian microregions, the decision-making units (DMUs). The method implementation returns efficiency scores for each DMU that are used to represent performance composite indicators. The formulation of a DEA model depends on a set of characteristics, such as the assumptions regarding the returns to scale (variable, VRS; or constant, CRS) or its orientation (output, input, or non-oriented) (Charnes et al., 1978; Banker et al., 1984). The slack-based measure (SBM) model (Tone, 2001) allows DMUs to simultaneously maximize their output and minimize their inputs (Cook et al., 2014). Several studies have used DEA models to solve agricultural and logistic problems (Gómez-Limón et al., 2012; Toma et al., 2017; Melo et al., 2018; Melo et al., 2020). Equations (10.1) through (10.6) represent the slack-based measure (SBM) model developed by Tone (2001), assuming variable returns to scale (VRS):

$$Minimize\, \tau = t - \left(\frac{1}{m}\right)\left(\sum_{i=1}^{m} \frac{S_i^-}{x_{i0}}\right) \tag{10.1}$$

$$t + \left(\frac{1}{n}\right)\left(\sum_{j=1}^{n} \frac{S_j^+}{y_{j0}}\right) = 1 \tag{10.2}$$

$$\sum_{k=1}^{z} {}_k x_{ik} + S_i^- - t x_{i0} = 0 \quad i = 1,2,\ldots,m \tag{10.3}$$

$$\sum_{k=1}^{z} {}_k y_{jk} - S_j^+ - t y_{j0} = 0 \quad j = 1,2,\ldots,n \tag{10.4}$$

$$\sum_{k=1}^{z} {}_k - t = 0 \tag{10.5}$$

$${}_k \geq 0, S_i^- \geq 0, S_j^+ \geq 0, \text{and}\, t > 0 \tag{10.6}$$

Where τ is the efficiency level, t is the model linearization variable, S_i^- is the slack variable for the i-th input, S_j^+ is the slack variable for the j-th output, Λ_k is the contribution of the k-th DMU to the DMU_0 (under analysis), x_{i0} is the amount of i-th input used by DMU_0 (under analysis), y_{j0} is the amount of the j-th output produced by DMU_0

(under analysis), x_{ik} is the amount of the i-th input used by the k-th DMU, y_{jk} is the amount of the j-th output produced by the k-th DMU, m is the number of inputs used in the model, n is the number of outputs considered in the model, and z is the number of DMUs in our analysis.

The model represented with Eqs. 10.1 through 10.6 is the standard SBM model under VRS assumptions and is formulated considering desirable inputs, simply deemed as inputs, and desirable outputs, simply deemed as outputs. However, according to Liu et al. (2010), in real-world situations, the DMUs also face both undesirable inputs (UI) and undesirable outputs (UO). Since the DMUs aim to minimize inputs and UO as negative factors (Rentizelas et al., 2019; Seiford & Zhu, 2002), we can consider UO as inputs in the model. Moreover, because the DMUs aim to maximize outputs and UI as positive factors (Liu et al., 2015), we can use UI as outputs in the model. In our approach, the reduction of potential yield due to water balance deficit and yield risk is an example of UO, and the average temperature is an example of UI.

We use our model to estimate individual scores for each DMU (region), which correspond to their relative efficiency levels. DMUs with estimated score equal to one ($\tau^* = 1$), the maximum value, are considered efficient. Because more than one DMU can be assigned with the maximum efficient score, we use the double frontier composite indicator (CI) tiebreaking method, suggested by Leta et al. (2005), to avoid ties in the first position of our rank. The CI (Eq. 10.7) is the arithmetic average calculated using the inverted (handling outputs as inputs and vice versa) and the standard efficiency scores. We standardize each score dividing it by the maximum value.

$$t_k^{\text{composite}} = \frac{\left[t_k^{\text{standard}} + \left(1 - t_k^{\text{inverted}}\right)\right]/2}{\max\left\{\left[t_k^{\text{standard}} + \left(1 - t_k^{\text{inverted}}\right)\right]/2\right\}} \quad k = 1, 2, 3, \ldots, z \tag{10.7}$$

Where τ_k^{inverted} is the inverted efficiency score estimated for the k-th DMU and τ_k^{standard} is the standard efficiency score calculated for the k-th DMU.

Based on the efficiency score rank obtained from the DEA model results, the DMUs (regions) were split into three groups: (i) high-performance regions, consisting of the first tertile of the most efficient production regions accordingly to DEA; (ii) regular-performance regions, consisting of the second tertile; and (iii) the low-performance regions, consisting of the last tertile. The estimated cost reflects the reduction of the maximum soybean potential yield due to the water balance deficit for each group of regions. This cost was included in the total cost of the NEM.

10.3.2 Data and Variable Description

Our analysis uses data from all Brazilian microregions (Brazilian Institute of Geography and Statistics – IBGE, 2020) which were considered suitable for soybean production and the expansion areas identified by the territorial intelligence decision-making support system provided by AGROIDEAL (2019). Figure 10.1 summarizes the analyzed microregions and their current soybean production. This approach indicates the available cropland areas, considering the locations that are currently used for pasture and could be used for agricultural production expansion. As all forest areas are out of the scope of the current investigation, the expansion necessary occurs only on pasture areas.

We consider the average soybean yield as a positive performance factor in our model, because this variable shows how efficiently a DMU converts land into soybean production. The standard deviation of yields is calculated to represent yield risk and is considered a negative performance factor. We used the soybean yield data, from 2014 through 2018, for each microregion, from the Brazilian Institute for Geography and Statistics (IBGE, 2020).

The potential reductions in soybean yield due to water balance deficit were obtained from the National Institute of Meteorology (INMET, 2019). The INMET (2019) developed

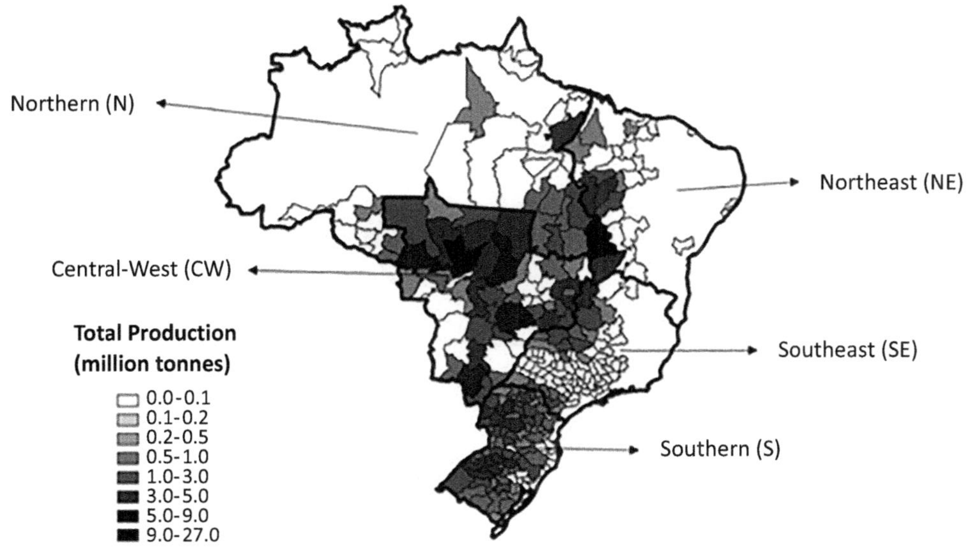

Fig. 10.1 Brazilian regions and current soybean production per microregion

a forecast model that uses weather information from many climate stations to estimate this potential reduction in production. The estimated values obtained from this model were used as negative performance factors. Our assumption is that producers located in areas with higher expected yield reduction will be more likely to use irrigation in their crop. The data set from INMET (2019) covers the period between 2014 and 2018.

According to EMBRAPA (2013), temperatures lower than 20 °C can reduce germination, and temperatures lower than 10 °C may severely stunt the growth of the soybean plant. The ideal temperature for soybean cultivation is around 30 °C. Despite temperatures higher than 40 °C also being harmful to the growth and flowering, the average temperature range of the selected microregions was between 20 °C and 30 °C. Therefore, within this temperature range, there is a positive relationship between temperature and soybean yield, justifying the use of the average temperature as a positive factor of performance in our model. The average temperature within microregions was obtained from (INMET 2019), between 2014 and 2018.

10.3.3 Network Equilibrium Model Formulation

The goals of our NEM model are to:

(i) Find the optimal distribution of corn and soybean production forecast for 2050.
(ii) Use results from (i) to minimize the total cost of a multimodal transport network used to model the transportation flows among producing and demand regions.
(iii) Calculate CO_2 emissions originated from the transportation flows in (ii).
(iv) Measure the possible impact of the inclusion of planned railways in our transport intermodal network, in terms of cost and CO_2 emission reduction.

Our model was built under the assumption that soybean and corn producers choose where to grow their crops, considering the distribution of transportation flows that provides the minimum total cost.

Figure 10.2 summarizes the transportation network and its connection between supply and demand regions, as well as the main variables, used in our model.

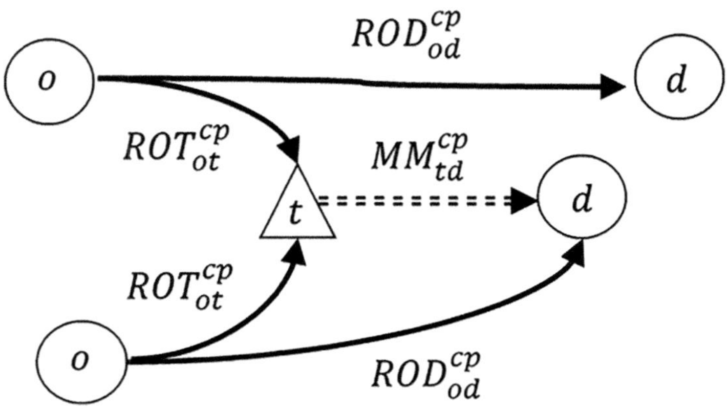

Fig. 10.2 Transportation network and main variables of the model. (Source: Adapted from Branco et al., 2020)

where:

o: Supply regions (origins).

d: Demand regions (destination) – domestic demand region or an export terminal.

c: Demand market, which can be a domestic market (*dm*) or international market (*im*).

t: Transshipment terminal, which loads cargo into different transportation modes including rail and inland waterways, but not roadways.

p: Product: soybean (*s*) or corn (*co*).

The transportation nodes are connected through network arcs, which represent the available transportation infrastructure, utilized in interregional freight flows between producing and demand regions. We use the following variables to represent interregional flows:

ROD_{od}^{cp}: Road transportation flow of product *p* between origin *o* and destination *d*, in the market *c*.

ROT_{ot}^{cp}: Road transportation flow of product *p* between origin *o* and transshipment point *t*, in the market *c*.

MM_{td}^{cp}: Multimodal transportation flow of product *p* between transshipment point *t* and destination *d*, in the market *c*.

Our goal is to minimize the total supply cost (*C*) represented in the following objective function (Eq. 10.8):

$$C = \sum_{o}\sum_{d}\sum_{c}\sum_{p}\mathrm{ROD}_{od}^{cp}.\left(YC_{o} + PC_{o}^{p} + TC_{od}\right)$$

$$+\sum_{o}\sum_{t}\sum_{c}\sum_{p}\mathrm{ROT}_{ot}^{cp}.\left(YC_{o} + PC_{o}^{p} + TC_{ot}\right) \quad (10.8)$$

$$+\sum_{t}\sum_{d}\sum_{c}\sum_{p}MM_{td}^{cp}.TC_{td}$$

where:

C: Total cost (*US\$/tonne*), including the maximum yield reduction cost and production and transportation costs.

TC: Transportation cost (US\$/tonne of cargo) between an origin and a destination ($\mathrm{TC_{od}}$), between an origin and a transshipment terminal ($\mathrm{TC_{ot}}$), and between a transshipment terminal and a destination ($\mathrm{TC_{td}}$).

PC_{o}^{p}: Production cost (*US\$*/tonne of corn and soybean) in each origin *o* (origins suitable for corn and soybean production) for product *p* (soybean or corn).

YC_o: Yield cost (*US\$/tonne* of soybean) representing the soybean maximum yield reduction of each suitable production region *o* due to the water balance deficit.

We only consider soybean yields in our analysis because we assume that corn is taken as a secondary crop in most of the country. Our assumption is that corn is mainly used for soil rotation after soybean harvest, planted as the winter crop. Moreover, the soybean crop represents the largest part of farming revenue, while corn revenue is secondary. With that said, we assume that the spatial distribution of the future production regions will be mainly guided by the soybean regional growing performance.

We use Eq. (10.9) to represent the total CO_2 emissions from transportation:

$$CO_2 = \sum_o \sum_d \sum_c \sum_p \mathrm{ROD}_{od}^{cp}.ECO_{2od} + \sum_o \sum_t \sum_c \sum_p \mathrm{ROT}_{ot}^{cp}.ECO_{2ot} + \sum_t \sum_d \sum_c \sum_p MM_{td}^{cp}.ECO_{2td} \quad (10.9)$$

where:

ECO_2: Emissions of CO_2 (tonnes of CO_2/tonne of cargo) by transportation between an origin and a destination (ECO_{2od}), between an origin and a transshipment terminal (ECO_{2ot}), and between a transshipment terminal and a destination (ECO_{2td}).

The variable "ECO_2" represents the CO_2emission from each network link. Emissions were calculated multiplying the amount of CO_2 emission per kilometer and tonne, by the distance between the nodes of the network.

Equation (10.10) represents a set of constraints regarding the total amount of soybeans and corn shipped to supply domestic demand and exports. The total amount of shipped oilseed and grains must be lower than or equal to the total production (PRO_o^p) of each product p and origin region o plus their respective future production. Future production is obtained multiplying the size (in hectares) of the future area (FA_o) used for corn and soybean production expansion, in each origin o, by the respective expected yield $\left(\mathrm{YL}_o^p\right)$ for each product p in the origin o (tonnes/hectare).

$$\sum_d \sum_c \mathrm{ROD}_{od}^{cp} + \sum_t \sum_c \mathrm{ROT}_{ot}^{cp} \leq \mathrm{PRO}_o^p + \left(FA_o.\mathrm{YL}_o^p\right), \forall o \text{ and } p \quad (10.10)$$

Equation (10.11) shows that the future production area in each origin o must be lower or equal than the maximum suitable area for corn and soybean production in that same region (SA_o).

$$FA_o \leq SA_o, \forall o \quad (10.11)$$

We use Eq. (10.12) to cover the supply of domestic demand ($c = dm$). The following equation states the total amount of products moved to each demand region must be equal to the domestic demand (DEM_d^p) for each product in that region:

$$\sum_o \mathrm{ROD}_{od}^{cp} + \sum_t MM_{td}^{cp} = DEM_d^p, \forall d, p \text{ and } c = dm \quad (10.12)$$

Equation (10.13) is used to determine that the total amount of each commodity transported to each destination node addressed as an exporting terminal ($d \in \{sp\}$) is equal to the international demand ($c = im$):

$$\sum_o \mathrm{ROD}_{od}^{cp} + \sum_t MM_{td}^{cp} = EXP_d^p, \forall d \in \{sp\}, p \text{ and } c = im \quad (10.13)$$

Equation (10.14) determines that the sum of transportation flows arriving in each transshipment terminal t must be equal to the sum of transportation flows that departs from each terminal:

$$\sum_o \mathrm{ROT}_{ot}^{cp} = \sum_d MM_{td}^{cp}, \forall t, c \text{ and } p \quad (10.14)$$

Equation (10.15) is used to add a limitation for the total quantity of cargo assigned to each transshipment terminal t. This total must be equal to or less than the load capacity of each terminal (TCAP_t):

$$\sum_o \sum_c \sum_p \mathrm{ROT}_{ot}^{cp} \leq \mathrm{TCAP}_t, \forall t \quad (10.15)$$

10.3.4 Transportation Network and Scenarios

We built the multimodal transportation network and determined the capacity for each node in the model based on information regarding the total amount of soybeans and corn shipped in 2017 (ANTAQ, 2017; ANTT, 2017).

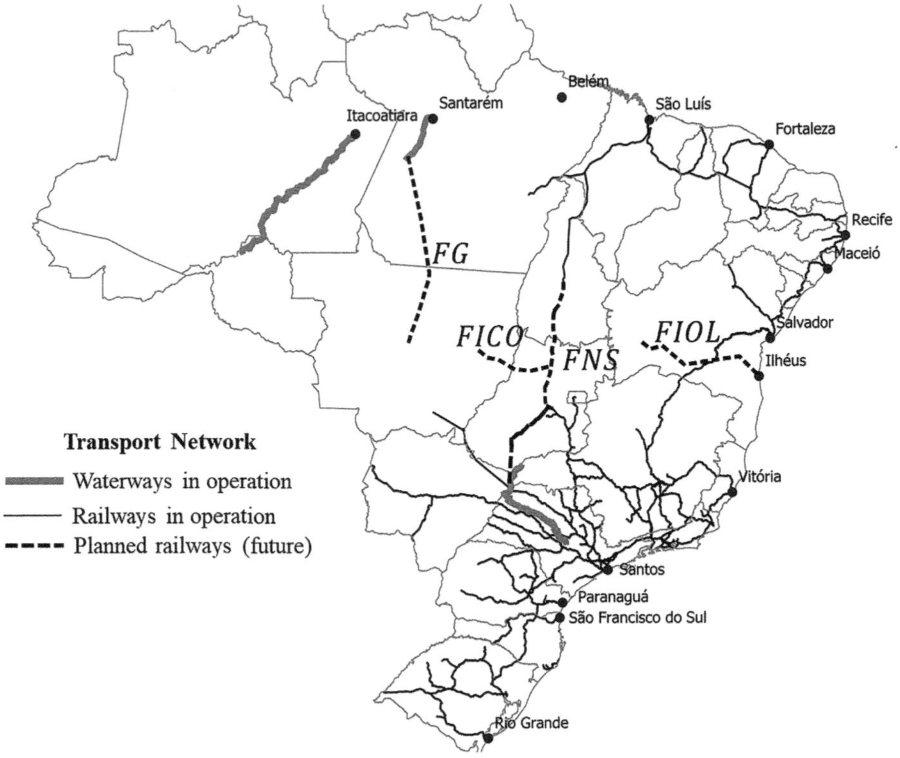

Fig. 10.3 Brazilian current transportation network, planned railways, and export ports. (Source: Elaborated by the authors)

The future scenarios consider the following planned projects: West-East Integration Railway (FIOL), South stretch of North-South railway (FNS), Center-West Integration Railway (FICO), and Ferrogrão railway (FG). We assumed unconstrained capacities. Figure 10.3 shows the current and planned infrastructure (railways, waterways, and ports).

We analyze the following scenarios in our model:

(i) Current production: The objective is to reach the minimum total cost using the current multimodal network – MTC (CP/CMN). The costs are minimized considering the 2018 corn and soybean production.
(ii) Forecasted production: The objective is to reach the minimum total future cost using the current multimodal network – MTC (FP/CMN). The costs in 2050 are minimized costs, using the constraints established in the current multimodal network, i.e., using the current availability for grain transportation and capacities.
(iii) Planned infrastructure and forecasted production: The objective is to reach the minimum total cost with future multimodal network – MTC (FP/FMN). The total cost in 2050 is minimized, considering that transportation will occur in the future multimodal network, including planned railways, and assuming no capacity constraints.
(iv) CO_2 emission minimization: Assuming the Forecasted Production with Future Multimodal Network – MCO_2 (FP/FMN). The total CO_2 emissions in 2050 are minimized, considering that transportation will occur in the future multimodal network, including planned railways, and assuming no capacity constraints.

10.4 Results and Discussion

10.4.1 DEA Approach

The relative efficiency scores for a total of 107 microregions were calculated using Eqs. (10.1) through (10.7). We only considered microregions that are suitable for soybean production in our analysis. Figure 10.4 shows all microregions used in our analysis, which are combined in nine groups of similar efficiency scores, according to the DEA model results. The more efficient the microregion, the darker the color in the map.

The FEW Nexus approach resulted in a spatial distribution with the best-ranked microregions in the central-west and northern regions. After applying the DEA model, we ranked microregions accordingly to their scores and analyzed them, considering three efficiency groups: high (position 1 through 35), average (position 36 through 72), and low position 73 through 107).

According to our results, the microregions (DMUs) in the central-west (CW) and in the north (N) presented relatively high scores. We found that 53.7% (22 out of 41) of the microregions in the CW and 75% (6 out of 8) in the north are classified within the high-efficiency group. Only 12.2% (5 out of 41) of the CW microregions and none in the north are within the low-efficiency group. On the contrary, our results reveal that microregions in the southeast (SE) and northeast (NE) have relatively low scores, with respectively 57.5% (23 out of 40) and 37.4% (6 out of 16) of the microregions within the low-efficiency group. Only 5% (2 out of 40) of the microregions in the SE and 31.3% (5 out of 16) in the NE are within the high-efficiency group. Because all high-efficiency areas in the south (S) were already occupied, we could analyze only two microregions that presented suitable areas to

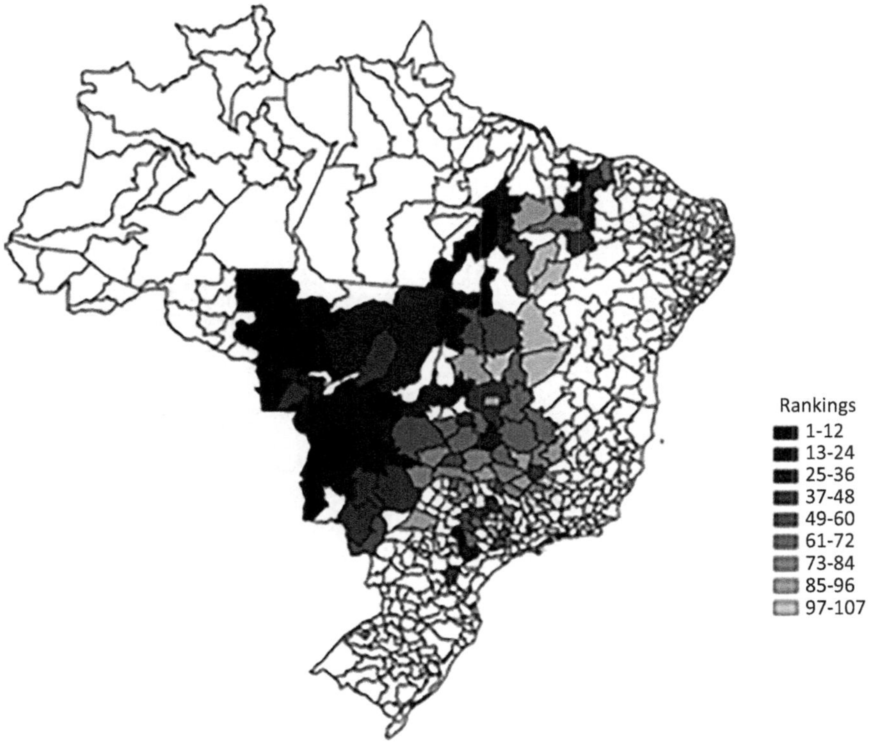

Fig. 10.4 Microregions suitable for soybean growing, classified according to their efficient scores

expand production: one is classified in the average-efficiency group and the other in the low-efficiency group.

We forecasted the future production in 2050 using the estimated efficiency scores to rank the availability of suitable areas for soybean production, in each microregion. We then assigned additional production to suitable areas, starting with the most efficient microregions. For those areas with no suitable land to expand the production, we used the same 2018/2019 production for 2050. We additionally assumed that the production expansion can only occur in released pasture suitable areas (PA). The expected soybean spatial production distribution, in 2050, is shown in Fig. 10.5.

The expansion of soybean production in Brazil (over PA) was calculated based on FAO (2017) and MAPA (2019). Our results indicate that the country can potentially produce a total of 192 million tonnes in 2050 (expanding only over PA). When compared to the 115 million tonnes of soybeans produced in the 2018/2019 crop year (IBGE, 2020), the expected production represents an increase of 67%.

The results in Fig. 10.5 show that deforestation for agriculture production can be avoided since Brazil does not need to use forest areas to achieve the expected production in 2050. Our findings suggest that the exclusive use of pasture areas is sufficient to meet 100% of the production goals. This outcome is achievable if future soybean production occurs only in areas located in the most efficient microregions in the country. This result is in line with previous studies in literature, such as Stabile et al. (2020), that demonstrated that it is possible to increase production (also in the Amazon region) and simultaneously decrease deforestation. The same authors suggested investments in increasing productivity and public policies to avoid deforestation.

According to the results obtained from our DEA models, the 2050 expected soybean production could be reached using part of the areas currently used as pasture and reducing the water balance deficit. With expansion occurring only in pasture areas (PA), our results indicate yield reductions due to water balance deficit in comparison to the current scenario from 35.1% (average between 2014 and 2018) to 31.7% (2050).

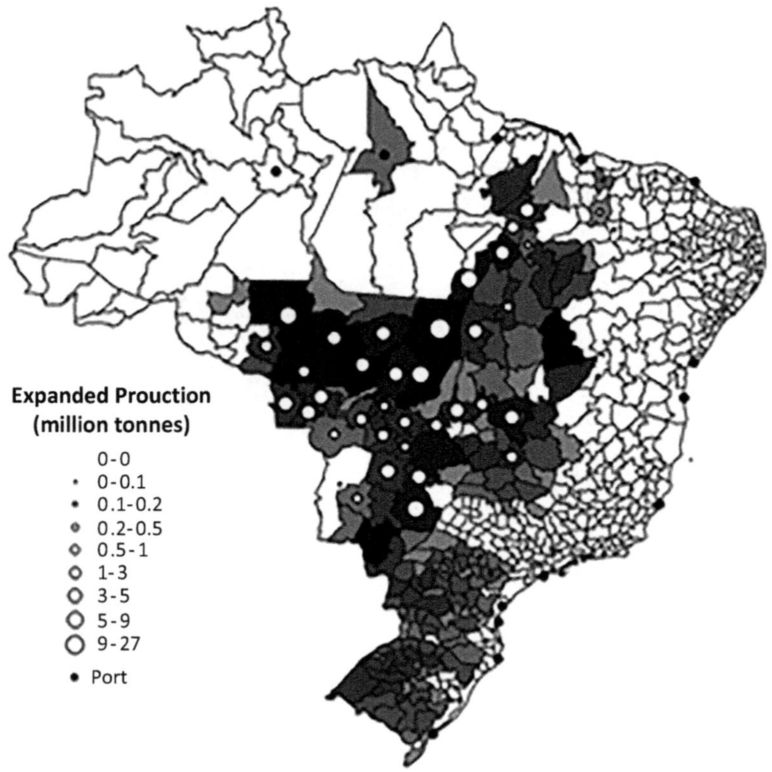

Fig. 10.5 Expected soybean production for 2050, considering the expansion only over PA

In summary, incentives aiming to improve production and transportation efficiency should encourage soybean production in areas with lower water deficit, where irrigation is less necessary, as well as where there are near routes connecting the production areas to export terminals. Since the most efficient regions (considering the number of microregions and the production volume) are located far from the main ports, in the central-west, policymakers are recommended to prioritize infrastructure investments.

Whether future investments, technological evolution, and public policies promote the improvement of crop yields and, at the same time, the increase of soybean production in areas with lower water deficit, it will be possible to reduce even more the projected need of irrigation, improve energy efficiency, and reduce CO_2 emissions transporting corn and soybeans. It is fundamental to highlight that, even for the current scenario, our findings demonstrate that it is possible to achieve a cleaner production for the next 30 years using only pasture area and no deforestation is needed. Moreover, the development of alternative routes to transport grains and oilseeds (including waterways and railroads in green transport corridors), the improvement of road pavement conditions, and the encouragement of cargo fleet renewal will also result in positive environmental and economic impacts.

10.4.2 NEM Approach

Table 10.1 shows the current and the future (2050) production by state. The results indicate that the spatial distribution of future production in the MTC (FP/CMN) scenario points out to an

Table 10.1 Current and future (2050) corn and soybean production by state (thousand tonnes)

	2018	2050		
Brazilian state	MTC (CP/CMN)	MTC (FP/CMN)	MTC (FP/FMN)	MCO_2 (FP/FMN)
Mato Grosso (MT)	60,763	69,149	75,754	68,454
Mato Grosso do Sul (MS)	18,937	54,157	49,438	49,891
Goiás (GO)	21,614	39,717	50,413	45,085
Minas Gerais (MG)	11,977	42,402	33,892	42,402
Paraná (PR)	36,777	41,875	41,875	41,875
Rio Grande do Sul (RS)	26,251	28,181	28,181	28,181
Maranhão (MA)	4056	21,830	20,384	21,830
Bahia (BA)	7431	14,545	14,545	14,545
São Paulo (SP)	8267	13,548	11,497	13,548
Tocantins (TO)	3432	12,721	12,299	12,721
Piauí (PI)	3537	6381	6227	6381
Santa Catarina (SC)	5424	5775	5775	5775
Pará (PA)	2596	2511	2511	2511
Rondônia (RO)	1849	1978	1978	1978
Distrito Federal (DF)	683	1268	1268	860
Sergipe (SE)	759	844	844	844
Ceará (CE)	336	374	374	374
Amazonas (AM)	8	298	298	298
Roraima (RR)	99	147	147	147
Acre (AC)	81	91	91	91
Pernambuco (PE)	54	60	60	60
Amapá (AP)	62	55	55	55
Espírito Santo (ES)	33	37	37	37
Alagoas (AL)	32	37	37	37
Paraíba (PB)	24	26	26	26
Rio de Janeiro (RJ)	7	8	8	8
Rio Grande do Norte (RN)	4	4	4	4
Total	215,092	358,018	358,018	358,018

increase in corn and soybean production in the states of Mato Grosso do Sul/MS (35 million tonnes of additional production), Minas Gerais/MG (30 million tonnes), Goiás/GO (18 million tonnes), Maranhão/MA (18 million tonnes), Tocantins/TO (9 million tonnes), Mato Grosso/MT (8 million tonnes), Bahia/BA (7 million tonnes), São Paulo/SP (5 million tonnes), and Paraná/PR (5 million tonnes). Together, these states can potentially represent about 87% of the expected production expected in 2050.

The results for the analysis of scenarios MTC (CP/CMN) and MTC (FP/CMN) are shown in Fig. 10.6a, b, respectively. Figure 10.6b shows where production should be located considering the total cost minimization and that the 2050 transport network will remain the same as in 2018. In this scenario, there is no change in the current railway network.

The spatial distribution of the expected 2050 production according to scenarios MTC (FP/FMN) and MCO_2 (FP/FMN) is shown in Fig. 10.7. The results do not indicate any significant difference between these scenarios in terms of improvements in the future spatial production distribution when we minimize costs and emissions. According to it, the future railway network will simultaneously improve the competitiveness and efficiency of transport in terms of CO_2 emissions.

Since we could only find relatively small changes in the production spatial distribution using the scenarios that aim to minimize total costs, we can conclude that the routes designed for the planned railways are efficient. The planned railways cross regions with high efficiency for grain production and are in line with the optimal corn and soybean interregional transportation flows for the expected production, in 2050.

When we focused on the cost minimization using the MTC (FP/FMN) scenario, our models indicate an additional production of 11 million tonnes in Goiás/GO and 7 million tonnes in Mato Grosso/MT. Our results also suggest a reduction in the production in Minas Gerais/MG (−9 million tonnes), Mato Grosso do Sul/MS (−5 million tonnes), São Paulo/SP (−2 million tonnes), Maranhão/MA (−1 million tonnes), and Tocantins/TO (−1 million tonnes). Therefore, we can conclude that the planned railways will play an important role in improving the competitiveness of grain production in the more central states, such as Mato Grosso/MT and Goiás/GO.

Considering the CO_2 emission minimization in scenario MCO_2 (FP/FMN), our results suggest

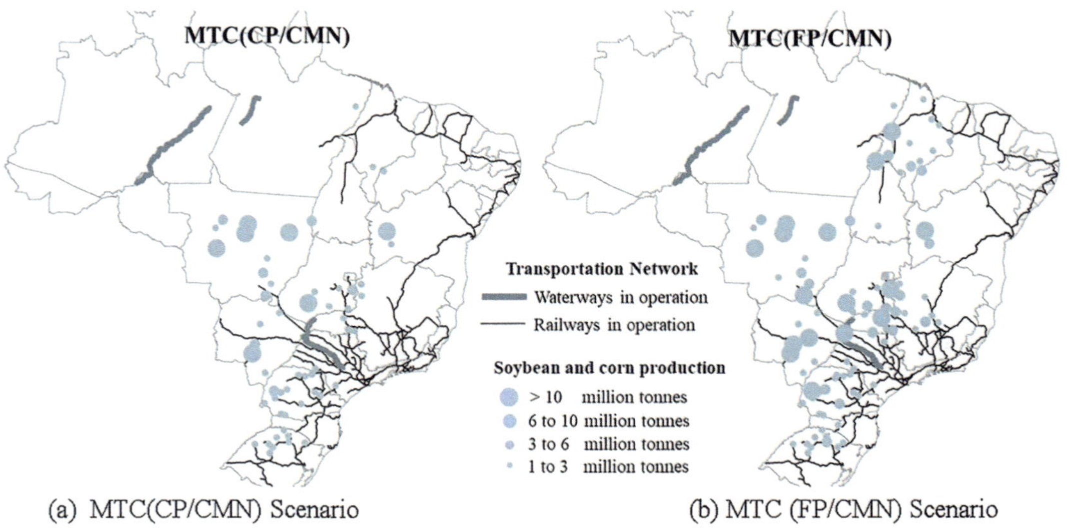

Fig. 10.6 Spatial distribution of current (2018) and future (2050) production maintaining the current multimodal network. (**a**) MTC(CP/CMN) Scenario. (**b**) MTC (FP/CMN) Scenario

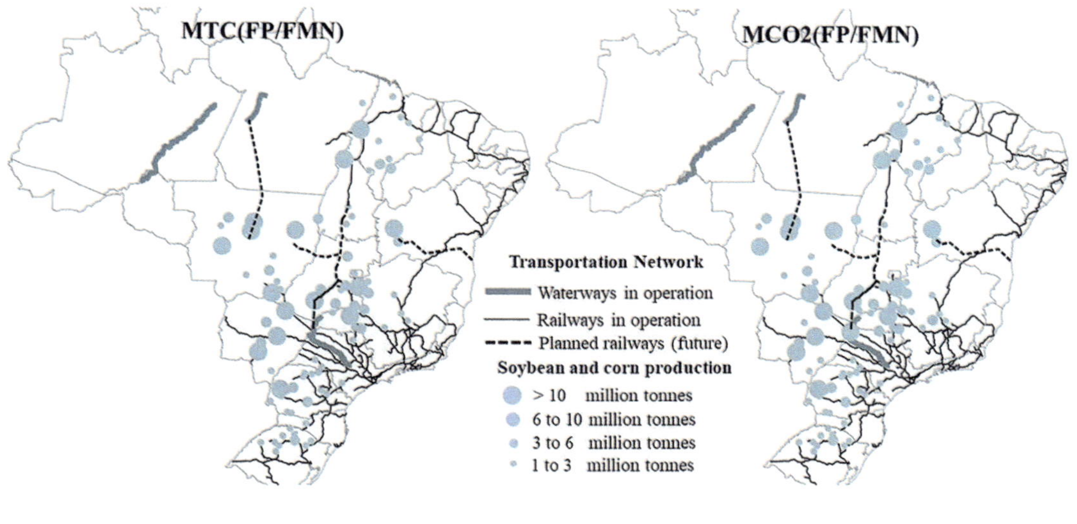

(a) MTC (FP/FMN) Scenario (b) MCO_2 (FP/CMN) Scenario

Fig. 10.7 Spatial distribution of future (2050) production including new planned railways. (**a**) MTC (FP/FMN) Scenario. (**b**) MCO_2 (FP/CMN) Scenario

Table 10.2 Total CO_2 emissions and total cost for each scenario

	Total CO_2 emissions		Total cost	
Scenario	Million tonnes	$kgCO_2$/tonne	Billion US$[a]	US$/tonne
MCO2 (FP/FMN) (2050)	5.9	17	239.4	668.6
MTC (FP/FMN) (2050)	6.3	18	237.2	662.6
MTC (FP/CMN) (2050)	7.2	20	241.9	675.8
MTC (CP/CMN) (2018)	4.7	22	144.1	669.9

Source: Results of the modeling
[a]It was considered the following exchange rate: @ US$ 4.3/R$ (April 2020, Brazilian Central Bank)

production increase in Minas Gerais/MG (9 million tonnes), São Paulo/SP (2 million tonnes), and Maranhão/MA (1 million tonne). Additionally, the model results suggest production reductions in Mato Grosso/MT (−7 million tonnes) and Goiás (−6 million tonnes). With these results, we can conclude that model was able to assign smaller shares of the additional production to further regions and higher shares to regions closer to the export terminals, aiming to minimize the CO_2 emissions with transportation.

As presented in Table 10.2, MTC (FP/FMN) scenario indicates a decrease of 0.9 million tonnes of CO_2 emissions compared to scenario MTC (FP/CMN). This result indicates that the implementation of the planned railways could decrease approximately 13% of the current CO_2 emissions, from 20 $kgCO_2$ by tonne of corn and soybean transported to 18 $kgCO_2$ by tonne. Using the minimization of CO_2 emission levels, the MCO2 (FP/FMN) scenario indicates a potential reduction of 18% on emissions, reaching 17 $kgCO_2$ by tonne.

The minimum total cost resultant from the spatial optimization of the future production and the interregional transportation flows is around US$ 663 per tonne in the MTC (FP/FMN) scenario and increases to US$ 669 per tonne when the model minimizes the CO_2 emissions, in the MCO_2 (FP/FMN) scenario. However, the latest scenario promoted a reduction in CO_2 emissions from 18 to 17 $kgCO_2$/tonne. It is important to highlight that when compared to the current total cost, scenarios MTC (FP/FMN) and MCO_2 (FP/FMN) present lower average total costs (US$). This result indicates that, under the following conditions, the planned railways could foster a cost reduction by 2050: (i) when agriculture production and planning of transportation flow goals are related to a minimum CO_2 emission and (ii) when agriculture expansion and planning of transportation flow planning reach the minimum total cost..

10.5 Main Conclusions

There is a lack of studies applying DEA to evaluate FEW Nexus systems. As shown before, some authors applied DEA to evaluate the efficiency of OECD countries, Chinese cities, or regions, based on FEW Nexus variables, and one of them showed that adequate use of resources is related to the Land-FEW-Nexus efficiency rather than simply using the minimum Land-FEW-Nexus resources. And the CO_2 emission is a common undesirable output in DEA applications in climate change studies that could be more explored in FEW Nexus in city studies. One of the advantages of applying DEA to evaluate FEW Nexus is that it does not need to assume a complex functional form to evaluate it, and it can be easily integrated to other methods, but one of the limitations is the lack of information on the dynamics of the system and its networks.

In this chapter, we showed how to apply a dual-step model, based on DEA and NEM, to minimize total crop production costs and CO_2 emissions with transportation; at the same time, we promote sustainable production expansion in Brazil. The analysis using our model considers four scenarios that account for the possible expansion of the transportation system in the country. The results indicate that it is possible to reach minimal costs and emissions by 2050, considering several constraints, including the production expansion only in pasture areas, with no need for additional deforestation. At last, our results proved to be the optimal configuration of the multimodal transportation network.

Our model could be used as an important tool, once is a powerful toll for policymakers planning the future agriculture production and the transportation network simultaneously, considering the mutual relation between transportation and land use. The results provided important thoughts regarding the environmental and economic impacts of the planned railways. Additionally, our model provided results regarding the dimension of transportation infrastructure capacities, such as roads, railways, waterways, and ports, for the location of transshipment terminals.

The results of this type of analysis are useful for guiding and coordinating public policies and private initiatives, aiming to optimize the selection and prioritization of future investments in transport infrastructure and, at the same time, establish incentives (programs, actions, rules) to conduct agricultural production to more interesting areas from an agronomic, economic, and environmental points of view.

The model can be replicated in other regions and countries, as well as considering other types of agricultural products.

Acknowledgments The authors thank the São Paulo Research Foundation (FAPESP) (Projects #2018/24353-3, #2018/20436-1, #2017/50420-7) and the Belmont Forum for the financial support to this research project.

References

AGROIDEAL. (2019). *Soy – Territorial Intelligence.*

Alexandratos, N., & Bruinsma, J. (2012). *World agriculture towards 2030/2050* (ESA working paper no 12-03). Food and Agriculture Organization of the United Nations – FAO.

Alves Junior, P. N., Melo, I. C., Branco, J. E. H., Bartholomeu, D. B., & Caixeta-Filho, J. V. (2021). Which Green Transport Corridors (GTC) are efficient? A dual-step approach using Network Equilibrium Model (NEM) and Data Envelopment Analysis (DEA). *Journal of Marine Science and Engineering, 9*, 247. https://doi.org/10.3390/jmse9030247

ANTAQ. (2017). *Estatístico Aquaviário*. http://web.antaq.gov.br/Anuario/

ANTT. (2017). *Declaração de Rede – Ferrovias*. www.antt.gov.br

Banker, R. D., Charnes, A., & Cooper, W. W. (1984). Some models for estimating technical and scale inefficiencies in data envelopment analysis. *Management Science, 30*(9), 1078–1092. https://doi.org/10.1287/mnsc.30.9.1078

Batten, D.F., & Boyce, D.E. (1986). Spatial interaction, transportation, and interregional commodity flow models, in: Nijkamp, P. (Ed.), Handbook of Regional and Urban Economics, Volume I. North-Holland, Amsterdam, pp. 357-406.

Branco, J. E. H., Bartholomeu, D. B., Alves Junior, P. N., & Caixeta Filho, J. V. (2020). Evaluation of the economic and environmental impacts from the addition of new railways to the Brazilian's transportation network: An application of a network equilibrium model. *Transport Policy*. https://doi.org/10.1016/j.tranpol.2020.03.011

Branco, J. E. H., Bartholomeu, D. B., Alves Junior, P. N., Caixeta-Filho, J. V. (2021). Mutual analyses of agriculture land use and transportation networks: The future location of soybean and corn production in Brazil. Agricultural Systems 194103264-10.1016/j.agsy.2021.103264

BRASIL (Ministério da Infraestrutura). (2020). *Política de Concessões: Rio, Terra, Céu e Mar*. https://infraestrutura.gov.br/concessoes/

BRASIL (Ministério dos Transportes). (2018). *Anuário Estatístico de Transportes, 2010–2017*. http://www.infraestrutura.gov.br

BRASIL and EPL (Ministério da Infraestrutura e Empresa de Planejamento e Logística). (2019). Diagnóstico Logístico - 2010 – 2018. https://www.ontl.epl.gov.br/diagnostico-logistico1

Caixeta-Filho, J. V.; & Macaulay, T. G. (1995). Interstate Movements: An Application of Spatial Equilibrium Models. Central European Journal of Operations Research, v. 3, n.3, p. 203–213. https://www.esalqlog.esalq.usp.br/upload/kceditor/files/1995/01/Interstate-movements-an-application-of-spatial-equilibrium-models.pdf

Carey, H. C. (1858). Principles Of Social Science. Kessinger Publishing, LLC.

Calil, D., Clements, Y., & Ribera, L. (2019). Brazil's agricultural production and its potential as global food supplier. *Choices Mag Food, Farm, Resour Issues, 34*, 1–12. https://doi.org/10.22004/ag.econ.292352

Crainic, T. D., Florian, M., Gue´lat, J., & Spiess, H. (1990). Strategic planning of freight transportation: STAN, an interactive-graphic system, Transportation Research Record, 1283, pp. 97–124.

Charnes, A., Cooper, W. W., & Rhodes, E. (1978). Measuring the efficiency of decision making units. *European Journal of Operational Research, 2*, 429–444.

Chen, C. M., Delmas, M. A., & Lieberman, M. B. (2015). Production frontier methodologies and efficiency as a performance measure in strategic management research. *Strategic Management Journal, 36*(1), 19–36. https://doi.org/10.1002/smj.2199

Cook, W. D., Tone, K., & Zhu, J. (2014). Data envelopment analysis: Prior to choosing a model. *Omega, 44*, 1–4. https://doi.org/10.1016/j.omega.2013.09.004

De Simone, L. D., & Popoff, F. (2000). *Eco-efficiency: The business link to sustainable development*. The MIT Press.

De La Cruz, B. C. B., Pizzolato, N. D., De La Cruz, A. B. (2010). An application of the spatial equilibrium model to soybean production in tocantins and neighboring states in Brazil. *Pesquisa Operacional 30*(2), 443–464 https://doi.org/10.1590/S0101-74382010000200011

EMBRAPA – Empresa Brasileira de Pesquisa Agropecuária. (2013). *Tecnologias de produção de soja*. Embrapa Soja.

European Comission. (2019). *Global food supply and demand: Consumer trends and trade challenges*. EU Agricultural Markets Briefs No 16.

FAO – Food and Agriculture Organization of the United. (2009). *How to feed the world*, 2050.

FAO – Food and Agriculture Organization of the United Nations. (2017). *The future of food and agriculture*.

Fare, R. S., Grosskopf, S., & Lovell, C. A. K. (1994). Production Frontiere. Cambridge University Press, London.

FAO - Food and Agriculture Organization of the United Nations. (2017). The future of food and agriculture: Trends and Challenges. https://www.fao.org/3/i6583e/i6583e.pdf

Farrell, M. J. (1957). The measurement of productive efficiency. *Journal of the Royal Statistical*.

Friesz, T. L., Tobin, R. L., & Harker, P. T. (1983). Predictive intercity freight network models: the state of the art. *Transportation Research Part A: General 17*(6) 409–417 https://doi.org/10.1016/0191-2607(83)90161-9

Friesz, T. L., & Harker, P. T. (1983). Multicriteria spatial price equilibrium network design: Theory and computational results. *Transportation Research Part B: Methodological 17*(5) 411–426 https://doi.org/10.1016/0191-2615(83)90007-3

Gédéon, C., Florian, M., & Crainic, T. G. (1993). Determining origin-destination matrices and optimal multiproduct flows for freight transportation over multimodal networks. *Transportation Research Part B: Methodological 27*(5) 351–368 https://doi.org/10.1016/0191-2615(93)90022-3

Guélat, J., Florian, M., & Crainic, T. G. (1990). A Multimode Multiproduct Network Assignment Model for Strategic Planning of Freight Flows. *Transportation Science 24*(1) 25–39 https://doi.org/10.1287/trsc.24.1.25

Gómez-Limón, J. A., Picazo-Tadeo, A. J., & Reig-Martínez, E. (2012). Eco-efficiency assessment of olive farms in Andalusia. *Land Use Policy, 29*, 395–406.

Holguin-Veras, J, et al. (2001). An assessment of methodological alternatives for a regional freight model. New York Metropolitan Transportation Council Report, New York, NY.

IBGE – Brazilian Institute of Geography and Statistics. (2020). *Produação Agrícola Municipal*. www.ibge.gov.br

Ibrahim, M. D., et al. (2019). Transnational resource generativity : Efficiency analysis and target setting of water, energy, land, and food nexus for OECD countries. *Science of the Total Environment, 697*, 134017.

INMET – Brazilian National Institute of Meteorology. (2019). *Cultivation water balance and loss of productivity*.

Kalirajan, K. P., Obwona, M. B., & Zhao, S. (1996). A decomposition of Total factor productivity growth: The case of Chinese agricultural growth before and after reforms. *American Journal of Agricultural Economics., 78*, 331–338. https://doi.org/10.2307/1243706

Kresge, D. T., & Roberts, P. O. (1971). Techniques of Transportation Planning: Systems Analysis and Simulation Models, Washington, DC, USA: Brookings Institution, Transport Research Program.

Kibirige, D., Singh, A. S., & Rugube, L. M. (2019). Small-scale irrigation and production efficiency among vegetable farmers in the eastern Cape Province of South Africa: The DEA approach. *Journal of Agricultural Studies., 7*, 149. https://doi.org/10.5296/jas.v7i1.14564

Labys, W. C., Yang, C. W. (1991). Advances in the Spatial Equilibrium Modeling of Mineral and Energy Issues. International Regional Science Review 14(1) 61-94 10.1177/016001769101400104

Lampe, H. W., & Hilgers, D. (2015, January). Trajectories of efficiency measurement: A bibliometric analysis of DEA and SFA. *European Journal of Operational Research, 240*(1), 1–21.

Leontief, W. W. (1936). Quantitative input and output relations in the economic systems of the United States. Review of Economics and Statistics 18(3), 105-125.

Leta, F. R., Soares de Mello, J. C. C. B., Gomes, E. G., & Meza, L. A. (2005). Métodos de melhora de ordenação em DEA aplicados à avaliação estática de tornos mecânicos. *Investigação Operacional, 25*(2), 229–242.

Li, G., Huang, D., & Li, Y. (2016). China's input-output efficiency of water-energy-food nexus based on the data envelopment analysis (DEA) model. *Sustainability, 8*(9), 16.

Liu, W. B., Meng, W., Li, X. X., & Zhang, D. Q. (2010). DEA models with undesirable inputs and outputs. *Annals of Operations Research, 173*(1), 177–194. https://doi.org/10.1007/s10479-009-0587-3

Liu, W., Zhou, Z., Ma, C., Liu, D., & Shen, W. (2015). Two-stage DEA models with undesirable input-intermediate-outputs. *Omega, 56*, 74–87. https://doi.org/10.1016/j.omega.2015.03.009

Mackett, R. L. (1985). Integrated land use — transport models. *Transport Reviews, 5*(4), 325–343. https://doi.org/10.1080/01441648508716610

MAPA – Ministério da Agricultura, Pecuária e Abastecimento (2019) *Projeções do Agronegócio*. Brasil 2018/19 a 2028/29: Projeções de Longo Prazo.

Masuda, K. (2016). Measuring eco-efficiency of wheat production in Japan: A combined application of life cycle assessment and data envelopment analysis. *Journal of Cleaner Production., 126*, 373–381. https://doi.org/10.1016/j.jclepro.2016.03.090

Melo, I. C., Alves Junior, P. N., Perico, A. E., Guzman, M. G. S., & Rebelatto, D. A. N. (2018). Benchmarking freight transportation corridors and routes with data envelopment analysis (DEA). *Benchmarking: An International Journal, 25*(2), 713–742. https://doi.org/10.1108/BIJ-11-2016-0175

Melo, I. C., Péra, T. G., Alves Junior, P. N., Rebelatto, D. A. N., & Caixeta-Filho, J. V. (2020). Framework for logistics performance index construction using DEA: An application for soybean haulage in Brazil. *Transportation Research Procedia, 48*, 3090–3106. https://doi.org/10.1016/j.trpro.2020.08.179

Mwambo, F. M., Fürstb, C., Nyarkoc, B. K., Borgemeistera, C., & Martiusd, C. (2020). Maize production and environmental costs: Resource evaluation and strategic land use planning for food security in northern Ghana by means of coupled energy and data envelopment analysis. *Land Use Policy, 95*, 104490.

Nijkamp, P. (2007). Ceteris paribus spatial complexity and spatial equilibrium. *Regional Science and Urban Economics 37*(4) 509–516 https://doi.org/10.1016/j.regsciurbeco.2007.04.006

Pierluigi, Toma Pier Paolo, Miglietta Giovanni, Zurlini Donatella, Valente Irene, Petrosillo (2017) A non-parametric bootstrap-data envelopment analysis approach for environmental policy planning and management of agricultural efficiency in EU countries. *Ecological Indicators 83*, 132–143 https://doi.org/10.1016/j.ecolind.2017.07.049

Raju, K. S., & Kumar, D. N. (2006). Ranking irrigation planning alternatives using data envelopment analysis. *Water Resources Management, 20*, 553–566. https://doi.org/10.1007/s11269-006-3090-5

Rentizelas, A., Melo, I. C., Alves Junior, P. N., Campoli, J. S., & Rebelatto, D. A. N. (2019). Multi-criteria efficiency assessment of international biomass supply chain pathways using data envelopment analysis. *Journal of Cleaner Production, 237*, 117690. https://doi.org/10.1016/j.jclepro.2019.117690

Rodriguez Diaz, J. A., Camacho, P. E., & Lopez Luque, R. (2004). Application of data envelopment analysis to studies of irrigation efficiency in Analusia. *Journal of Irrigation and Drainage Engineering, 130*, 175–183.

Seiford, L. M., & Zhu, J. (2002). Modeling undesirable factors in efficiency evaluation. *European Journal of Operational Research, 142*(1), 16–20. https://doi.org/10.1016/S0377-2217(01)00293-4

Stabile, M. C. C., Guimarães, A. L., Silva, D. S., RibeiroV, M. M. N., CoeMT, P. E., Moutinho, P., & Alencar, A. (2020). Solving Brazil's land use puzzle: Increasing production and slowing Amazon deforestation. *Land Use Policy, 91*, 104362. https://doi.org/10.1016/j.landusepol.2019.104362

Teodor Gabriel, Crainic Gilbert, Laporte (1997) Planning models for freight transportation. *European Journal of Operational Research 97*(3) 409–438 https://doi.org/10.1016/S0377-2217(96)00298-6

Tone, K. (2001). A slacks-based measure of efficiency in data envelopment analysis. *European Journal of Operational Research, 130*(3), 498–509. https://doi.org/10.1016/S0377-2217(99)00407-5

USDA (United States Department of Agriculture). (2018). Foreign Agricultural Service Data & Analysis. https://www.fas.usda.gov/data

United Nations. (2019). *World Population Prospects 2019.*

VALEC. (2020). *Ferrovias*. www.valec.gov.br/ferrovias

Yilmaz, B., Yurdusev, M. A., & Harmancioglu, N. B. (2009). The assessment of irrigation efficiency in Buyuk Menderes Basin. *Water Resources Management, 23*, 1081. https://doi.org/10.1007/s11269-008-9316-y

Zhang, T., & Xu, Y. (2019). Evaluation on the efficiency of water-energy-food nexus based on data envelopment analysis (DEA) and Malmquist in different regions of China. *International Journal of Computational Intelligence Systems, 12*(2), 1649–1659.

Part III

Lessons in Urban Resilience

11 Learning About the Nexus from Vulnerable Urban Communities

Leandro Luiz Giatti, Susanne Börner, and Carolina Monteiro de Carvalho

Abstract

Urban communities in socio-environmental vulnerable conditions constitute places of constant processes of creation and reproduction of social practices and knowledge, related to finding alternatives to mitigate the scarcity of essential resources for survival. The water-energy-food nexus thinking, in turn, consists of a fundamental perspective for reducing vulnerabilities, contributing to equating interdependent scarcities, and expanding the supply of resources and social inclusion. In this context, social practices and common sense knowledge in local communities are self-organized and in constant dialogue with water, energy, and food scarcity. These practices can point to legitimate alternatives for synergies that mitigate nexus trade-offs or even contradictions in the form of situations that corroborate unsustainability scenarios. This chapter explores community-based studies conducted in vulnerable urban peripheries of the Global South in the Sao Paulo Metropolitan Region, Brazil. We find that there are forms of isolation of social practices synergistic to the nexus, which must be understood and incorporated in interaction with management promoted by local governments. Such structural coupling can make public policies more assertive, efficient, and legitimate from the point of view of local to global relations. Hence, we recognize the need for cognitive inclusion through participatory approaches within nexus thinking to achieve transformations to sustainability.

Keywords

Cognitive exclusion · Transformation to sustainability · Nexus thinking · Urban nexus · Urban vulnerable communities

11.1 Introduction

The search for sustainable development cannot be successful through mechanisms analogous to those that have put the world on an unsustainable path. For that matter, we highlight the necessary

L. L. Giatti (✉)
Department of Environmental Health, School of Public Health, University of Sao Paulo, Sao Paulo, Brazil
e-mail: lgiatti@usp.br

S. Börner
Department of Environmental Health, School of Public Health, University of Sao Paulo, Sao Paulo, Brazil

School of Geography, Earth and Environmental Sciences, University of Birmingham, Birmingham, UK

C. M. de Carvalho
Institute of Energy and Environment, University of Sao Paulo, Sao Paulo, Brazil

L. L. B. Lazaro et al. (eds.), *Water-Energy-Food Nexus and Climate Change in Cities*, Sustainable Development Goals Series, https://doi.org/10.1007/978-3-031-05472-3_11

criticism against the idea of massification of propositions for sustainable development. Solutions based on hegemonic thinking are often oppressive and impermeable to the conditioning elements of local contexts and, therefore, doomed to failure due to a lack of legitimacy (Santos, 2007, 2009; Giatti, 2019). In this context, uncertainties emerge as a threat of failure, and they are moreover related to different levels of complexity that play an essential role in restricting simplification or reductionism in the search for alternatives (Jasanoff, 2007; Scoones & Stirling, 2020).

As the world is composed of a myriad of socio-environmental contexts, there is no single recipe for sustainability. Thus, local conditions, social practices, and common sense must play a determinant role in the search for balancing intertwined challenges within the social, environmental, and economic dimensions (Leff, 2017).

Thinking along these lines, we assume that the water-energy-food nexus thinking provides an important learning exercise, especially in local contexts such as vulnerable urban peripheries. The water-energy-food nexus (or simply, the nexus), as a perspective for searching synergies to reduce trade-offs, is not only a technical approach but also allows for a new rationality on complex trade-offs between scarcity of resources in contexts permeated by sociopolitical dimensions (Artioli et al., 2017; Cairns & Krzywoszynska, 2016). Thus, we understand that interdependencies associated with water-energy-food in urban environments may allow for essential insights based on community-based social practices, agency, and knowledge (Magnani, 2002; Schatzki, 2015). Marginalized social groups, such as young residents of these peripheries, can also provide important insights into their everyday experiences and confrontation with problems associated with the nexus. Moreover, local concerns and practices can be considered essential in promoting legitimate transformation processes to sustainability (Scoones et al., 2018).

Conventionally, knowledge and practices of peripheral communities are segregated through a cognitive exclusion barrier, a situation that disables dialogue with hegemonic knowledge and the construction of legitimately inclusive public policies. Thus, a diversity of possibilities and its creative potential of solutions is rendered unviable, reproducing the hegemonic knowledges and actions conducted vertically (Santos, 2007). In this respect, we emphasize the importance of developing collaborative studies and reflections capable of promoting interaction and transformative social learning (Souza et al., 2019). Thus, we emphasize that it is possible and necessary to dialogue and ecologize knowledge through studies that allow us to learn from these peripheral and vulnerable communities. We consider, therefore, that the diversity of ideas and practices produced marginally has a fundamental role to play in the search for nexuses aimed at sustainable development.

This chapter explores and recognizes the relevance of knowledge and social practices for transformations to sustainability in urban communities. It focuses on local capacities, agency, and perceptions as necessary inputs for mitigating nexus' trade-offs with legitimacy and cognitive inclusion. The lessons presented relate to two research projects conducted in vulnerable urban communities in the Sao Paulo Metropolitan Region. In the second section, we introduce the idea of the nexus as a dynamic inherent to the contexts of multiple and intertwined scarcities of urban peripheries. The studied cases are presented in Sects. 11.3 and 11.4 as citations of research conducted by our research group. The research experiences refer to participatory action research (PAR) approaches (Baum et al., 2006; Thiollent, 2011), in which community-based research activities occurred through the production of hybrid knowledge in the search for local problems and concerns. For that, researchers and local subjects collaborated through workshops, application of participatory tools (Giatti, 2019), and adaptations for online collaborative activities, like using WhatsApp, in the context of COVID-19 (Auerbach et al., 2022) – the pandemic context affected the case related in Sect. 11.4. In Sect. 11.5, we discuss and conclude considering the related experiences as a perspective for learning and coupling through different knowledges, sectors, and social actors.

11.2 The Urban Nexus from Urban Peripheries

The applicability of the nexus to urban contexts – the urban nexus – requires holistic analyses starting from resource and energy demands for cities, considerations of local resource possibilities, cross-scale connections, interdependencies, and the need for optimization of local resource use (Amaral et al., 2021; Lehmann, 2018; Perrone et al., 2011). Considering the urban nexus is fundamental to finding sustainable solutions to meet the growing global urbanization trends and its respective expansion of demands for ecosystem services. By their nature, urban areas centralize the demands arising from broad territories affecting availability and interdependence relations between water, energy, food, and other resources. As a result, urban scenarios present significant challenges and also opportunities for integration between the different sectors as well as for the socio-material governance of the nexus (Covarrubias, 2019; Lazaro & Giatti, 2021).

The urban nexus seeks to respond to the urgent need for policies and actions that create alternatives to the conventional siloed thinking, i.e., patterns of self-centered structures in their own domains. Thus, integration among sectors is also necessary to transcend policy areas such as land use, social inclusion, waste management, and transport. It refers to achieving more efficient and effective use of resource cycles in urban and peri-urban areas or beyond. This integration should provide benefits across sectors, enabling tailored solutions to govern resource interdependencies through a comprehensive multi-scale spatial perspective (ICLEI, 2014).

Nexus thinking is an irrefutable contribution to the quest for sustainable urban development since the magnitude and trends of urban growth accelerate the planetary environmental crisis (Steffen et al., 2015). However, cities are deeply unequal, and these conditions are made more complex by the relationship between the need to reduce vulnerability and the demand for resources such as water, energy, and food. Worldwide, millions of people in urban poverty demand these resources to alleviate their vulnerability. However, as we know, increasing access to water, energy, and food requires the necessary understanding of the constraints imposed by the trade-offs inherent to the nexus. Nevertheless, these vulnerable urban populations are often underserved and often rely only on their own creative, self-organizing, and local cooperative capacities to cope with the challenges of inequities, resource scarcity, and environmental injustices.

Therefore, resource scarcities within the nexus are closely connected to ongoing creative processes in the daily life of people in vulnerable urban communities. Social practices created and reproduced in those communities can tell us a lot about the search for everyday needs for resources and ways of dealing with intertwined scarcities. For instance, in Kampala, Uganda, city dwellers face complicated and expensive access to energy for cooking associated with water scarcity and food availability. In this context of irregular settlements, people's food choices are conditioned by the need to save water and energy. Thus, people buy charcoal when they have money or, alternatively, search, buy, or borrow charcoal, deadwood, or biomass briquettes. The search for alternatives also relates to food; instead of cooking beans, they opt for bananas because of faster cooking and resource-saving. This nexus interplay in their practices makes it possible to understand alternatives to trade-off mitigation, to avoid generating a harmful interference due to a lower nutritional quality. The complex relations mediating the cooking activities in Kampala requires extensive effort and agency in finding nexus-related alternatives highlighting a form of precarious consumption. Moreover, the exact relation of scarcities expands upscale, connecting the local constraints with regional nexus issues. For instance, the local insufficient water supply is under pressure from the ongoing wood overexploitation required to meet urban energy demands (Mguni et al., 2020).

Basically, nexus thinking has emerged from the complex challenges of interdependencies between resources, and, therefore, it requires integration between sectors with their pre-existing governance structures (Benson et al., 2015). At its core, the urban nexus requires integration between municipal sectoral strategies in the search for synergies and the reduction of trade-offs, which can be challenging in practice. The urban nexus should also use interactions between different organizational levels through arrangements that favor stakeholder engagement and a convergence of behaviors as well as social patterns in a context of urban complexity. In urban contexts, nexus thinking and respective innovation must occur through public participation (Wahl et al., 2021). However, there is a need to overcome a marked gap of cognitive exclusion, considering that other rationalities and values from diverse communities must be integrated. In this context, key issues are not only the promotion of participation but also the promotion of legitimate participation in the co-production and application of hybrid and democratic knowledge (Santos, 2007).

11.3 Social Practices Related to Intertwined Scarcities

Once the focus is on nexus thinking in the context of resource-related social practices in vulnerable urban communities, adequate tools are needed to understand this relationship at the local level. Participatory Geographic Information Systems (PGIS) is one of these tools. According to Kahila-Tani et al. (2016), PGIS integrates collaborative mapping technologies and promotes social participation in generating and managing spatial information collaboratively. Applying PGIS and participatory mapping in vulnerable communities can promote empowerment and raise environmental awareness among local dwellers and contribute to the urban nexus thinking understanding since it is possible to map local flows and trade-offs at local, regional, and also global levels (Eftelioglu et al., 2016).

PGIS was applied in the community of Novo Recreio (Carvalho et al., 2021), inside Guarulhos municipality (Sao Paulo state, Brazil), a city with about 1.3 million inhabitants that suffers from several social and environmental challenges on a daily basis, especially peripheric and vulnerable areas. Novo Recreio has approximately 4500 family and borders on Cantareira State Park, a protected rainforest area that is a relevant source of water resources for the Sao Paulo Metropolitan Region. However, due to the unplanned urbanization, Novo Recreio has limited access to water, energy, and fresh food, such as fruits and vegetables, the mentioned nexus resources. Guarulhos also has a socio-environmental vulnerability index (SVI) of 0310, medium vulnerability, and Novo Recreio has an SVI index of 0374, indicating that it has a higher vulnerability in comparison to the whole municipality (Atlas Brasil, 2018).

The research in Novo Recreio was developed jointly through the Resnexus Project,[1] which aimed to assess the local urban nexus resources through interviews with local residents that had been pre-selected with the support of the local primary healthcare unit. The project's first stage included local observation, informal conversations, and interviews with 14 residents, guided by questions related to people's access to water, energy, and food, based on a methodology already reported by Giatti et al. (2019). The local interviews were carried out during February and March of 2017. Project members transcribed and analyzed the information.

The interviews provided a detailed picture of the food and eating habits of the interviewees, according to the local conditions, as it was possible to observe such as the following: Local markets or street markets had little variety of food, especially with regard to fruits and vegetables. Purchases were made with local producers or at street markets and at bigger markets in other neighborhoods or even the center of Guarulhos. Thus, residents depended on public transport or on owning a car. Public transport was scarce, and on rainy days public transport was suspended, according to reports from some of the interviewed residents. The neighborhood itself had no

[1] International cooperation funded in Brazil by FAPESP, involving University of São Paulo (Brazil), University of Sussex (UK) and University of Wageningen (Netherlands) – https://bv.fapesp.br/en/auxilios/92515/resilience-and-vulnerability-at-the-urban-nexus-of-food-water-energy-and-the-environment-resnexus/

street market, only small local markets, mainly selling ultra-processed products.

People from Novo Recreio make constant choices related to food acquisition, given the scope of their local conditions. Thus, local social practices are mediated by the described material constraints (Schatzki, 2015). The results are complex forms of dialogue with the urban nexus and can negatively impact their health conditioning. The context of being in the urban periphery has led to a low possibility of acquiring healthier and fresher food, thereby worsening trade-offs within the nexus.

On the other hand, some good lessons can be learned from social practices that generate conditions to preserve resources. For example, people from Novo Recreio develop ways to save water, recycle materials, and produce small amounts of food through urban gardening. These practices and knowledge make possible some synergies within the nexus, representing possibilities to reduce the trade-offs among water, energy, and food. However, these proceedings are not necessarily recognized by local government policies, and it characterizes a kind of disruption among communitarian competencies and a higher organizational level of public management (Giatti et al., 2019).

In the second stage of research in Guarulhos, participatory mapping was applied in an extension course for the community, as part of the postdoctoral research project "PGIS for socio-environmental governance in Guarulhos, Brazil (FAPESP 2015/21311-0)," offered to adolescents aged 14–17 years, who had previously attended the local NGO "Clube de Mães" ("Mothers' Club"), in the period between April and August 2017. The main objective was to introduce participants to the fundamentals of mapping and local urban planning, to promote awareness and socio-environmental reflections and also stimulate dialogue. In the progressive development of the course, there was a preparatory mapping phase, which provided basic information on mapping and also on the surroundings of the area, including a group exercise.

Participants identified rivers, the environmental protection area nearby, their homes, and their daily routine in the neighborhood. In this step, participants produced a so-called sketch map (Fig. 11.1) with an initial diagnosis of the main problems and their expectations for changes in the neighborhood over the next few years. The next phase provided more accurate geoinformation as well as the location of the socio-environmental problems mentioned by the participants in the sketch map phase. Through this mapping process, places with socio-environmental problems were mapped, such as those with potential for landslides/flooding and inadequate waste disposal, as well as areas where these problems overlapped, representing a risk to the health and safety of local dwellers. Furthermore, the discussion around marking the different points on the map was a rich process, which promoted empowerment and awareness among the participants. The post-mapping phase reflected the new discoveries and mapped points and their impacts on their lives. In addition, suggestions for addressing these problems were discussed using other participatory methods, such as integrated panels and the development of community journals (Fig. 11.1), in which the nexus resource problems discussed were reported along with possible solutions. Community journals can play an important role in communicating results of the mapping process to the wider neighborhood. The material produced during the participatory process was disseminated to the community and the wider society of Guarulhos through project workshops, stakeholder meetings, articles, and other publications, such as blogs and social media (Carvalho et al., 2021).

With the interviews and the mapping process, it was possible to recognize new and legitimate proposals for the Novo Recreio community related to the nexus thinking water-energy-food, such as the need for closer fresh food markets with a variety of fresh food, and solutions like new urban gardens implementation for food security and health improvement. Renewable energy sources, such as solar panels, were also discussed, along with other environmental education measures (to avoid deforestation, to preserve local streams).

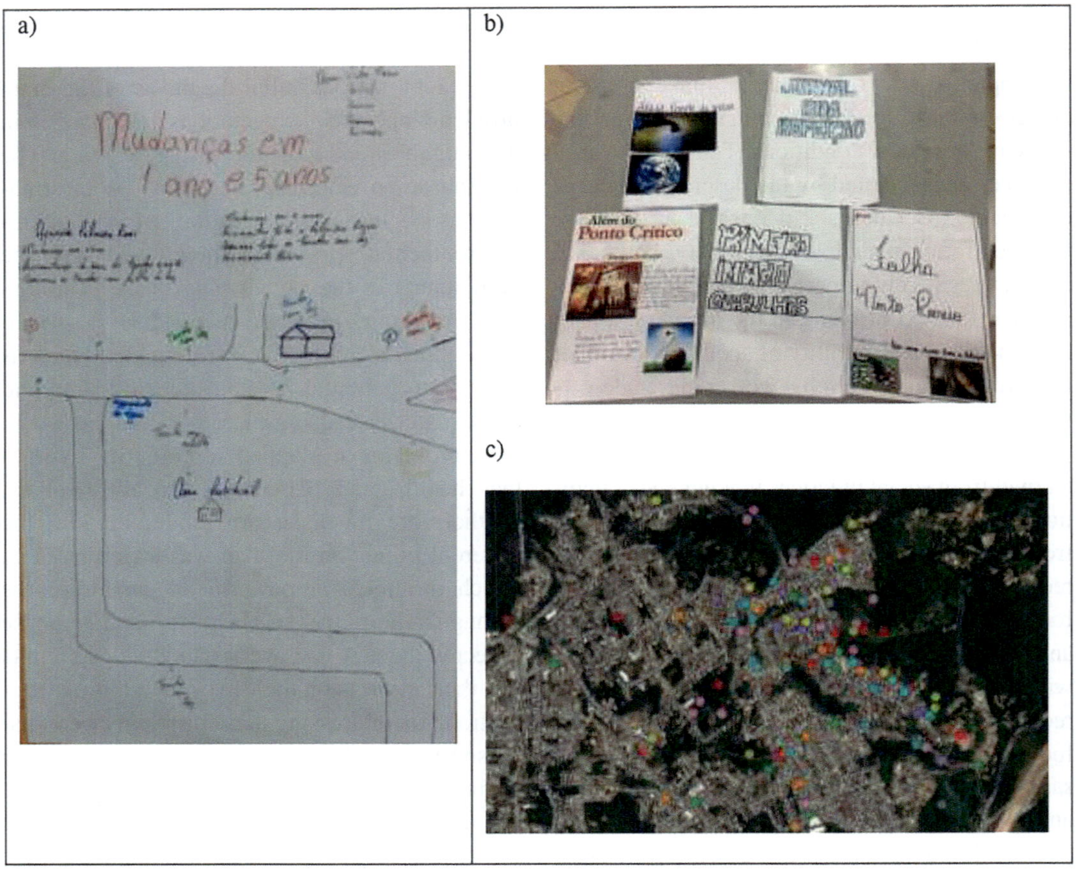

Fig. 11.1 (**a**) Sketch map developed by participants, (**b**) community journals containing proposals and solutions for the mapped issues, and (**c**) local planning mapped with online platform Maptionnaire. (Carvalho & Giatti, 2021)

The mapped themes and proposals brought up in the activities with the youth were motivators of discussions and reflections on the interdependence of resources. It enabled ways to approach nexus thinking from the participants' perspective, considering their daily activities and related concerns. As a result, the dialogue process helped develop alternatives and innovations for young people, allowing them to discuss and reflect on themes initially absent from their daily activities and concerns. However, the approaches through participative methodologies – including GIS – provided legitimacy to the discussions to the concerns of youth.

Social participation is key to achieving efficient and fair governance and sustainable urban development, integrating the complexities and nexus' flows of metropolitan regions with citizens' demands. To this end, the study of the urban nexus and its particularities, associated with social protagonism, can promote robust support for integration, providing a more complete view of the access to and flows of basic resources.

11.4 Youth Agency and the Nexus

The debate on social practices and interdependent resource scarcities intersects with debates on agency and the nexus. The debate on agency acquires relevance especially in contexts of social exclusion and power imbalances as they are present in Brazil's urban peripheries. Moreover, by focusing on youth as the main protagonists in nexus dialogues, we seek to challenge "dominant forms of nexus thinking" (Kraft et al., 2019) which derive from adult voices that create top-down and technocratic understandings of the nexus and focus on the nexus essentially as part of high-level policy agendas (Kraftl et al., 2019; Walker, 2019).

These dominant and adult-centered approaches overlook the importance of a better understanding and analysis of "young people's everyday, embodied engagements with water, food and energy" (Kraftl et al., 2019) which call for "bottom-up" approaches. Hence, Walker (2019), for instance, explores the "mundane and life-sustaining ways in which food, water and energy are used, imagined and contested in situated and embodied ways in children and young people's lives" (Walker, 2019, 360). In this contribution, we seek to introduce youth agency as a hybrid concept that integrates young people's experiences with the nexus through an understanding of "everyday agency" rather than inscribing ourselves in an approach that sees youth exclusively as agents of change and radical societal transformation. At the same time, we do not want to abolish the juxtaposition of young people's voice and agency (Kraftl, 2013), since there are numerous impressive examples of youth leadership in urban hubs and also in the urban peripheries that seek social inclusion, equality, and climate justice. Examples from Sao Paulo's periphery include initiatives such as *Perifa Sustentavel* or *Festival Favela em Casa* that dialogue with the transformative potential of these peripheral spaces. Yet, not all young people actually "speak out" as voices of change but nonetheless have everyday knowledge of their own reality and of what matters to them, based on their everyday experiences. We feel that these "silent expressions" of experiencing and engaging with the nexus on a daily basis matter in themselves, especially in contexts of vulnerability, resource scarcity, and social (and cognitive) exclusion. Hence, we seek to broaden the concept of youth agency to include understandings of agency in mundane, everyday spaces outside politicized arenas (Börner et al., 2020), thereby introducing a more pluralistic and multifaceted concept of youth agency.

Drawing from insights of ongoing research from the project: "Building resilience in the face of nexus threats: local knowledge and social practices of Brazilian youth" (NEXUS-DRR)[2] with young people aged 12–18 in Sao Paulo's urban periphery in the municipality of Franco da Rocha, we inscribe youth agency not only in an understanding as *contest and transformation* as in most social youth movements pressuring for transformational change. Rather, we argue that youth agency can and must also be understood as *living with* and *adapting to* existing conditions of resource scarcity through local knowledge(s) and adaptive social practices as discussed above.

Conceived pre-COVID, the research was originally designed as participatory, face-to-face research, including participatory youth workshops (e.g., for the production of a community journal), youth-led thematic community walks, photo-voice, participatory mapping, and focus groups. Participants had been recruited through youth groups in two social assistance reference centers in Franco da Rocha in the Sao Paulo urban periphery and introductory youth workshops took place in February 2020. The research had been conceived in the format of a university extension course offered through the School of Public Health at the University of Sao Paulo to benefit approximately 40 young people. With the onset of COVID-19 and the subsequent displacement of the researcher from the field, the project was temporarily suspended before shifting the research activities to an online format using mainly WhatsApp groups (see case study Börner, in Auerbach et al., 2022). The online university extension course was delivered between January and June 2021, including weekly (a)synchronous activities such as discussion groups, photo-voice activities, participatory video, as well as practical activities with a focus on access to and use of food, water, and energy as well as disaster risk and climate change (case study Börner in Auerbach et al., 2022). Approximately 30 young people participated regularly in the online activities, of which 15 successfully completed the extension course.

[2]Research funded by the European Union's Horizon 2020 Research & Innovation Programme under the Marie Sklodowska-Curie Grant Agreement No. 833401

The research uses a participatory action research methodology to elicit youth protagonism in critically reflecting on resource interdependencies and scarcities in contexts of disaster risk, a process during which young people become empowered as co-creators of knowledge to understand and transform their environment through a reflection-action approach (Freire, 1972). Some of the participants indicated issues related to flooding and access to clean drinking water or to healthy food, while others named air pollution or lack of access to leisure areas as key concerns (Kraftl et al., 2021). Solution-oriented reflections on living with and adapting to scarcity included discussions on access to fruits and vegetables, e.g., by planting fruit trees (Fig. 11.2) and vegetable patches in the backyards of houses. Young people also engaged in discussions on ways of integrating water and energy-saving practices into everyday lives or adapting to extreme weather conditions, for instance, by unplugging electric appliances before a storm.

Lessons learned from young people's engagement with the nexus in a context of disaster risk in the municipality of Franco da Rocha in the Sao Paulo Greater Metropolitan Area furthermore showed the importance of exploring young people's emotions regarding their vulnerabilities but also their potentials (Börner et al., 2020; Kraftl et al., 2021). Although not always visible, young people engage in various "ways of dialoguing with and adapting to resource scarcity through everyday adaptive actions" (Kraftl et al., 2021).

As young people – especially in vulnerable conditions – are not always aware of the value of their everyday experiences and narratives as a transformational force, the project NEXUS-DRR seeks to stimulate critical awareness of everyday realities through a process of social learning based on reflection and action, based on the participatory philosophy of Paulo Freire. Based on this understanding, everyday youth agency becomes more than "living with" the everyday environment. Rather, we understand youth agency as a dynamic concept where the process of knowledge creation and reproduction goes hand in hand with a critical awareness of social adaptation practices to mitigate the scarcity of essential resources and to live "well" – or as well as possible – with resource scarcity (Börner et al., 2020; Kraftl et al., 2021). One example is the

Fig. 11.2 Photograph of her backyard with fruit trees taken by female participant aged 16

way in which youth in Franco da Rocha dialogue with their vulnerability related to mobility in urban spaces. Youth in the project mentioned their needs for better connectivity and a reduced dependency on and better quality of public transport. Moreover, they argued for the importance of receiving better value for money, for instance, based on an integrated transport scheme for public transport to reduce costs. Furthermore, they advocated for alternative modes of transport, encouraging cycling including a proposal for new cycling lanes. They were also aware of the impacts of the intersection of transport with environmental hazards such as flooding as this exacerbated the precarity of access to safe and reliable transport.

The transformational force of young people's micro-realities with the nexus comes from their critical awareness of the complex challenges posed by resource scarcity and interdependency as well as the solutions that they propose for more sustainable futures. As youth narratives of their everyday agency and the nexus unfold, we need to be careful to give space to a diversity of individual youth identities which create a mosaic of experiences with different micro-realities (Kraftl et al., 2021). Young peoples' stories "reflect manifold realities which cannot be reduced to a single 'one size fits all' portrait of their situation" (ibid).

Finally, we should also be careful with the expectations that we put on youth as drivers of change. Although youth can play an important role both as everyday and transformational agents, there is a need for a set of actions and actors to enable comprehensive and systemic transformational change. This requires integrating youth knowledge into public policies for sustainable urban planning to create more youth-friendly and resilient urban spaces. To achieve this, navigating youth everyday interaction with their local environments and resource scarcities from a nexus perspective may help acquire a better understanding of how young people understand, negotiate, and overcome multi-scalar sociopolitical, economic, and ecological processes (Walker, 2019).

11.5 Learning and Coupling with Local Contexts and Knowledges

The situations explored in the two urban communities show that, despite conditions of marginality and vulnerability, there are possibilities for association and convergence between social practices, perceptions, and local actions regarding the scarcity associated with the water-energy-food nexus. The local self-organizing capacities can generate synergies that improve access to resources while alleviating the trade-offs among the different nexus dimensions. They may, however, generate contradictions where local social practices exacerbate trade-offs within the nexus. Furthermore, a tendency of disconnection can be observed between aspects of daily life in urban communities and the ways in which communities name, plan, and act on attributes of interconnected scarcities within the global crisis of unsustainability. Such disconnections can be described as a nexus of exclusion, represented by "an abysmal frontier that disconnects peripheral residents from sustainable and healthy choices and interaction with other organizational levels" (Giatti et al., 2019, 8). Thus, people from poor urban communities are not only excluded based on conditions of vulnerabilities and lack of opportunities; they are also excluded by the lack of appropriate translations of different knowledge and lack of cognitive inclusion (Santos, 2007).

The ethnographic investigation in Guarulhos confirms that some social practices can represent synergies within the nexus. Moreover, the participatory mapping shows that besides perceiving their marginalization, young people engaged in a process of self-reflection that enabled them to integrate their daily lives' experiences into urban planning approaches for sustainability and health. As the same, the research in Franco da Rocha depicts how young people in urban peripheries can adapt within conditions of resource scarcity. They created alternatives like planting fruit trees, and youth narratives around the nexus unfolded in ways that acquired new meanings and opened up alternatives for youth everyday agency. Finally, youth narratives opened new pathways for cognitive inclusion of young people in generating alternatives.

Based on these reflections and experiences, we suggest an integrated model for overcoming the marginalization of social practices and cognitive exclusion (Fig. 11.3). The model must target misconnections among the opportunities of local communities and upper organizational levels where the debate about sustainability and environmental crises regularly occurs. Figure 11.3 represents how the reproduction of contradictions towards the nexus and the related lack of interaction with public policies feeds back on the social and cognitive exclusion scenario, maintaining inertia and the lack of alternatives for mitigating vulnerabilities, unsustainability, and trade-offs within the nexus. Such contradictions are also reflected in unsatisfactory dialogues among local social actors and decision-makers from local governments.

In the suggested model, the scenario of social and cognitive exclusion is represented as being impervious to changes and interactions with higher organization levels and public policies. On the other hand, it is a dialogical model to face a void of interactions and the absence of translations. Implementation and operation of this proposal remit to the promotion of collective conversations through the application of participatory research tools, like in workshops involving laypeople, practitioners from public administration, policy-makers, academics, and other possible partners like nongovernmental organizations and the private sector. All the social actors must reflect from the local contexts, analyzing constraints and potentialities, identifying possible cooperation among different sectors and actors, and envisaging practical and collaborative interventions and attainable outcomes (Oetzel et al., 2018; Wallerstein et al., 2017). The green arrows in Fig. 11.3 represent ruptures of isolation and alternatives to be built considering social practices and the possibility of collaborative alternatives with the local studied social practices.

We recognize the importance of understanding local constraints, social practices, and local knowledge as central in developing new struc-

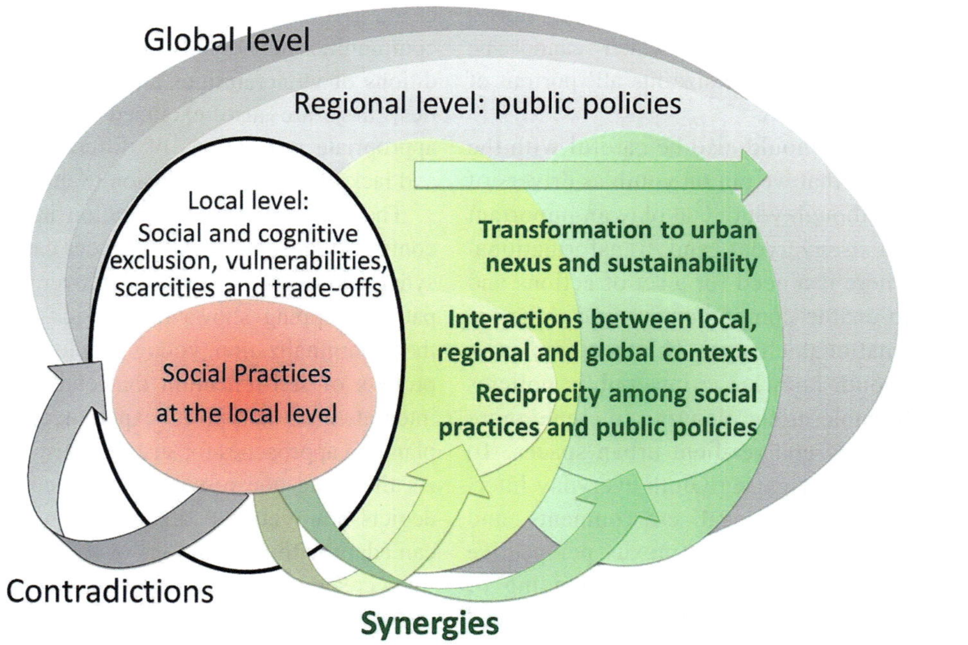

Fig. 11.3 Local social practices and perspectives for transformation to sustainability and urban nexus. (Source: Produced by the authors)

tures for integrated responses to nexus challenges. Necessary changes are more likely to arise from horizontal collaborations in a bottom-up manner, thereby shaping reflexive processes that connect global-local dimensions within public policies at the local community level (Bott et al., 2019; Harwood, 2018; Stirling, 2015). This means, for example, that social practices synergistic to the nexus of people in Guarulhos, such as ways to save water and urban gardening, must encounter and make sense in new arrangements through the actions of public policies. At the same, the empowering property of participatory activities must create space for youth engagement and agency.

Therefore, the aimed transformations are conditioned by the accomplishment of structural coupling, which relates to breaking the isolation of the communities to facilitate a higher level of self-organization through dialogue and social learning (Maturana & Varela, 1992; Souza et al., 2019). In practical terms, it means pushing the natural ability of such systems to interact, learn, cooperate, and, thus, transform themselves acquiring new possibilities and adaptations (Preiser et al., 2018). The rupture with the conventional isolation and cognitive exclusion also relates to allowing autonomy through self-organization, all of this as an emancipatory process (Giatti, 2019). In this regard, a policy decision of providing a street fair based on the suggestion of the young people from Guarulhos could be more effective in determining what and how the best foodstuffs could be sold. Moreover, by taking the subjects' concerns into account, local needs can be respected. This may include choosing appropriate weekdays or specific places for a street fair or taking account of monetary constraints for local people to acquire fresh and healthier food.

The cross-fertilization provided by more symmetric interactions involving diverse knowledge can be favorable to sustainability and the management of scarce resources (Wutich et al., 2018). Consequently, cognitive justice, hybridization, and horizontal collaboration are useful for delivering innovation aimed at urban transitions to sustainability and resilience (Ernstson et al., 2010). Also, the legitimacy of dialogical and participatory approaches contributes to equalizing power relations in the process of making hybrid knowledge into action for effective change (Wallerstein & Duran, 2010).

Considering the fundamental cross-sectoral and multilevel approach, the urban nexus requires substantial empirical grounding. In this sense, communities as particular spaces of practices represent interesting choices for study and for the creation of bottom-up initiatives. For instance, when women from vulnerable urban communities learn about nutritious food and cooking, they can focus on well-being possibilities, which may lead to dynamics beyond the original limited focus on the nexus between climate, water, food, and energy. However, as the local level alone cannot account for all trade-offs, the nexus thinking must provide upper-scale interactions and cross-border synergies (Boas et al., 2016). In this respect, we reinforce the importance of the urban nexus towards reflexive interactions with local social practices. It can represent the nurturing diversity of socio-technical practices for transformative engagement (Arora, 2019).

Concerning the nexus, we conclude that there are many things to learn from urban communities, and it relates to local social practices, knowledges, and agency. By recognizing these learnings and with the promotion of cognitive justice, we can make possible new, legitimate, and collaborative nexus-related alternatives that stimulate transformations towards sustainability in urban contexts. In the expectation of broad changes comprehending global-local intrinsic relationships within the nexus, transformation must be preceded by a shift to overcome the absence of dialogue between social actors through different levels and sectors. Community members must learn from public policies as well practitioners, and policy-makers must learn better about communities, their creative capacities, and agency. Approaching the nexus from the community level can promote a nexus of broad inclusion, which can then contribute to optimally mitigating the interdependent scarcities of water, energy, and food, from a perspective of social and cognitive justice. This rationality is based on the consideration that interdependency and unsustainability are not only limited to resources. It is also subordinated by social concerns, local capacities, and necessary reflexivity. Finally, the challenge of sustainability for urban communities is also a challenge for democracy and social inclusion from a cognitive perspective.

Acknowledgments The authors acknowledge funding by the National Council for Scientific and Technological Development (CNPq – proc. n. 309840/2018-0) and by the Sao Paulo Research Foundation (FAPESP – proc. n. 2015/03804-9, 2015/21311-0, and 2021/07399-2). Dr. Susanne Börner's research is financed by the European Union's Horizon 2020 Research & Innovation Programme under the Marie Sklodowska-Curie Grant Agreement No. 833401 [NEXUS-DRR].

References

Amaral, M. H., Benites-Lazaro, L. L., de Almeida Sinisgalli, P. A., da Fonseca Alves, H. P., & Giatti, L. L. (2021). Environmental injustices on green and blue infrastructure: Urban nexus in a macrometropolitan territory. *Journal of Cleaner Production*, 125829.

Arora, S. (2019). Admitting uncertainty, transforming engagement: Towards caring practices for sustainability beyond climate change. *Regional Environmental Change, 19*, 1571–1584.

Artioli, F., Acuto, M., & Mcarthur, J. (2017). The water-energy-food nexus: An integration agenda and implications for urban governance. *Political Geography, 61*, 215–223.

Atlas do Desenvolvimento Humano no Brasil. (2018). Available in: http://www.atlasbrasil.org.br/. Accessed 24 April 2018.

Auerbach, J., Muñoz, S., Affiah, U., de la Barrera Torre, G., Börner, S., Cho, H., Cofield, R., CM, D. E., Graddy-Lovelace, G., Klassen, S., Limeberry, V., Morse, A., Natarajan, L., & Walsh, E. A. (2022). *Displacement of the Scholar? Participatory Action Research Under COVID-19*. Frontiers in Sustainable Food Systems. Section: Social Movements, Institutions, and Governance. https://doi.org/10.3389/fsufs.2022.762065

Baum, F., MacDougall, C., & Smith, D. (2006). Participatory action research. *Journal of Epidemiology and Community Health, 60*, 854.

Benson, D., Gain, A. K., & Rouillard, J. J. (2015). Water governance in a comparative perspective: From IWRM to a nexus' approach? *Water Alternatives, 8*.

Boas, I., Biermann, F., & Kanie, N. (2016). Cross-sectoral strategies in global sustainability governance: Towards a nexus approach. *International Environmental Agreements: Politics, Law and Economics, 16*, 449–464.

Börner, S., Kraftl, P., & Giatti, L. L. (2020). Blurring the "-ism" in youth climate crisis activism: Everyday agency and practices of marginalized youth in the Brazilian urban periphery. *Children's Geographies, 19*(3), 275–283.

Bott, L. M., Ankel, L., & Braun, B. (2019). Adaptive neighborhoods: The interrelation of urban form, social capital, and responses to coastal hazards in Jakarta. *Geoforum, 106*, 202–213.

Cairns, R., & Krzywoszynska, A. (2016). Anatomy of a buzzword: The emergence of 'the water-energy-food nexus' in UK natural resource debates. *Environmental Science & Policy, 64*, 164–170.

Carvalho, C. M., & Giatti, L. L. (2021). Sistema de informação geográfica participativo (SIGP) para a integração da governança socioambiental da Região Macrometropolitana Paulista. In P. R. Jacobi & L. L. Giatti (Eds.), *Inovação para governança da macrometrópole paulista face à emergência climática* (pp. 99–119). CRV.

Carvalho, C. M., Giatti, L. L., Fagerholm, N., Bedran-Martins, A. M., & Kytta, M. (2021). Participatory Geographic Information Systems (PGIS) to assess water, energy and food availability in a vulnerable community in Guarulhos (Brazil). *International Journal of Urban Sustainable Development, 13*, 516–529.

Covarrubias, M. (2019). The nexus between water, energy and food in cities: Towards conceptualizing socio-material interconnections. *Sustainability Science, 14*, 277–287.

de Santos, BS. (2007). Para além do pensamento abissal: Das linhas globais a uma ecologia de saberes. *Novos estudos-CEBRAP*, 71–94.

de Santos, BS. (2009). Una epistemología del sur: la reinvención del conocimiento y la emancipación social. *Siglo, XXI*.

Eftelioglu, E., Jiang, Z., Ali, R., et al. (2016). Spatial computing perspective on food energy and water nexus. *Journal of Environmental Studies and Sciences, 6*, 62–76. https://doi.org/10.1007/s13412-016-0372-y

Ernstson, H., Der Leeuw, S. E. V., Redman, C. L., Meffert, D. J., Davis, G., Alfsen, C., & Elmqvist, T. (2010). Urban transitions: On urban resilience and human-dominated ecosystems. *Ambio, 39*, 531–545.

Freire, P. (1972). *Pedagogy of the oppressed*. Herder and Herder.

Giatti, L. L. (2019). *Participatory research in the post-normal age: Unsustainability and uncertainties to rethink Paulo Freire's pedagogy of the oppressed*. Springer.

Giatti, L. L., Urbinatti, A. M., de Carvalho, C. M., Bedran-Martins, A. M., de Santos, I. P. O., Honda, S. O., Fracalanza, A. P., & Jacobi, P. R. (2019). Nexus of exclusion and challenges for sustainability and health in an urban periphery in Brazil. *Cadernos De Saude Publica, 35*, 00007918.

Harwood, S. A. (2018). In search of a (WEF) nexus approach. *Environmental Science and Policy, 83*, 79–85.

ICLEI, GIZ. (2014). *Operationalizing the Urban NEXUS towards resource-efficient and integrated cities and metropolitan regions Case Studies*. Disponível em: http://www2.gtz.de/urbanet/opencommunity/focus/urbannexus.asp

Jasanoff, S. (2007). Technologies of humility. *Nature, 450*, 33–33.

Kahila-Tani, M., Broberg, A., Kytta, M., & Tyger, T. (2016). Let the citizens map—Public participation GIS as a planning support system in the Helsinki master plan process. *Planning Practice & Research, 31*(2), 195–214. https://doi.org/10.1080/02697459.2015.1104203

Kraftl, P. (2013). Beyond 'Voice', Beyond 'Agency', Beyond 'Politics'? Hybrid Childhoods and Some Critical Reflections on Children's Emotional Geographies. *Emotion, Space and Society, 9*, 13–23. https://doi.org/10.1016/j.emospa.2013.01.004.

Kraftl, P., Perrella Balastieri, J. A., Marques Campos, A. E., Coles, B., Hadfield-Hill, S., Horton, J., Valladares Soares, P., Nogueira Vilanova, M. R., Walker, C., & Zara, C. (2019). (re)thinking (re)connection: Young people, 'natures' and the water–energy–food nexus in Sao Paulo state, Brazil. *Transactions of the Institute of British Geographers, 44*, 299–314.

Kraftl, P., Hadfield-Hill, S., & Börner, S. (2021). Building better, safer, healthier cities for children and young people. In *Addressing the climate challenges*. COP 26 Essay Collection. Forum for Global Challenges, University of Birmingham.

Lazaro, L. L. B., & Giatti, L. L. (2021). O Nexo água-energia-alimentos—Uma abordagem para cidades sustentáveis e o desenvolvimento sustentável. In P. R. Jacobi & L. L. Giatti (Eds.), *Inovação para governança da macrometrópole paulista face à emergência climática* (pp. 79–98). CRV.

Leff, E. (2017). Power-knowledge relations in the field of political ecology. *Ambiente & Sociedade, 20*, 225–256.

Lehmann, S. (2018). Implementing the Urban Nexus approach for improved resource-efficiency of developing cities in Southeast-Asia. *City, Culture and Society, 13*, 46–56.

Magnani, J. G. C. (2002). De perto e de dentro: notas para uma etnografia urbana. *RBCS, 14*.

Maturana, H. R., & Varela, F. J. (1992). *The tree of knowledge* (Rev. ed.). Shambala.

Mguni, P., van Vliet, B., Spaargaren, G., Nakirya, D., Osuret, J., Isunju, J. B., Ssekamatte, T., & Mugambe, R. (2020). What could go wrong with cooking? Exploring vulnerability at the water, energy and food Nexus in Kampala through a social practices lens. *Global Environmental Change, 63*, 102086.

Oetzel, J. G., Wallerstein, N., Duran, B., Sanchez-Youngman, S., Nguyen, T., Woo, K., Wang, J., Schulz, A., Kaholokula, K., & Israel, B. (2018). Impact of participatory health research: A test of the community-based participatory research conceptual model. *BioMed Research International*, 2018.

Perrone, D., Murphy, J., & Hornberger, G. M. (2011). *Gaining perspective on the water- energy nexus at the community scale*.

Preiser, R., Biggs, R., De Vos, A., & Folke, C. (2018). Social-ecological systems as complex adaptive systems. *Ecology and Society, 23*.

Programa das Nações Unidas para o Desenovolvimento – PNUD. Atlas do Desenvolvimento Humano no Brasil. Disponível em: http://atlasbrasil.org.br/2013/. Acesso em: 23 ago. 2017.

Schatzki, T. (2015). Spaces of practices and of large social phenomena. *Espaces Temps, 24*.

Scoones, I., & Stirling, A. (2020). *The politics of uncertainty: Challenges of transformation*. Taylor & Francis.

Scoones, I., Stirling, A., Abrol, D., Atela, J., Charli-Joseph, L., Eakin, H., Ely, A., Olsson, P., Pereira, L., & Priya, R. (2018). *Transformations to sustainability*.

Souza, D. T., Wals, A. E., & Jacobi, P. R. (2019). Learning-based transformations towards sustainability: A relational approach based on Humberto Maturana and Paulo Freire. *Environmental Education Research*, 1–15.

Steffen, W., Broadgate, W., Deutsch, L., Gaffney, O., & Ludwig, C. (2015). The trajectory of the Anthropocene: The great acceleration. *The Anthropocene Review, 2*, 81–98.

Stirling, A. (2015). *Developing 'Nexus Capabilities': Towards transdisciplinary methodologies*.

Thiollent, M. (2011). *Metodologia da Pesquisa-Ação* (18th ed.). Cortez.

Wahl, D., Ness, B., & Wamsler, C. (2021). Implementing the urban food–water–energy nexus through urban laboratories: A systematic literature review. *Sustainability Science*, 1–14.

Walker, C. (2019). Nexus thinking and the geographies of children, youth and families: Towards an integrated research agenda. *Children's Geographies, 18*, 351–363.

Wallerstein, N., & Duran, B. (2010). Community-based participatory research contributions to intervention research: The intersection of science and practice to improve health equity. *American Journal of Public Health, 100*.

Wallerstein, N., Duran, B., Minkler, M., & Oetzel, J. G. (2017). *Community-based participatory research for health: Advancing social and health equity*. Wiley.

Wutich, A., Pahl-Wostl, C., Rauschmayer, F., Schleyer, C., Srinivasan, V., Suhardiman, D., Tallis, H., & Zwarteveen, M. (2018). Integrating sustainability, justice and diversity?: Opportunities and challenges for inclusively framing water research. In S. Lele, S. Brondizio, J. Byrne, G. Mace, & J. Martinez-Alier (Eds.), *Rethinking environmentalism: Linking justice, sustainability and diversity*. MIT Press.

12 Urban Gardens and Composting: Effective Government for Strengthening Urban Resilience and Community Waste Management

Clarice Araújo Carvalho, Suzana Más-Rosa, and Andréa Cardoso Ventura

Abstract

Organic waste management is considered an effective alternative for treating solid urban waste. Implementing organic waste management may greatly contribute to global environmental governance, focused on meeting the 17 Sustainable Development Goals (SDGs) of the 2030 United Nations Agenda, especially when added to urban agriculture programs. However, these projects require a good governance system in order to succeed. This study aims to define the key factors in urban garden governance for implementing an effective organic waste management. To identify these key factors, the involved actors and their responsibilities, besides establishing the possible networks and their respective benefits and conflicts, we studied the case of the Hortas Urbanas Salvador Project – an urban community garden in one of the main neighborhoods in the city of Salvador, Bahia, Brazil. This case study was conducted based on a bibliographic research, a field analysis, and interviews with the actors responsible for the project. Civil society participation and the support of the municipal sphere were essential for the project performance. As a result, we proposed a governance model aiming to enhance urban garden contribution and decentralize organic waste management, thus promoting sustainable development.

Keywords

Urban green commons · Municipal solid waste · Environmental governance · Urban resilience

12.1 Introduction

Brazil is the fifth most populous country in the world, and 91% of its population is expected to live in urban areas by 2030 (UN-Habitat, 2014). This population growth and the disordered occupation of the urban space will lead to the reduction of available nonrenewable resources, the loss of biodiversity, and climate change. Moreover, the issue of food losses and waste has recently raised serious concern, spurring worldwide mobilization (Santos et al., 2020). These data point to the urgent need to rethink the sustainability of cities considering the inevitable growth of urban populations and their demands (Food and Agriculture Organization – FAO,

C. A. Carvalho (✉) · S. Más-Rosa · A. C. Ventura
Universidade Federal da Bahia [Federal University of Bahia] (UFBA), Salvador, Brazil
e-mail: andreaventura@ufba.br

L. L. B. Lazaro et al. (eds.), *Water-Energy-Food Nexus and Climate Change in Cities*, Sustainable Development Goals Series, https://doi.org/10.1007/978-3-031-05472-3_12

2012), whereby the efficient management of urban solid waste (MSW) plays a fundamental role

According to the Intergovernmental Panel on Climate Change (IPCC), the waste sector is among the most polluting and the largest emitters of greenhouse gases (GHG). As a result of the decomposition of organic materials such as food and pruning wastes, inadequate waste disposal contributes to GHG emissions, shortening landfill's lifespan, incurring social and economic losses, contaminating groundwater, endangering population health, and spreading diseases (Van Elk, 2007; Gouveia, 2012). In that sense, several projects around the world are developing approaches to adequately manage municipal solid waste (MSW) (Alves, 2017).

In Brazil, organic waste represents about 45% of all MSW generated annually (Brazilian Association of Public Cleaning Companies and Special Waste – ABRELPE, 2018). The country is also among the ten countries with the highest food wastage in the world, with approximately 35% of its production being wasted every year (FAO, 2015).

Zago and Barros (2019) state that reducing food waste is one of the most sustainable ways to reduce natural resource depletion, besides being one of the United Nations (UN) Sustainable Development Goals (SDGs). The agenda also includes the goals of eradicating poverty (SDG1) and hunger (SDG2). According to the Food and Agriculture Organization (FAO, 2014), the nexus approach considers the interdependence and equality among three dimensions – water, energy, and food, – whose challenges are also addressed in the SDGs, especially in SDG2, which seeks to end hunger, achieve food security and improve nutrition, and promote sustainable agriculture (Un-habitat, 2020). In turn, SDG11 includes themes intrinsically related to urbanization, such as mobility, solid waste management, and sanitation, seeking to achieve sustainable communities and cities, as well as the planning and resilience of human settlements, while considering the different needs of the rural, peri-urban, and urban areas (UN, 2018). The so-called nexus concept (water, energy, food) has been gaining strength in scientific research aiming to achieve the SDGs, as it seeks to integrate the three resources, creating synergies and avoiding trade-offs (Märker et al., 2018).

In Brazil, the management of MSW is regulated by Federal Law No. 12,305 (2010), which determines the Brazilian National Policy on Solid Waste (PNRS), and by Decree No. 7404/2010. According to the PNRS, only tailings should be directed to landfills, whereas the waste organic fraction must receive proper disposal, being applied to a more noble purpose, such as composting. One of the main instruments employed by the PNRS is the National Plan for Solid Waste, which recommends the use of composting units already installed and the implementation of new ones (primarily accompanied by selective collection of organic waste). The plan also includes decentralized and local strategies that stimulate home composting and its modalities (earthworms and compost heaps), besides encouraging large-quantity generators of waste to allocate specific areas within their establishments for the practice of composting. Another measure proposed in the National Plan is the implementation of school gardens and the use of compost in urban agriculture (Law No. 12,305, 2010). Despite the existence of PNRS, 53% of the Brazilian cities still do not comply with the legal determination, and the composting index for organic waste in the country is low (Instituto de Pesquisa Econômica Aplicada – IPEA, 2020).

In this context of noncompliance with PNRS legislation and the incipient use of organic waste, few studies approach the management of MSW as part of the food-water-energy nexus (Garcia & You, 2017). However, in a study conducted by Borghi, Moreschi, and Gallo on the importance of composting for the circular economy, the authors establish a strong connection between the nexus at stake and systems associated with MSW, addressing on themes strictly related to the growing scarcity of natural resources and climate change (Bhaduri et al., 2015). Thus, the proper management of MSW, especially the organic ones, has the potential for promoting synergies between systems associated with food, water, and energy.

Just as Brazil, underdeveloped countries have encountered difficulties in transposing national strategies to local contexts, so that several composting projects have failed in their development and implementation (Grimble, 1998). To tackle such difficulties, Borghi, Moreschi, and Gallo point to the need for considering the perceptions, interests, responsibilities, and functions of all involved actors while establishing an integrated governance system.

Zurbrügg and Rothenberger (2013) present decentralized composting as a low-cost method capable of generating a high-quality organic compost. In developing countries, several small-scale pilot projects have demonstrated the benefits of this method. When associated with urban agriculture programs, this type of composting can be easily operated by the community itself, creating employment opportunities and income and strengthening urban resilience. In this context, urban green commons emerge as an alternative to institute an efficient organic waste management system (Swagemakers et al., 2018) and encourage community organization in favor of a collective activity, increasing environmental awareness, creating local jobs, and promoting biodiversity (Siqueira & Assad, 2015). According to Dennis and James (2016), implementing urban green commons and fostering population engagement help strengthen local resilience.

Before the need to implement sustainable practices (FAO, 2012), especially those regarding urban planning aimed at environmental conservation and quality of life, one must understand the aspects of environmental governance (Lemos & Agrawal, 2006). Environmental governance refers to interventions related to the environment, knowledge, institutions, decision-making, and behavior. At a global level, this concept focuses on meeting the SDGs and the 169 targets that make up the 2030 Agenda of the United Nations (UN, 2018; Seixas et al., 2020).

The literature on the subject postulates that several stakeholders must participate, get involved, and negotiate in order to achieve better governance. Multi-actor partnerships are collaborative processes established among different stakeholders to achieve a common goal, cooperate to resolve problems and disagreements related to them, and develop joint actions in relation to perceived problems (Warner, 2007). At a local level, the municipalities – including governments and citizens – are responsible for planning and executing measures that help achieve national objectives (Grimberg & Blauth, 2005). Studies on municipal solid waste management report that the main difficulty in implementing a sustainable model is its practical application (Le et al., 2018).

The Municipality of Salvador, located in the State of Bahia, Brazil, has been developing and implementing plans, policies, and actions focused on climate change and sustainable development (Salvador City Hall – PMS, 2020a). Among the various measures aimed at confronting climate change, the municipality has increased the registration and implantation of urban green commons (PMS, 2020a, b), excelling as a means of achieving benefits at both environmental and social levels.

Although the municipality currently includes 54 urban gardens (PMS, 2020a), only one plans to implement an organic waste composting system. In this context, this chapter presents a case study on the implementation of the first composting yard in Salvador, conducted by the Urban Green Salvador Project – here referred to as UG. By identifying the key factors, the main authors involved in the process, as well as their respective responsibilities, our main objective is to propose a governance model for urban green commons capable of motivating the success of the project and strengthening urban resilience. With this study, we hope to assist in the effective implementation of other composting yards in Salvador and other Brazilian cities and, to a greater extent, in other countries with similar social and environmental contexts.

12.2 Theoretical Framework

12.2.1 Environmental Governance

The concept of governance consists of a set of rules, norms, and attitudes that maintain and control the activities of a given system. When applied to the environmental scope, it refers to the management of natural resources aimed at sustainable development (Santos & Bacci, 2017). Governance implies the analysis of all actors involved in the process of planning and implementing a system, understanding their values and interests. This concept is structured by formal (defined and explicit public policies, rules, and managers) and informal bias (agreements, social movements, and spontaneous leaders), which must all be considered for a broad and assertive understanding of management. In turn, environmental governance is concerned with developing programs and projects focused on the environment (Seixas et al., 2020).

The concept of environmental governance varies according to the context in which it is analyzed, gaining new nuances with the purpose of each institution or project (Grimberg & Blauth, 2005). When put into practice, environmental governance allows the creation of development models that incorporate the local social and environmental needs to strengthen urban resilience (Santos & Bacci, 2017).

Environmental governance has been used to achieve the 17 SDGs proposed by the UN, but the success of the 2030 Agenda depends on the local efforts of each participating country (UN, 2018). Although international agreements motivate countries to adopt sustainable development, they are not enough to induce the implementation of the necessary and urgent changes, for transformations must occur within each territory with the participation of the political sphere and civil society upon local governance arrangements (Seixas et al., 2020). The 2030 Agenda of the United Nations is a large-scale program, thus requiring management, monitoring, support, and communication of all countries involved (UN, 2018), as well as a clearly defined and explicit governance and effective and assertive management (Le et al., 2018).

At the local level, authorities must not only implement new environmental projects but also develop those already in effect, whereby governance initiatives – associated with the water-food-energy nexus – play a key role for them to achieve success (Weitz et al., 2017). In a study conducted by Le et al. (2018), the authors report that several composting programs were discontinued in underdeveloped countries due to management mistakes. They also state that governance practices applied to the environmental sector require a holistic approach that considers all stakeholders, that is, such projects (especially those related to MSW composting) depend on actors' integration and a clear division of roles and responsibilities. According to the authors, municipal support proved to be essential for project development in developing countries.

In our study, we will analyze the concept of governance focusing on an urban green common in Salvador that plans to implement the city's first composting yard. To recommend possible adjustments and progress in the implementation of the composting yard, we will identify the actors involved and their respective responsibilities, as well as the challenges encountered in the project. These analyses are encompassed by the concept of multi-stakeholder platform, in which different actors interact and seek solutions to situations or shared resources, mainly considered in natural resource management contexts such as urban gardens (Thiele et al., 2011).

12.2.2 Urban Green Commons

Urban green commons are characterized as "areas that allow residents and citizens to actively rework urban nature in ways that support ecological processes, while allowing for a collective caring of pieces of land in the city" (Colding et al., 2013, p.1). The appropriation of urban spaces and the establishment of urban gardens by the local community tend to alleviate stresses caused by urbanization, besides providing a series of benefits such as personal well-being, improvement in food resources, community empowerment and autonomy, citizenship practice, green spaces preservation, and strengthening of local food security (Dennis & James, 2016).

According to the Brazilian Agricultural Research Corporation [EMBRAPA] (2015), lands underutilized for the irregular disposal of waste may also serve for the implementation of composting systems, favoring local sustainability and organic waste management, complying with article 36, paragraph V of the PNRS.

Urban green commons are also related to food and nutrition security (FNS), whose goal constitutes a global challenge and motivated the 2030 Agenda. Thus, programs encouraging the creation of urban green commons at the municipal level can be seen as an alternative for meeting the 2030 Agenda (Navarro et al., 2019).

Being spaces for small-scale agricultural cultivation, urban green commons are mostly maintained by the voluntary work of local residents (Colding & Barthel, 2012; FAO, 2012). However, the governance models applied to the management of this initiative are diverse (Fox-kämper et al., 2018). Urban greens can be private (when the activities relative to them are assigned or executed by the owner) and maintained either under a top-down or bottom-up model. Whereas the orders and rules in the top-down model are determined by a structured hierarchy, from municipal authorities to the community, in the bottom-up model, activities are managed by active members of the community. In each of these models, the hierarchy is either fixed or flexible, varying as to the autonomous participation of members and support levels of municipal or external professionals.

As observed by Fox-kämper et al. (2018), urban green commons are mostly regulated by bottom-up governance models with political or administrative assistance. In this model, gardens are implemented as a government initiative, but maintained by the community, without further financial aid from government authorities. They can also be initiated by the community with government support, which includes not only financial, legal, or professional matters but also resources such as water. When the actors involved in the project sought for a common goal, the application of this model succeeded. Depending on the development stage and actors involved in each process, combining different models in project management is a common doing (Fox-kämper et al., 2018).

However, the reduction and lack of access to green and public areas, intensified by the privatization in urban development, impact the quality of life and weaken urban resilience (Colding & Barthel, 2012; FAO, 2012). Thus, projects such as urban gardens require government support and policies aimed at their development.

12.2.3 Organic Waste Management in Urban Gardens

Composting is a controlled biological process that consists of converting organic substances (plant and animal by-products) into a stable, sanitary product, similar to solid humus. Being the largest source of carbon and energy for microorganisms, the organic compost incorporated into the soil constitutes an indispensable substrate for the development of biological life (Robert, 1996).

Composting has been practiced in rural areas for centuries as a means of improving soil fertility. However, urban waste composting has another motivation: reducing and recycling waste (Ali, 2004).

The highest percentage of waste produced in Brazilian cities (51.4%) refers to organic waste (Instituto Brasileiro de Geografia e Estatistica – IBGE, 2010). When disposed in landfills or dumps, these residues cause high environmental impacts, shorten landfill's lifespan, and generate expenses that could have been avoided. A viable and sustainable alternative for recycling such a large volume of organic waste is to process it through composting and use it as fertilizer in urban and rural agriculture. Yet, only 1.6% of the organic waste produced in Brazil receives this kind of treatment (IPEA, 2012).

In Brazil, Federal Law No. 12,305 (2010) recommends composting as destination and treatment for organic waste. For being a low-cost technique that produces agricultural inputs, creates employment opportunities and income, and encourages urban agriculture, composting can be used as an instrument to strengthen urban resilience (Más-Rosa & Carvalho, 2021).

According to Ali (2004), despite presenting a history of bankruptcy in Brazil and in several low-income countries, large composting plants have been showing good results in developed countries (Fehr, 2006). This data points to the need for identifying and characterizing different types of MSW composting experiences to encourage new technological processes and the diversification of waste management systems in the municipalities.

Several countries around the world have adopted the decentralization of composting (Ali, 2004). In Brazil, community-based and decentralized composting systems have reached successful outcomes in different regions of the country (Fehr, 2009; Inácio & Miller, 2009; Moura, 2012; Abreu, 2013). Decentralized activities are efficient when it comes to diverting waste from the final disposal and mobilizing and sensitizing people (Siqueria & Assad, 2015), besides guaranteeing autonomy at the community level. Moreover, the use of compost in agriculture and the consequent generation of income are beneficial to the success of urban greens, strengthening urban resilience (Swagemakers et al., 2018).

Implementing urban green commons and MSW management systems within these spaces is also crucial for supporting the municipal public power (Fox-kämper et al., 2018; Grimberg & Blauth, 2005). Mainly, due to the nature of solid waste management as a public service, institutions and legislation must support initiatives as a conditional factor for their development.

12.3 Methodology

This study aims to propose a governance model for community urban gardens with decentralized MSW management through composting. The case study reported here focused on the urban green common and the implementation of the composting yard. For that, we identified the main actors involved in the process and their respective responsibilities, that is, the key factors for the implementation and success of projects of this nature.

The key factors in decentralized MSW management projects through composting applied in urban agriculture were determined using the Stakeholder Analysis (SA), which consists in identifying the involved actors, categorizing them according to their function and networks, and investigating their relationships (Grimble, 1998). Considering that the analyzed composting yard is still at the planning stage, we adapted the method to this reality.

We chose this method for providing a broad view of groups involved in initiatives that include the use and conservation of natural resources. When considering the system as a whole, one can understand its main structures and actors, as well as their respective interests. All parties involved are contemplated in the analysis, even the most vulnerable or those with less power (Grimble, 1998). The method comprises projects that consider the following points: (i) multiple economic, social, and environmental objectives among the actors; (ii) property rights, referring to the use of the area for implementing urban gardens; (iii) generation of nonmarketable products/services, as a result of the waste management process; and (iv) least favored actors, generally the poorest, who have their political and economic needs undervalued in other methodologies. Considering the context of the Urban Green Salvador Project (UG), points (i), (ii), and (iii) are pertinent, indicating the usefulness of the methodology for creating a multi-stakeholder platform to guide the project governance practices (Thiele et al., 2011). The SA was applied into five phases adapted to the reality at stake, as described below.

12.3.1 Clarify the Objectives of the Analysis

By consulting the records on urban gardens made available by the State of Salvador, we verified the existence of 50 urban gardens and 4 community orchards registered by the Municipal Secretariat of Sustainability, Innovation and Resilience [SECIS] (PMS, 2020a). Among these, only one garden plans to implement a structured management project aimed at municipal solid waste (MSW) within its space: the urban green common entitled Projeto Hortas Urbanas Salvador. After acknowledging the project for developing the first composting yard in the city, this study was oriented toward understanding the essential factors for implementing this initiative. Thus, our study was focused on analyzing the actors (current and potential) involved in the UG project, as well as the necessary relationships for its achievement. As the composting yard in question is still at the planning phase, we considered both the context of the project and information presented in the literature.

12.3.2 Create a System That Encompasses All Objectives

Besides considering our research objective, the system also included the objective of implementing the project. In that sense, it focused on analyzing the existing relationships to contextualize the current reality – which comprises an urban garden in operation – and creating an interaction model between the involved actors and their responsibilities, thus enabling the development and success of the composting yard as an efficient and decentralized system of MSW management.

12.3.3 Identify Decision-Makers and Stakeholders

We identified a single individual acting as both space administrator and creator of the UG and the composting yard. Before the reduced number of volunteers due to the COVID-19 pandemic, the manager acted as the sole decision-maker in the project development. To understand the developed activities and the role of other potential or previously involved actors, a semi-structured interview and a questionnaire were applied to the manager, and a field research was conducted in the UG to observe the *in loco* reality.

The questionnaire provided data on other actors who should be involved in the project implementation, maintenance, and development. These actors were contacted and also underwent a semi-structured interview, as shown in Table 12.1.

12.3.4 Investigate Stakeholder Interests and Agendas

The applied SA enabled us to understand the function and interaction between each key actor identified in the previous step. Based on a bibliography review on practices presented in other case studies, the key responsibilities of each actor were also identified.

Carvalho (2020) and Colding and Barthel (2012) reported the main characteristics of urban community gardens and their positive effects on social, environmental, and urban resilience, indicating the benefits of this model in response to crises and as an alternative model for the management of public areas. According to Colding and Barthel (2012), also stress the importance of legislation addressing the topic and the predominance of projects with bottom-up management. When monitoring the implementation of an urban garden with bottom-up management, Camps-Calvet et al., (2015) likewise highlight the influence of this model in strengthening local resilience.

The review also included case studies analyzing the relevance of community participation in the development of urban gardens such as those conducted by Dennis and James (2016), Honda (2018), and Nagib (2019), besides focusing on issues related to project implementation, as promotion of biodiversity (Dennis & James, 2016), function of government participation, and monitoring of management and composting processes (Nagib, 2019; Honda, 2018). In a study conducted by

Table 12.1 Interviewed actors

Identified actors			
Reference name	Position/function	Methodological technique	Information obtained
Interviewee 1	UG manager	Semi-structured interview questionnaire	Implementation of the urban green common
			Performed activities
			Future projects
			Volunteer management
			Partnerships made and estimated
			Management and logistics of the composting yard
Interviewee 2	Former SECIS secretary	Semi-structured interview	SECIS support for the implementation of the UG
			Communication with the manager
			Performed activities
Interviewee 3	Former SECIS secretary		Management of MSW in Salvador and selective collection
Interviewee 4	Vegan restaurant management		Was unaware of the composting yard
Interviewee 5	Vegetarian restaurant management		Knows the project and will participate, but does not know about the logistics
Interviewee 6	Neighboring commercial establishments		Unknown to the UG project
Interviewee 7			
Interviewee 8			
Interviewee 9			
Interviewee 10			
Interviewee 11			
Interviewee 12			
Interviewee 13			
Interviewee 14			
Interviewee 15			

Source: Own elaboration

Nagib (2019), the author suggests the creation of a statute/manifesto that supports the establishment of urban green commons and possible legislation on the topic.

The methodology used in our study also considered the studies developed by Le et al. (2018) and Thampi and Rao (2015). Decentralized MSW management, the stakeholders involved in the processes, and their respective relationships are central points in our research. The studies conducted by Villa et al. (2020) and Swagemakers et al. (2018) focused on the necessary logistics and management for implementing a decentralized solid waste management from composting, indicating the relevance of community participation in the entire process and the consequent strengthening of urban resilience.

The analyzed studies enabled us to identify the key actors and their respective responsibilities in different scenarios. Based on these data and on the information regarding the stakeholders of the UG in question (presented in Topic 3.1.3), we created a model for the composting yard implementation aiming to prevent obstacles and failures during its execution.

12.3.5 Identify Patterns of Interaction and Dependency, Such As Conflicts and Compatibilities

From the bibliographic review previously described in the fourth step, adapted to the context of UG, and data on the current and potential stakeholders, we created a stakeholder map, identifying the partnership or conflict relationships established among actors (Fig. 12.1).

After identifying the actors and their main activities, they were categorized according to their nature into government, enterprises, educational institutions, and civil society – following the model created by Le et al. (2018). After analyzing each category individually, existing and potential relationships were traced to identify common interests or conflicts, characterizing an analysis of multi-stakeholder processes to guide governance practices.

Thus, the model created covers the key factors, including actors and responsibilities considered essential to the governance of the urban garden in question, the implementation of MSW management to guarantee the development of the composting yard, and the implementation of further initiatives of the same nature.

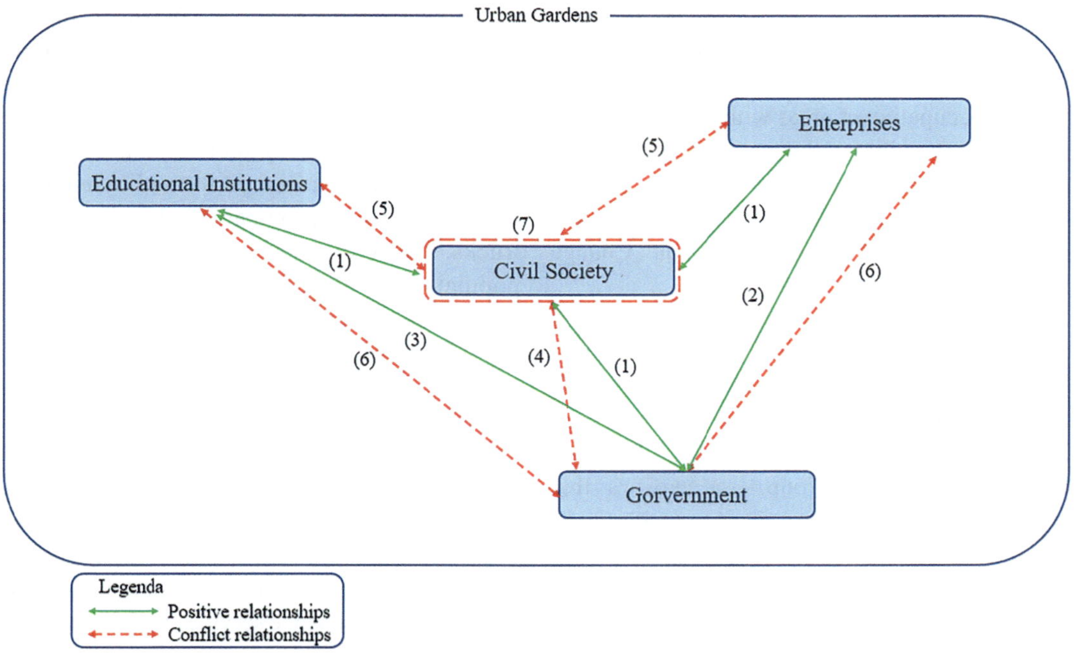

Fig. 12.1 Network of interactions between the main actors involved in the development of urban community gardens. Source: Visual network adapted from Le et al. (2018) with actors and relations from own elaboration

12.4 Results

12.4.1 Characterization of the Study Area

12.4.1.1 The Municipality of Salvador and the Direction to Sustainable Development

Bahia is a Brazilian state located in the northeast region of the country. The state borders eight other Brazilian states and is limited by the Atlantic Ocean to the east. Its territory occupies 564,760.427 km^2, with a coastline of about 932 km, the largest in Brazil, and its capital is Salvador (Brazilian Institute of Geography and Statistics - IBGE, 2021).

The Brazilian Institute of Geography and Statistics [IBGE] 2020 census estimated that the population of Salvador was 2,886,698 inhabitants, comprising the most populous municipality in the northeast and the third largest in Brazil (IBGE, 2021), despite being and the most economically developed municipality in the state (PMS, 2020b).

In Salvador, the urban network development was strongly marked by the accelerated and disordered occupation of areas with poor infrastructure and susceptible to the occurrence of natural and social disasters. In that sense, the state population is highly vulnerable to the adverse impacts of climate change (Brazilian Climate Change Panel – PBMC, 2016).

In recent years, the state of Salvador has developed and implemented plans, policies, and fundamental actions aimed at the sustainable development of the city (PMS, 2020a). Since 2015, the city has been part of the C40 Cities Climate Leadership Group – a network that brings together cities committed to tackling climate change and reducing its impacts. The municipality also participates in the Global Pact of Mayors for Climate and Energy (GCoM), considered the largest global alliance of cities committed to fighting global warming. In 2019, Salvador published its *Resilience Strategy* (Salvador's Resilience Plan) – a document prepared with the direct support of the Rockefeller Foundation's 100 Resilient Cities program (R100) (PMS, 2019), which was later transformed into the Global Resilient Cities Network (GRCN, 2020). In 2020, the municipality launched the Second Greenhouse Gas Inventory and the Plan for Mitigation and Adaptation to Climate Change (PMAMC).

12.4.1.2 The Municipal Solid Waste Management System of Salvador

In Bahia, as well as in other Brazilian states, landfills are still an issue of serious political and social concern. In total, 359 of the 417 municipalities in the state (86.09%) contain landfills within their territories (Urban Development Secretariat – SEDUR, 2016).

Similar to the largest Brazilian metropolises, the municipality of Salvador presents problems related to basic sanitation and MSW generation that directly impacts the health and well-being of the population (Azevedo, 2004).

In Salvador, the operational system of waste management is conducted by the Urban Cleaning Company (LIMPURB), a mixed-capital public company responsible for defining policies and inspecting and controlling services (LIMPURB, 2020; Pinto, 2018). The adopted collection system is the "door-to-door" waste collection, performed by environmental agents accompanied by compactor trucks, collecting the waste disposed by the population at their homes (SOUZA, 2018). According to the LIMPURP, 3000 tons of household waste and 2500 rubble on average are collected daily in the municipality.

All MSW produced in Salvador is disposed at the Metropolitan Landfill Centro (AMC), which serves two other municipalities, thus receiving an average of 3000 tons/day of waste. The Bahia Transfer and Treatment of Waste LTDA (BATTRE) company is responsible for the operations and maintenance services of the transshipment station, which serves the city of Salvador, and for the implementation, operation, and maintenance of the landfill (BATTRE, 2020).

Although the landfill employs different technologies for pollution control, we verified that Salvador does not have any type of differentiated collection system and use of organic waste, as determined in the Brazilian National Policy on Solid Waste (PNRS). In 2015, the municipality published Decree No. 26,916 of December 16, 2015 (PMS, 2015), regulating the mandatory collection, transportation, treatment, and disposal of solid waste and final disposal of tailings from large-quantity generators of waste. However, an interview conducted with the Municipal Secretariat of Sustainability, Innovation and Resilience (SECIS) in July 2020 revealed that large waste generators are not yet held responsible for the destination and treatment of their organic waste.

The Municipality of Salvador has not yet adapted to the solid waste management model postulated in Law No. 12,305, so they adopt a logic of removal and non-recovery of the waste generated in the city (PMS, 2020a). Until March 2021, the municipality did not present a Municipal Plan for Integrated Solid Waste Management (PMGIRS), and we found no record of the planning or execution of specific municipal projects or legislation on the subject. Likewise, the municipality lacks a municipal selective collection system, but we identified the existence of some initiatives mainly aimed at the collection of recyclable materials. In 2015, the SECIS implemented the Salvador Selective Collection Program, whose main goal is to install voluntary delivery points for recyclable materials (Pinto, 2018). In a study conducted by Sousa (2018), the author identified that commercial establishments of "in natura and processed" food, such as fruits and vegetables from restaurants and supermarkets, have a great demand for the disposal of organic matter residues.

The author points that the city also faces other administrative and financial problems, such as the lack of qualified technicians and little money directed to this sector, hindering compliance with the legislation. At times, the municipal public management transfers its responsibility and legal obligation toward waste collection to private management through public tenders, concession contracts, and public-private partnerships, possibly resulting in the loss of control over service inspection. The deficiency in collection inspection by the public management entails low-quality services in low-income neighborhoods to the detriment of the noble neighborhoods.

12.4.1.3 Overview of the Urban Green Salvador Project

The Urban Green Salvador Project (UG) was created in 2019 as the city's first urban green common. At the time, the creator's proposal (Interviewed 1, 2021) was to revitalize an abandoned land in a residential area with irregular disposal of waste, as a way to manage public areas (Dennis & James, 2016; Colding et al., 2013). The project was presented to the City Hall, which collaborated with land cleaning and donated useful materials for the garden maintenance, such as land, fertilizer, and plant seedlings (Interviewed 1; Interviewed 2, 2021). Approximately, 60 tons of irregularly dumped rubble were removed during the land cleaning process.

The project creator (who is also the current manager and responsible for the UG in partnership with SECIS) initiated the activities alone, receiving volunteers interested in the processes only afterward (Interviewed 2, 2021). The volunteers autonomously organized their activities, without any hierarchical control or imposition of obligations; however, the project development experienced some challenges regarding the frequency of their presence and the diversity of the necessary tasks and participants' profiles. Prior to the coronavirus pandemic, the garden relied on the support of 20 volunteers, but the situation obliged them to interrupt their activities, so that the manager was left to perform all of them alone (Interviewee 1, 2021).

The main activities performed in UG are maintaining vegetable beds, cultivating bees, and maintaining fruit trees. A sustainable joinery was also created at the site, aiming to train community members and produce artifacts to generate income. The gardening pruning waste undergoes composting by the vermicomposting technique. The UG also has HomeBiogas, a device that converts organic matter into biogas in a short period of time to be used in the kitchen.

Currently, the project is implementing a composting yard, a pioneering initiative in the municipality, which still lacks a selective collection system and composting plants, thus comprising the main focus of our study.

12.4.2 Stakeholder Analysis (SA): Identification of the Actors and Their Characteristics

Community urban gardens is a cross-sectional initiative that includes social and environmental themes approaching environmental education, volunteering, urban health, and local biodiversity, thus strengthening urban resilience (Le et al., 2018). In this sense, understanding the involved actors and their functions and other initiatives associated with space (such as the decentralized MSW management) is essential for the development of urban gardens.

For expanding the scope of these initiatives, effectively strengthening urban resilience (Colding & Barthel, 2012; Dennis & James, 2016; Carvalho, 2020), and contributing to the achievement of SDGs (FAO, 2012), governance is a necessary structure that guides the interactions between involved and potential actors and their main activities. Based on the UG context and other case studies on urban gardens performed in Brazilian cities such as São Paulo (Honda, 2018; Nagib, 2019) and Rio de Janeiro (Villa et al., 2020), as well as in other countries, such as Vietnam (Le et al., 2018), India (Thampi & Rao, 2015), Barcelona (Camps-Calvet et al., 2015), Sweden, Germany, and South Africa (Colding et al., 2013), we investigated the actors considered key for the successful implementation of the first composting yard in Salvador.

The subsequent topics characterize the stakeholders and their respective functions, according to data presented in Table 12.2. Categorization was conducted based on the fourth step of the methodology proposed by Grimble (1998), presented in Topics "3.1.3. Identify Decision-makers and Stakeholders" and "3.1.4. Investigate the Interests and Agendas of the Interested Parties" of this article. The nomenclatures used were adapted from the case study presented by Le et al. (2018).

12.4.2.1 Government

All studies on urban green commons and composting projects considered in our review relied on the support of government agencies, regardless of the divergences in the activities performed. Government agencies offered political-administrative support for urban community garden implementation as to the use of public space, material donation, technical support (Colding & Barthel, 2012; Dennis & James, 2016; Fox-kämper et al., 2018), and legislation

Table 12.2 Categorization of stakeholders

Key actors		
Category	Function	Actors
Government	Technical support	City hall
	Services provision	Departments
	Legislation	Councilors
Enterprises	Financing	Private enterprises
	Technical support	Restaurants
	Services provision	
Educational institutions	Technical support	Public and private schools
	Publicity/disclosure	Colleges/universities
		Researchers
Civil society	Planning and deployment	Local residents
	Management	Volunteers
	Publicity/disclosure	NGOs
		CSOs

Source: Own elaboration adapted from Le et al. (2018)

that supports and recognizes these initiatives (Nagib, 2019; Santos & Bacci, 2017). Secretariats' participation associated with the public power is demonstrated through the promotion of informational events and employees' participation in the activities developed at the urban greens (Honda, 2018; PMS 2020a). Due to the dimension of urban green common performance, the municipal public power is presented as its most important sphere, translated into the City Hall, associated secretariats, and councilors. Honda (2018) reported a case in which an event held in a local urban green common by the health secretariat presented the initiative as associated with the 2030 Agenda and the SDGs. Besides that, secretariat staff assisted in the practical activities related to the garden, which stimulated knowledge between global and local actions in the municipalities (Honda, 2018). Secretariats can also act as important mobilizers through promoting employees' participation, providing maintenance services, and donating necessary tools – even if sporadically and without legal obligation.

Although the municipal government provided no financial contribution to any of the analyzed cases, their contribution proved to be essential for the implementation of the gardens, especially at the beginning of the project, through the provision of materials, plant seedlings, and professional support for teaching cultivation techniques (Colding et al., 2013; Colding & Barthel, 2012; Dennis & James, 2016; Nagib, 2019; Interviewee 1, 2021; Interviewee 2, 2021).

As public authorities at the municipal level are responsible for the administration of urban public areas and the management of MSW, their support in urban green common implementation consists of enabling land use (Colding et al., 2013; Colding & Barthel, 2012; Dennis & James, 2016) and in implementing decentralized MSW management systems. Implementing laws that support these initiatives tends to strengthen their proposals and motivate their expansion (Hettiarachchi et al., 2018; Villa et al., 2020). Law 16.212 (2015) is a great example of legislation that supports the establishment of urban green commons and the practice of composting in this space, providing for the participatory management of spaces, as pointed by Fox-Kämper et al. (2018), Nagib (2019), and Santos and Bacci (2017), in the Municipality of São Paulo, Brazil:

> Art. 19. The squares hosting organic community gardens may have composters built and maintained by those responsible for the respective urban garden, according to the principles of urban permaculture.
> § 1 The installation of composters must be authorized by the respective Subprefecture, after hearing the users' committee, when applicable.
> § 2 It will be up to those responsible for the urban garden to inform visitors of the square about the correct handling of composters, to develop educational campaigns and actions in the square and surrounding areas, involving the users' committee, when applicable. (City Hall of São Paulo, 2015, p. 1)

The Municipality of Salvador lacks a specific legislation regulating these practices. Currently, the main action of the city regarding this theme is registering existing urban green commons, which is conducted by SECIS (PMS, 2020a). Thus, drafting legislation addressing this topic is an important step toward the advancement of these initiatives. Such a legislation is particularly important for determining the division of responsibilities in the maintenance of urban green commons and composting plants (Nagib, 2019). In this context, the City Hall is responsible for regulating the processes and release of the execution of activities in a public environment, it is the community's duty to cherish and maintain urban green commons and the composting process, ensuring management autonomy on the part of volunteers (Santos & Bacci, 2017). Elaborating legal norms aimed at these initiatives makes the involvement of councilors in these projects essential (Nagib, 2019), bridging the municipal power and civil society.

Our data show that public authorities contributed to the composting yard implementation by making the site available, as the composting will be installed at the same site as the UG. According to the UG's manager, although he has attempted contact with SECIS for 3 years now looking for support to make the composting feasible, no contribution on their part has been provided yet. When the urban green common was created, SECIS provided the space, cleaned the land, and donated the necessary inputs for the garden operation (Interviewed 1, 2021; Interviewed 2, 2021).

Many municipal laws regulate the issue of solid waste and the responsibility of large-quantity generators in Salvador. However, we identified no specific public policies or legislation stimulating the implementation of composting projects during the course of this study (PMS, 2020a). Although SECIS has collaborated with several urban green commons in the municipality, initiatives aimed at implementing composting as a tool for decentralized waste management are still lacking.

12.4.2.2 Enterprises

The participation of private organizations in activities performed at community urban gardens generates benefits for commercial disclosure (Andrade et al., 2018), tax incentives (Brazil, 2000), and employees' integration (Colding & Barthel, 2012; Jagt et al., 2017). Moreover, the support arising from these organizations excels when it comes to material donation, provision of maintenance and technical services related to garden cultivation and construction, dissemination of the initiative, and funding.

Environmental companies also tend to offer their services, but partnerships are not restricted to this sector. Besides donations and financial support, organizations can participate in projects promoted by urban gardens, as well as in the composting process. In that sense, the vegan and vegetarian restaurants located in the Pituba neighborhood will be relevant actors for the maintenance of the composting yard. Being located around the Hortas Urbanas Project and due to the predominance of foods suitable for composting, such as vegetables and fruits (Souza, 2018), they may be responsible for the supply of organic waste, contributing to their prioritization (Interviewed 1, 2021). Moreover, Federal Law No. 9985 (2000) stipulated that private organizations have an incentive for environmental compensation, guaranteeing tax exemptions upon the provision of support to environmental projects. Finally, the partnerships established between community gardens and private organizations allow the dissemination of these brands as supporters of environmental causes, thus occupying a positive position in the society (Schenini et al., 2006; Andrade et al., 2018).

As stated by Interviewee 1 (2021) regarding the management and logistics of the composting yard, participating restaurants must pay a monthly fee for waste collection, which will be performed by bicycle and then packed in containers. Initially, the manager expects that the collection will gather about 360 kg of waste weekly with the collaboration of four restaurants (Interviewee 5, 2021). However, as shown in Table 12.1, only one establishment confirmed its interest in participating. The composting yard managers must train and qualify all staff from establishments that agree to participate in the project, besides holding campaigns aimed at raising consumers' awareness (Interviewee 1, 2020).

The composting yard will be managed by the Compartiquali – a company headquartered in Salvador that operates in the area of environmental consultancy and waste management. Considering the above-exposed waste, managers expect to receive between 600 kg and 1500 kg of waste weekly.

12.4.2.3 Educational Institutions

This category comprises institutions of primary, secondary, and tertiary education (Honda, 2018), of either public or private nature, as well as scientific researchers (Le et al., 2018). Although not considered relevant in the decision-making process of projects such as urban community garden implementation and decentralized waste management system, the participation of these institutions within these spaces enables the development of teaching, research, and extension (Le et al., 2018; Interviewed 1, 2020). The presence of students and the development of academic work on biodiversity and activities performed on these projects guarantee the professionalization of these practices and stimulate social participation in this environment (Colding & Barthel, 2012; Honda, 2018; Nagib, 2019). Promoting environmental education is one of the most prominent factors in community urban gardens, especially when it arouses the support of public authorities.

The support potential of each educational sector is related to its nature and level of knowledge. Institutions of tertiary education tend to contribute in technical issues and management practices and with the efficacy of cultivation methods. In turn, institutions of primary and secondary education provide for the education of children and adolescents, promoting participation and knowledge related to sustainable development. Moreover, the production of scientific knowledge such as the creation of bibliographic references cited in this research unveils the relevance of the project, disseminating it not only through the scientific community but for the overall society, awakening individuals as potential volunteers (Colding & Barthel, 2012; Honda, 2018).

Currently, UG provides space for research development in partnership with the Institute of Biology of the Universidade Federal da Bahia (Interviewed 1, 2021). Yet, it also shows great potential for scientific research in different areas of knowledge and extension projects. With composting implementation, the possibilities for studies will be expanded.

12.4.2.4 Civil Society

In urban green commons, the participation of civil society included active volunteers, local residents/community, nongovernmental organizations (NGOs), and civil society organizations (CSOs) (Grimble, 1998). This category includes the main actors responsible for developing and ensuring the success of the community gardens (Colding & Barthel, 2012; Dennis & James, 2016; Honda, 2018; Nagib, 2019). The categories aforementioned function in partnership with civil society, enabling the performance of activities and the maintenance and expansion of the project described by Fox-kämper et al. (2018) as a bottom-up model. Thus, the population strata that executes the project is the one that keeps it alive.

In our research, we verified that volunteers are distinguished from local residents: whereas volunteers are active participants of the project, local residents live nearby and tend to sporadically encourage its development by participating and publicizing actions promoted in this environment (Nagib, 2019). Information regarding the role of NGOs and CSOs (Grimble, 1998; Lassa & Nugraha, 2015) are scarce in previous studies on the topic, but these organizations mainly contribute in task forces (events created to stimulate collective work) for the maintenance or construction of urban gardens (Nikolaïdou et al., 2016; Le et al., 2018; Nagib, 2019).

In an environmental governance analysis conducted by Grimberg (2005), the author stresses that civil society is responsible for implementing participatory management processes, including planning urban garden projects, proposing them to the public authorities, and implementing them. Although other categories do provide support to the management and execution of activities, it is up to volunteers to develop self-management skills.

Self-management requires the establishment of hierarchical relationships, but no determination has been made as to which structure is the most beneficial to the urban green common development (Fox-kämper et al., 2018; Nagib, 2019). Whereas some gardens function based on horizontal relationships, waiving a management nucleus controlling each member's attendance or activities, others rely on a group responsible for the managing processes, organizing participation, and planning events and programs (Fox-kämper et al., 2018). Determining the most suitable model depends on the origin of each project and on the community in which it is inserted. Our results indicate that precursors to garden implantation tend to become leaders in front of the other volunteers (Nagib, 2019).

Despite self-management, voluntary work in these spaces is characterized by the freedom and autonomy of participating members. In this scenario, the requirement for discipline, planning, and compliance with schedules and tasks becomes inefficient, as activities are performed according to individuals' physical and emotional availability. This points to the need for publicizing the project in order to expand the number of volunteers, as well as to the need for creating a statute/manifesto that strengthens the primary purpose of the community garden and reverberates its creation as a collective movement.

Currently, the UG relies on a limited number of volunteers, most of whom are older women. As aforementioned, before the coronavirus pandemic, around 20 people collaborated with the activities required by the garden; however, few of them were young people. Interviewee 1 (2021) indicated the need for the participation of young volunteers due to the manual work related to cultivation and composting practices. Besides volunteers for maintaining the space, the UG also requires the participation of individuals for dissemination of the project, for managing and formalizing the practices, and for assuming undertakings that assist in financing, material donations, and technical knowledge.

For the composting system maintenance, a group of people must actuate on receiving waste, cleaning and controlling packaging containers, and handling composting rows. Thus, social involvement will be fundamental to the success of the project. UG is conducting a campaign through social networks and commercial establishments in the Pituba neighborhood to raise funds and invite volunteers to participate in the project. The population can collaborate both by working on composting and by contracting the collecting service for their household-generated residues for a fee, which will provide individuals with buckets for storing waste (Interviewee 1, 2021).

12.4.3 Network of Interactions Between Key Actors

After characterizing each key actor participating in the implantation and development of the community garden, one must understand the relationships established between them. These relationships can be either positive or conflicting, depending on each actor's specific interests. According to Grimble (1998), stakeholders participate in situations that address their own interests, so that all activities must meet their own expectations, always being bilateral. A support activity may possibly supply the interest of only one side of the relationship, but in this case, the relationship is unlikely to last. Due to the necessity of seeking alternatives that meet the interests of all actors involved in each situation, this scenario is characterized as multi-stakeholder platform (Thiele et al., 2011). Thus, urban green commons will witness a favorable scenario for their progress when valuing and devising strategies that stimulate mutual support.

Figure 12.1 shows the interactions between the key actors categorized in Topic 4.2.4, which expatiates on the necessary network for community garden development. These relationships were classified into positive or conflictual and numbered to describe the characteristics of each interaction.

Based on Le et al. (2018), we understand "civil society" as a key actor of this process, being central to all relationships and activities performed in this context, such as the responsibility for planning and proposing the garden project to government authorities, besides maintaining it and assuring its progress (Colding et al., 2013; Dennis & James, 2016; Nagib, 2019; Santos & Bacci, 2017; Villa et al., 2020). The other actors can implement urban green commons without the participation of civil society, but in this case, it would be classified as of non-community nature – that is, civil society participation is seminal for characterizing the community nature of urban green commons (Fox-kämper et al., 2018).

12.4.3.1 Positive Relationships Between Key Actors

Positive relationships were observed in (1), (2), and (3) as well as mutual benefits between key actors. The positive relationship observed between "civil society" and "government" is characterized by the implantation of urban green commons, with the population being responsible for the revitalization of a public urban area (FAO, 2012; Nikolaïdou et al., 2016). Furthermore, it encourages environmental education of those involved and the neighboring community, because of the contact with the environment generated by the green areas and sustainable development practices. In the case of the implementation of decentralized MSW management by composting, civil society supports the municipality to comply with Federal Law No. 12,305 (2010) and reduces the impacts caused by irregular solid organic waste management, when carrying out the necessary actions for the implantation of this model in the urban green space. On the other hand, it is up to the government to use the public area for the implementation of the urban green common, as well as the full realization of its activities based on technical support, service provision, and institution of laws that support and regulate it.

The positive relationship indicated in (1), between "civil society" and "enterprises," provides for the provision of technical knowledge and services of the latter, which receives as a counterpart of tax exemptions derived from the tax compensation law in force throughout the national territory. Moreover, the partnership between the urban green commons and the companies promotes a positive position in the society (Schenini et al., 2006; Andrade et al., 2018).

In the case of UG, the implementation of the composting yard will occur in association with a private company of the Environmental Consultancy and Management field. In addition to the company publicizing its service among project participants, the company will receive a financial return from the maintenance fees paid by residents and participating restaurants. This is a case of mutual beneficial relationship between the activities present in an urban green common. Finally, the relationship described in (1), between "educational institutions" and "civil society," corresponds to the existence of a place that conducts the creation of new technical, scientific, and educational knowledge by the academy. At the same time, it allows for greater adaptation of species as well as practical and management activities carried out in gardens.

The relationship pointed out in (2), between "government" and "enterprises," occurs, mainly, due to the legislation favorable to the support provided by private institutions. Note that, in all the relations presented, the public power emerges as a regulatory agent, facilitating partnerships. By instituting measures that benefit enterprises, the government makes them allies in meeting their obligations, as those related to the maintenance of the local environment and, potentially, to the decentralized management of MSW.

The relationship (3) occurs between the "government" and "educational institutions," who represent the public municipal education departments and municipal schools – referring to basic and elementary education – respectively. Municipal educational institutions need the release and encouragement of public management to act together in urban green commons. By promoting this interaction, the government indirectly encourages the development of urban gardens in urban spaces with the presence of students and the promotion of environmental education. In this scenario, only educational institutions associated with the municipal public power were considered, as urban green commons comprehend this governmental sphere.

12.4.3.2 Conflict Relationships Between Key Actors

The relationships presented in (4), (5), (6), and (7) demonstrate conflicts of interest between actors. In these cases, the expectations of only one actor were met, configuring a weak and probably punctual relationship (Grimble, 1998; Le et al., 2018). This type of relationships can be useful in specific situations, such as events and extraordinary activities (Nagib, 2019). However, they are not recommended for the application of long-term activities, which define the development of urban green common.

The conflict presented in (4) stems from the possible lack of assistance from the public authorities in making the operation of urban green commons feasible, even after the organization of the civil society (Santo & Bacci,

2017). The delay in evaluating the proposals and putting them into practice can disperse the volunteers and lead to the abandonment of the implementation, since the occupation of public areas requires an endorsement from the City Hall (Nagib, 2019). The government must analyze, review, and enable the implementation of this type of projects. However, rights and obligations must be well defined so that the expectations of civil society do not exceed the limits of the government's responsibilities (Santos & Bacci, 2017). All activities developed, as well as the management and execution of these activities, are the responsibility of the volunteers (Nagib, 2019). However, there are some records that indicate the active participants' feeling of helplessness related to the public power, mainly because of the lack of support in routine activities (Interviewed 1, 2021).

In (5), the disparity of expectations may occur due to the lack of knowledge or interest of enterprises and educational institutions in participating in urban green commons (Hettiarachchi et al., 2018; Le et al., 2018). For the same reason described in (4), it can generate a feeling of helplessness by the volunteers in relation to the society (Nagib, 2019; Interviewed 1, 2021). It is up to the active members to disseminate and to search for these actors so that they actively participate in the project (Santos & Bacci, 2017), demonstrating its potential benefits.

The conflict presented in (6) occurs between "government" and "educational institutions" or "enterprises." It happens when the public power does not provide the appropriate environment for conducting the activities of the other two actors (Zurbrügg & Rothenberger, 2013). In this case, the control and enforcement of laws – as well as the drafting of laws that support the establishment of urban green commons – are shown to be essential (Nagib, 2019; Santos & Bacci, 2017), because it enables the fluidity of support from companies and educational institutions.

Finally, the conflict (7) illustrates the possible dissent between individuals and groups which the civil society is composed of. As shown, this key actor is primarily responsible for the development of the gardens, so that all beneficial relationships or conflicts affect them, both directly and indirectly. Self-management becomes essential to deal with this context, but it is a complex path to be followed (Fox-kämper et al., 2018; Nagib, 2019). Each individual (residents, volunteers, NGOs, and CSOs) has their own interest and condition. When these interests spontaneously need to be discussed and agreed upon, conflict scenarios tend to arise (Le et al., 2018; Nagib, 2019). Thus, the maintenance of this environment must be the main goal as it is the greatest interest of civil society – when working in urban green common – and is able to provide benefits for society as a whole (Nagib, 2019). It is important to create regulations that guide the management and maintenance practices of each urban green common. Furthermore, the support of other actors (enterprises, academia, and government) is also more likely with these regulations, who will have defined the purpose of the project and how they can help (Nagib, 2019).

12.5 Discussion

By applying the SA methodology in the identification and categorization of key actors, as well as their relationships (Fig. 12.1), it was possible to present a multi-stakeholder platform that drives governance structure feasible for the implementation of urban green commons and in the decentralized management of MSW. By analyzing the current relationships, as well as the data collected about UG, as a result of this study, it was possible to identify the key responsibilities to the development of these projects. These responsibilities, when executed by the main actors, are identified here as the key factors for the governance of an urban green common associated with a composting yard, as described in Table 12.3.

The key factors represent the quintessential pieces for the progress of urban green commons

Table 12.3 Set of key factors observed in the implementation of urban community gardens and decentralized MSW management

Key factors						
		Responsibilities				
Category	Actors	Legislation	Statute/manifest	Popular participation	Publicity/disclosure	Technical knowledge
Government	City hall	X				
	Departments				X	X
	Councilors	X				
Civil society	Local residents			X		X
	Volunteers		X	X	X	X
	NGOs			X		X
	CSOs			X		X

Source: Own elaboration

and decentralized MSW management, despite other useful contributions to this context. It is noted that the actors "educational institutions" and "enterprises" emerge as potential supporters, but they are not essential. On the other hand, the performances of "civil society" and "government" are the central points in the implementation of this type of projects, according to Fig. 12.1.

For the aim of UG's expansion and the implementation of the composting yard, it is necessary to regulate and to manage the UG activities. Promoted by the key actor "government," the creation of specific legislation is a means to support its execution and development and tends to enhance partnerships, presented in Table 12.2, relationship (1) (Santos & Bacci, 2017). With this measure, UG gains relevance in the municipal context and is no longer considered just a punctual community action.

The creation of regulations addressing the objective and rules of the project emerges as essential. Also, such regulations may motivate the population to participate, and the "educational institutions" and the "enterprises" may note the significance of the survival of such spaces. The creation of the manifest reflects the implantation of the urban green common and the decentralized management of MSW as a social movement (Nagib, 2019). This manifest must be created by the civil society – which is composed of volunteers working in the garden – based on the interests agreed between each group. The conflict of interests tends to generate conflicts, but common goals focused on the progress of the project should be sought, as shown in Fig. 12.1, relationship (7) (Le et al., 2018).

The population's participation is what makes these spaces feasible (Colding & Barthel, 2012; Dennis & James, 2016; Honda, 2018; Nagib, 2019), as shown in Fig. 12.1 in which the key actor "civil society" is presented in the center of relations. It is up to the population (represented in NGOs, CSOs, and citizens) to recognize the significance and benefits of working in urban green commons; they should also propose themselves to dedicate part of their time to participate (Santos & Bacci, 2017). Sustainable development must be a collective action generated from an individual choice (UN, 2018). The existence of regulation and its future disclosure tend to bring the population closer to this environment (Nagib, 2019). Beyond the volunteers, the disclosure finds great repercussion when carried out by the public authorities, as proposed in Fig. 12.1, with positive relationship (1). This act characterized urban green commons and decentralized management of MSW as something relevant in the society.

The last key factor observed is offered essentially by the "civil society" and the "government" (translated by the secretariats) regarding the technical knowledge necessary to carry out such activities. The activities are composed of the construction and maintenance of urban gardens, food cultivation practices, logistical processes for the storage and collection of MSW, training for those involved in the process of segregating waste at partner commercial establishments, handling of the composting system, and management of internal processes. Only with the mastery of these techniques, the project may be successful. This knowledge must essentially be provided by the government, as well as by the civil society, between volunteers and citizens – who act punctually – sharing their knowledge (Hettiarachchi et al., 2018; Thampi & Rao, 2015; Interview 1, 2021; Interview 2, 2021).

The implementation of these factors, together, summarizes what is essential in the effectiveness of urban community gardens and in the composting carried out in these spaces. The participation of other institutions or groups is beneficial, by enabling the expansion of this initiative. The proposal of urban green commons, as well as the decentralized management of MSW, serves the society by promoting a better quality of life and achieving sustainable development, and these aspects must be clear to the entire society.

12.6 Conclusions

The creation of a model for the implementation and development of urban green commons associated with the decentralized management of MSW is possible by identifying the main actors, as well as their most relevant responsibilities. Without an adequate model, the execution of these projects tends to follow common sense, causing the failure or stagnation of their original proposals. The active presence of civil society and public authorities is essential to promote the expansion of urban green commons and the effective composting practice. The residues composted are reverted into organic fertilizer that may be used in their own crops or even for sale or donation.

Considering the model presented, the UG and the implementation of the composting yard would tend to strengthen the whole project, attracting new volunteers and expanding the participation in decentralized MSW management. For this, it is suggested to review the actions currently underway and plan the yard based on the model proposed in this work.

The UG has its governance model with bottom-up characteristics with political/administrative help, a factor often pointed out by the gardens that have achieved great progress. This model eases the definition of common goals among active participants. With the understanding of multi-stakeholder platform and the identification of the aforementioned key factors, cooperation between potential actors can be strengthened. To this end, the creator of the UG should use the model proposed in this work, in order to interact proactively with each of the key actors mentioned in the Network of Interactions proposed in Fig. 12.1. Also, the observance of the identified key factors, as shown in Table 12.3, will be necessary to seek positive relationships and to minimize existing conflicts.

This study has limitations. It was not possible to interview members of the "civil society," both voluntary actors involved in the process and others who should be included in the future governance structures. It would be important that the implementation of the composting yard had already started to verify the effective potential of positive interactions between key actors, concrete existence of conflicts, and real involvement and dedication to enforce key responsibilities. These issues were identified as key factors for a successful governance model in urban green commons with composting yards. The limitations found were mainly related to the measures arising from the COVID-19 pandemic, in which fieldwork, as well as meetings with different actors, was hampered. Subsequent works can be developed focusing on the practical validation of the model presented here, through its application in projects of this nature.

The creation of specific municipal legislation for composting yards connected to urban green commons, definition of a garden manifest, involvement of the neighborhood as well as other members of local civil society, propagation of the project among priority stakeholders, and holding of technical knowledge workshops were presented as the key factors of the governance process. These aspects should lead to the achievement of sustainable urban development. It is expected that other projects of this nature will be developed in Salvador and in other Brazilian municipalities as well as in the world, aiming at the promotion of biodiversity, tackling climate change, increasing the population's quality of life, empowering communities, and strengthening urban resilience, thus contributing to the achievement of the SDGs.

References

Abreu, M. J. (2013). Gestão comunitária de resíduos orgânicos: o caso do Projeto Revolução dos Baldinhos (PRB), Capital Social e Agricultura Urbana. Universidade Federal de Santa Catarina. Florianópolis, Brazil. https://repositorio.ufsc.br/handle/123456789/107404

Ali, M. (2004). *Sustainable composting. Case studies and guidelines for developing countries*. Water, Engineering and Development Centre. Loughborough University.

Alves, C. G. C. (2017). *Inventários municipais de emissões de gases de efeito estufa (GEE) no Brasil: Uma análise de sua prática, potencialidades e desafios*. Thesis. Federal University of Bahia (UFBA), Bahia, Salvador.

Andrade, N. A., Rainatto, G. C., de Faria, J. M. B., & da Silva, A. L. (2018). Jornada de engajamento do consumidor por meio do Marketing 4.0 e o Big Data. *PMKT – Revista Brasileira de Pesquisas de Marketing, Opinião e Mídia, São Paulo, 11*(3), 272–287.

Azevedo, G. O. D. (2004). *Por menos lixo: A minimização dos resíduos sólidos urbanos na cidade do Salvador/BA*. Thesis. Federal University of Bahia (UFBA), Bahia, Salvador.

Bahia Transfer and Treatment of Waste LTDA (BATTRE). (2020). https://www.battre.com.br

Bhaduri, A., Ringler, C., Dombrowski, R. M., & Scheumann, W. (2015). Sustainability in the water–energy–food nexus. *Water International, 40*(5–6), 723–732.

Brazil. (2000). Federal Law No. 9,985. Institutes the National System of Nature Conservation Units and makes other provisions. http://www.planalto.gov.br/ccivil_03/leis/l9985.htm

Brazil. (2010). Federal Law No. 12,305. Institutes the National Solid Waste Policy (PNRS). http://www.planalto.gov.br/ccivil_03/_ato2007-2010/2010/lei/l12305.htm

Brazilian Agricultural Research Corporation (EMBRAPA). (2015). Hortas Comunitárias. Observatório de Comunicação Pública. EMBRAPA. https://www.embrapa.br/documents/10180/6542367/Observat%C3%B3rio+de+Comunica%C3%A7%C3%A3o+P%C3%BAblica-+Hortas+Comunit%C3%A1rias/e351af38-3376-40a2-9c32-43b3eed3af84

Brazilian Climate Change Panel (PBMC). (2016). Mudanças Climáticas e Cidades: Relatório Especial do Painel Brasileiro de Mudanças Climáticas. http://www.pbmc.coppe.ufrj.br/documentos/Relatorio_UM_v10-2017-1.pdf

Brazilian Association of Public Cleaning and Special Waste Companies (ABRELPE). (2018). *Panorama dos Resíduos Sólidos no Brasil*. https://abrelpe.org.br/panorama/

Brazilian Institute of Geography and Statistics (IBGE). (2010). Sanitation Atlas. http://www.ibge.gov.br/home/estatistica/populacao/atlas_saneamento/default_zip.shtm.

Brazilian Institute of Geography and Statistics (IBGE). (2020). Cidades e Estados - Salvador, code 2927408. https://www.ibge.gov.br/cidades-e-estados/ba/salvador.html

Brazilian Institute of Geography and Statistics (IBGE). (2021). Portal de Cidades. Bahia. https://www.ibge.gov.br/cidades-e-estados/ba.html

Camps-Calvet, M., *et al.* (2015). Sowing resilience and contestation in times of crises: The case of urban gardening movements in Barcelona. *Partecipazione e Conflitto, 8*(2), 417–442. https://doi.org/10.1285/i20356609v8i2p417

Carvalho, C. A. (2020). Hortas urbanas e resiliência: Benefícios da implantação. *CONVIBRA*. ISSN 2179-5967.

City Hall of São Paulo. (2015). Law 16.212. Provides for the participatory management of squares in the city of São Paulo, and other provisions.http://legislacao.prefeitura.sp.gov.br/leis/lei-16212-de-10-de-junho-de-2015

Colding, J., & Barthel, S. (2012). The potential of ' Urban Green Commons ' in the resilience building of cities. Ecological Economics, 86, 156–166. https://doi.org/10.1016/j.ecolecon.2012.10.016

Colding, J., Barthel, S., Bendt, P., Snep, R., van der Knaap, W., & Ernstson, H. (2013). Urban green commons: Insights on urban common property systems. *Global Environmental Change, 23*(5), 1039–1051. https://doi.org/10.1016/j.gloenvcha.2013.05.006

Dennis, M., & James, P. (2016). Urban Forestry & Urban Greening User participation in urban green commons: Exploring the links between access , voluntarism , biodiversity and well being. *Urban Forestry & Urban Greening, 15*, 22–31. https://doi.org/10.1016/j.ufug.2015.11.009

dos Santos, V. M. N., & de Bacci, D. L. C. (2017). Proposta para governala ambiental ante os dilemas socioambientais urbanos. *Estudos Avancados, 31*(89), 199–212. https://doi.org/10.1590/s0103-40142017.31890017

Fehr, M. (2006). A successful pilot project of decentralized household waste management in Brazil. The Environmentalist. 26(1): 21-29. https://doi.org/10.1007/s10669-006-5355-1

Fehr, M. (2009). Setting A Benchmark For Landfill Diversion Of Domestic Waste In Brazil. In: ISWA – APESB World Congress. Lisbon. 377 (1). 1-12.

Food and Agriculture Organization (FAO). (2012). *Criar cidades mais verdes*. http://www.fao.org/ag/agp/greenercities/pt/hup/index.html

Fox-Kämper, R., Wesener, A., Münderlein, D., Sondermann, M., McWilliam, W., & Kirk, N. (2018). Urban community gardens: An evaluation of governance approaches and related enablers and barriers at different development stages. *Landscape and Urban Planning, 170*(April 2016), 59–68. https://doi.org/10.1016/j.landurbplan.2017.06.023

García, D. & You, F. (2017). Systems engineering opportunities for agricultural and organic waste management in the food-water-energy nexus, Current Opinion in Chemical Engineering, 18, 23- 31.

Gouveia, N. (2012). Resíduos sólidos urbanos: impactos socioambientais e perspectiva de manejo sustentável com inclusão social. Ciência e Saúde Coletiva, [s.l.], v. 17, n. 6, p. 1503-1510.

GRCN. (2020). Chief Resilience Officers from around the world announce the evolution and expansion of the Global Resilient Cities Network. United Arab

Emirates: The Rockefeller Foundation.https://www.rockpa.org/project/global-resilient-cities-network/

Grimberg, E. (2005). Governança democrática e um novo paradigma de gestão de resíduos sólidos São Paulo. Instituto Pólis. São Paulo.http://www.polis.org.br/artigo_interno.asp?codigo=24

Grimberg, E., & Blauth, P. (2005). *Governança democrática e um novo paradigma de gestão de resíduos sólidos*. Instituto Polis. http://www.polis.org.br/uploads/753/753.pdf

Grimble, R. (1998). Stakeholder methodologies in natural resource. *Socio-Economic Methodologies Best Practice Guidelines, 12*. http://www.nri.org/publications/bpg/bpg02.pdfk

Hettiarachchi, H., Meegoda, J. N., & Ryu, S. (2018). Organic waste buyback as a viable method to enhance sustainable municipal solid waste management in developing countries. *International Journal of Environmental Research and Public Health, 15*(11), 1–15. https://doi.org/10.3390/ijerph15112483

Honda, S. O. (2018). *Pesquisa-ação na implantação de horta comunitária: empoderamento e sustentabilidade na periferia de guarulhos*. Universidade de São Paulo (USP).

Inácio, C. T. & Miller, P. R. M. (2009). Compostagem: ciência e prática para gestão de resíduos orgânicos. EMBRAPA Solos. Composting Technologies and Automation. ISBN: 85-85864-31-6.

Institute of Applied Economic Research (IPEA). (2012). Diagnosis of solid waste from health services. IPEA Research Report. https://www.ipea.gov.br/portal/images/stories/PDFs/relatoriopesquisa/120806_relatorio_residuos_solidos.pdf

Institute of Applied Economic Research (IPEA). (2020). Resíduos sólidos urbanos no Brasil: desafios tecnológicos, políticos e econômicos. Centro de Pesquisa em Ciência, Tecnologia e Sociedade. https://www.ipea.gov.br/cts/pt/central-de-conteudo/artigos/artigos/217-residuos-solidos-urbanos-no-brasil-desafios-tecnologicos-politicos-e-economicos

Jagt, et al. (2017). Cultivating nature-based solutions: The Governance of Communal Urban Gardens in the European Union. *Environmental Research, [s.l.], 159*, 264–275, Elsevier BV. https://doi.org/10.1016/j.envres.2017.08.013

Lemos, M.C. & Agrawal, A. (2006) Environmental Governance. Annual Review of Environment and Resources, 31, 297-325.https://doi.org/10.1146/annurev.energy.31.042605.135621

Lassa, J. A., & Nugraha, E. (2015). From shared learning to shared action in building resilience in the city of Bandar Lampung, Indonesia. *Environment and Urbanization, 27*(1), 161–180. https://doi.org/10.1177/0956247814552233

Law no 9.985, 18 of July of 2000. (2000). Esta Lei institui o Sistema Nacional de Unidades de Conservação da Natureza – SNUC, estabelece critérios e normas para a criação, implantação e gestão das unidades de conservação. https://legislacao.presidencia.gov.br/atos/?tipo=LEI&numero=9985&ano=2000&ato=77ck3aq1kMNpWTfc9

Law no 12.305, 02 of August of 2010. (2010). *Institui a Política Nacional de Resíduos Sólidos: altera a Lei no 9.605, de 12 de fevereiro de 1998; e dá outras providências*. http://www.planalto.gov.br/ccivil_03/_Ato2007-2010/2010/Lei/L12305.htm

Le, N. P. *et al.* (2018). Understanding the stakeholders' involvement in utilizing municipal solid waste in agriculture through composting: A case study of Hanoi, Vietnam. *Sustainability (Switzerland), 10*(7), 1–32. https://doi.org/10.3390/su10072314

Märker, C., Venghaus, S. & Hake, J. F. (2018). Integrated governance for the food-energy-water nexus - The scope of action for institutional change. Renewable and Sustainable Energy Reviews 97:290-300. doi: 10.1016/j.rser.2018.08.020

Moura, J. (2012). Minicentral de compostagem completa 5 anos de operação ininterrupta em Visconde de Mauá. Comissão de educação ambiental da Associação Pró-Bem-Viver Visconde de Mauá. http://agriculturaurbana.org.br/boas_praticas/compostagem/central%20de%20compostagem.htm

Más-Rosa, S., & Carvalho, C. A. (2021). A compostagem como estratégia de gestão de resíduos orgânicos: Fortalecendo a resiliência urbana em Salvador. *Congresso Virtual UFBA, 2021*.

Nagib, G. (2019). Processos e materialização da agricultura urbana como ativismo na cidade de São Paulo: o caso da Horta das Corujas. *Cadernos Metrópole, 21*(46), 715–740. https://doi.org/10.1590/2236-9996.2019-4602

Navarro, R. S., Marques, A. F., Nunes D. S., Quadros A. (2019). *Hortas comunitárias e os objetivos da agenda 2030 para o desenvolvimento sustentável*. Desenvolvimento Regional: Processos, Políticas e Transformações Territoriais. RS, Brasil. ISSN: 2447-4622

Nikolaïdou, S., et al. (2016). Urban gardening and green space governance. *Cogitatio: Towards New Collaborative Planning Practices*, [s.i], *1*(1), 5–19, 29 https://www.cogitatiopress.com/urbanplanning/article/view/520

Pinto, A. P. S. (2018). Gestão de resíduos sólidos urbanos: estudo sobre a atuação do programa de coleta seletiva salvador na cooperativa dos agentes ambientais da nova república (canore) no município de Salvador – Bahia. Universidade Federal da Bahia. Instituto de Geociências.http://www.geografia.ufba.br/TCC%20Ana%20Paula%20Pinto.pdf

Robert. M. (1996). Le Sol: Interface dans l'environnement ressource pour le développement. Masson, Paris, 244 p.

Salvador City Hall (PMS). (2015). Municipal Decree No 26.916, 16/12/2015. Mandatory tocollection, transport, treatment and disposal of waste solids and final disposal of tailings from big generators. https://www.sefaz.salvador.ba.gov.br/Documento/ObterArquivo/1426

Salvador City Hall (PMS). (2019). Salvador Resilient Plan. The Rockefeller Foundation. http://salvadorresiliente.salvador.ba.gov.br/

Salvador City Hall (PMS). (2020a). *Primeira horta urbana de Salvador completa quatro anos de atividades*. http://www.comunicacao.salvador.ba.gov.br/index.php/todas-as-noticias-4/57362-primeira-horta-urbana-de-salvador-completa-quatro-anos-de-atividades

Salvador City Hall (PMS). (2020b). *Plano de mitigação e adaptação às mudanças do clima em Salvador*. http://sustentabilidade.salvador.ba.gov.br/wp-content/uploads/2020-/12/Versao_Completa_PMAMC.pdf

Santos, K. L. dos, Panizzon, J., Cenci, M. M., Grabowski, G., Jahno, V. D. Food losses and waste: reflections on the current brazilian scenario. *Brazilian Journal of Food Technology* [online]. 2020, 23, e2019134. Epub Mar 13, 2020. ISSN 1981-6723. https://doi.org/10.1590/1981-6723.13419

Schenini, P. C., Schmitt, V., Silva, F. A., & Pereira, M. F. (2006). Marketing verde como uma abordagem estratégica frente ao novo perfil de consumo. In *XLIV Congresso da Sober. Congress* (pp. 1–13). Federal University of Santa Catarina (UFSC). http://www.sober.org.br/palestra/5/670.pdf

Seixas, C. S., Prado, D. S., Joly, C. A., May, P. H., Costa Neves, E. M. S., & Teixeira, L. R. (2020). Governança ambiental no Brasil: rumo aos objetivos do desenvolvimento sustentável (ODS)? *Cadernos Gestão Pública e Cidadania, 25*(81), 1–21. https://doi.org/10.12660/cgpc.v25n81.81404

Siqueira, T. M. O. & Assad, M. L. R. C. (2015). Compostagem de resíduos sólidos urbanos no Estado de São Paulo (Brasil). Anppas. Revista Ambiente e Sociedade. doi: https://doi.org/10.1590/1809-4422ASOC1243V1842015

Souza, R. C. T. A horta agroecológica: instrumento de inclusão social e deeducação ambiental na escola Maria Montessouri, Anápolis, Estado de Goiás. 2018.39f. Monografia - Faculdade de Agronomia, Centro Universitário de Anápolis –UniEvangélica, 2018.

Swagemakers, P., Garcia, M. D. D., & Wiskerke, J. S. C. (2018). Socially-inclusive development and value creation: How a composting project in Galicia (Spain) "Hit the Rocks". *Sustainability (Switzerland), 10*(6). https://doi.org/10.3390/su10062040

Thampi, A., & Rao, B. (2015). Application of multi-criteria decision making tools for technology choice in treatment and disposal of municipal solid waste for local self government bodies – A case study of Kerala, India. *Journal of Solid Waste Technology and Management, 41*(1), 84–95. https://doi.org/10.5276/JSWTM.2015.84

Thiele, G., Devaux, A., Reinoso, I., Pico, H., Montesdeoca, F., Pumisacho, M., Andrade-Piedra, J., Velasco, C., Flores, P., Esprella, R., Thomann, A., Manrique, K., & Horton, D. (2011). Multi-stakeholder platforms for linking small farmers to value chains: Evidence from the Andes. *International Journal of Agricultural Sustainability, 9*(3), 423–433. https://doi.org/10.1080/14735903.2011.589206

UN-Habitat. (2015). Global Activities Report 2015. Increasing synergy for greater national ownership. United Nations. https://sustainabledevelopment.un.org/content/documents/1726Habitat%20Global%20Activties%202015.pdf

UN-Habitat. (2014). Global Activities Report 2014. Agência Brasil.https://agenciabrasil.ebc.com.br/geral/noticia/2016-10/mais-de-90-da-populacao-brasileira-vivera-em-cidades-em-2030

Un-habitat. (2020). World Cities Report: The value of sustainable urbanization. Kenya: United Nations Human Settlements Programme. 418 p. https://unhabitat.org/World%20Cities%20Report%2020204

United Nations (UN). (2018). Transforming our world: The 2030 agenda for sustainable development. *A New Era in Global Health*. https://doi.org/10.1891/9780826190123.ap02

Urban Cleaning Company (LIMPURB). (2020). Carta anual de políticas públicas e governança corporativa. Prefeitura Municipal de Salvador.http://limpurb.salvador.ba.gov.br/images/Aspla/Carta_de_Governana_2020.pdf

Urban Development Secretariat (SEDUR). (2016). Relatório de atividades 2015: um capítulo para o desenvolvimento da Bahia. Salvador.

Van Elk, A. G. H. P. (2007). Redução de emissões na disposição final. Mecanismo de Desenvolvimento Limpo Aplicado a Resíduos Sólidos. Rio de Janeiro: IBAM. https://www.ibam.org.br/media/arquivos/estudos/03-aterro_mdl_1.pdf

Villa, F., Arcidiacono, A., Causone, F., Masera, G., Tadi, M., & Grosso, M. (2020). Entering rocinha: A gis approach for the improvement of solid waste management in a slum in Rio de Janeiro (Brazil). *Detritus, 9*(March), 221–231. https://doi.org/10.31025/2611-4135/2020.13900

Warner, J. (2007). *Multi-stakeholder platforms for integrated water management*. Aldershot.

Weitz, N., Strambo, C., Kemp-Benedict, E., & Nilsson, M. (2017). Closing the governance gaps in the water-energy-food nexus: Insights from integrative governance. *Global Environmental Change, 45*(July 2017), 165–173.

Zago, V. C. P., & Barros, R. T. V. (2019). Gestão dos resíduos sólidos orgânicos urbanos no Brasil: Do ordenamento jurídico à realidade. *Eng Sanit Ambient, 24*(2), 219–228.

Zurbrügg, C., & Rothenberger, S. (2013). Determinants of resilience in community-led waste management. *Proceedings of Institution of Civil Engineers: Waste and Resource Management, 166*(2), 93–100. https://doi.org/10.1680/warm.12.00006

WEF Nexus and Sustainable Investments in West Africa: The Case of Nigeria

13

Damilola S. Olawuyi and Alex Oche

Abstract

Water, energy and food (WEF) resources are vital to all aspects of human existence and survival. Recent studies have therefore increasingly underlined the need for a nexus governance approach as an important vehicle through which countries can holistically achieve sustainable development and the Sustainable Development Goals. Existing studies have analysed the guiding principles of emerging law and governance frameworks across Europe and North America that implement WEF nexus approach in practice. However, an assessment of the legal and governance frameworks on WEF nexus in the African context, and the challenges that limit their successful implementation, remains sparse. This chapter fills a gap in this regard.

Drawing lessons from Nigeria in West Africa, this chapter discusses the importance of WEF nexus governance for advancing sustainable investments in these important and interdependent sectors. It examines the preconditions for WEF nexus integration, the scope of application in a developing country context, the current barriers to implementation and the emerging solutions for addressing such practical challenges. The study suggests that elaborating sustainable development and human rights safeguards in investment instruments in the WEF sector, reforming existing laws and institutions to ensure coherent implementation of WEF programmes, improving the availability and reliability of statistical data needed to drive integrated WEF planning and designing capacity development programmes for WEF actors are significant steps towards improving the development and successful implementation of WEF programmes in West African countries, such as Nigeria.

Keywords

Sustainable investment · Developing countries · WEF nexus · Climate change · Human rights

13.1 Introduction

The water, energy and food (WEF) nexus governance approach has become increasingly recognized as an important framework through which countries can holistically implement and achieve coherent planning in these three important sectors,

D. S. Olawuyi (✉)
College of Law, Hamad Bin Khalifa University (HBKU), Doha, Qatar
e-mail: dolawuyi@hbku.edu.qa

A. Oche
College of Law, Afe Babalola University, Ado Ekiti, Nigeria

L. L. B. Lazaro et al. (eds.), *Water-Energy-Food Nexus and Climate Change in Cities*, Sustainable Development Goals Series, https://doi.org/10.1007/978-3-031-05472-3_13

in order to avoid overlap.[1] However, despite the emergence of several discourses on WEF nexus governance approach and its law and governance requirements, the practical challenges of implementing it in practice in the African context have yet to receive in-depth analysis. Drawing lessons from Nigeria in West Africa, this chapter aims to fill this gap by exploring the significance of, and the preconditions for, a successful WEF nexus governance integration in the African contexts. Several African countries face pre-existing governance challenges that make interoperability, rule linkage, systemic integration and inter-ministerial coordination more complex and challenging. Without addressing such practical challenges, a WEF nexus governance approach becomes difficult. This chapter illustrates practical governance challenges that must be addressed by African countries in order to advance WEF governance as a tool for promoting sustainable development in local contexts.

The chapter is divided into five sections with this introduction being the first. The second section discusses the drivers of WEF nexus and benefits of a nexus governance as a tool for promoting sustainable development. Drawing lessons from Nigeria, the third section discusses the barriers to integrating WEF nexus governance for developing countries. Building on the third section, the fourth section makes recommendations on how pre-existing and practical challenges to WEF nexus governance can be addressed through innovative legal solutions. The fifth section is the concluding section.

13.2 Drivers and Significance of WEF Nexus in West Africa

The search for sustainable development requires integrated social, economic and environmental progress in key sectors.[2] Whilst social development focuses on the provision of basic infrastructure and services needed for human development (such as water, energy, food, motorable roads, healthcare, education and the fulfilment of human rights including the right to water, food and decent living), economic development focuses on creation of jobs, growth in government revenue, inflow of foreign direct investments (FDIs), growth in gross domestic product (GDP) and poverty eradication (including eradicating water, energy and food poverty), whilst environmental development aims to achieve the protection of natural capital (e.g. land, air, water, minerals, etc.) from degradation and pollution control through sustainable consumption of the earth's resources, including water, energy and food.[3] Given the importance of water, energy and food resources, none of the pillars of sustainable development can be achieved without water, energy and food security.[4]

The central tenet of the WEF governance model therefore is the idea of integrating WEF institutions, policies and rules in order to ensure coherence in their implementation.[5] Over the last decades, and as far back as the 1980s, researchers and policymakers began to point out the need for a cross-sectoral and integrative approach to managing WEF resources.[6] In developed country contexts, the need for integration is driven largely by efforts to reduce fragmentation, role duplication and incoherence in the implementation of government policies and programmes relating to WEF resources.[7] Whilst these reasons have also

[1]Olawuyi (2020) 1–9.

[2]See para. 1.2 of the ILA New Delhi Declaration on Sustainable Development and the 2012 Sofia Guiding Statements on the Judicial Elaboration of the 2002 New Delhi Declaration of Principles of International Law Relating to Sustainable Development, *Resolution* No. 7/2012 noting in para. 3 that 'the sustainable use of all natural resources represents an emerging rule of general customary international law, with particular normative precision identifiable with respect to shared and common natural resources'.

[3]See Schrijver (2017) 99–102; Olawuyi (2021) 1–15.

[4]Olawuyi (2020)1–9.

[5]See Olawuyi (n.5).

[6]Calls for integrated water resources management (IWRM) and integrated coastal zone management (ICZM) have featured prominently in the literature since the 1980s. However, they were seen as too narrow and inadequate to respond to emerging challenges in food and energy sectors. For a detailed overview, see Hayley Leck et al. (2015) 445.

[7]Pavoni (2010) 641; See also the International Law Association (2018), stating that the principle of mutual supportiveness between trade and environmental policies or agreements may inspire solutions to a conflict of norms.

driven the WEF discourse to prominence in developing countries, a number of additional local and governance challenges make WEF nexus integration an urgent and essential prerequisite for sustainable development, especially for developing countries where progress on the SDGs has been relatively slow. In order to address WEF scarcities, and to advance the UNSDGs (United Nations Sustainable Development Goals) relating to water, energy and food, many developing countries in Africa and Latin America, where many of the WEF poorest people live, have scaled up investments in infrastructure development projects aimed at expanding WEF access (WEF projects). A good example is the recently approved Mambilla hydropower project, Africa's largest. Estimated to cost US$5.7 billion, the Mambilla hydropower project aims to expand electricity access in Nigeria.[8]

However, the implementation of several WEF projects needed to advance sustainable development, such as the Mambilla hydropower project in Nigeria, renewable energy projects in Senegal and large energy pipeline projects such as the West African Gas Pipeline project in Ghana, amongst others, has been stifled by sector-based planning, human rights concerns, delays in approval processes and lack of integrated planning and governance.[9] The tendency of infrastructure development activities in one sector, to result in overlap, human rights challenges and policy reversal in other sectors, has resulted in calls for a more coherent and less fragmented approach to the design, approval, planning and implementation of WEF projects, through a WEF governance approach.[10]

Generally, calls for a nexus governance approach to WEF resources as a tool for promoting sustainable investments have become prominent over the last decade due to the following four main reasons[11]:

13.2.1 Unprecedented Increase in Demand for Water, Energy and Food in Cities

Population growth, rapid urbanization and economic development have caused demand for water, energy and food to increase unprecedented levels in developing countries.[12] Demand for food, water and energy has grown over the last decade and is projected to grow further by approximately 35%, 40% and 50%, respectively, by 2030, owing to an increase in the global population and rise in consumption patterns across the world, especially in developing countries.[13] In the West African region, for example, demand for energy, food and water is on the rise, spurred by unprecedented economic and population growth and urbanization.[14] A recent study on Ghana has shown that rural-urban migration caused a massive increase in demand for water, energy and food, on the already strained infrastructure struggling to feed the former inhabitants.[15] Furthermore, efforts to actualize SDG7 (energy) for the over 1.7 billion people lacking access to modern energy sources, majority of which reside in developing countries, will surely cause further

[8]Labelled as Africa's Three Gorges Dam project, this expansive project will include the construction of four dams and 700 km of electricity transmission lines in Africa's most populous nation; see *Premium Times* (2017).

[9]Ecowas Centre for Renewable Energy and Energy Efficiency (2010) 5.

[10]Olawuyi (2018).

[11]Cairns and Krzywoszynska (2016) trace the recent focus on the WEF nexus to the year 2008, where business leaders of the World Economic Forum issued a call to engage with nexus issues between economic growth and water, energy and food resource systems.

[12]Terrapon-Pfaf (2018).

[13]According to UN estimates, the world population of 7.4 billion (in 2016) will increase by 1 billion over the next 10 years and reach 9.6 billion by 2050. See United Nations (2015).

[14]International Energy Agency (IEA) (2019).

[15]The 2010 national census in Ghana recorded a population of 22.7 million, of this number, the Ghana Statistical Service reported in 2010 that more than 80% of Ghanaians migrate with about 70% going to the urban areas. For context, 70% of 22.7 million is 15,890,000 which is more than the population of Monaco and Liechtenstein combined. Ghana Statistical Service (2010); also Tanle (2020) 1–18.

increase in demand for energy through energy expansion projects, which will have an impact on water resources and land, given their interdependencies.[16] Similarly, a number of countries have intensified efforts to reduce the current electricity demand rate by promoting energy efficiency in water, food and industrial sectors and eliminating waste. Addressing energy demand, without understanding the implications of energy efficiency measures and activities on food and agricultural productivity, may result in counterproductive policies. Also, advancing agricultural expansion projects without addressing the impacts of such projects on electricity usage and demand could provide counterproductive effects on energy efficiency and security. For example, the food and agricultural sector which is currently the largest consumer of water (70%) and one of the largest users of energy (30%) is prone to impact as food price will increase as energy, fertilizer, water and transportation costs rise.[17] Also, addressing energy demand through expansion of alternative and renewable energy projects could affect both land access for agricultural purposes and food security.

To meet the needs of people with limited access to these resources, there is a need to address water, energy and food issues jointly so that actions in one sector do not affect the other negatively. The WEF nexus model allows policymakers to understand the implications of increasing energy demand on food and water supply and the implications of rising food production and agriculture sector demand for water and energy.[18]

For example, by implementing efficiency programmes in water- and energy-intensive industries, such as the oil and gas, agriculture and manufacturing sectors, West African countries can reduce waste and promote conservation of available water and energy resources. Similarly, by promoting administrative coordination in the approval and implementation of WEF projects, the design and implementation of critical infrastructure investments needed to advance WEF security can proceed efficiently thereby reducing risk levels.

13.2.2 Climate Change

One of the key challenges to sustainable development across the world is the problem of climate change.[19] Climate change increases the pressure on availability of water, energy and food resources at local and regional levels in diverse ways.[20] For example, a rise in the frequency of droughts due to drier summers, wild fires, increase in number of pests and insects such as locusts,[21] migrant grazing and changes in climate variables such as temperature, precipitation and humidity could significantly affect food production, distribution, storage and transportation.[22] For example, in the Economic Community of West African States (ECOWAS) region, climate change has caused a change in pattern and duration of rainfall, with shorter raining seasons lately, whilst desertification and drought have become more persistent.[23] Consequently, countries in the region are considered highly vulnerable to climate change as they are prone to experiencing impacts on agricultural productivity and food and energy security.[24] Studies indicate that a temperature change of two degree Celsius could result in 40–80% loss of cropland conducive for farming and agriculture, which could further worsen food insecurity in already poor West African countries.[25] In order to effectively limit and address the catastrophic

[16] Olawuyi and Nsoh (2020).

[17] Food and Agricultural Organization (FAO) (2011).

[18] Oguamanam (2015) 238–4.

[19] Olawuyi (2022)1–10.

[20] Ibid.

[21] Recently, in 2018, climate change-related disaster – particularly cyclone – dumped heavy rain on the Arabian Desert which caused locust to breed unnoticed. Strong winds in 2019 blew the growing swarms into Yemen's inaccessible conflict zones and then across the Red Sea into Somalia, Ethiopia and Kenya, causing several severe locust invasions on these countries who were caught unprepared. See McConnell (2021).

[22] Ellen Wall et al. (2007); Budong Qian et al. (2012).

[23] World Bank (2013).

[24] Ibid ECREEE (n 20) 5; Oche (2020).

[25] World Bank (2013).

impacts of climate change, the international climate change regime has emphasized the importance of cooperating across relevant sectors in both the preparation and implementation of the National Adaptation Plan (NAP).

Addressing climate change impacts in one domain, without addressing trade-offs and impacts in other domains, may result in maladaptation and ineffectiveness. An approach which addresses the impacts of climate change in one domain whilst neglecting trade-offs and impacts in other domains might prove ineffective. Also, given the cross-cutting implications of climate change for water, energy and food infrastructure, promoting the design of smart and climate-resilient infrastructure can enhance efficiency and reduce emissions across entire WEF domains.[26] Infrastructure for climate adaptation could be designed for multifunctional purposes across the different WEF domains, for example, dams for electricity, irrigation and drainage systems.[27] Similarly, given the urgency of climate change mitigation and adaptation in vulnerable cities of West Africa, there is a need for a coordinated climate action across all key sectors. The WEF nexus approach provides an opportunity for key stakeholders in water, energy and food sectors to coordinate governance efforts and streamline regulatory procedures in a manner that fosters the rapid development and implementation of climate-smart infrastructure investments across the three key domains.

13.2.3 The Need for Holistic Implementation of Sustainable Development Goals (SDGs)

The nexus governance approach provides a systemic understanding and view of the trade-offs and synergies between the SDGs, and this helps to avoid overlapping and duplication of sector-specific actions and programmes. Through information sharing and cooperation between the different institutions responsible for the various SDGs, a country can formulate and implement a multiscale, holistic and integrated plan for achieving the SDGs.

SDGs include a wide spectrum of topics and issues, including poverty, food security, health, climate change and gender equality, amongst others. Progress towards at least 12 of these 17 goals is directly linked with the sustainable use of resources such as land, water, energy and food. However, progress made in several developing countries, in terms of advancing several of the SDGs, remains comparably low.[28] According to the UN, one key reason is that a nexus approach has not been adequately considered in framing the targets of the SDGs, leaving out the interdependencies, trade-offs and synergies amongst the SDGs and their targets.[29] For example, in Nigeria, practical challenges such as lack of intergovernmental coordination and lack of holistic planning across relevant sectors continue to stifle progress on all of the SDGs.[30] Sector-based and piecemeal responses to SDG implementation therefore remain a key reason for slow progress recorded in several of the SDGs in Nigeria and across West Africa.[31] Poor information exchange, data gaps and lack of coordination amongst key stakeholders in the sectors often result in poor planning and decision-making, which ultimately results in conflicts amongst stakeholders and poor outcomes in SDG delivery.[32] Given the interconnectivity of the targets of goals 2, 6 and 7 on food, water and energy, respectively, there is a need to coherently implement the applicable legislation and policies in investment, human rights, environment and agricultural law regimes in order to achieve any of the SDGs.

The WEF governance approach provides a framework for West African countries to coherently track and monitor progress made in advancing the SDGs relating to water, energy and food

[26] Olawuyi and Nsoh (n.19).

[27] Ringler et al. (2013).

[28] Olawuyi and Olusegun (2018) 37–60.

[29] United Nations (2018).

[30] Olawuyi and Olusegun (2018) 37–60.

[31] Ibid.

[32] Stockholm Resilience Centre (2018) 39–40.

and to promote holistic responses to challenges that impede progress in each sector.

13.2.4 Addressing the Interactions Between WEF Projects and Human Rights

The need to respect, protect and fulfil human rights is a vital aspect of the social pillar of sustainable development.[33] Consequently, efforts and projects aimed at promoting development should not stifle or affect fundamental human rights. However, over the last decade, with several climate and energy expansion projects being associated with negative impacts on human rights, especially the right to food, property and water, the dangers of sector-based planning have become so apparent.[34] Several WEF projects have been linked with complex human rights violations in developing countries.[35] These human rights impacts take several forms and manifest in different ways especially compulsory acquisition of property as project sites, forced displacement, marginalization, exclusion, concentration of energy access projects in poor and vulnerable communities and governmental repression in developing countries.[36] In many cases, these human rights challenges have been exacerbated by lack of integrated planning by key institutions responsible for project design and delivery. For example, in granting approvals for energy projects, failure to consider the implications of such projects on agricultural lands and food access has decelerated progress on food security programmes.[37] Similarly, failure to consider the implications of new investments on existing projects and programmes in other sectors has resulted in land grabs and human rights violations.[38] For example, as seen in the Cameroon BIT (bilateral investment treaty) as well as in a number of other BITs negotiated by African countries, there have been terms included in investment agreements which allow investors to utilize as much land, forest and water resources in a given area for their investment activities during the term of the treaty.[39] Such provisions not only conflict with pre-existing rights of landowners in such communities but could also result in transborder disputes over shared water resources.[40] Similarly, whilst stabilization clauses are important from an investor's standpoint to mitigate legal and political risks that are typically commonplace in emerging markets due to change of laws,[41] there have been instances of overly broad stabilization clauses that may make improvements to domestic laws on labour, land and human rights to be inapplicable to an investment.[42]

The gaps and prevalent human rights violations associated with WEF expansion projects have highlighted the need for a more integrative, transparent and human right approach to WEF governance.[43] As West African countries increasingly seek to attract investments in WEF sectors, there is a need to ensure that human rights safeguards are integrated into project approval requirements and standards across the sectors in order to address adverse human rights impacts. For example, approving an agricultural expansion investment may result in land grabs and water scarcity and could worsen access to energy in poor communities who rely on biomass for cooking and heating. Such an outcome may result in the violation of several human rights relating to environment, water, food and human life.[44] To address such impacts, the African Union's 2012

---

33 Olawuyi (2018) 73–104.

34 Ibid.

35 Oguamanam (2015) 237–238.

36 Olawuyi (2016) 1–25.

37 Cordes and Bulman (2016) 53–67.

38 Richards (2013).

39 Mbengue and Waltman (2015) 24–30.

40 Cordes and Bulman (2016) 53–67.

41 Maniruzzaman (2008) 127; Kriebaum (2007) 717–744.

42 Gehne and Brillo (2014) 3–5; Olawuyi (2015) 247–265.

43 Olawuyi and Nsoh (n 19) at 7–8.

44 See, for example, the decision of the African Commission in the case of *Social and Economic Rights Action Centre and the Centre for Economic and Social Rights v Nigeria* (2001) African Commission on Human and People's Rights, Comm. No. 155/96, which recognizes the obligation of governments to prevent human rights violations resulting from business and commercial activities.

Resolution on a Human Rights-Based Approach to Natural Resources Governance (hereafter the African Union Resolution) calls on governments to 'set up independent monitoring and accountability mechanisms that ensure that human rights are justiciable and extractive industries and investors legally accountable in the country hosting their activities and in the country of legal domicile'.[45] The WEF governance approach provides an accountability framework that allows regulators to undertake a holistic and systemic assessment and monitoring of the human rights WEF projects and investments. By reviewing the human rights impacts of a project in a wide range of interrelated sectors, efforts to advance water, energy or food security will reduce human rights problems in other sectors. Key challenges that limit mutual supportiveness and responsible investment in WEF systems must be comprehensively addressed in order to achieve sustainable development in these key sectors. The next section discusses some of such challenges.

13.3 Law and Governance Barriers to Implementing WEF Nexus in Nigeria

Drawing lessons from Nigeria, this section develops a profile of key barriers and challenges that limit the implementation of WEF nexus governance in the African context. Nigeria offers a classic example of how WEF governance may be stifled by pre-existing sustainability challenges which make interoperability and integration complex and near unattainable.

13.3.1 Methodology

This study relies primarily on published government reports and data in Nigeria that document the current scope and status of WEF nexus initiatives being implemented to advance integrated and coordinated delivery of the SDGs in Nigeria. These include publicly accessible legislation, policy documents and periodic reports collated and released by the Federal Government of Nigeria.[46] A qualitative analysis of these documents has allowed conclusions to be drawn on how WEF governance programmes and initiatives are progressing in Nigeria and the challenges to their implementation. Furthermore, an analytical review of published literature is adopted because existing research in the field has satisfactorily compiled the importance of WEF for advancing the SDGs.[47] However, a detailed examination of practical challenges facing the development and implementation of WEF programmes in the context of Africa, and specifically Nigeria, has remained absent.

This article moves the discussion forward by examining the current challenges and future opportunities for mainstreaming WEF governance into sustainable development planning in Nigeria. Owing to the scope of the study and the nature of the methodological approach, the survey can by no means be regarded as representative. However, since this survey is combined with a review of the literature, this section provides an analytical profile of, and insights on, the key challenges facing the integration of WEF governance into the design and implementation of WEF projects in the context of Nigeria and how lessons from Nigeria can guide other African countries to overcome law and governance barriers to WEF planning.

[45] African Commission on Human and Peoples' Rights, Resolution on a Human Rights-Based Approach to Natural Resources Governance (2 May 2012) ACHPR/Res.224 (LI) (hereafter the African Union Resolution).

See also the Niamey Declaration on Ensuring the Upholding of the African Charter in the Extractive Industries Sector, ACHPR/Res.367 (LX) (2017), which requires all African countries to 'put in place regulatory bodies vested with the relevant powers for ensuring that human rights, environmental and labour standards are duly respected and environmental and social impacts are mitigated'.

[46] Government of Nigeria (2017), Government of Nigeria (2020a, b, c).

[47] See Olawuyi (2020).

13.3.2 Results

Despite Nigeria's abundant water, energy and food resources and one of the world's largest exporters of oil and gas resources, it has been ranked as one of the world's poorest countries.[48] A wide range of complex social, economic and environmental challenges also mean that progress in advancing the SDGs remains critically slow in Nigeria.[49] For example, due to inadequate water, sanitation and hygiene (WASH) facilities, Nigeria has one of the highest rates of open defecation in the world.[50] Similarly, despite Nigeria's vast oil and natural gas resources, electrification rate remains critical at 45%, whilst more than 55% of Nigerians do not have access to electricity and still rely on non-commercial fuels like coal dust, wood, animal dung and crop residues for cooking, lighting, heating and commercial activities.[51] Furthermore, food insecurity remains a critical challenge in Nigeria with about nine out of ten Nigerians being unable to afford a healthy diet.[52] Furthermore, Nigeria ranks as number two in the world in terms of the population of stunted and malnourished children.[53] In addition to other governance challenges such as governmental corruption, conflicts, insecurity and poor maintenance of WEF infrastructures, amongst others, Nigeria's poor sustainable development outcomes are worsened by lack of integrated and holistic implementation of WEF programmes. Problems of fragmented laws and institutions, weak multi-stakeholder engagement and participation in WEF projects, inadequate infrastructure and technology for knowledge sharing and weak capacity for implementing the WEF nexus approach continue to limit governance coordination across the three sectors. These problems are discussed below:

13.3.2.1 Fragmented Laws and Institutions

At the core of the WEF nexus governance discourse is the idea of integration. Integration of management and governance across WEF systems is crucial to reducing trade-offs and maximizing the synergies between them.[54] However, attempts to integrate water, energy and food systems still have a long way to go in Nigeria.[55] In Nigeria, as it is in several African countries, most institutions – like the various water, energy and agriculture departments or agencies – operate under a complex system of unclear mandates and obscure funding and do not consult with each other.[56] This is because they operate under separate laws which do not make reference to intersecting WEF resources. In Nigeria, for example, the laws and institutions governing WEF resources are scattered about several different statutes, with several departments, water boards and basin authorities responsible for licensing different projects without considering the interactions and pressures existing between the different domains.[57] For example, the Petroleum Industry Act (2021) which empowers the minister of petroleum resources to grant prospecting and mining licences makes no provisions to consider impacts on water bodies. Similarly, the several regulatory bodies created by subsidiary regulations to the act do not consider the trade-offs on water resources caused by energy production processes. At most, what is considered are pollution-related issues which are typically reactionary than proactive.[58] This increases the ten-

[48]Adebayo (2018); Government of Nigeria (2020a), stating that 40.1% of Nigerians live in poverty.

[49]Olawuyi and Olusegun (2018) 37–60; Government of Nigeria (2017), Government of Nigeria (2020a).

[50]Government of Nigeria (2020c).

[51] Olawuyi (2022).

[52]In the 2020 Global Hunger Index, Nigeria ranked 98th out of the 107 countries and was described as having 'a level of hunger that is serious' Global Hunger Index (2021).

[53]Bloomberg (2020).

[54]Olawuyi (2020) 1–9.

[55]Government of Nigeria (2017) 70–72. See also Government of Nigeria (2020a) 71–75, which documents poor governance and weak regulatory structures as a barrier to the integration and coordination of SDG efforts in Nigeria.

[56]Government of Nigeria (2017) 70–72; Government of Nigeria (2020a) 71–75; also Maupin and Mwanikah Ojoyi (2017).

[57]Olawuyi and Tubodenyefa (2018).

[58]See Petroleum Act (1969), S. 9 (1) para. i.

dencies for red tape and 'turf wars' by ministries who should ordinarily work together for the common goals of advancing WEF programmes, especially where their respective enabling statutes leave doubts about their jurisdictions.[59] Cooperation between WEF ministries has been largely on an informal basis, without any clear legislative framework for such integration. This has resulted in role duplication, limited information exchange and the tendency of actors to stay in their own confines and become unwilling to cooperate with other key sectors that have complementary roles.[60] This raises the need for clear, comprehensive and specific legislation on coordination and integration by key WEF actors, in order to promote interoperability. The absence of linked rules, procedures and obligations on WEF nexus governance will not only make it unduly difficult to evolve and implement the guiding principles of WEF; it will also make regulatory supervision difficult and complex.

As West African countries develop strategies to address climate change and promote the SDGs, a holistic reassessment of existing laws and regulations in WEF sectors will be very important so as to address existing gaps. To implement a nexus approach to WEF governance, a comprehensive linkage of legislation, rules and procedures across the WEF domains is essential.[61] Recent legislative developments in the UK and Alberta, Canada, suggest that rule linkage across the WEF sectors is indeed possible and emerging. For example, the UK's Environmental Permitting (England and Wales) Regulations (EPEWR 2016) offers an example of innovative legislation that integrates the administration of a range of environmental regimes. The system requires regulators to control certain activities that could harm the environment or human health and covers facilities such as the energy industry, waste operations, food industry and water discharge and groundwater activities. The system also allows a single permit to cover more than one regulated facility if they are on the same site. Similarly, in Alberta, Canada, Section 2 of the Responsible Energy Development Act (REDA 2012) empowers the Alberta Energy Regulator (AER) to regulate water conservation and management, environmental protection and public lands management with respect to energy resource activities in Alberta. Section 67(1) also creates an obligation on the AER to ensure that its activities are consistent with programmes, policies and work of the government in respect of energy resource development, public lands management, environmental management and water management. This provision aims to ensure systemic coherence and integration of activities and programmes related to water, energy and public lands management in Alberta. As can be learned from Canada and the UK, implementing the WEF nexus approach will require West African countries to update existing laws and regulations in WEF sectors to promote integration of standards and regulatory requirements. Whilst country-specific assessments will be required to identify how simplifying measures can be designed and applied in WEF sectors and the possible political barriers, rule integration and the elimination of duplicative procedures are essential steps towards integrative governance of WEF sectors.

13.3.2.2 Weak Multi-Stakeholder Engagement and Participation in WEF Projects

The WEF nexus governance approach aims to promote multi-stakeholder partnership, through the development of an integrated regulation system that allows members of the public and different regulatory stakeholders to come together and tackle a common issue that no stakeholder would have been able to tackle alone.[62] Considering the essential nature of water, energy and food to the human existence, projects that may impact or limit access to these resources should involve the active participation and engagement of members of the public that could be adversely impacted in a coordinated and integrated manner. International law has increasingly emphasized the need for

[59] Olawuyi and Tubodenyefa (2018).

[60] Ibid.

[61] Olawuyi (2020).

[62] Beisheim, N S (2016) 6.

free, prior informed consents (FPIC) of local communities that may be affected by an investment activity or project prior to its approval or implementation.[63] Yet in many parts of Africa, including Nigeria, the enforcement and implementation of procedural human rights safeguards on participation, access to information, non-discrimination and accountability in the design and implementation of development projects remain weak.[64] Whilst virtually all African countries are signatories to international human rights instruments that guarantee these procedural rights, the level of domestic implementation remains less robust, especially in politically sensitive sectors such as water, energy and food sectors.[65]

Nearly all cases of conflict over land-based water, energy and agricultural investments and projects arise from projects executed without proper consultations with the local communities.[66] For example, one of the key concerns expressed with the Mambilla hydropower project is the limited consultation of local communities by the Nigerian government. Furthermore, the dam is expected to displace 100,000 people from their ancestral homes and lands which could worsen food insecurity in those communities.[67] Similar concerns on lack of consultations, land grabs and dislocation of indigenes and residents have also been raised with the Zungeru Hydropower Dam in Niger State, Nigeria.[68] These communities are not generally opposed to such projects or investment, but they know and insist upon their rights, even if governments and supervisory bodies are unwilling or unable to protect them.[69] Given the critical and common issues that arise with WEF projects, there is a need for responsible administrative authorities to develop common and coordinated frameworks that will allow local communities to take part in, and influence, decision-making processes relating to WEF infrastructure projects and how possible short- and long-term impacts to the local communities that rely on land, forest, food and water resources in the investment area are to be mitigated.[70] Over their entire life cycle of WEF projects, the channels through which the community and key stakeholders who may be affected by such projects will access information and make inputs must be clear, credible and reliable.

13.3.2.3 Weak Infrastructure and Technology for Knowledge Sharing

Implementing holistic responses to common and multiscale threats to WEF security will require effective sharing of data and information, in open linked databases and interlinked information and communication technology (ICT) systems by WEF institutions and actors. Such interlinked technological systems will aim to deliver an inventory of WEF programmes, facilities, stakeholders and hazards that are accessible in real time all levels of governance in order to facilitate timely and integrated decision-making.[71] However, due to lack of modern and interoperable ICT systems, the sharing of vital information relating to the nature, scale, overall impact, budgeting and statistical data on the scope of WEF programmes is often difficult to access across sectors in Nigeria.[72] The digitalization of government processes and records in these and other key sectors has progressed very slowly in Nigeria, which makes it difficult to develop coherent response measures and frameworks that cut across sectors. Furthermore, as a recent mapping indicates, the reluctance of various government actors to disclose relevant information, due to internal confidentiality measures, means that analogue modes of storing and processing relevant

[63] Ruggie (2011) paras. 8–12; UNEP (2012) p. 7.

[64] Government of Nigeria (2020a) 71–75; Olawuyi (2018) 73–104.

[65] Olawuyi (n.13). pp. 1–15.

[66] Ibid.

[67] Lammers (2020).

[68] Ibid.

[69] Olawuyi (2018) 73–104.

[70] Dada (2012) 67–90.

[71] Olawuyi (2020) 1–9.

[72] Government of Nigeria (2017) 70–72; also Government of Nigeria (2020a) 71–75, which notes insufficient data as a barrier to the coordination of SDG efforts in Nigeria including in WEF sectors.

information remain prevalent.[73] The end result is that upscaling ICT systems to enhance system modernizations and interoperability has not received sustained funding over the last several years. The COVID-19 pandemic has particularly highlighted the low level of ICT deployment in key government functions in Nigeria and the need to address this gap in order to improve the abilities of stakeholders in WEF sectors to better respond to future pandemics and disasters.[74]

The WEF nexus governance model provides a framework for mandating interoperability and information sharing across WEF sectors in order to enhance the timely and effective planning and delivery of urgently needed WEF projects. This will require interlinked information sharing platforms and databases which will allow regulators across the three domains to access relevant information on ongoing programmes and projects in a timely manner. Improving data sharing and interoperability across the WEF sectors should begin with an investigation of system modernization opportunities that exist to improve data collection and information sharing in the most effective and efficient way. Furthermore, governments and regulators will need to urgently allocate the required budgetary allocation and funding to accelerate investment in modern ICT infrastructure across the WEF domains.

13.3.2.4 Capacity Questions

Underlying the concerns relating to lack of open and interlinked information sharing system is dearth of technical capacity needed to coherently apply and interlink investment, human rights and environment and agricultural law regimes.[75] In Nigeria, for example, agriculture, water and human rights institutions continue to function separately without clear and in-depth consideration of the intersections and linkages of how water and agricultural projects may affect human rights and vice versa.[76] The lack of institutional coordination is often fuelled by the absence of a formal agenda or obligations to collaborate between actors, the lack of fluid programmatic activity between agencies and ministries, the lack of a coherent agenda between human rights and investment interest groups, the unwillingness by actors to collaborate across sectors and the absence of training and capacity to do so.[77] In Nigeria, for example, investment treaties are negotiated by the Ministry of Foreign Affairs which is largely comprised of diplomats, with little or no expertise on agriculture, water, environment, justice and human rights issues. Without cross-sectoral participation or linkage with relevant stakeholders, the end result is that the resulting investments may fail to adequately respond to complex land allocation and human rights puzzles that may result across the WEF sectors.

There is a need for increased training and capacity development across key sectors in order to enhance the abilities of responsible authorities to understand and implement WEF thinking. In addition to exposing key stakeholders to training on water, energy, food and human rights, there is a need for staff exchange and development of complimentary roles and functions in order to increase interoperability and coordination.[78] Fostering institutional coordination and constructive engagement between stakeholders in foreign investment, food, agriculture, energy and human rights ministries and agencies, can advance coherent implementation of programmes and projects to advance the SDGs. A multistakeholder approach to WEF management can enable relevant agencies and communities to identify joint initiatives and projects that can be effectively designed, financed and implemented together is a holistic way.

The above challenges underscore practical and on-the-ground challenges that stifle WEF nexus governance and overall progress in the

[73] Olawuyi and Tubondenyefa (2018) pp. 1–25.

[74] Olawuyi and Nalule (2020).

[75] Manrique Gil and Bandone (2013) pp. 3–12 identifying weak institutional capacity as a key reason for the violation of human rights across Africa pp. 3–12.

[76] Government of Nigeria (2017) 70–72; Government of Nigeria (2020a) pp. 43–45; also Olawuyi and Tubondenyefa (2018) pp. 1–25.

[77] Ibid.

[78] See Olawuyi (n.1).

realization of the SDGs on water, energy and food in Nigeria, as is the case in many other developing countries. The next section offers recommendations on how challenges to WEF nexus governance can be addressed through dynamic and innovative governance reforms.

13.4 Advancing Responsible Investment in Food, Water and Energy Sectors

The above gaps that hinder the advancement of WEF governance, especially human rights concerns, capacity and infrastructure deficit, discordant legislations and institutional isolation on energy, food and water investments, continue to exacerbate fragmentations and overlap in the implementation of WEF investments in Nigeria and in several West African countries.[79] Without a nexus governance approach, with human rights at its core, investment and projects designed to advance SDGs on water, energy and food may exacerbate scarcity of these resources and produce complex human rights violations. Lack of policy integration will not only compound WEF security problems on a grand scale but also frustrate concerted efforts to attain associated SDGs and increase other protracted WEF-related crises at regional and international levels caused by migration. A nexus governance approach requires that when implementing WEF development projects and investments, due care must be taken to ensure a holistic consideration of the implications of such projects in the context of promoting sustainable development across all sectors. The core philosophy behind a nexus governance approach is that the long-term feasibility and profitability of any investment activity are tied to multiscale social, environmental and economic advancement and to address adverse impacts; one must consider those factors at an initial stage so as to avoid shortfalls.[80]

Accordingly, clear and comprehensive regulatory frameworks are required to address the multidimensional implications of WEF investments and projects on human rights, local communities and livelihoods. Firstly, appropriate attention is required by investors and host states alike to ensure that water, energy and agricultural expansion projects and activities aimed at advancing WEF security clearly integrate and reference sustainable development standards in areas such as business and human rights, corporate social responsibility (CSR), labour, environment and anti-corruption, amongst others. For example, a new generation of international investment agreements (IIAs) and bilateral investment treaties (BITs) is emerging that clearly reference and integrate these often-neglected social standards.[81] By incorporating international standards and norms on human rights into investment treaties in WEF sectors, parties can avoid complex disputes and protests associated with isolated investment treaties that neglect the livelihoods and rights of local communities. Furthermore, as international law relating to business and human rights continues to evolve, reforms designed to better improve local conditions in investment communities are bound to emerge. It is therefore imperative for countries to integrate fair and equitable contractual mechanisms that can allow the state to update domestic WEF laws, especially on human rights, labour and CSR, without triggering treaty violations. Equilibrium clauses that allow both parties to renegotiate the agreement, in cases of human rights violations, will need to be integrated into

[79] See Lammers (2020) and also UNCTAD (2017), stating that there is fragmentation of the international legal governance system for investment more broadly.

[80] See PRI (2020), stating that 'investment policies should promote and facilitate the observance by investors of international best practices and applicable instruments of responsible business conduct and corporate governance'.

[81] For example, Burkina Faso-Canada BIT (2015); Colombia-Panama FTA (2013) which clearly reference 'internationally recognized standards' on labour, environment, human rights and anti-corruption. See also Morocco-Nigeria BIT (2016), EFTA-Georgia FTA (2016), CETA (2016) and Bosnia and Herzegovina-EFTA FTA (2013) which reference SDGs and corporate governance.

BITs and IIAs in order to avoid costly project delays, protests and investor-state arbitrations.

Secondly, steps must be put in place to ensure that all stakeholders that may be impacted by an investment activity are given transparent and adequate opportunities to take part in, and influence, decision-making processes leading to the adoption of investment treaties. BITs and IIAs can require the host state to demonstrate that the free, prior, informed consent of the stakeholders that can be impacted by the treaty has been obtained. Furthermore, investors should expressly commit to providing frequent, timely and detailed reporting on how environmental standards are applied in project approval, construction and operations and closure or decommissioning all through the entire life cycle of the project. In addition to conducting EIAs and HRIAs prior to the commencement of the investment activity, such assessments should be conducted with the active engagement of the public concerned. Besides environmental impact assessments (EIAs), social impact assessments (SIAs) should be conducted so as to foster acceptance and trust between local communities and WEF-related projects. Doing so will forestall the likelihood of community protests, sabotage and other security-related problems as the host communities are more likely to see themselves as key stakeholders in the WEF projects.[82]

Thirdly, ensuring coherence in the design and implementation of investment treaty provisions on energy, food and water will require the integration of relevant bodies and agencies. To advance institutional coordination and cooperation in the design and implementation of programmes, projects and policies across relevant sectors, gaps to information exchange, resource sharing and interoperability must be carefully examined and addressed. This can be done by establishing a joint committee or task force that will coordinate knowledge, expertise and information sharing across the sectors.[83] For example, such a coordinating body with a wide range of sectoral experts can be involved in the process of treaty negotiation, as well as in key phases of project approval and implementation. Similarly, enhancing cross-sectoral access to relevant project data through data sharing platforms can help the relevant actors to detect conflicting projects, rules and procedures.

Furthermore, to reduce administrative bureaucracies and bottlenecks that delay project implementation, WEF institutions at domestic levels will need to pool resources together through multi-stakeholder planning approaches. Achieving this will require the establishment and empowerment of compliance committees and review teams who would be responsible for approving projects and monitoring compliance across WEF sectors. Such monitoring institutions must be independent and should not be easily influenced by governmental interests or pressure, for example, through a monitoring committee or expert review team who would inspect project locations to ensure that projects are being executed based on approved guidelines and methodologies. Such review institutions should be equipped with the resources to perform spot assessments, fact-findings and investigations, such that they can gather first-hand information on the true impacts of the investment activity on water, food and energy projects and how they affect human rights and other sustainability issues of host communities. One issue that must be considered is the problem of governmental repression through withholding of funds required to enable such institutions carry out their functions at an optimal level. It is recommended that funding for such institutions should be separate from state executive budgets in order to ensure independence.

Fourthly, investment in research and development on water, energy and food infrastructure and technologies is required to fill the infrastructure deficit across WEF sectors. Across West Africa, infrastructure for gathering statistical data on water, energy and food is mostly lacking and, where present, is in dilapidated conditions. This makes management and accountability of WEF resources difficult or almost impossible. This problem is typically institutionalized in African countries which relegate basic human rights needs such as water and food from the

[82] Lindsay and Kirkpatrick (2019) 113.

[83] See Olawuyi, note 1.

realm of human rights to making them mere directive principles of state policy which are non-justiciable, thereby creating room for poor government funding and lack of sustained implementation.[84] Hence, there is the need for deliberate and committed funding for infrastructure development as well as sustained legal and institutional framework insulated from regime change and neglect. However, this can only be achieved if foundational legislation relating to WEF needs is viewed from the human rights lens and constitutionally recognized, such that citizens can have greater legal authority and safeguard to demand for greater investment in WEF infrastructure projects. Sustained funding on research in WEF sectors can improve the availability and reliability of statistical data needed to improve policymaking and interoperability across WEF sectors which will accelerate progress on the SDGs.

13.5 Conclusion

The attainment of sustainable development, and all of the SDGs, will be impossible without water, energy and food security. There is therefore an urgent need for improved governance of WEF resources in order to coherently address multi-scale threats to their availability, affordability and accessibility. The WEF governance approach provides a framework for interlinking investments, legislation and policies relating to WEF to accelerate sustainable development and reduce adverse human rights impacts. To ensure that WEF governance approach moves from theory to successful practical integration and adoption, fragmented legal structures, inadequate institutional coordination, weak technological infrastructure and capacity gaps that stifle the development and application of hybrid and linked rules, procedures and processes across the sectors will have to be comprehensively addressed.

Partnership is required across all levels of governance in order to meet the urgent sustainability threats in critically important WEF sectors. Governments, institutes and consumers must build relationships that are based on mutual trust, resource exchange and development of scalable programmes that reduce role duplication and conflictive overlaps. One common factor is that investment in innovation and technical support is vital so that a range of solutions can be developed by WEF actors. However, this can only happen if the approach is turned into actions through sustained funding and investment in research, development and innovation programmes to accelerate the digitization and interoperability of key institutions.

[84] See, for example, Chapter II, Constitution of the Federal Republic of Nigeria (1999) (as amended).

References

Adebayo, B. (2018). *Nigeria overtakes India in extreme poverty ranking (Cable News Network, June 26, 2018).* https://edition.cnn.com/2018/06/26/africa/nigeria-overtakes-india-extreme-poverty-intl/index.html

Beisheim, N. S. (2016). *Multistakeholder partnerships for implementing the 2030 Agenda: improving accountability.*

Bloomberg. (2020). *Congo, Nigeria are among worst hotspots for hunger*, https://www.bloomberg.com/news/newsletters/2020-09-18/supply-chains-latest. Accessed 12 Mar 2022.

Cairns, R., & Krzywoszynska, A. (2016). Anatomy of a buzzword: The emergence of 'the water-energy-food nexus' in UK natural resource debates. *Environmental Science & Policy, 64*, 164–170.

Constitution of the Federal Republic of Nigeria 1999 (as amended).

Cordes, K., & Bulman, A. (2016). Corporate agricultural investment and the right to food: Addressing disparate protections and promoting rights-consistent outcomes. *UCLA Journal of International Law and Foreign Affairs, 20*(1), 87–162.

Dada, J. (2012). Impediments to human rights protection in Nigeria. *Annual Survey of International & Comparative Law, 18*(1), 67.

ECREEE, Ecowas Centre for Renewable Energy and Energy Efficiency. (2010). *Managing Water, Food and Energy (WEF) Resources for Sustainable Development in the ECOWAS Region.*

Food and Agricultural Organization (FAO). (2011). *'Energy-Smart' food for people and climate, Rome.* www.fao.org/3/i2454e/i2454e00.pdf. Accessed 25 Feb 2021.

Gehne, K., & Brillo, R. (2014). *Stabilization clauses in international investment law: Beyond balancing and fair and equitable treatment* (Working paper no 2013/46) (pp. 3–5).

Ghana Statistical Service. (2010). *Population and housing census report*. Ghana Statistical Service.

Gil, M., & Bandone, A. (2013). *Human rights protection mechanisms in Africa: Strong potential, Weak Capacity*. European Parliament.

Global Hunger Index. (2021). *Nigeria*. https://www.globalhungerindex.org/nigeria.html. Accessed 17 Mar 2021.

Government of Nigeria. (2017). *Implementation of the SDGs: A national voluntary review*. https://www.thecommonwealth.io/wp-content/uploads/2020/08/NigeriaVNR2017.pdf

Government of Nigeria. (2020a). *Nigeria: Integration of the SDGs into National development planning: A second voluntary national review*. The Presidency. Available online: https://sustainabledevelopment.un.org/content/documents/2630 9VNR_2020_Nigeria_Report.pdf. Accessed 15 Sept 2021.

Government of Nigeria. (2020b). *Nigeria Economic Recovery and Growth Plan (ERGP), 2017–2020*. Available online: https://statehouse.Gov.ng/policy/economy/economic-recovery-and-growth-plan/. Accessed 15 Sept 2021.

Government of Nigeria. (2020c). *Making Nigeria open-defecation-free by 2025: A national road map (Federal Ministry of Water Resources)*. https://www.unicef.org/nigeria/media/1491/file/Nigeria-making-Nigeria-open-defecation-free-by-2025.pdf.pdf

International Energy Agency (IEA). (2019). *African Energy Outlook 2019*. https://www.iea.org/reports/africa-energy-outlook-2019. Accessed 21 Mar 2022.

International Law Association. (2018). *Report of the Committee on sustainable development and the green economy in international trade law*.

Kriebaum, U. (2007). Regulatory taking: Balancing the interests of the investor and the state. *Journal of World Investment and Trade, 8*, 717–744.

Lammers. (2020). *A just energy transition for Africa: Mapping the impacts of ECAs active in energy sectors in Ghana*. Nigeria. https://www.bothends.org/uploaded_files/document/A_Just_Energy_Transition_for_Africa_November_2020.pdf

Leck, H., et al. (2015). Tracing the water–energy–food nexus: Description, theory and practice. *Geography Compass, 9*(8), 445–460.

Lindsay, R., & Kirkpatrick, A. (2019). *Global arbitration review: The guide to mining arbitrations* (J. Fry & L.-A. Bret, Eds.) (p. 113). Law Business Research Ltd.

Maniruzzaman, A. F. M. (2008). The pursuit of stability in international energy investment contracts: A critical appraisal of the emerging trends. *The Journal of World Energy Law & Business, 1*(2), 121–157.

Maupin, A., & Mwanikah Ojoyi, M. (2017). *Africa // Africa needs to manage food, water and energy in a way that connects all*. Africa // Africa needs to manage food, water and energy in a way that connects all three | Nexus – The Water, Energy & Food Security Resource Platform (water-energy-food.org). Accessed 17 Mar 2021.

Mbengue, M., & Waltman, S. (2015). *Farmland investments and water rights: The legal regimes at stake*.

McConnell, T. (2021). A locust plague hit East Africa. *The pesticide solution may have dire consequences*. https://www.msn.com/en-xl/africa/other/a-locust-plague-hit-east-africa-the-pesticide-solution-may-have-dire-consequences/ar-BB1fgsKx?ocid=msedgntp. Accessed 25 Feb 2021.

Oche, A. (2020). Comparative analysis of green bond regimes in Nigeria and China. *Journal for sustainable Development Law and Policy, 11*(1), 1–2.

Oguamanam, C. (2015). Sustainable development in the era of bioenergy and agricultural land grab. In S. Alam et al. (Eds.), *International environmental law and the global south*. Cambridge University Press.

Olawuyi, D. (2015). Legal strategies and tools for mitigating legal risks associated with oil and gas Investments in Africa. *Opec Energy Review (Blackwell Wiley Publishers), 39*(3), 247–265.

Olawuyi, D. (2016). *The human rights-based approach to carbon finance* (pp. 1–25). Cambridge University Press.

Olawuyi, D. (2018). Energy (and human rights) for all: Addressing human rights risks in energy access projects. In R. Salter et al. (Eds.), *Energy justice: US and international perspectives (Edward Elgar)* (pp. 73–104).

Olawuyi, D. (2020). Sustainable development and water-energy-food nexus: Legal challenges and emerging solutions. *Journal of Environmental Science & Policy, Elsevier, London, 103*(1).

Olawuyi, D. (2021). *Local content and sustainable development in global energy markets*. Cambridge University Press.

Olawuyi, D. (2022). *The role of natural gas in a just and equitable energy transition. In Olawuyi D & Pereira, E., The palgrave handbook of natural gas and the global energy transition (pp. 77–95)*. London, Palgrave Macmillan.

Olawuyi, D., & Nalule, V. (2020). *Ensuring universal access to modern energy services in times of pandemic related disruptions: Key considerations for policymakers*.

Olawuyi, D., & Nsoh, W. (2020). Water energy food nexus. In S. Bagg, P. Galvao, B. Tileman, & A. Lucas (Eds.), *Environmental law and policy* (4th ed.). Emond Montgomery Publishing.

Olawuyi, D., & Olusegun, O. (2018). Achieving the United Nations sustainable development goals on biological diversity in Nigeria: Current issues and future directions. *Global Journal of Comparative Law (Brill), 7*(1), 37–60.

Olawuyi, D., & Tubodenyefa, Z. (2018). *Review of the environmental guidelines and standards for the petroleum industry in Nigeria* (commissioned by the DFID Facility for Oil Sector Transformation (FOSTER) Project; and published by the Institute for Oil, Gas, Energy, Environment and Sustainable Development).

Pavoni, R. (2010). Mutual supportiveness as a principle of international and law-making: A watershed for the "WTO-and-competing-regimes" debate?. *European Journal of International Law, 21*(3), 649–679. https://doi.org/10.1093/ejil/chq046.

Premium Times. (2017). Nigeria Approves Construction of 3,050MW Mambilla Electricity Project at $5.7 billion (*Premium Times*, September 4, 2017).

PRI. (2020). *What are the principles for responsible investment.* https://www.unpri.org/pri/an-introduction-to-responsible-investment/what-are-the-principles-for-responsible-investment. Accessed 7 July 2020.

Qian, B., et al. (2012). Changing growing season observed in Canada. *Climatic Change, 112*(2), 339.

Richards, M. (2013). *Social and environmental impacts of agricultural large-scale land acquisitions in Africa—With a focus on West and Central Africa.* 24–27 (Rights & Resources Initiative). http://www.rightsandresources.org/documents/files/doc_5797.pdf. Accessed 27 Feb 2021.

Ringler, C., et al. (2013). The nexus across water, energy, land and food (WELF): Potential for improved resource use efficiency? *Current Opinion in Environmental Sustainability, 5*(6), 617.

Ruggie, J. (2011). United Nations guiding principles on business and human rights: Implementing the United Nations. *'Protect, Respect and Remedy' Framework.*

Schrijver, N. (2017). Advancements in the principles of international law on sustainable development. In M. C. CordonierSegger & C. G. Weeramantry (Eds.), *Sustainable development principles in the decisions of international courts and tribunals, 1992–2012.* Routledge.

Stockholm Resilience Centre. (2018). *Transformation is feasible: How to achieve the sustainable development goals within planetary boundaries.* https://www.stockholmresilience.org/download/18.51d83659166367a9a16353/1539675518425/Report_Achieving%20the%20Sustainable%20Development%20Goals_WEB.pdf

Tanle, A. (2020). Rural-urban migration and household livelihood in the Agona West Municipality, Ghana. *Journal of Geography and Regional Planning, 13*(1), 1–18.

Terrapon-Pfaf, J. (2018). Energising the WEF nexus to enhance sustainable development at local level. *Journal of Environmental Management, 223*, 409–416.

UNEP. (2012). UN-REDD programme guidelines on free prior informed consent, p. 7. https://www.unccelearn.org/wp-content/uploads/library/un-redd05.pdf. Accessed 22 Mar 2022.

UNCTAD. (2017). *Phase 2 of IIA reform: Modernizing the existing stock of old-generation treaties.* https://unctad.org/en/PublicationsLibrary/diaepcb2017d3_en.pdf. Accessed 7 July 2020.

United Nations. (2015). *Transforming our world: The 2030 Agenda for Sustainable Development, Resolution adopted by the General Assembly on 25 September 2015, a/res/70/1.*

United Nations. (2018). Water-Energy-Food Nexus for the Review of SDG 7 (Policy Brief 9). https://sustainabledevelopment.un.org/content/documents/17483PB9.pdf. Accessed 22 Mar 2022.

Wall, E., et al. (Eds.). (2007). *Farming in a changing climate: Agriculture adaptation in Canada* (p. 273). UBC Press.

World Bank. (2013). *What climate change means for Africa, Asia, and the coastal poor.* World Bank. https://www.worldbank.org/en/news/feature/2013/06/19/what-climate-change-means-africa-asia-coastal-poor

The Food-Water-Renewable Energy Nexus Resource Security Examples for Asia-Pacific Cities

14

K. M. Nazmul Islam and Steven J. Kenway

Abstract

Population growth, economic development, and climate change pose critical challenges to water, energy, and food security of the cities in emerging nations of the Asia-Pacific region. The unprecedented urbanization growth in the Asia-Pacific region during the past century has triggered the need for food-water-energy nexus thinking in cities. The biggest 100 cities of the Asia-Pacific region together are contributing annually around 20% of global gross domestic product (GDP) and 29% of GDP growth. Over half of the populace of this Asia-Pacific region will be living in cities by 2025. Asia-Pacific region cities will need to increase their food supply by 77% by 2050, and consequently, water and energy consumption will increase. This chapter deals with the food-water-renewable energy nexus in some Asia-Pacific cities. The interaction of Sustainable Development Goals (SDGs) and food-water-energy nexus with a focus on renewable energy is mentioned first. The status and progress of renewable energy use for water, energy, and food are then reviewed. Recent research is highlighted, and key challenges and opportunities such as fragmented and inconsistent policies are discussed. We conclude that cities need to make decisions to ensure integrating renewable energy for the city-level water-energy-food nexus, and achievement of SDGs, such as sustainable cities and communities (SDG 11) and climate action (SDG 13).

Keywords

Sustainable investment · developing countries · WEF nexus · Climate change · Human rights

14.1 Introduction

The food system is an interconnected process of agricultural production, food processing and transportation, preparation and cooking, and disposal. Throughout this system, water and energy are interconnectedly consumed. For example, energy is used for (a) irrigation in the agriculture phase; (b) water heating, cooling, sterilization, softening, and wastewater treatment in the food processing phase; and (c) cooking and dishwashing at the household (Islam et al., 2021).

K. M. N. Islam (✉)
School of Chemical Engineering, the University of Queensland, St Lucia, Australia

Agriculture and Food, Commonwealth Scientific and Industrial Research Organisation (CSIRO), St Lucia, Australia
e-mail: kmnazmul.islam@uq.net.au

S. J. Kenway
Advanced Water Management Centre (AWMC), the University of Queensland, St Lucia, Australia

L. L. B. Lazaro et al. (eds.), *Water-Energy-Food Nexus and Climate Change in Cities*, Sustainable Development Goals Series, https://doi.org/10.1007/978-3-031-05472-3_14

Why a nexus approach is needed?

"Water development and management programmes, if planned properly, can serve multiple functions, from contributing to energy and food production to helping communities adapt to climate change. A nexus approach to sectoral management, through enhanced dialogue, collaboration and coordination, is needed to ensure that co-benefits and trade-offs are considered and that appropriate safeguards are put in place"

UN World Water Development Report 2014 Vol.1.

Awareness about this linkage among water, energy, and food system is growing; it is now widely established that one sector-oriented policy would inadvertently affect others (say, national desalination policy can increase the national energy consumption) (Kenway et al., 2019a). Hence, to ensure strong sustainability along the food system globally, there is a need to act and take decisions through integrated approaches due to rising food demand from population, economic, and consumption growth (Yuan et al., 2021). We need to ensure a balance among the water-energy-food systems and different goals and interests of associated stakeholders.

The notable growth in urbanization around the world during the past century has provoked the need for water-energy-food nexus thinking in cities. The number of people living in cities has increased by four times during the last 60 years. Around 57% of the global population now live in the urbanized area compared to around 15% in 1900 and 34% in 1960. By 2050, it is projected to reach 68% (Ritchie & Roser, 2018). It took 25 years (1960–1985) to reach the urban population of 1–2 billion, while it took 12 years (2003–2015) for reaching 3–4 billion urban population (Satterthwaite, 2020). This growth of urbanized people is a major challenge to ensure environmental sustainability, making the study of the water-energy-food nexus important for climate-friendly food systems in cities and increasing resource efficiency (Islam et al., 2021).

Understanding the renewable energy use in the context of the water-energy-food nexus in cities is important for addressing the global challenge of climate change mitigation (Kennedy et al., 2015). Utilizing renewable energy use for food systems (agriculture, industrial food processing, and cooking as residential end use) is particularly critical for large cities, due to their sheer size and complexity (Scott et al., 2015). Here, we have discussed the water, energy, and food flows of some selected Asia-Pacific cities. City-scale correlations are discussed, followed by renewable energy perspective in Sustainable Development Goals (SDGs). To show the progress of renewable energy use for water, energy, and food, we discussed the renewable energy use progress and targets, as well as supporting targets at the city scale in the selected Asia-Pacific cities. Finally, the challenges and opportunities of water, energy, and food nexus focusing on renewable energy for holistic resource security are highlighted.

14.2 Food-Water-Energy Flows

The selected 18 cities covering high-income, upper-middle-income, and lower-middle-income emerging nations in the Asia-Pacific region had a combined population of 241.5 million in 2016, equal to 3.25% of the global population (7.424 billion). Their combined gross domestic product (GDP) was USD 4.063 trillion in 2016, which was higher than the population percentage, at 5.33% of global GDP (USD 76.16 trillion) (Table 14.1).

The per capita food consumption in the cities of high-income countries of the Asia-Pacific regions ranged from 11,720 to 13,470 kJ/person/day, in comparison with the global average of 12,300 kJ/person/day in 2015 (FAO, 2003). In the cities of the upper-middle-income category, the range was 10,630–12,510 kJ/person/day, and for the lower-middle-income category, the range was 9500–11,630 kJ/person/day (Table 14.1). The estimated total water use in 2016 of the

Table 14.1 Population, contribution to national GDP, food, water, and energy consumption of selected major cities of Asia-Pacific countries

Economic status[a]	Major cities	Country	2016 estimated population (mill)[b]	Contribution to national GDP (%)[c]	Daily dietary consumption (kJ/person/day)[d]	2016 estimated total water use (GL)[e]	2016 estimated total energy use (PJ)[e]
High income	Tokyo	Japan	38.2	19	11,720	4325	2647
	Sydney	Australia	5.2	24	13,470	603	738
	Melbourne	Australia	4.9	19	13,470	403	1165
	Auckland	New Zealand	1.4	38	11,760	157	300
	Singapore	Singapore	5.7	100	13,000	314	1221
	Seoul	South Korea	9.9	21	12,720	1134	804
Upper middle income	Shanghai	China	24.5	4	12,510	5423	2776
	Beijing	China	21.3	3.5	12,510	4508	1519
	Kuala Lumpur	Malaysia	7.1	17	12,090	746	369
	Jakarta	Indonesia	10.5	18	10,670	330	410
	Bangkok	Thailand	9.4	48	10,630	1086	364
Lower middle income	Delhi	India	26.5	5	9870	3383	553
	Mumbai	India	21.4	7	9870	3550	184
	Dhaka	Bangladesh	18.3	40	9500	785	446
	Karachi	Pakistan	15.2	42	9540	623	373
	Manila	Philippines	13.1	33	10,790	1354	604
	Ho Chi Minh City	Vietnam	7.5	23	11,630	528	64
	Kathmandu	Nepal	1.4	45	9790	43	22

Note:
[a]Economic status of the country is based on the World Bank classification taken from World Bank open database
[b]Population data is based on *The World's Cities in 2016* published by the United Nations. https://www.un.org/en/development/desa/population/publications/pdf/urbanization/the_worlds_cities_in_2016_data_booklet.pdf
[c]Contribution to national GDP is based on the national statistics of the respective country
[d]The country-level daily dietary consumption (kJ/person) data is used because of the absence of uniform data at the city level. The data is based on the FAO Food Consumption Nutrients spreadsheet (2008)
[e]For the cities Tokyo, Delhi, Shanghai, Mumbai, Beijing, Dhaka, Karachi, Manila, Jakarta, and Seoul, the total estimated water and energy use was sourced from Kennedy et al. (2015). For all other cities, data are based on national/regional statistics and news reports

selected Asia-Pacific cities was 29,296 GL, which is about 1.5% of global water use (2.6 million GL) (FAO, 2021). The highest water consumption was in Shanghai (5423 GL with a population of 24.5 million), followed by Beijing (4508 GL with a population of 21.3 million), Tokyo (4325 GL with a population of 38.2 million), Mumbai (3550 GL with a population of 21.4 million), and Delhi (3383 GL with a population of 26.5 million). Although Tokyo is the largest megacity, with around 38 million people, its water consumption is surpassed by Shanghai and Beijing. To give those numbers a perspective, let us consider the Wivenhoe Dam (Queensland, Australia), which can hold 1165 GL of water at the full supply level. So, the annual water consumed in Shanghai, Beijing, and Tokyo is around 4–5 times higher than the water-holding capacity of Wivenhoe Dam (Queensland, Australia) at the full supply level.

The direct energy consumption of these 18 cities was 14,560 PJ, which is ~4% of global energy consumption (393,892 PJ) (IEA, 2021), whereas the embodied energy consumption can be 3–4 times higher than the direct energy consumption (Long et al., 2017). Annual energy consumption in these cities for 2016 ranges from 22 PJ for Kathmandu (population 1.4 million) to ~2776 PJ for Shanghai (population 24.5 million). Similar to water, energy consumption in Tokyo is surpassed by Shanghai and Beijing. Three other cities – Melbourne, Singapore, and Seoul – consumed annually higher than 800 PJ energy. As an example, an oil supertanker can hold about 12.2 PJ of oil (Kennedy et al., 2015). So, Shanghai consumes the energy equivalent to one supertanker in around 1.8 days, whereas for Melbourne, the energy consumption is equivalent to one supertanker in about four days. Direct and indirect energy inputs at different stages of the food sector globally are around 24% (95,000 PJ/year) of total global energy consumption in 2015 (Ferroukhi et al., 2015). So, considering that in these 18 Asia-Pacific cities, direct and indirect energy inputs at different stages of the food sector were around 3495 PJ.

14.3 City-Scale Analysis of Food-Water-Energy Flows

Per capita water use in the selected cities of the Asia-Pacific region ranges from 31,025 liters/person/year for Kathmandu to 221,337 liters/person/year for Shanghai. It is notably higher in Shanghai, Beijing, Mumbai, and Delhi when we compared water use with dietary inputs. Given the total water use, the highest user is in Tokyo, Shanghai, Beijing, Mumbai, and Delhi than in the other selected cities of the Asia-Pacific region when we compared the water use with dietary inputs (Fig. 14.1a). Interestingly, cities of the high economic country, such as Sydney, Melbourne, Singapore, and Seoul, are comparatively high water efficient than in the other selected cities of the Asia-Pacific region, when we compared the water use with dietary inputs (Fig. 14.1a). The high water use efficiency is perhaps because these cities are highly invested in water governance, monitoring, and associated efficiency schemes. For energy, the opposite pattern is overserved. Cities of high-economic countries, such as Sydney, Melbourne, and Singapore, are comparatively less energy efficient than in the other selected cities of the Asia-Pacific region when we compared the energy use with dietary inputs (Fig. 14.1d).

There is a positive correlation among food, water, and energy consumption at the city scale for the selected cities of the Asia-Pacific region. The causes of cities of high economic countries such as those of Sydney, Melbourne, Tokyo, and Singapore suggest that energy consumption with the increase in food consumption will likely to increase in the cities of the low-income and middle-income country tremendously.

14.4 Food-Water Renewable Energy Perspective on SDGs

The Sustainable Development Goals (SDGs) were introduced during the Rio + 20 Conference of the United Nations in 2012 (Dincer, 2000). SDG 7 is focused on access to affordable, reli-

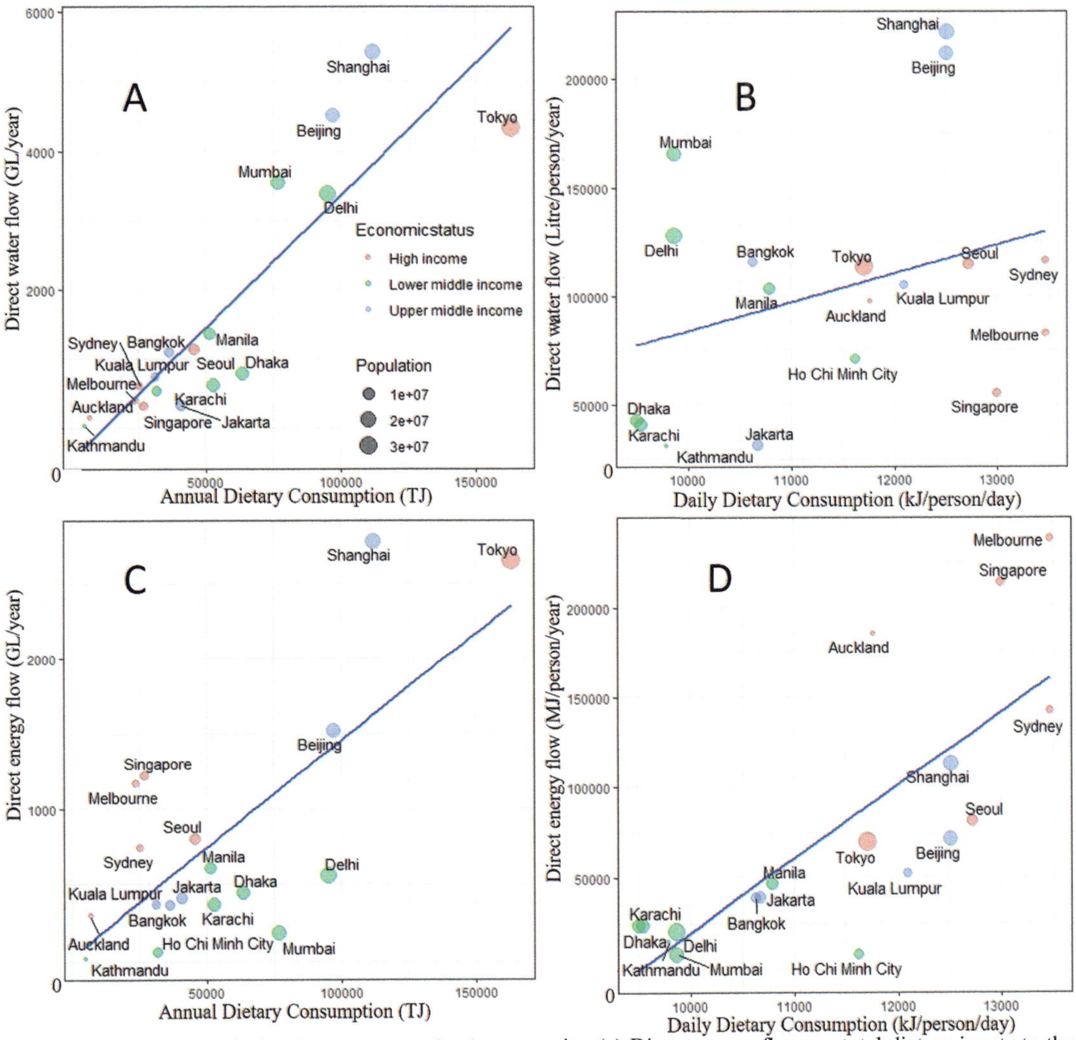

Fig. 14.1 City-scale analysis of water-energy-food consumption in the selected cities of the Asia-Pacific region. (**a**) Direct water flow vs. total dietary inputs to the city. (**b**) Per capita water flow vs. per capita dietary inputs to the city. (**c**) Direct energy flow vs. total dietary inputs to the city. (**d**) Per capita energy flow vs. per capita dietary inputs to the city. (Source: Authors based on the Table 14.1 data)

able, sustainable, and modern energy with three targets: (i) ensuring universal access to energy services (SDG 7.1), (ii) increasing the share of renewables in the energy mix (SDG 7.2), and (iii) improving energy efficiency (SDG 7.3). While SDG 6 is focused on ensuring availability and sustainable management of water and sanitation for all, SDG 2 is focused on ending hunger, achieving food security, improving nutrition, and promoting sustainable agriculture. These three SDGs are interconnected, for example, to meet the food security target (SDG 2); our society also needs to comply with the sustainable water target (SDG 6) and clean energy target (SDG 7). The same interconnection exists between SDG 6 and SDG 7 (Fig. 14.2). Hence, to achieve those SDGs, nexus consideration is obvious (Gupta, 2017).

One of the key aspects of SDG 7 is renewable energy, and the use of renewable energy to ensure food security (SDG 2) and sustainable water and sanitation (SDG 6) is critically important for SDG 11 (sustainable cities and communities), SDG 12 (responsible consumption and production), and

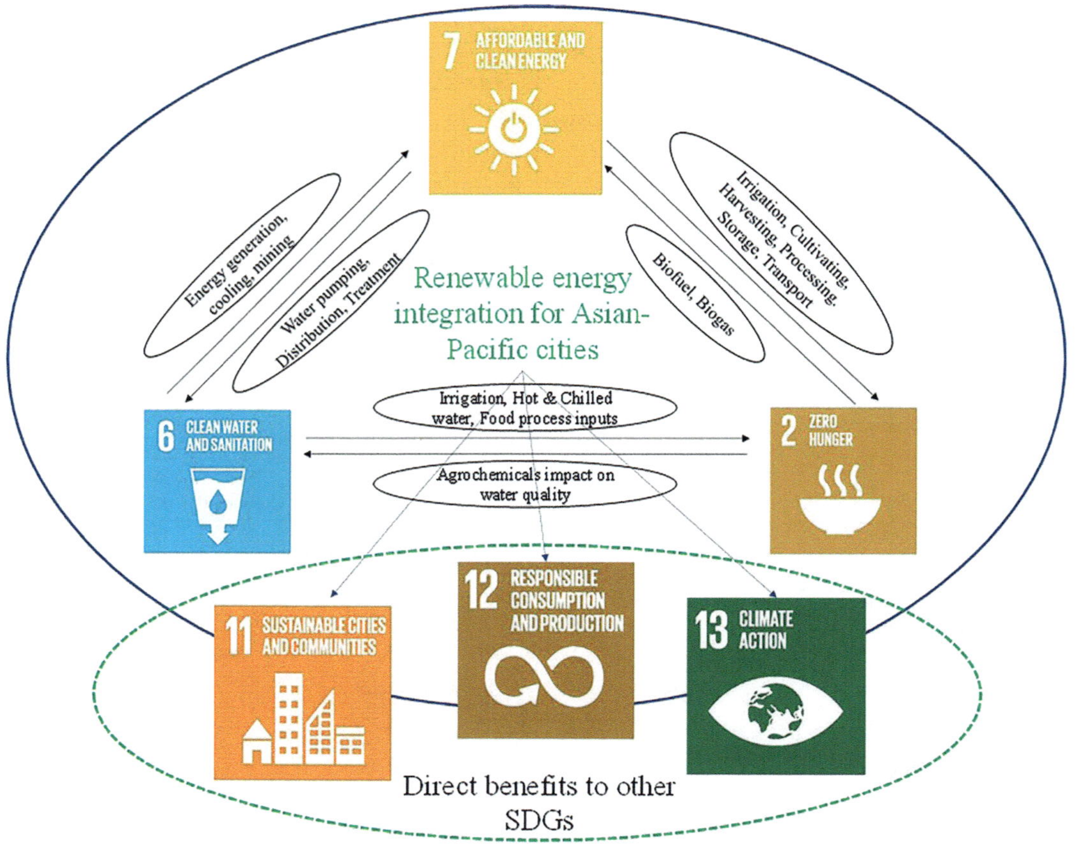

Fig. 14.2 Interconnection of food, water, and energy and direct benefits of renewable energy integration for this interconnection on SDGs. (Source: Authors)

SDG 13 (climate action). Globally, energy demand will be doubled by 2050, while water and food demand rise will be over 50%. Most of this growth will be in the cities, and in particular, the growth will be in Asia and African regions (FAO, 2014). Given these challenges, cities around the world should make a balance among the competing needs for limited resources considering the greatest global threat of climate change. Renewable energy utilization can play a central role to overcome these challenges and in achieving the SDGs mentioned above.

The available renewable energy technologies can provide a cost-effective and sustainable supply of energy. The rapid growth of renewable energy technologies can also initiate sustainable changes in the water and food sectors due to substantial spillover effects. Eventually, such changes will influence SDGs 11, 12, and 13. For example, the diffusion of renewable energy technologies will reduce 20% of power generation-related water withdrawals globally by 2030. Power generation-related water withdrawals will be reduced by about 50% in the UK from solar photovoltaic- and wind energy-based electricity; more than 25% in the USA, Germany, and Australia; and over 10% in India by 2030 (Ferroukhi et al., 2015). Water will be also accessible for domestic and agricultural purposes because of renewable energy technologies (say, renewable energy use for irrigation), thus decoupling the growth in the water and food sector avoiding fossil fuels (FAO, 2017). Renewable integration along with the food system can improve efficiency and productivity, and the agri-food sector can contribute further to bioenergy development from agricultural residues and food wastes (Berardy and Chester, 2017; FAO, 2017). This in turn will help toward reducing poverty (SDG 1) by transforming rural economies, enhance energy security (SDG 7), and contribute to other SDGs, such as SDGs 11, 12, and 13.

14.5 Food-Water-Renewable Energy Nexus: Status and Progress of Renewable Energy Use

The population of the Asia-Pacific region is now about 4.5 billion, and urbanization is increasing with on average 3% annual growth of urban population. This urbanized population growth leads to the increase in energy demand in the cities of this region due to associated growth in food consumption, industrial production, and transport. In the Asia-Pacific regions, countries under the high-income category showed higher levels of per capita energy consumption (UNESCAP, 2018). In this region, fossil fuel still dominates the final energy consumption (more than 50%). When traditional biomass is excluded, the modern renewable share is growing, but still low (Fig. 14.3). Electricity generation in this region is largely from coal-based power plants, which represents more than 50% of the total power generations (UNESCAP, 2019).

Traditional cooking fuel and technology are still dominant in the Asia-Pacific region because still around 2.1 billion people (around 47% of the total population) rely on it. Traditional cooking fuels such as wood, dung, and charcoal are dominant in the rural areas of this region (UNESCAP, 2018). In 2016, countries such as Australia, Japan, Malaysia, New Zealand, Singapore, and South Korea had 100% clean cooking (electricity, LPG (liquefied petroleum gas), natural gas, and biogas) access rates. The regional rate of access to clean cooking was around 53%. In some of the middle-income countries (say, China and Indonesia), clean cooking access rate was around 60%, but for other low-income countries in this region, the access rate is around 30–40%, well below the pace required to achieve universal access by 2030 (Fig. 14.4). In the Asia-Pacific region, which does not have 100% access to clean cooking fuel (electricity, LPG, natural gas, and biogas) at the country level, clean cooking fuel comparatively dominates in the urban areas (Fig. 14.5).

Cities of the high-income countries in this region demonstrated priority for energy efficiency and the use of renewable energy, as indicated by the declaration of national- and city-level

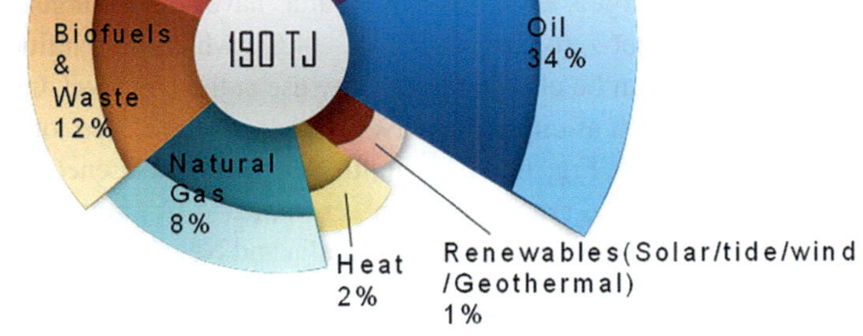

Fig. 14.3 Final energy consumption (in TJ) in Asia-Pacific region by product in 2018. (Data source: IEA, 2021)

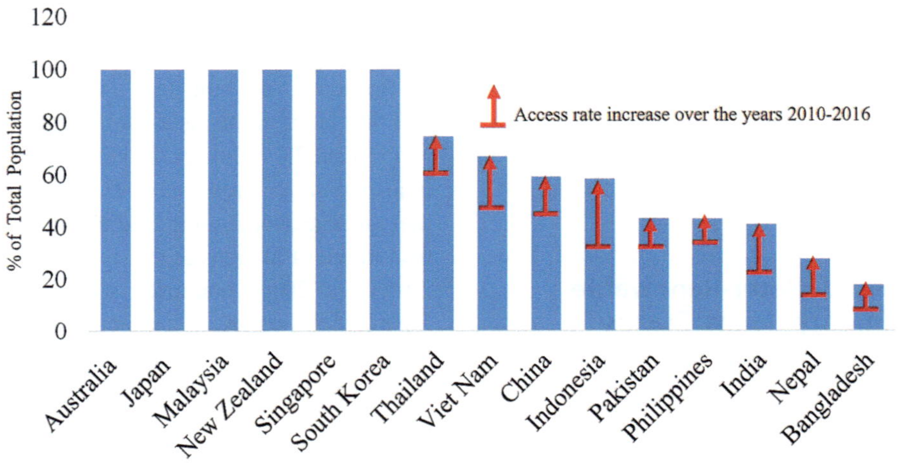

Fig. 14.4 Population percentage with access to clean cooking in selected Asia-Pacific countries, 2016. (Data source: UNESCAP, 2018)

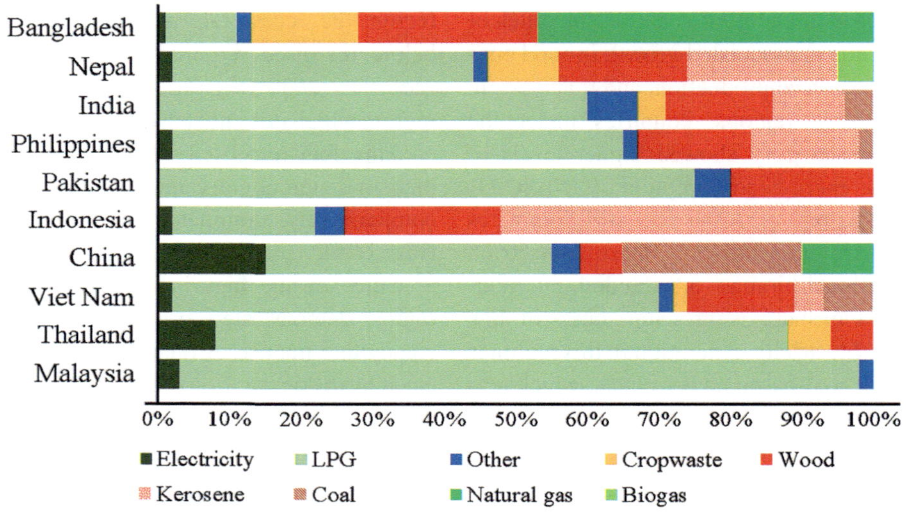

Fig. 14.5 Cooking fuel mix in the urban areas of selected Asia-Pacific countries, 2016. (Data source: UNESCAP, 2018)

targets (UNESCAP, 2019). Those targets have a different perspective, such as some of the cities have declared renewable energy targets and policies and some other declared net-zero targets (Fig. 14.6). Again, these targets can be classified based on their scale and focus, such as city-wide and city authority operation or both (Fig. 14.7).

Thirteen cities of Australia declared renewable energy targets and policies, and 11 of them have 100% renewable energy targets. Out of these, Adelaide, Sydney, and Newcastle have achieved the targets of 100% of renewable energy for municipal operations as of 2020. In comparison with renewable energy targets, nine cities in Australia have other targets such as net-zero (Fig. 14.6), which can influence the renewable energy use at the city level. Cities in Australia and New Zealand have also other enabling targets influencing renewable energy at the city level. As an example, 114 municipal governments in Australia and New Zealand had declared a climate emergency by the end of 2020 (REN21, 2021).

Net-zero targets in 103 cities of Japan are declared by the end of 2020 in comparison

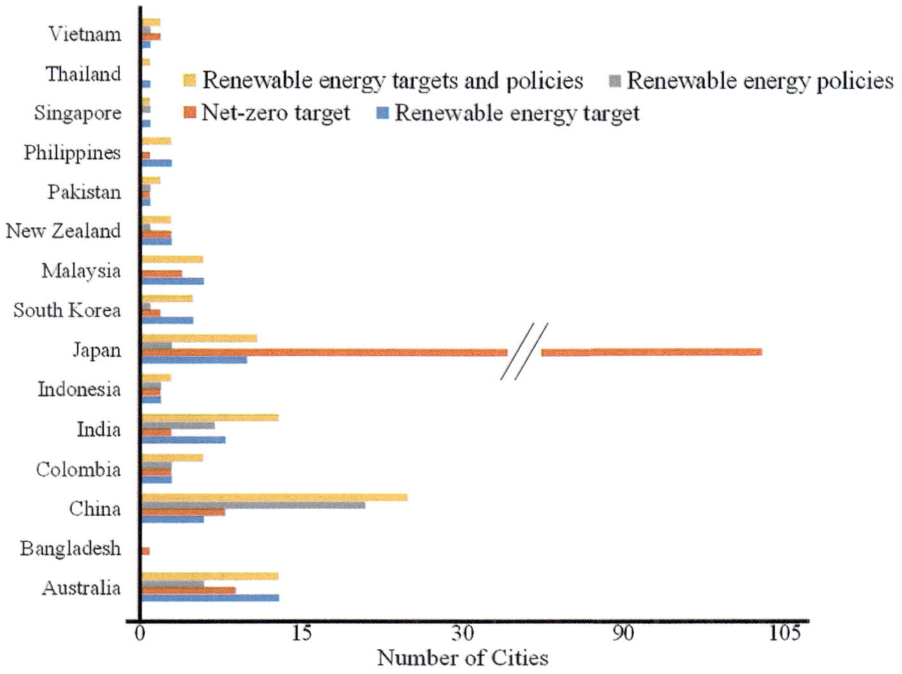

Fig. 14.6 Number of cities with the net-zero target, renewable energy target, and policies in 2020 of selected Asia-Pacific cities. (Data source: REN21, 2021)

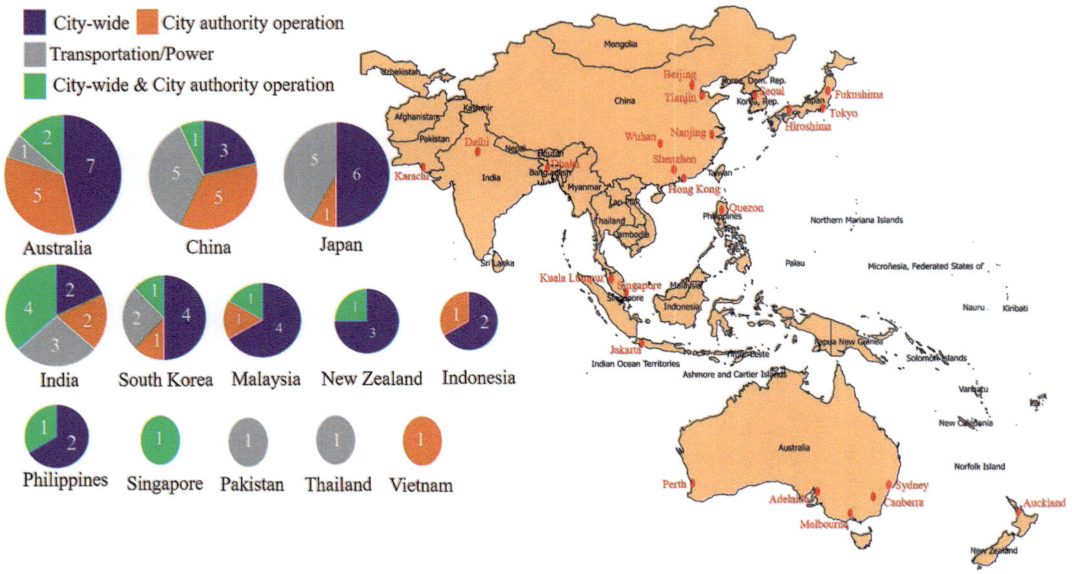

Fig. 14.7 Number of cities with renewable energy target and their scale in 2020 of selected Asia-pacific cities. Cities of Table 14.2 are shown with red color on the map. (Data source: REN21, 2021)

with the 800 cities around the world (REN21, 2021). So, compared to other cities in the Asia-Pacific region, Japanese cities are more focused on the enabling environment that will influence renewable energy use at the city level. However, 11 cities have also renewable energy targets and policies, and 6 of them have a 100% renewable target (e.g., Fukushima, Kuzumaki, Takarazuka, Tokyo, Yokohama, and Yusuhara) (Fig. 14.6).

Table 14.2 City-level renewable energy targets and other enabling targets influencing renewable energy use in selected cities of Asia-Pacific countries

Country	City	Renewable energy target sector and scale	City-level renewable energy/electric vehicle (EV) targets		National renewable power targets	Other targets influencing renewable energy use			Source
			Renewables	EV		Net-zero	Emission reduction	Carbon neutral	
Australia	Adelaide	Power (municipal operation and city-wide)	100%	–	23% electricity by 2020	(2025)	–	(2025)	Adelaide City Council (2016)
	Canberra	Power (municipal operation and city-wide)	100%	–		(2045)	–	(2045)	MASON (2020)
	Perth	Power (city-wide)	25%	–			32% (city-wide)	–	Perth City Council (2018)
	Sydney	Power (municipal operation and city-wide)	50%	–		(2050)	70% (2030)	–	Sydney City Council (2013)
	Melbourne	Power (municipal operation)	100%	–		2050) Net-zero buildings	45,242 t/year	–	Melbourne City Council (2016)
China	Beijing	Transport and power (city-wide)	11.47 GW (solar), 15% (wind)	50% of vehicle	35% by 2030	–	–	–	REN21 (2021)
	Shenzhen	Power (municipal operation and city-wide)	15%	100% of public bus		–	–	–	
	Tianjin	Power (municipal operation)	1115 MW generation capacity	–		–	–	–	
	Hong Kong	Not available	–	–		(2050)	–	–	
	Nanjing	Not available	–	–		(2050)	–	–	
	Wuhan	Not available	–	–		(2050)	–	–	
Japan	Fukushima	Power (city-wide)	100%	–	22–24% by 2030	(2050)	–	–	REN21 (2021)
	Hiroshima	Power (city-wide)	50%	–			–	–	
	Tokyo	Power (municipal operation)	100%	–		(2050)	–	–	
India	Delhi	Power (city-wide)	2000 MW generation capacity	–	350 GW by 2030	(2050)	–	–	REN21 (2021)
Indonesia	Jakarta	Power (city-wide)	30%	–	26% by 2025	(2050)	–	–	UNFCCC (2021)

Country	City	Renewable energy target sector and scale	City-level renewable energy/electric vehicle (EV) targets		National renewable power targets	Other targets influencing renewable energy use			Source
			Renewables	EV		Net-zero	Emission reduction	Carbon neutral	
South Korea	Seoul	Transport (city-wide)		100% bus fleet	35% by 2040 and net-zero by 2050		–	–	Climate Change News (2020)
Malaysia	Kuala Lumpur	–	–	–	20% by 2025	(2050)	–	–	UNFCCC (2021)
Pakistan	Karachi	Power (city-wide)	5%	–	20% by 2025 and 30% by 2030	(2050)	–	–	REN21 (2021)
Philippines	Quezon City	–	–	–	23% by 2025	(2050)	–	–	REN21 (2021)
Singapore	Singapore	Power (municipal operation and city-wide)	550 MW generation capacity	–	2GW of solar energy by 2030		–	–	REN21 (2021)
New Zealand	Auckland	Power (city-wide)	100%	–		(2050)	–	–	UNFCCC (2021)
Bangladesh	Dhaka	–	–	–	10% by 2020	(2050)	–	–	UNFCCC (2021)

In China, most of the city-level targets and actions are following the country's commitment to carbon neutrality by 2060. Six cities (e.g., Dalian) have net-zero targets as of 2020; there are also other targets such as climate neutrality by 2050 in Rizhao. In Chinese cities, there are some measures to popularize solar heating systems. As an example, suitable new residential buildings with less than 12 floors in Luanzhou need to install a solar water heating system. Building higher than 12 floors can install solar water heating system up to the 12th floor. In addition, solar district heating systems are also commissioned in some of the cities (e.g., Shenzhen, Zhongba, and Saga). The country also introduced a geothermal heating system at the city level (e.g., Fengxi City) (REN21, 2021).

Jakarta in Indonesia is also committed to deploying 30% renewables for its power mix by 2030 (UNFCCC, 2021). Jakarta and Balikpapan cities have also net-zero target by 2050 (Table 14.2). Renewable energy use in the transport sector in the cities (five cities) of South Korea is highlighted with renewable energy targets and policies (Figs. 14.6 and 14.7). Only two cities in South Korea have a net-zero target (e.g., Dangjin-si and Seoul) (REN21, 2021).

The low-income countries of the Asia-Pacific region are also progressing toward the use of renewable energy in the cities, but not at the pace of high−/middle-income countries. For example, India launched a smart city program in 2015. The target is to increase solar power generation capacity in an urban area. Solar PV and solar water heater mandates are growing in several cities, for example, in 2009, the Bangalore municipal utility company started to deny electricity grid access to households not equipped with solar water heaters. As a result, 1234 million m^2 of solar collector area was installed across the city as of 2017. A similar approach is followed in cities across Karnataka, and now it is the leading Indian state for solar water heaters (REN21, 2021). Only three cities (e.g., Kolkata, Chennai, and Delhi) in India set net-zero targets for 2050. In comparison with India under the low-income country of the Asia-Pacific region, some other countries are also introducing enabling environment like net-zero targets at the city level, which will influence the renewable energy in their cities. For example, Dhaka (Bangladesh), Karachi (Pakistan), and Quezon City (Philippines) already set net-zero targets (Table 14.2).

14.6 Food-Water Renewable Energy Nexus: Research Gaps in the Asia-Pacific

China and Australia in the Asia-Pacific region are the leading countries of water-energy implication assessment within the water, energy, and food nexus research arena. China and Australia have conducted most studies in this research arena representing 17% and 15% of the total nexus studies, respectively, after the USA (Islam et al., 2021). The drivers that influenced such nexus research in China are (i) analyzing the challenges and opportunities of WrE consumption reduction for agricultural water supply (Zhao et al., 2018), (ii) analyzing the sustainable development challenges due to uneven energy and water distribution (Wang et al., 2018), and (iii) analyzing the integrated water and energy sectoral planning opportunities (He et al., 2019). In China, among the nexus research, energy assessment for water use, such as for irrigation in the agriculture phase (Zhao et al., 2018) and industrial process (Gao et al., 2019), is notable.

The drivers of water, energy, and food nexus assessment in Australia are (i) achieving water security rather than energy security because of drought conditions (Kenway et al., 2011) and (ii) analyzing the energy consumption from desalination (Kenway & Lam, 2016) and rainwater harvesting (Talebpour et al., 2014). Energy implication of water use in Australia was much focused on residential hot water use (Kenway et al., 2019b), urban water supply (Kenway et al., 2015), and rainwater harvesting (Siems & Sahin, 2016). A few studies also quantified irrigation in the agriculture phase (Tyson et al., 2012).

A range of water, energy, and food nexus studies considered energy assessment and associated GHG (greenhouse gas) emissions particularly for agriculture focusing on irrigation (Benbi, 2018), urban water service (Valek et al., 2017), residential end use (Chhipi-Shrestha et al., 2017), and manufacturing sector (Chinese et al., 2017). Technology utilization focusing on water provisioning services by exploring the different combinations of technology at the city level is also explored (e.g., desalination plant, hydropower generation plant, water reclamation, algae production in wastewater pond, and gray water recycling) (Valek et al., 2017).

Historical change in energy consumption under food-water-energy nexus studies was mostly conducted for agriculture focusing on irrigation (Benbi, 2018) and urban water supply (Xie et al., 2018). Very few studies addressed the inter-regional flow of energy through the food trade (Ramaswami et al., 2017). However, the economic implications of energy consumption were considered by only a few studies focusing on the residential user (Chini et al., 2016), urban water service (Walker et al., 2014), and irrigation water supply (Cremades et al., 2016).

So, China and Australia in the Asia-Pacific region have more focus on food-water-energy nexus research. The key driver that influenced such nexus research in these two countries is exploring integrated water and energy sectoral planning opportunities to solve the resource scarcity problem. Of the nexus research, most focus is given on analyzing energy implications for irrigation water, urban water supply, and residential hot water use. There exists a major research gap considering the technological (e.g., renewable energy use) and economical aspects of energy assessment of the food system focusing on the food industry and cooking at the city scale. This gap needs to be explored because significant financial implications are associated with water and energy saving (Keairns et al., 2016). Also, the historical evolution of energy consumption in the food system at the city level and changes in interregional flow need to be explored to assess the impacts of water- or energy-oriented policy on sustainable food system development.

14.7 Food-Water Renewable Energy Nexus: Challenges and Opportunities

The key challenges are (i) inherent complexity of interaction; (ii) lack of knowledge about feedback loop, rather than an integrated target; (iii) lack of partnerships; (iv) city governance system; and (v) individual sectoral target. All of these issues are briefly discussed here:

(i) *The inherent complexity of interaction*: It is difficult to manage the interactions among water, energy, and food due to their inherent complexity and different stakeholder and regulatory authority involvement at different levels. Attempt to deal with one component of this complex nexus taking fragmentary approaches may lead to failure or unintended consequences in other components (Kenway et al., 2019a).

(ii) *Lack of knowledge about feedback loop*: In addition, there still exists a large uncertainty about the feedback loop (say, respond to change in renewable energy use in the food system) of the involved systems and across the sectors (Islam et al., 2021). The multi-level feedbacks across the water, energy, and food sectors for renewable energy use are difficult to identify and manage.

(iii) *Individual sectoral target*: In addition, each sector (water, energy, and food) has its own goals, targets, and national compliance procedures and key performance indicators (KPI). The whole water, energy, and food nexus system-oriented policy and strategy are still absent (Islam et al., 2021). Hence, there is a need to identify how one sector goal (say, vegetable-based diet recommendation) can either facilitate or hinder the achievement of other sectors (say, water and energy efficiency improvement target) at the city level.

(iv) *Lack of partnerships*: Fragmented, and sometimes inconsistent, policies often arise due to functional mismatches across sectors and stakeholders (Artioli et al., 2017). The key reason for such functional mismatches is often the narrow focus of the policymaker on optimizing the benefits of one sector and overlooking the risks related to other sectors. Integrated water, energy, and food nexus strategies are currently absent because policy development and implementation are fragmented among different government entities in the Asia-Pacific regions. Say, the agriculture department is focused on agriculture activities, the industrial department is focused on food industry-related activities, and the water and energy departments have their mandates. That is why, operationalization of the water, energy, and food nexus approach and renewable energy integration for this nexus will need streamlining interdepartmental strategies and policies (Ololade et al., 2017).

(v) *City governance system*: Another major challenge is governance. The electoral cycle of the city-level government is often short enough to deal with complex issues of nexus management at the city level (Simpson & Jewitt, 2019). If there is no continuity of one city-level government, the successive governments have their ideologies and commitment (e.g., funding policies and institutional resource utilization) that hinder the advances of renewable energy use for water-energy-food nexus at the city level.

There is a huge opportunity to develop nexus thinking at the city level by promoting integrative and interdisciplinary research. The water-energy-food nexus at the city level is unique because of the inherent difference between cities. So, such research will inform the policymaker about the potential opportunities and synergies of renewable energy integration for the water-energy-food nexus at the city level. The research should focus on bridging this knowledge gap at the city level for integrated assessment and decision support systems by combining different data (say, water, energy, and food) and models to assess cross-sector impacts and analyzing the feedback loop.

Collaborative planning and multi-stakeholder dialogue should be encouraged to explore the synergies for renewable energy use for the food-water-energy nexus. For such planning, issues such as (i) risk of one sector policy to other sectors (water, food, and energy), (ii) regulatory reformation for coherent policy, and (iii) water and energy footprints of food system business enterprises and solutions to minimize impacts need to be explored.

The city authority should create and strengthen partnerships between national, state, and local actors (say, water, energy, and food sector actors) to develop long-term urban resilience. In this process, the city government system should utilize the big data sources (say, city-level input-out data, water balance, and energy balance data) to connect the hinterlands and the city communities. This will ultimately improve the local government and city inhabitant's sustainability literacy.

Finally, a holistic governance system needs to develop with shared strategic direction for a coherent cross-cutting policy at the city scale. The public needs to be engaged to drive the change and political commitment for nexus thinking at the city level. Hence, a reporting/accounting system covering the full value and costs of water and energy across the city food system is needed. As discussed in Sect. 14.4, renewable energy use for the food-water-energy nexus can play a major role in achieving the SDGs through reduction of negative consequences and externalities (e.g., economic, social, and environmental) and by increasing resource use efficiency. Though renewable energy use for food-water-energy nexus at the city level is still evolving, Asia-Pacific cities can benefit from an integrated strategy at the city level as a pilot basis, rather than dealing with the food-water-energy nexus challenges individually.

14.8 Conclusion

The economy, society, and environment of the Asia-Pacific region will be under tremendous pressure and challenges because of the additional 1.2 billion urban residents by 2050 (UN, 2019). Cities should drive the action for climate change mitigation and sustainable development to ensure resource security. The sustainability of cities in Asia-Pacific will determine the future development pathways in this region. More strategic approaches as mentioned in the previous section are needed to overcome the mismatch between short-term and long-term planning for integrating renewable energy for food-water-energy nexus. Urban challenges in each city of the Asia-Pacific are unique. Hence, it is important to recognize them by analyzing the challenges mentioned in the previous section. Accordingly, their solutions should be developed by analyzing country- and local-specific scenarios. We conclude that city authorities should plan to finance and deploy resources by analyzing the key challenges mentioned in the previous section from local contexts to materialize multiple benefits, such as achieving clean energy target (SDG 7), while contributing to sustainable cities and communities (SDG 11) and climate action (SDG 13).

References

Adelaide City Council. (2016). *Carbon Neutral Adelaide Action Plan 2016–2021* (p. 51). Government of South Australia and Adelaide City Council.

Artioli, F., Acuto, M., & Mcarthur, J. P. G. (2017). The water-energy-food nexus: An integration agenda and implications for urban governance. *Political Geography, 61*, 215–223.

Benbi, D. (2018). Carbon footprint and agricultural sustainability nexus in an intensively cultivated region of Indo-Gangetic Plains. *Science of the Total Environment, 644*, 611–623.

Berardy, A., & Chester, M. V. (2017). Climate change vulnerability in the food, energy, and water nexus: Concerns for agricultural production in Arizona and its urban export supply. *Environmental Research Letters, 12*(3), 035004.

Chhipi-Shrestha, G., Hewage, K., & Sadiq, R. (2017). Impacts of neighborhood densification on water-energy-carbon nexus: Investigating water distribution and residential landscaping system. *Journal of Cleaner Production, 156*, 786–795.

Chinese, D., Santin, M., & Saro, O. (2017). Water-energy and GHG nexus assessment of alternative heat recovery options in industry: A case study on electric steelmaking in Europe. *Energy, 141*, 2670–2687.

Chini, C. M., Schreiber, K. L., Barker, Z. A., & Stillwell, A. S. (2016). Quantifying energy and water savings in the US residential sector. *Environmental Science & Technology, 50*(17), 9003–9012.

Climate Change News. (2020). *South Korea formally commits to cutting emissions to net zero by 2050.*

Cremades, R., Rothausen, S. G., Conway, D., Zou, X., Wang, J., & Li, Y.e. (2016). Co-benefits and trade-offs in the water–energy nexus of irrigation modernization in China. *Environmental Research Letters, 11*(5), 054007.

Dincer, I. (2000). Renewable energy and sustainable development: A crucial review. *Renewable and Sustainable Energy Reviews, 4*(2), 157–175.

FAO. (2003). *World agriculture: Towards 2015/2030 An FAO Perspective*. Food and Agriculture Organization of the United Nations (FAO).

FAO. (2014). *The Water-Energy-Food Nexus: A new approach in support of food security and sustainable agriculture*. Food Agriculture Organization of the United Nations (FAO).

FAO. (2017). *The future of food and agriculture–Trends and challenges* (p. 163). Food and Agriculture Organization of the United Nations.

FAO. (2021). AQUASTAT Database. Nations (FAO), T.F.a.A.O.o.t.U. (ed).

Ferroukhi, R., Nagpal, D., Lopez-Peña, A., Hodges, T., Mohtar, R. H., Daher, B., Mohtar, S., & Keulertz, M. (2015). *Renewable energy in the water, energy and food nexus* (p. 124). The International Renewable Energy Agency (IRENA).

Gao, C., Song, K., Na, H., Tian, F., & Zhang, S. (2019). Comprehensive evaluation on energy-water saving effects in iron and steel industry. *Science of the Total Environment, 670*, 346–360.

Gupta, A. D. (2017). *Water-Energy-Food (WEF) Nexus and Sustainable Development.*

He, G., Zhao, Y., Wang, J., Li, H., Zhu, Y., & Jiang, S. (2019). The Water–Energy Nexus: Energy use for water supply in China. *International Journal of Water Resources Development, 35*(4), 587–604.

IEA. (2021). IEA Sankey Diagram, International Energy Agency (IEA).

Islam, K. N., Kenway, S. J., Renouf, M. A., Lam, K. L., & Wiedmann, T. (2021). A review of the water-related energy consumption of the food system in nexus studies. *Journal of Cleaner Production, 279*(123), 414.

Keairns, D., Darton, R., & Irabien, A. (2016). The energy-water-food nexus. *Annual review of chemical and biomolecular engineering, 7*, 239–262.

Kennedy, C. A., Stewart, I., Facchini, A., Cersosimo, I., Mele, R., Chen, B., Uda, M., Kansal, A., Chiu, A., & Kim, K.-g. (2015). Energy and material flows of megacities. *Proceedings of the National Academy of Sciences, 112*(19), 5985–5990.

Kenway, S. J., & Lam, K. L. (2016). Quantifying and managing urban water-related energy use systemically: Case study lessons from Australia. *International Journal of Water Resources Development, 32*(3), 379–397.

Kenway, S., Lant, P., Priestley, A., & Daniels, P. (2011). The connection between water and energy in cities: A review. *Water Science & Technology, 63*(9), 1983–1990.

Kenway, S., Binks, A., Lane, J., Lant, P., Lam, K., & Simms, A. (2015). A systemic framework and analysis of urban water energy. *Environmental Modelling & Software, 73*, 272–285.

Kenway, S. J., Lam, K. L., Stokes-Draut, J., Sanders, K. T., Binks, A. N., Bors, J., Head, B., Olsson, G., & McMahon, J. E. (2019a). Defining water-related energy for global comparison, clearer communication, and sharper policy. *Journal of Cleaner Production. 236,* 117502.

Kenway, S. J., Scheidegger, R., & Bader, H.-P. (2019b). Dynamic simulation of showers to understand water-related energy in households. *Energy and Buildings, 192*, 45–62.

Long, Y., Yoshida, Y., & Dong, L. (2017). Exploring the indirect household carbon emissions by source: Analysis on 49 Japanese cities. *Journal of Cleaner Production, 167*, 571–581.

Mason, B. (2020). *The ACT is now running on 100 renewable electricity, SBS.*

Melbourne City Council. (2016). *Emissions reduction plan for our operations 2016–2021* (p. 34). Melbourne City Council.

Ololade, O. O., Esterhuyse, S., & Levine, A. D. (2017). The water-energy-food nexus from a South African perspective. *Water-Energy-Food Nexus: Principles and Practices, 229*, 129.

Perth City Council. (2018). *Towards an energy resilient city strategic directions paper* (p. 19). Perth City Council.

Ramaswami, A., Boyer, D., Nagpure, A. S., Fang, A., Bogra, S., Bakshi, B., Cohen, E., & Rao-Ghorpade, A. (2017). An urban systems framework to assess the trans-boundary food-energy-water nexus: Implementation in Delhi, India. *Environmental Research Letters, 12*(2), 025008.

REN21. (2021). *Renewables in cities 2021 global status report* (p. 201). REN21 Secretariat, UN Environment Programme.

Ritchie, H., & Roser, M. (2018). *Urbanization.* OurWorldInData.org

Satterthwaite, D. (2020). *An urbanising world, International Institute for Environment and Development (IIED).*

Scott, C. A., Kurian, M., & Wescoat, J. L. (2015). *Governing the nexus* (pp. 15–38). Springer.

Siems, R., & Sahin, O. (2016). Energy intensity of residential rainwater tank systems: Exploring the economic and environmental impacts. *Journal of Cleaner Production, 113*, 251–262.

Simpson, G. B., & Jewitt, G. (2019). The development of the water-energy-food nexus as a framework for achieving resource security: A review. *Frontiers in Environmental Science, 7*, 8.

Sydney City Council. (2013). *Decentralised energy master plan: Renewable energy* (p. 140). Sydney City Council.

Talebpour, M., Sahin, O., Siems, R., & Stewart, R. (2014). Water and energy nexus of residential rainwater tanks at an end use level: Case of Australia. *Energy and Buildings, 80*, 195–207.

Tyson, A., George, B., Aye, L., Nawarathna, B., & Malano, H. (2012). Energy and greenhouse gas emission accounting framework for groundwater use in agriculture. *Irrigation and Drainage, 61*(4), 542–554.

UN. (2019). *The future of Asian & Pacific Cities: Transformative pathways towards sustainable urban development* (p. 183). United Nations.

UNESCAP. (2018). *Statistical perspectives 2018: Sustainable energy in Asia and the Pacific* (p. 43). United Nations Economic and Social Commission for Asia and the Pacific (ESCAP).

UNESCAP. (2019). *Statistical Perspectives 2019: Sustainable Energy in Asia and the Pacific* (p. 77). United Nations Economic and Social Commission for Asia and the Pacific (ESCAP).

UNFCCC. (2021). *Climate Ambition Alliance: Net Zero 2050*. UNFCCC.

Valek, A. M., Sušnik, J., & Grafakos, S. (2017). Quantification of the urban water-energy nexus in México City, México, with an assessment of water-system related carbon emissions. *Science of the Total Environment, 590*, 258–268.

Walker, R. V., Beck, M. B., Hall, J. W., Dawson, R. J., & Heidrich, O. (2014). The energy-water-food nexus: Strategic analysis of technologies for transforming the urban metabolism. *Journal of Environmental Management, 141*, 104–115.

Wang, S., Liu, Y., & Chen, B. (2018). Multiregional input–output and ecological network analyses for regional energy–water nexus within China. *Applied Energy, 227*, 353–364.

Xie, X., Jia, B., Han, G., Wu, S., Dai, J., & Weinberg, J. (2018). A historical data analysis of water-energy nexus in the past 30 years urbanization of Wuxi city, China. *Environmental Progress Sustainable Energy, 37*(1), 46–55.

Yuan, M.-H., Chiueh, P.-T., & Lo, S.-L. (2021). Measuring urban food-energy-water nexus sustainability: Finding solutions for cities. *Science of the Total Environment, 752*(141), 954.

Zhao, R., Liu, Y., Tian, M., Ding, M., Cao, L., Zhang, Z., Chuai, X., Xiao, L., & Yao, L. (2018). Impacts of water and land resources exploitation on agricultural carbon emissions: The water-land-energy-carbon nexus. *Land Use Policy, 72*, 480–492.

15 Urban Living Labs and the Water-Energy-Food Nexus: Experiences from the GLOCULL Project in São Paulo, Brazil

Fabiano de Araújo Moreira,
Michele Dalla Fontana, Ana Maria Bertolini,
Patrícia Marra Sepe, Tadeu Fabrício Malheiros,
and Gabriela Marques Di Giulio

Abstract

Cities are prominent in the process of transition to sustainability due to their high concentration of people and economic activities, which make urban centers large consumers of resources and large producers of waste. On the other hand, cities are centers of knowledge and fertile environments for innovation and creativity. The complexity of São Paulo's megacity offers fertile ground for developing experiments and advancing the transition to sustainability through the establishment of an innovative system of governance, such as the Urban Living Lab (ULL). The ULL can be applied, for example, to identify local problems and knowledge gaps concerning the water, energy, and food nexus in cities, a concept that addresses the complex interrelationships between those elements to identify synergies and minimize trade-offs to ensure more efficient, equitable, and fair use of natural resources. In addition, the ULL promotes the co-production and creative dissemination of knowledge to drive change and achieve transformative paths to urban sustainability, while strengthening the relation between academia and society. In this chapter, we address the relevance of creating a space for the interaction of academic and nonacademic actors in cities to advance in the process of transition to sustainability and the insertion of the concept of the nexus in the policy sector as an important element for the sustainability of cities. To do so, we approach the experience of the ULL in the GLOCULL project, a collaboration between scientists and actors from the municipality of São Paulo, Brazil, exploring sustainable actions related to the nexus.

Keywords

Transdisciplinarity · Sustainability · Urban laboratory

F. de Araújo Moreira (✉) · M. Dalla Fontana · A. M. Bertolini · G. M. Di Giulio
School of Public Health, University of São Paulo (ESP/USP), São Paulo, Brazil
e-mail: fabiano.moreira@usp.br

P. M. Sepe
Connect the Dots Project, Urban Development Municipal Secretariat of São Paulo City (SMDU/PMSP), São Paulo, SP, Brazil

T. F. Malheiros
School of Engineering of São Carlos, University of São Paulo, São Carlos, SP, Brazil

L. L. B. Lazaro et al. (eds.), *Water-Energy-Food Nexus and Climate Change in Cities*, Sustainable Development Goals Series, https://doi.org/10.1007/978-3-031-05472-3_15

15.1 Introduction

The socio-environmental challenges of the contemporary era are spreading in different domains and at different scales. Characterized by uncertainties and complexities, these challenges demand responses and interventions that take into account the need for profound changes in the socioeconomic system, in the modes of production and consumption in areas such as water, energy, agriculture, health, transport, and housing (Coenen et al., 2012; Markard et al., 2012). These necessary changes, interpreted as "socio-technical transitions," involve multisectoral changes that require the participation of different actors. Such transitions do not just include technological innovations; they also require a general reconfiguration of policies, infrastructure, market dynamics, production, and consumption practices (Geels, 2004, 2011).

Cities have played an important role in the transition to sustainability due to their high concentration of people and economic activities, which make urban centers great consumers of resources and great producers of waste. On the other hand, cities are centers of knowledge and a fertile environment for innovation and creativity (Brescia & Marshall, 2016). The 2030 Agenda for Sustainable Development, for example, emphasizes the critical role of cities, with their diversity of projects and forms of management, as real physical spaces for the implementation of the 17 objectives proposed to address the most difficult problems that threaten the planet and humanity. It is no coincidence that SDG 11 is dedicated to cities, with the prospect of "making cities and human settlements inclusive, safe, resilient, and sustainable." Moreover, the implementation plan for the New Urban Agenda establishes transformative commitments based on the social, economic, and environmental dimensions of sustainable urban development and is considered a tool to take advantage of urbanization as a transforming element (Frantzeskaki et al., 2017).

Particularly in megacities such as São Paulo, Brazil, the transition to urban sustainability requires different transformations that can be drivers of "radical" changes and contribute to sustainable "multidimensional" urban structures (McCormick et al., 2013). Governance and planning, innovation and competitiveness, lifestyle, and consumption are considered factors of change (McCormick et al., 2013). On the other hand, management of resources such as water, energy, and food, mitigation and adaptation to climate change, and transport and accessibility to housing and public space are understood as sustainable urban structures. A variety of studies have addressed the relationships between transport, mobility, environmental issues, public health, and climate change (Banister, 2005; Dulal et al., 2011; Jacyna et al., 2017; Moriarty & Honnery, 2013) and, more recently, the water, energy, and food nexus (WEF nexus) – crucial themes in the discussion on the transition to urban sustainability. The WEF nexus, for example, is directly related to the increase in demand for resources by the population, mainly in the cities. This concept was conceived as an attempt to address the complex interrelationships that arise between water, energy, and food, at different scales, strengthening synergies between sectors and minimizing trade-offs to ensure more efficient, equitable, and fair use of natural resources (Bazilian et al., 2011; Hoff, 2011). Therefore, promoting actions addressing water, energy, and food sectors would be of great benefit for the sustainability of cities (EC, 2006; Yuan et al., 2021).

The scholarship recognizes that the transition to urban sustainability requires dialogue and cooperation between academic and nonacademic actors. In this sense, cities are seen as ideal places to develop this relationship, since this jurisdiction is where the closest proximity between the state and civil society takes place. The way in which these nonacademic actors (government and civil society) organize, interact, and make decisions – the so-called governance process – is a central element in the discussion on the transition to sustainability. It is this interaction, in fact, that allows transition experiments to be proposed and tested in order to promote transformational changes. Transition experiments also include those interventions based on existing efforts and actions, which make it possible to create new actions or guide transitions (Luederitz et al.,

2017). Although the idea of promoting transition processes through experiments is not completely new, the concept of spaces (real or virtual) that facilitate experimentation and learning based on the participation and involvement of different actors can bring new possibilities both for academia, which focuses on different ways of producing knowledge, and for other actors who benefit from the process of building solutions to concrete problems.

Some approaches, known as urban laboratories (ULs), have been developed to foster the integration of different actors in the process of identifying problems and solutions for the sustainability of cities. There are different types of ULs, including the Urban Transition Labs (Nevens et al., 2013), City Labs (Culwick et al., 2019), Real-World Labs (Wanner et al., 2018), and Urban Living Labs (Bulkeley et al., 2016). All these UL approaches are temporary or permanent think tanks that understand society as a laboratory capable of identifying relevant solution paths. These labs share a common methodology that promotes interactions between different groups of stakeholders and researchers with a focus on societal problem-solving in order to help municipalities to find multisectoral solutions in a dynamic and complex urban context, which significantly increases the chances for successful and sustainable urban development (Hansson & Polk, 2018; Lehmann, 2018; Wahl et al., 2021). Belcher et al. (2016) conceptualized four key aspects of effective UL in order to produce transferable, useful information bridging different disciplines and science and society, namely, salience, credibility, legitimacy, and effectiveness.

The Urban Living Lab (ULL) is an instrument of urban governance co-created by public institutions, civil society, research organizations, and the private sector, which facilitates experimentation and, therefore, the transition to sustainability (Bulkeley et al., 2016). The setting of a ULL provides an interactive platform to identify key issues affecting the locality and knowledge gaps, foster bilateral cooperation, and combine co-design, co-production, and co-dissemination on the process of co-creation of knowledge and information to solve fundamental issues to achieve sustainability (Bulkeley et al., 2016; Mauser et al., 2013). The use of the ULL has enormous potential, especially as it accommodates the idea of an experimental approach that aims to respond to the challenges of urban sustainability and provide necessary spaces that facilitate experimentation. However, the transition to urban sustainability through experiments requires that all participants in the experiments recognize the uncertainties as to the expected results, avoid unrecognized risks, be open to taking advantage of new learning opportunities, and prepare for change (Luederitz et al., 2017).

In this chapter, we seek to advance the discussion on (a) the relevance of creating a space for the interaction of academic and nonacademic actors in cities to advance in the process of transition to sustainability and (b) the insertion of the concept of the WEF nexus in the policy sector as an opportunity for the sustainability of cities. To do so, we address the experience of engagement of academic and nonacademic actors that took place between the years 2018 and 2021 in the GLOCULL project – "Globally and LOCally-sustainable food-water-energy innovation in Urban Living Labs" (Forum Belmont/Sugi/JPI) – a transdisciplinary research in seven ULL (Austria, Brazil, Germany, South Africa, Sweden, Netherlands, United States), which combined an integrated, model-supported assessment of local-global interactions in the WEF nexus. More specifically, we focus on the actions and dynamics in the ULL of the city of São Paulo, Brazil. The complexity of the megacity of São Paulo offers fertile ground for developing urban sustainability experiments, especially to move forward with the transition to sustainability through the establishment of an innovative governance system, as proposed by the ULL. This system enables dialogue between actors who, in general, still have few opportunities to debate and build a "safe space," proposing and testing solutions that would not normally be considered. The proposal for a ULL can be extended, for example, to the discussion on the WEF nexus, expanding the possibilities of identifying key issues and knowledge gaps that affect the locality.

15.2 The GLOCULL Project in São Paulo, Brazil

Recent analyses indicate a set of challenges in the city of São Paulo that still seriously affect communities' quality of life, such as river and soil pollution, lack of sanitation, traffic and noise, soil sealing with consequent impact on urban drainage, changes in microclimates (e.g., urban heat islands), loss of biodiversity, and impacts on precipitation and water production and storage capacity by reservoirs, among others (Bonduki, 2011; Franco et al., 2015; Jacobi, 2006; Sepe & Pereira, 2015). These challenges are closely related to the inability of local governments to develop public policies and implement urgent actions to sustain the city's growth process and carry out the most appropriate territorial planning (D'Almeida, 2016; Di Giulio & Vasconcellos, 2014). This city is a major importer of resources, which results in the production of a large amount of waste and pollutants that put great pressure on the local environment, with high ecological footprint (Becker et al., 2012). In this sense, São Paulo, as well as other large urban agglomerations, relies on an unsustainable linear metabolism and needs suitable strategies to guarantee water, energy, and food supply.

The Brazilian team of the GLOCULL project in São Paulo is based in the School of Public Health/University of São Paulo (FSP/USP), in collaboration with the São Carlos School of Engineering (EESC/USP), and integrates an inter- and transdisciplinary group, with postdoc, graduate, and undergraduate students as well as institutional actors from the local municipality. The aim of the project is to produce a participatory assessment of the sustainability of ongoing actions in the rural south zone of São Paulo, analyzing the factors that determine the interdependence of the nexus systems and the dynamics of the territory. This peri-urban and rural area is considered strategic for water supply and energy and food production for the region, and it is home to more than 400 productive agricultural units with conventional and, more recently, organic production (Ligue os Pontos, 2020). It contains several environmental protection areas, municipal natural parks, and indigenous lands, denoting the municipal effort for the environmental protection of this region.

A particular case of interest of the ULL, and one of the main initiatives in progress in the southern rural area of the city, is the project "Ligue os Pontos" (LoP or "Connect the Dots"). This initiative was conceived and coordinated through the Municipal Secretariat for Urbanism and Licensing of the city of São Paulo and supported by the Bloomberg Philanthropies since 2016, when the LoP received the Mayors Challenge Latin America and the Caribbean awards, and officially ended in July 2021. The main goals of the LoP were to (i) contain urban sprawl in the south zone of São Paulo by maintaining the farmers in the area (ii) promote sustainable uses and agriculture, and (iii) preserve natural ecosystems. The project operated mainly on three fronts: (1) "data and evidence," collecting data and information about the farmers and agricultural properties in the region through a census and mapping; (2) "strengthening agriculture," bringing technical assistance to farmers to promote the transition to organic and agroecological agriculture and improve traditional techniques; and (3) "value chain," improving market access for farmers and seeking alternatives for better logistics between food production and distribution.

The ULL, specifically, worked with a focus on the activities supporting local agriculture, and contributed to these actions from a WEF nexus perspective, developing sustainability indicators that can capture the context of existing conflict and the potential responses to be given in the interface of these systems. The decision to develop the sustainability indicators was made by the actors involved at the beginning of the project, especially as a necessity revealed by the practitioners involved in the municipality and in the LoP in producing new data to support local actions toward the promotion of better practices in agriculture and environmental protection. The connection of the researchers and the practitioners came from previous partnerships in building indicators for the municipality, which helped the team to build this proposal collaboratively since the beginning of the project (see Sect. 15.3 for further explanations in this sense).

Our ULL aims to contribute to these actions by bringing together several actors from academic and nonacademic sectors (e.g., representatives from the Municipal Secretariat of Urbanism and Licensing, the Municipal Secretariat of Environment, the House of Ecological Agriculture of the District of Parelheiros, the Environmental Protection Areas, and the staff hired by the municipality for the LoP project). With a transdisciplinary approach (Bulkeley et al., 2016; Mauser et al., 2013), the ULL engaged actors through four main strategies: (i) interactive workshops, (ii) participation in meetings of the main municipal and local councils, (iii) scientific/technical meetings, and (iv) fieldwork in the study area.

The methodological procedure for the development of the sustainability indicators included three main phases: (i) identification and mobilization of local actors with interests in the southern region of the city of São Paulo and meetings for executive planning and identification of previous knowledge and common goals, (ii) workshops to raise awareness of the WEF nexus and to build a model and set of indicators, and (iii) validation.

Overall, four activities/workshops were performed to co-produce the sustainability indicators. Workshops 1 ("WEF nexus") and 2 ("flow and causal loop diagram") were held to raise awareness of the WEF nexus concept among all the partners of the ULL and to identify the main opportunities and barriers regarding water, energy, and food in the south zone of São Paulo. For the water, energy, and food systems, aspects of sociopolitical, infrastructure, uses, logistics, source, natural resources, and production categories were identified. Workshop 3 ("first indicators' list") was performed to work on the flows and causal loop diagram, which integrates the previously identified elements. Through the flows and causal loop diagram, the participants (researchers and practitioners) were able to identify a number of possible indicators. After this, the process of filling the indicators forms took into consideration several quality criteria to exclude some indicators. In workshop 4 ("sustainability indicators"), participants added the indicators in a broader sustainability model (Pressure-State-Impact-Response – PSIR), which was chosen by the participants according to their own previous experience in using this model. This model was then presented to all the participants, who discussed if there was any missing information and if there were overlapping indicators.

15.3 Reflecting on the ULL Experience

15.3.1 Methodological Process

Data used to inform this chapter is based on reports of four workshops performed to co-produce sustainability indicators that took place between the years of 2019 and 2021 in São Paulo, transcripts of five personal interviews with practitioners who work on urban, environmental, and agricultural issues in the municipality of São Paulo, and notes regarding experiences of the ULL's researchers during the project. The collected data was compiled and organized in five main topics, guided by themes in the discussions in this chapter: (i) features that facilitate the integration process, (ii) process of creation of space for interaction between actors, (iii) inclusion of the WEF nexus approach within the policy sector, (iv) barriers and difficulties of the integration process, and (v) opportunities created in the process for future collaboration among the actors.

15.3.2 Features that Facilitate the Integration Process

While transdisciplinary studies acknowledge and incorporate multi-stakeholders' perspectives and values, we recognize that aligning purposes, expectations, and efforts in a time frame is quite challenging, requiring substantial cooperation and ongoing adaptation. Previous experiences of cooperation among the stakeholders and their institutions, face-to-face practice time, and a shared understanding of principles and challenges regarding co-production favor researchers and stakeholders to work together and ultimately reshape academic and management structures and operational procedures (Roux et al., 2010). In this sense, the longtime and solid relationship between the University of São Paulo and the São Paulo City Hall, in fact, was a critical aspect that facilitated the initial negotiations and dialogues among the ULL participants. Many of the practitioners who work at the City Hall and in the LoP have already attended or have been involved in several training courses provided by the university. The School of Public Health, in particular, has a well-consolidated experience in offering applied courses focused on public health, sanitation, and environmental and urban planning aspects. Some of the ULL participants have already attended a few of these training courses and have engaged in past research projects. Their experience of training, research, and learning was critical to the formation of a network, engaging academic and nonacademic actors in a range of experiments, which encouraged the practitioners' engagement in the ULL, and might make possible the process and the required flexibility to incorporate new approaches and research findings into their decision-making and strategic planning.

City Halls of large cities, like São Paulo, normally combine institutional, technological, and financial capacity to undertake actions regarding urban planning, environmental protection, and climate adaptation, for example, and accelerate innovative processes. They also usually integrate boundary organizations that can connect policy to academics, and transnational interactions that help to establish regulatory framework, in order to boost sustainable actions (Serrao-Neumann et al., 2020; Setzer & de Macedo, 2015). The existence of policy entrepreneurs is also a critical factor that facilitates the engagement of practitioners in and around policymaking venues to promote policy change (Mintrom & Luetjens, 2017; Uittenbroek et al., 2014). The City Hall of São Paulo combines all these aspects, including policy entrepreneurs who have the key role of identifying knowledge gaps and needs for interventions and actions, mobilizing change processes in the policy system. The inclusion of these actors in an action research project (e.g., GLOCULL) is decisive for the development of information that is credible, salient, and legitimate for the local governments (Wahl et al., 2021).

Engaging stakeholders in transdisciplinary studies through ULL is easier when ongoing integrated projects and actions focused on enabling capacity building and learning are developed. That is the case of the LoP, based on an integrated approach that sought to increase local sustainable agricultural production to preserve the landscape, while promoting local social and economic development. The network built around the LoP, and the engagement of the practitioners with local farmers, as well as their commitment to the goals of this project, invoked a genuine interest in the GLOCULL approach, on the WEF nexus perspective, and facilitated their involvement in the ULL. Sufficiently, long duration and requested funding are also critical aspects that facilitate the integration process in transdisciplinary studies (Bergmann et al., 2021). In this sense, the three years of the GLOCULL project, funded by the São Paulo Research Foundation (Fapesp) – which has a positive reputation as a public foundation, funded by the taxpayer in the State of São Paulo with the mission to support research projects – and the Belmont Forum positively helped the commitment of the stakeholders. The highly competitive nature of such funding scheme certainly led the stakeholders to understand the innovative perspective of this research project, addressing critical and timely problems. Having three years to produce and analyze data, reach outcomes, and co-develop products and results that may have the potential of short-term impact was certainly an aspect that encouraged the practitioners to integrate this action research and be part of this ULL.

15.3.3 Creating Space for Interaction Between Actors

Dealing with WEF nexus issues, and particularly with projects such as LoP or any other action aimed at developing sustainable agricultural practices in a metropolitan context, requires the collaboration of a great number of actors. Scholars recognize that integrated approaches are crucial when facing such comprehensive and complex topics (Blay-Palmer et al., 2018; Dubbeling et al., 2017). Advocates of environmental integrated policies have been advancing in new models of governance for decades, and the importance of cooperation between public authorities, private sectors, and civil society is widely recognized to be an important point for sustainable development (Scalia et al., 2018). However, the cross-sectoral coordination and collaboration within the public sector are often overlooked.

Internal coordination in the public sector, both horizontally and vertically, is a fundamental step to build coherent and integrated strategies. We recognize that internal coherence and coordination can be challenging within local authorities, no less than the dialogue with representatives of the private sectors and civil society. In reality, whatever the context, public administrations tend to be organized in compartmentalized sectors (e.g., ministries, departments, secretariats). Each of these sectors is specialized in specific topics and has distinct areas of expertise, specific mandates, and objectives to be achieved, and they generally compete for the same scarce resources (Benson, 2011). This is no different in the case of São Paulo. On the contrary, things are amplified by the size of the metropolis, the extension of its territory, and the diversity of its challenges. This is matched by an equally massive administrative machine of 25 secretariats, each responsible for managing a set of actions, programs, and public policies related to a specific management theme (e.g., health, education, social assistance, culture, transportation, natural environment, urban planning), 32 subprefectures, and hundreds of civil servants.

Difficulty in coordination and dialogue between different entities is an endemic problem in bureaucratic systems of this size. Benson (2011, p. 146) distinguishes three main, often coexisting, problems that make cooperation between government sectors difficult: "i) the differing worldviews and mandates of sectors; ii) the resource allocation and planning processes within government; iii) capacity constraints within sectors for generating necessary information." These might be due to institutional and bureaucratic mechanisms that have been developed over time and established a way of working in silos in the public sector.

In our case, all the actors that took part in the ULL, despite working on common issues in the same territory, usually do not have many opportunities, for various reasons, to sit at the same table and have regular conversations. In addition, they all have very complicated agendas, with pressing deadlines and objectives to be achieved, which do not allow much time for self-reflection. Some of the participants in the ULL reported that the GLOCULL project workshops and meetings were one of the few opportunities for them to discuss and get updates about what each one was doing. The project was seen as an opportunity to strengthen internal relations and adopt a more horizontal way of working. The ULL setting in particular was considered useful in its offering an alternative to the common way of functioning of a public administration, in which tasks are separated and oriented to obtain specific goals with low levels of articulation. Every opportunity to reflect on what has been done and exchange ideas and impressions with a larger group of people allows to broaden the perspective and see new possibilities to deal with the municipalities' challenges. The chance to be exposed to different views and thoughts is critical to avoid making recurring mistakes.

Although doubts remain as to whether these relationships between actors can be maintained after the project ends, we can argue that ULL offered the space and time for different actors to meet, discuss, and reflect. This is outside the normal working routine and detached from specific and mandatory objectives required by institutional roles. It is in these opportunities, in which participants discuss more freely, that creative thinking can occur and different solutions can be found.

We do not claim that working in ULL is the ultimate solution, but even if the only purpose of a ULL was to offer this opportunity to improve dialogues and exchange ideas, this would already be a successful result. This may also suggest that research institutions and researchers could play as facilitators by providing these spaces for interaction, "breaking down the walls of the university," as one of the ULL partners claimed, being more present on the territory and engaged with local actors. This is very much in line with what some scholars say about the fundamental role of researchers in bringing different actors, from different sectors together to co-produced solution-focused research (Bréthaut et al., 2019; Wahl et al., 2021), improving the relations between actors and strengthening institutional capacity (Bazilian et al., 2011). Researchers are in the position to organize and lead ULL. In particular, social scientists, who are able to reveal and take into consideration power dynamics in knowledge co-production processes, are supporting the provision of a space where all actors can discuss and negotiate in a fair way (Allouche et al., 2015; Williams et al., 2014).

15.3.4 Including the WEF Nexus Within the Public Sector

In order to achieve the necessary transformation for the WEF nexus and implementation of innovative solutions, the nexus approach should be included in the policy sector through the engagement of institutions and actors that mediate environmental outcomes in a multi-governance style (Venghaus & Hake, 2018). This includes the definition of clear roles for different sectors and actors involved and the identification of specific challenges, objectives, and policy integrations, so the actions might be more effective and efficient (Venghaus & Hake, 2018). One entry point to mainstream the WEF nexus in the policy sector would be the inclusion of this rationale, for instance, in national development plans led by ministries and authorities of water, agriculture, energy, environment, infrastructure, and urban/rural planning, which could ultimately influence other local plans (Hoff & Kasparek, 2016).

In the GLOCULL project, we sought to maintain a permanent contact with the ULL participants, trying to bring the WEF nexus in the discussions, so the nonacademic actors could improve their knowledge on the approach and, then, could use this material for other plans and projects. While the participants recognized that the subdivision of responsibilities within municipal bureaucracies is a critical aspect, causing difficulties for effective collaboration, information sharing, and coordinated action among municipal staff, they appreciated that the WEF nexus brings a possible path to improve relations within different secretariats and departments. They also revealed that this project helped them to have a deeper look at some sectoral issues, including public policies related to water and their relation with energy and rural food production, indicating that the nexus approach helps to identify connections previously unconsidered. This also allowed them to have a closer look into the interactions with other plans and actions that are currently undergoing in the territory, which indicates that one of the most important features of the nexus concept is raising new questions and interactions between compartmentalized projects that might be improved with an integrative rationale.

The sustainability indicators co-produced by the ULL members were also an entry point for the nexus in the local policies. This activity allowed the actors to evaluate the current progress and sustainability of their undergoing projects. While the use of the WEF nexus concept was considered as an innovative element by the participants, they also understood that the work on a new concept may be challenging for the stakeholders. Nevertheless, they expressed their enthusiasm in engaging with this project and co-developing the sustainability indicators, suggesting that it is also important to work with new logics, improving their actions in the public sector.

Another effort to make the WEF nexus clearer was the use of systems thinking. The creation of flows and causal loop diagrams helped the participants to better visualize the elements and interactions of the nexus in the south zone of São Paulo. They stated that the flows and causal loop diagrams were extremely useful tools to understand the concept of the nexus, to see the interactions, and to locate their activities in a broader system/context. Many of them mentioned how it was useful to visualize interactions that until then were very theoretical and unclear. However, they also noted that the exercise can easily become very complex, and it requires more rounds of reading, discussion, analysis, and readjustments to be fully understood. Some participants also pointed out that these tools would be highly complex to be used with other social actors (e.g., farmers), recommending the adoption of simpler systems, while maintaining the visual feature.

15.3.5 Barriers and Difficulties

Participants indicated that there were some difficulties at the beginning of the project to comprehend their role in the GLOCULL project. They noted that after the first connections, executive meetings and interactive workshops helped them to better understand their roles. On the other hand, participants also revealed that they could have had a more proactive participation to determine responsibilities and objectives, but this did not happen mainly due to the lack of time availability in their agenda.

Some participants recognized that there were flaws in the communication process in the ULL, since the first discussed goals of the group were adjusted along the process and not all the participants were notified. This makes clear the importance of establishing transparent modes of communication between all the actors. One important aspect for the improvement of trust is the establishment of formal agreements between the actors. In our case, an agreement between the university and the Municipal Secretariat of Urban Development and Licensing was developed and signed, but due to changes in the public administration and bureaucracy, this agreement was never fully formalized. This might have some impact on the information sharing from the municipality and, consequently, on the compilation of the sustainability indicators forms. Nevertheless, some participants pointed out that a clear establishment of shared responsibilities between the actors at the beginning of the process was reached, which was helpful for building trust within the group.

Other issues were noted by the participants regarding time lag in the process. For example, stakeholders stated that the timing difference between their actions/activities and the academic process resulted in less participation in the project than they would expect; this was mainly due to their busy routine and engagement in many other projects. The participants also noticed that more interactive activities along the project would be helpful to absorb the new concepts and approaches. Another critical issue refers to the assessment of the results of sustainability projects, which usually takes longer than the project itself. While the assessment of the impacts of the LoP project in the territory using the sustainability indicators might need some years, since social and economic aspects may not have an immediate change, the projects (LoP and GLOCULL) were concluded in 2021/2022. How to assess sustainability impacts of new solutions after projects are over is, then, a challenge.

External factors might also have some impacts on the ULL and the process of engagement between the participants, especially due to the Covid-19 crisis. The pandemic limited face-to-face interactions, including fieldworks, meetings, and workshops in the region and in the municipality. Instead, all interactions were done virtually, undermining informal dialogues among the actors. This fact was considered a relevant aspect by the participants, because informal interactions help to build trust between the actors involved, while virtual meetings make the interactions less spontaneous. At the same time, the pandemic made it difficult to involve other important local actors, including the farmers, whose participation was limited to fieldwork, interviews conducted by the researchers, and one workshop. Although the initial work plan included a set of workshops with the local farmers, they did not happen due to social distancing rules and difficulties to access the Internet by the farmers. The lack of inclusion of local actors in the process was reported as a downfall by the ULL participants, who recognized the importance of the insights that might have arisen from the close interactions with the farmers.

15.3.6 Opportunities for Future Research

Working in the format of a ULL can bring short-term results regarding the own research process. However, there are outcomes that might be developed over time, even after the end of the project. These results are more closely connected to the ULL dynamics rather than specific research findings. Indeed, our experience confirms that at the foundation of a laboratory, there must be a network of actors with common goals, willing to collaborate, and with a solid trust relationship. While it can be difficult to maintain active collaboration after the end of a project, mainly due to the lack of funding, the institutional and interpersonal relationships that have been built and the experiences that have been accumulated are a critical capital that can lead to the development of new collaborations and projects.

The experience of a ULL can then lead to future projects, giving continuity to the previous one, or it can create the foundation for developing new projects that fit the participants' interests. Discussion and reflection on ULL results are important moments that lead to the formulation of new research questions and the beginning of new cycles of transdisciplinary research (Mauser et al., 2013).

In our case, for example, the "PEGASuS Take-It-Further Grant" was a clear opportunity to go further in this partnership. The goal of this grant, promoted by the Future Earth in partnership with the Belmont Forum and with funding from the Gordon and Betty More Foundation's Science Program, is to enhance and accelerate the existing Belmont Forum-funded projects, including the GLOCULL project. For 12 months, participants of the ULL have the opportunity to interact and exchange experiences with other 5 cities from the Global South to increase knowledge of decision-making and sustainability and to promote innovation and solutions around WEF nexus issues. This was an unexpected development of the ULL that was not planned at the beginning of the project. Certainly, the funding played a key role, but it is equally true that the network of actors that has been established with the ULL and the agreement to continue working together on sustainability issues have been critical factors to apply and win the grant. We see no reason for other similar opportunities not to occur in the future under other funding circumstances.

The development of other studies, which run parallel to the initial project objectives, is also an important unfolding of a ULL. The discussion between academic and nonacademic actors can in fact uncover new problems and new research questions that were not necessarily conceived in the initial proposal (e.g., food deserts, local accessibility to healthy food, food security, and nutrition of local farmers). In the case of the GLOCULL in São Paulo, an undergraduate and a PhD research project have been designed and developed taking into consideration the discussions and the insights from the ULL. It is important to highlight that, while these studies focus on issues that are important for the scientific debate in the academic community, they also have concrete importance for the local reality testified by the interest shown by nonacademic actors in the ULL. We can say that, at least as far as the definition of the general objectives of the research is concerned, the confrontation and exchange of feedback between different actors in the ULL have largely contributed to the design of these studies.

As a final point, it is worth highlighting how the Covid-19 pandemic has prompted researchers and other ULL participants to question research methods (e.g., how to adapt participatory methods to socially distanced contexts) and to reflect on research questions to be addressed in future projects (e.g., impact of the pandemic on local food systems, post-pandemic sustainability).

15.4 Conclusions

This chapter presents the experience of a transdisciplinary project in which academics and nonacademics co-produced sustainability indicators related to the WEF nexus in the rural area of São Paulo, Brazil. The focus was on the relevance of creating a space for the interaction of multiple actors in cities, in this case, the ULL approach, to advance in the process of transition to sustainability, and the insertion of the concept of the nexus in the policy sector as an important element for the sustainability of cities.

Results revealed the importance of research institutions and researchers in creating a space for interaction between different actors, since practitioners usually have a busy agenda that undermines this kind of partnership. Previous experiences of collaboration, the existence of policy entrepreneurs, and sufficiently long duration and requested funding for the projects are important components that induce a better dialogue between researchers and nonacademics. The ULL, in this sense, was considered beneficial to establish better communication between different sectors in the municipality, especially when horizontal, and not only vertical, coordination within the public system is crucial to develop coherent and integrated strategies for the city. The WEF nexus approach was also beneficial since it calls for an integrative rationale in actions and plans of the city while it helps to raise new questions and to identify interactions between compartmentalized areas of the public sector.

Nonetheless, experiences in transdisciplinary projects are not free of challenges. In this case, more attention should have been given to communication since the very beginning of the project in order to establish a more transparent dialogue between actors in the definition of their roles and the common goals of the project. Time lag between academics and practitioners' activities posed another challenge, which impacted the nonacademic involvement throughout the process. The assessment of sustainability innovations' impacts in a longer term, after the end of the project, is also a critical aspect. It reveals the importance of maintaining a permanent contact between the different actors in order to continue collaboration in new projects and looking for new types of funding, both in the university and in the municipality. In our ULL, the collaboration unfolded new opportunities, with new projects and studies, which can promote new cycles of transdisciplinary researches. The Covid-19 pandemic resulted in several downfalls for this project, such as lesser participation of local actors, revealing the necessity for any project to be adaptable and dynamic throughout the process.

The experience analyzed in this chapter provides critical insights on how the use of a transdisciplinary methodology, such as the ULL, and the use of the WEF nexus in the policy sector could improve the ability of nonacademic actors to better integrate their ongoing actions in the city when dealing with social, economic, and environmental aspects in their compartmentalized agenda. While both perspectives – ULL and WEF nexus – may improve the sustainability of cities, they also provide room to bring together academic and nonacademic actors, revealing new questions and opportunities for interactions and initiating new cycles of transdisciplinary researches. This interactive experience also sheds light on the importance of continuing to work on collaborative projects in order to improve the expertise and skills among the participants, and facilitate future co-production projects, engaging a range of actors and knowledge in a mutual learning process.

Acknowledgments This research has been funded by FAPESP – São Paulo Research Foundation (Proc. 2018/21362-1; Proc. 2018/21249-0) within the GLOCULL project, "Globally and LOCally-sustainable food-water-energy innovation in Urban Living Labs" (Proc. Fapesp 2017/50423-6). The authors would like to thank all the practitioners that voluntarily took part in the research and contributed to the results.

References

Allouche, J., Middleton, C., & Gyawali, D. (2015). Technical veil, hidden politics: Interrogating the power linkages behind the nexus. *Water Alternatives, 8*(1).

Banister, D. (2005). *Unsustainable transport: City transport in the new century*. Taylor & Francis.

Bazilian, M., Rogner, H., Howells, M., et al. (2011). Considering the energy, water and food nexus: Towards an integrated modelling approach. *Energy Policy, 39*(12), 7896–7906.

Becker, M., Martins, T., Campos, F. D., et al. (2012). *A Pegada Ecológica de São Paulo-Estado e Capital e a família de pegadas*. WWF-Brasil.

Belcher, B. M., Rasmussen, K. E., Kemshaw, M. R., et al. (2016). Defining and assessing research quality in a transdisciplinary context. *Research Evaluation, 25*(1), 1–17.

Benson, T. (2011). In S. Fan & R. Pandya-Lorch (Eds.), *Cross-sectoral coordination in the public sector: A challenge to leveraging agriculture for improving nutrition and health* (pp. 145–152). International Food Policy Research Institute (IFPRI).

Bergmann, M., Schäpke, N., Marg, O., et al. (2021). Transdisciplinary sustainability research in real-world labs: Success factors and methods for change. *Sustainability Science, 16*(2), 541–564.

Blay-Palmer, A., Santini, G., Dubbeling, M., et al. (2018). Validating the city region food system approach: Enacting inclusive, transformational city region food systems. *Sustainability, 10*(5), 1680.

Bonduki, N. (2011). O modelo de desenvolvimento urbano de São Paulo precisa ser revertido. *Estudos Avançados, 25*(71).

Brescia, R., & Marshall, J. T. (2016). *How cities will save the world: Urban innovation in the face of population flows, climate change and economic inequality*. Routledge.

Bréthaut, C., Gallagher, L., Dalton, J., et al. (2019). Power dynamics and integration in the water-energy-food nexus: Learning lessons for transdisciplinary research in Cambodia. *Environmental Science & Policy, 94*, 153–162.

Bulkeley, H., Coenen, L., Frantzeskaki, N., et al. (2016). Urban living labs: Governing urban sustainability transitions. *Current Opinion in Environmental Sustainability, 22*, 13–17.

Coenen, L., Benneworth, P., & Truffer, B. (2012). Toward a spatial perspective on sustainability transitions. *Research Policy, 41*(6), 968–979.

Culwick, C., Washbourne, C. L., Anderson, P. M., et al. (2019). CityLab reflections and evolutions: Nurturing knowledge and learning for urban sustainability through co-production experimentation. *Current Opinion in Environmental Sustainability, 39*, 9–16.

D'Almeida, C. H. (2016). Desafios, hipóteses e inovação na gestão da política urbana. In R. Balbim & C. Krause (Eds.), *Eixos de estruturação da transformação urbana: inovação e avaliação em São Paulo*. Ipea.

Di Giulio, G. M., & Vasconcellos, M. D. P. (2014). Contribuições das Ciências Humanas para o debate sobre mudanças ambientais: um olhar sobre São Paulo. *Estudos Avançados, 28*(82), 41–63.

Dubbeling, M., Santini, G., Renting, H., et al. (2017). Assessing and planning sustainable city region food systems: Insights from two Latin American cities. *Sustainability, 9*(8), 1455.

Dulal, H. B., Brodnig, G., & Onoriose, C. G. (2011). Climate change mitigation in the transport sector through urban planning: A review. *Habitat International, 35*(3), 494–500.

European Commission (EC). (2006). Targeted summary of the European sustainable cities report for local authorities.

Franco, F. M., D'Almeida, C. H., & Abreu, G. K. M. (2015). A macroárea de estruturação metropolitana de São Paulo: o projeto urbano como instrumento de transformação do território. *Revista Iberoamericana de Urbanismo, 12*.

Frantzeskaki, N., Broto, V. C., Coenen, L., et al. (2017). Urban sustainability transitions: The dynamics and opportunities of sustainability transitions in cities. In N. Frantzeskaki, V. C. Broto, L. Coenen, et al. (Eds.), *Urban sustainability transitions* (pp. 1–20). Routledge.

Geels, F. W. (2004). From sectoral systems of innovation to socio-technical systems: Insights about dynamics and change from sociology and institutional theory. *Research Policy, 33*(6–7), 897–920.

Geels, F. W. (2011). The multi-level perspective on sustainability transitions: Responses to seven criticisms. *Environmental Innovation and Societal Transitions, 1*(1), 24–40.

Hansson, S., & Polk, M. (2018). Assessing the impact of transdisciplinary research: The usefulness of relevance, credibility, and legitimacy for understanding the link between process and impact. *Research Evaluation, 27*(2), 132–144.

Hoff, H. (2011). *Understanding the nexus background paper for the Bonn 2011 conference*. The Water, Energy and Food Security Nexus.

Hoff, H, & Kasparek, M. (2016). The water – energy – food security Nexus. German Technical Cooperation: Analysis of the Project Portfolio and Assessment of Opportunities for Nexus Mainstreaming. Report of the GIZ Project "International Water Policy" (PN: 2014.2264.1).

Jacobi, P. (2006). Dilemas socioambientais na gestão metropolitana: do risco à busca da sustentabilidade urbana. *Política & Trabalho. Revista de Ciências Sociais, 25*, 115–134.

Jacyna, M., Wasiak, M., Lewczuk, K., et al. (2017). Noise and environmental pollution from transport: Decisive problems in developing ecologically efficient transport systems. *Journal of Vibroengineering, 19*(7), 5639–5655.

Lehmann, S. (2018). Implementing the Urban Nexus approach for improved resource-efficiency of developing cities in Southeast-Asia. *City, Culture and Society, 13*, 46–56.

Ligue os Pontos. (2020). Quem são os produtores agrícolas da Zona Sul de São Paulo. Informes Urbanos, Prefeitura Municipal de São Paulo, n°45. Available at: https://www.prefeitura.sp.gov.br/cidade/secretarias/upload/desenvolvimento_urbano/arquivos/45_IU_PRODUTORES-AGRICOLAS_2020_final.pdf. Accessed 28 July 2021.

Luederitz, C., Schäpke, N., Wiek, A., et al. (2017). Learning through evaluation – A tentative evaluative scheme for sustainability transition experiments. *Journal of Cleaner Production, 169*, 61–76.

Markard, J., Raven, R., & Truffer, B. (2012). Sustainability transitions: An emerging field of research and its prospects. *Research Policy, 41*(6), 955–967.

Mauser, W., Klepper, G., Rice, M., et al. (2013). Transdisciplinary global change research: The co-creation of knowledge for sustainability. *Current Opinion in Environmental Sustainability, 5*(3–4), 420–431.

McCormick, K., Anderberg, S., Coenen, L., et al. (2013). Advancing sustainable urban transformation. *Journal of Cleaner Production, 50*, 1–11.

Mintrom, M., & Luetjens, J. (2017). Policy entrepreneurs and problem framing: The case of climate change. *Environment and Planning C: Politics and Space, 35*(8), 1362–1377.

Moriarty, P., & Honnery, D. (2013). Greening passenger transport: A review. *Journal of Cleaner Production, 54*, 14–22.

Nevens, F., Frantzeskaki, N., Gorissen, L., et al. (2013). Urban transition labs: Co-creating transformative action for sustainable cities. *Journal of Cleaner Production, 50*, 111–122.

Roux, D. J., Stirzaker, R. J., Breen, C. M., et al. (2010). Framework for participative reflection on the accomplishment of transdisciplinary research programs. *Environmental Science & Policy, 13*(8), 733–741.

Scalia, M., Barile, S., Saviano, M., et al. (2018). Governance for sustainability: A triple-helix model. *Sustainability Science, 13*(5), 1235–1244.

Sepe, P. M., & Pereira, H. M. S. B. (2015). *O conceito de Serviços Ambientais e o Novo Plano Diretor de São Paulo: Uma nova abordagem para a gestão ambiental urbana?* Anais do XVI ENANPUR: Espaço, planejamento & insurgências: Belo Horizonte, MG.

Serrao-Neumann, S., Di Giulio, G., & Choy, D. L. (2020). When salient science is not enough to advance climate change adaptation: Lessons from Brazil and Australia. *Environmental Science & Policy, 109*, 73–82.

Setzer, J., & de Macedo, L. V. (2015). Combining local and transnational action in the adoption and implementation of climate policies in the city of São Paulo. In C. Johnson, N. Toly, & H. Schroeder (Eds.), *The urban climate challenge: Rethinking the role of cities in the global climate regime* (pp. 101–118). Routledge.

Uittenbroek, C. J., Janssen-Jansen, L. B., Spit, T. J., et al. (2014). Political commitment in organising municipal responses to climate adaptation: The dedicated approach versus the mainstreaming approach. *Environmental Politics, 23*(6), 1043–1063.

Venghaus, S., & Hake, J. F. (2018). Nexus thinking in current EU policies–the interdependencies among food, energy and water resources. *Environmental Science & Policy, 90*, 183–192.

Wahl, D., Ness, B., & Wamsler, C. (2021). Implementing the urban food–water–energy nexus through urban laboratories: A systematic literature review. *Sustainability Science*, 1–14.

Wanner, M., Hilger, A., Westerkowski, J., et al. (2018). Towards a cyclical concept of real-world laboratories: A transdisciplinary research practice for sustainability transitions. disP-the. *Planning Review, 54*(2), 94–114.

Williams, J., Bouzarovski, S., & Swyngedouw, E. (2014). *Politicising the nexus: Nexus technologies, urban circulation, and the coproduction of water-energy*. Nexus Network Think Piece Series, Paper 001.

Yuan, M. H., Chiueh, P. T., & Lo, S. L. (2021). Measuring urban food-energy-water nexus sustainability: Finding solutions for cities. *Science of the Total Environment, 752*, 141954.

The Challenges of the Food, Water, and Energy Nexus and Potential Interlinkages with Instruments to Tackle Climate Change: Cases of Brazilian Cities

16

Pedro Roberto Jacobi, Sophia B. N. Picarelli, Larissa Miranda Heinisch, and Paulo Antonio de Almeida Sinisgalli

Abstract

The key actions to advance knowledge about sustainable and resilient urban governance are identifying and analyzing innovative initiatives for managing the food, water, and energy nexus (FWEN) in cities. Rapid urban changes in the world's cities are placing unprecedented demands on energy, water, food, and other systems, as each of them offers multiple life-supporting services. Identifying intersections across a diverse set of social actors and institutions represents a relevant agenda that can influence priority outcomes to urban communities, city regions, and their supporting ecosystems. It can be seen as a relevant strategy to advance toward a more integrated territorial planning in cities to strengthen actions to promote sustainable use of natural resources. These drivers have the potential to foster collaborative, functional, and transformative responses in the contexts of an institutional interplay, aiming to encourage co-management, bridging organizations, and social entrepreneurship among other stakeholders. From this background research, the chapter then presents the cases of the Brazilian cities of São José dos Campos and Florianópolis within the context of the "IFWEN – Understanding Innovative Initiatives for Governing Food, Water and Energy Nexus in Cities" project – as a basis to

P. R. Jacobi (✉)
Environmental Science Program of the Institute of Energy and Environment, Universidade de São Paulo (PROCAM-IEE/USP), São Paulo, SP, Brazil

Institute of Advanced Studies at the University of São Paulo (IEA/USP), São Paulo, SP, Brazil

S. B. N. Picarelli
Environmental Science Program of the Institute of Energy and Environment, Universidade de São Paulo (PROCAM-IEE/USP), São Paulo, SP, Brazil

ICLEI Local Governments for Sustainability, South America Secretariat (ICLEI-SAMS), São Paulo, SP, Brazil

L. M. Heinisch
ICLEI Local Governments for Sustainability, South America Secretariat (ICLEI-SAMS), São Paulo, SP, Brazil

P. A. de Almeida Sinisgalli
Environmental Science Program of the Institute of Energy and Environment, Universidade de São Paulo (PROCAM-IEE/USP), São Paulo, SP, Brazil

Institute of Advanced Studies at the University of São Paulo (IEA/USP), São Paulo, SP, Brazil

School of Arts, Sciences and Humanities, University of São Paulo (EACH/USP), São Paulo, SP, Brazil

L. L. B. Lazaro et al. (eds.), *Water-Energy-Food Nexus and Climate Change in Cities*, Sustainable Development Goals Series, https://doi.org/10.1007/978-3-031-05472-3_16

discuss this integrated perspective. The cases of São José dos Campos and Florianópolis illustrate, respectively, how actions on the FWEN for environmental protection and organic food production can contribute to the debate regarding innovation in urban governance and potential nexus enhancement. Due to the need to keep improving municipal planning, governance, and implementation of effective measures, a discussion is conducted regarding the potential of Local Climate Action Plans to increase the integration among sectors and actors. These insights arise from the cases of Recife and Fortaleza and their experience to tackle climate change and indicate that it might be possible to also foster the FWEN and enhance integrated governance at the municipal level. Moreover, this chapter also addresses insights from the cities of Recife and Fortaleza in Brazil regarding climate change adaptation and mitigation, focusing on their Local Climate Action Plan update under the "Urban-LEDS Phase II" project. These plans might be a strategic tool to foster FWEN and more integrated governance.

Keywords

Local Climate Action Plans · Integrated governance · Adaptation · Mitigation · FWEN

16.1 Introduction and Context

According to the World Economic Forum's (WEF) Global Risks Report (2021), climate action failure continues to be one of the main highest impact risks of the next decade, together with environmental risks, as well as weapons of mass destruction, livelihood crises, debt crises, and IT infrastructure breakdown, even though infectious diseases are in the top spot (WEF, 2021).

The current COVID-19 global pandemic, which is placing unprecedented impacts in so many countries and cities (Deslatte et al., 2020; Douglas et al., 2020; Harari, 2021), clearly reinforces planetary interdependence processes that can result in distinct scales of risks around the globe (Fortes & Ribeiro, 2014). Issues, such as the rapid loss of biodiversity in cities, should be considered as part of the need to connect variables that link water- and foodborne diseases and zoonosis (Dasgupta, 2021). The Organization for Economic Co-operation and Development (OECD) in "building back better" pandemic recovery plan highlights the need to invest in sustainable environmental management practices and conservation and restoration projects, as it can help prevent future pandemics and foster resilience and socio-environmental innovation in cities.

Fragmentation, technical-administrative barriers, political disputes, social polarization, and economic interests, among others, are components that bring even more complexity to the development and implementation of strategies to reduce potential negative risk impacts (Moran, 2011; Beck, 2018). According to Beck (2018, p. 15) "we live in a world that is not only changing but is metamorphosing, and this implies that some things change while others remain static." Thus, the concept of metamorphosis implies the destabilization of certainties of modern society. In current times, global risks create new forms of communities, and the public is increasingly intertwined by globalization. These in turn make global risks visible and political, as is the case of the COVID-19 pandemic. Within this changing reality, complexity demands the strengthening of hybrid scientific objects; the breaking down of knowledge boundaries, and hierarchies of knowledge, which is effective through cross sections and collaborative dynamics, as it challenges advances in disciplinary boundaries; and fertilizing exchanges, to promote awareness for change and uncertainty (Jacobi et al., 2020).

Urban areas can account for around 70% of CO_2 emissions from global energy use (Seto et al., 2014). Hosting more than half of the world's population, cities concentrate most of the built assets and economic activities, reinforcing their vulnerability to climate change. Impacts caused by climate changes are already being felt in urban areas and have been increasing in recent years, such as rising temperatures, rising sea levels, heat islands, floods, water and food shortages, and extreme events. Most Brazilian cities also face challenges associated with patterns of development and urban expansion (Klug et al., 2016). Changes in the hydrological cycle by global warming tend to accentuate these existing challenges and risks, such as floods, landslides, heat waves, and drinking water supply (PBMC, 2016).

Evaluations of these interrelationships between systems and complex problems, such as climate change, pandemics, and so many social challenges, are not simple, especially for the management of urban areas. However, some approaches reinforce that to seek better solutions to complex problems, it is important to consider broad, systemic, interdisciplinary, multisectoral, and participatory aspects, such as the concept of the water-energy-food nexus (Schulterbrandt-Gragg et al., 2018).

Under the Brazilian context, the Ministry of Regional Development is currently developing the National Policy of Urban Development, which aims to provide a more strategic vision of the diverse Brazilian territory, adopting a multilevel, inter-federative, inter-sectoral, and interinstitutional approach, and a systemic perspective, which incorporates cross-cutting themes and dimensions of economic development (Bruno, 2020). It is also important to mention some of the main planning and management instruments already established by national laws, which have potential linkages with water-energy-food nexus, such as the Municipal Master Plan (Plano Diretor), Municipal Sanitation Plan, Municipal Urban Mobility Plan, Municipal Plan for Disaster Risks Reduction, *Mata Atlantica* Municipal Plan, and Municipal Housing Plan of Social Interest. Even though national laws do not establish them, some Brazilian municipalities further approve specific municipal laws or instruments to develop and implement plans that can support to tackle local challenges, such as the Municipal Sustainability Development Laws, Food Security Laws, Urban Forestry Plan, Water Security Plan, Payment for Environmental Services Municipal Laws, Energy Efficiency Plan, and Local Climate Action Plans (LCAP), among others (CNM, 2014; Costa & Favarão, 2016). Regarding the LCAP, this chapter considers insights from two Brazilian cities' recent experience – Recife and Fortaleza – related to the development of their Local Climate Action Plans, especially in terms of the potential to strengthen the possible synergies between municipal sectors and increase co-benefits and nexus logics. All initiatives present a direct engagement of ICLEI – Local Governments for Sustainability – which provided access to the content development.

The interlinkage reflections consider if the FWEN can be increased through climate change planning instruments at the local level – since these instruments might present a potential to overcome some sectorial barriers and find innovative and more inclusive approaches to increase an effective and more efficient overlap between sectors aiming to enhance the co-benefits for social-environmental aspects.

In the section that follows, we present the study cases of two Brazilian cities – Florianópolis and São José dos Campos – under the nexus logic. Both cases present a stronger focus in one specific nexus' sector and also great potential to scale up the integration with other sectors and especially under a more integrated municipal governance. Then a discussion, based on the reflections and insights from the Local Climate Action Plans' experiences and background research and analysis, is presented. This chapter is focused on presenting and investigating specific empirical examples, focusing on the nexus governance in resource provision at city levels, if potential synergies can be fostered by the climate change agenda, considering different scales, from the global and national into regional and city levels.

This chapter contributes to scientific investigations which aim to identify ways to improve the nexus at the Brazilian municipal level and the potential enhancement of results that can arise from the "nexus approach," according to the description of the editors of this book – as multidimensional means that seeks to understand the interrelationships and interdependencies between water, energy, and food with climate change, taking into account complex and nonlinear aspects.

16.2 Evaluation of Two Brazilian Nexus Case Studies

This chapter was based mainly on the case review, analyses, and cross-check with literature. In this section, especially the cases of Florianópolis and São José dos Campos developed under the IFWEN project were further described to evaluate if the water-energy-food nexus is already being applied at their specific initiatives.

The "IFWEN – Understanding Innovative Initiatives for Governing Food, Water and Energy Nexus in Cities" project – is a two-year project (2019–2021) financed by the Belmont Forum, JPI Urban Europe e EU Horizon 2020. And it is implemented by the Fundação Getúlio Vargas (FGV); Yale University; ICLEI, Local Governments for Sustainability; The Nature of the Cities (TNOC); Stockholm Resilience Centre (SRC); Ming-Chuan University (MCU); and MIT CoLab.[1]

The IFWEN project has the following key objectives:

1. Assessing the changes in trade-offs of food-water-energy nexus (FWEN) in the Green and Blue Infrastructure (GBI) and their association with spatial planning and governance in cities using empirical cases.
2. Understanding the barriers that hinder innovative and integrated FWEN approaches using GBI at different scales and specifically looking for the standard features of diverse interventions, as well as how successful GBI-based innovations that changed the nexus took place, which capabilities cities had to innovate and how they develop those capabilities and approaches used to overcome the barriers that make the implementation more difficult in practice.
3. Designing a framework, a guide of best practices, and tools to foster IFWEN using GBI with better urban interventions and decision-making processes (Fig. 16.1).

The project has case studies in eight cities across three regions (South America, Africa, and Asia): Antananarivo (Madagascar), Dodoma (Tanzania), Florianópolis (Brazil), Gangtok (India), Johannesburg (South Africa), Lilongwe (Malawi), Nagpur (India), and São José dos Campos (Brazil). This chapter will focus on the discussion in the Brazilian cities – Florianópolis and São José dos Campos cases – especially concerning the work led by ICLEI South American Secretariat. To analyze these contexts and consolidate study cases, the methods applied included surveys, interviews, field visits, literature reviews, and desktop research.

The surveys were applied in an initial phase of the research, before the conduction of interviews. They included questions regarding the panorama of the city, such as its jurisdiction, geographic, socioeconomic, demographic, financial aspects, and legal and institutional matters that affect the FWEN. The surveys were answered separately by each city through an appointed representative who acted as a focal point within the municipality to gather the requested information from different departments and areas of expertise.

Regarding the interviews, these were conducted with the focal point and representatives for each municipality after the survey application to deepen the knowledge on the city's context and innovation initiatives on the FWEN. Questions included inquiries on the municipal departments involved with the nexus and GBI, key local challenges, identifying the innovative initiatives and enabling capabilities that allowed for the nexus, and knowledge management mechanisms on the thematic. The interviews were conducted both virtually and in person, in the latter case enabling site visits to better understand the city context and the specific innovative initiatives identified.

The description and evaluation that follows from both Brazilian cases consider that the nexus concept challenges existing structures, sector policies, and procedures to innovate traditional silo thinking and promote new dynamics of coordination. It is necessary to modify the prevalent logic based on "divided responsibilities that often result in poorly coordinated investments, increased costs and underutilized infrastructure

[1] https://ifwen.org/what-it-is-ifwen/

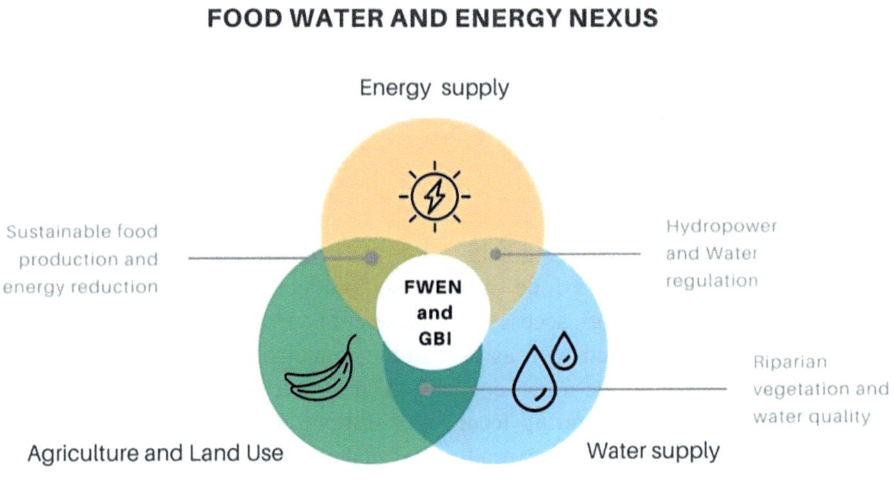

Fig. 16.1 Examples of GBI in the FWEN in the City of São José dos Campos. (Source: Elaborated by the authors within the IFWEN project work)

and facilities" (BMZ, 2014). The great challenge is to connect policies, practices, organizations, and institutions to work together horizontally and vertically to optimize resource use (Hezri, 2016). This requires capacity and institution-building on multiple levels to strengthen articulated governance with connectivity and integration of resource systems by shedding light on the possible synergies or trade-offs that might exist between different resources (C. Zhang et al., 2018).

16.2.1 Florianópolis, State of Santa Catarina, Brazil

16.2.1.1 Context

Florianópolis is located in the southern region of Brazil and is the capital of the State of Santa Catarina. Its population is mainly urban (96.21%) estimated at 492.977 inhabitants in 2018, the 17th densest capital of Brazil with 627.24 hab./km^2 (IBGE, 2010a, b). The city is highly dependent on resources from the surrounding municipalities in the continent, which implies less efficiency and more complex processes, especially regarding FWEN and the trade-offs between these services, the nexus. Dependence from metropolitan/regional and federal spheres of government to support the supply of FWEN in Florianópolis is a central challenge faced by the city. For instance, the leading electric energy supplier of the municipality is the *Centrais Elétricas de Santa Catarina* (CELESC), the company responsible for powering the State of Santa Catarina and 92.7% of the households in Florianópolis (BID et al., 2015). As well as power, water resources are also majorly provided to the city from continental sources by the Catarinense Company for Water and Sanitation (Companhia Catarinense de Água e Saneamento – CASAN). Food provision also heavily depends on continental sources – food production makes up less than 1% of the local economic activity, and the 22 municipalities that constitute the Florianópolis' Metropolitan Nucleus provide Florianópolis and for that are known as the "Green Belt." Tourism contributes to the economy of the city; this activity pressures the FWEN distribution net and accelerates the importation rates of resources to the city, as well as waste production.

16.2.1.2 Florianópolis Case Study

The case for the Florianópolis context was organized through semi-structured exploratory interviews and analysis with the city technical focal points to identify innovative approaches related to the FWEN (at least one sector as strong components). Also, the case should demonstrate a potential for GBI positive impacts and should be associated with municipal planning instruments and governance within the city. It consists of initiatives especially regarding the Municipal Urban Agriculture Program (MUAP – Programa Municipal de Agricultura Urbana). This program was established in Florianópolis in June 2017 through a decree,[2] to promote urban agroecological practices to produce and process food efficiently and locally, incentivizing access to healthy and low-cost food produced with agroecological methodologies within the urban context. Also, the MUAP aims at supporting local organic food producers to commercialize products directly, reducing production chains and transportation requirements. By doing so, the project envisions enhancing the co-benefits from these practices, such as repurposing land use within the urban context, maintaining clean and illegal littering of accessible urban areas, and promoting community participation and engagement (CEPAGRO, 2016).

The *Revolução dos Baldinhos*, "Bucket Revolution," was also key for triggering this initiative. In 2008, professionals from the Health Center in the Monte Cristo neighborhood, one of the most vulnerable and poor areas in the continental portion of Florianópolis, assembled with the *Centro de Estudos e Promoção da Agricultura de Grupo* (CEPAGRO) and local school and nursery representatives to discuss strategies for facing disease-spreading pests and illegal littering (of mainly organic materials) that caused it. This initiative has generated a net of workers responsible for collecting organic matter from houses, processing it, and distributing the composted matter in the urban gardens of the region.

Municipality representatives have identified these successful efforts as key for developing MUAP and addressing citywide challenges mentioned above, such as independence of food production, food security, and waste management. MUAP is funded through specific project notices and receives support from the project coordinators, such as the *Companhia de Melhoramentos da Capital* (COMCAP), responsible for providing technical support for the maintenance of the gardens as wood chips for covering the composting piles and seedling. Also, the program supports the Street Fair at the *Alfandega* Square at the center of the city. This allows food producers to sell products directly to consumers, shortening production circuits and enhancing the autonomy of the municipality in food production.

The initiative has provided the city with the possibility to reuse organic waste by composting the material and repurposing it as fertilizer for organic urban gardens. By doing so, the city has achieved (1) greater independence for food production at the local scale, increasing food safety and security, (2) greater independence for waste management, and (3) less waste destined to landfills (Florianópolis was a pioneer city on this aspect, being the first Brazilian city to approve a municipal law in 2019[3]). These direct impacts could potentially generate associated co-benefits, such as the following:

1. Reduction of GHG emissions due to a smaller volume of food input to landfills and reduced transportation of food from the continental to the insular portion of the island.
2. Increased water quality and quantity due to reduction on landfill by-products reaching aquifers:
   2.1. More permeable surfaces on urban centers reducing flooding events.
3. Reduced energy use due to the shorter production chain.
4. Qualities provided by urban green areas, such as better air quality, water infiltration, and micro-climate regulation, and provision of habitat for urban fauna.

[2] https://leismunicipais.com.br/a/sc/f/florianopolis/decreto/2017/1769/17688/decreto-n-17688-2017-dispoe-sobre-a-criacao-do-programa-municipal-de-agricultura-urbana?q=17688

[3] https://leismunicipais.com.br/a/sc/f/florianopolis/lei-ordinaria/2019/1051/10501/lei-ordinaria-n-10501-2019-dispoe-sobre-a-obrigatoriedade-da-reciclagem-de-residuos-solidos-organicos-no-municipio-de-florianopolis

In Florianópolis, the most substantial FWEN sector is related to the food sector. The case study is already an initiative focusing on food security and the improvement of local agroecological practices. More linkages were identified between water and energy sectors, such as related sustainable local food production leading to water quality maintenance and reduction in energy demand due to optimization of production chains and transportation. A direct link with all the three nexus sectors was not identified. It is also important to highlight that waste, although not a sector of the FWEN, appears as an important component, in this case, fostering the use of organic compost for gardens and agriculture under circular development principles and avoiding organic waste in landfills.

The image below represents the interlinkages and preponderance of nexus sectors related to Florianópolis analysis (Fig. 16.2).

16.2.2 São José dos Campos, State of São Paulo, Brazil

16.2.2.1 Context

São José dos Campos (SJC) is located in the eastern portion of the State of São Paulo, and it is the main municipality in the Metropolitan Region of *Vale do Paraíba*. The Paraíba Valley is situated between the cities of São Paulo and Rio de Janeiro and stands out for concentrating a considerable part of the regional GDP (gross domestic product). It is also known for being the most important aeronautical and aerospace hub in Latin America and hosting important federal scientific research institutes, technology companies, universities, colleges, and training centers to develop labor skills. Due to these aspects, the city is considered a hub of innovation in the region. In the Brazilian context, even though SJC is the fifth most populous city in the State of São Paulo with 629.921 people (2010) (IBGE, 2010a, b), the population density of the municipality is small compared to other large urban centers in Brazil. SJC is situated in a privileged area within the State of São Paulo that allows access to water sources – both for supplying the population and with hydroelectric potential – and to floodplains used for food production. Still, the development and urbanization history of the city have been heavily dependent on the nearby centers – São Paulo and Minas Gerais – and resulted in a profound dependence of São José dos Campos on important food, water, and energy. Also, enlightening the nexus between these elements and tracing more efficient technologies and policies to do so are a central challenge.

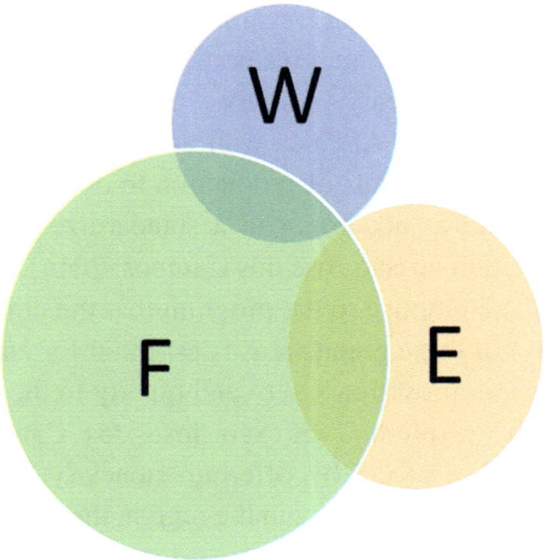

Fig. 16.2 Current integration and preponderance of nexus sectors from the Florianópolis study case. (Source: Elaborated by the authors)

Although the city does not have programs or policies explicitly aimed at the FWEN, SJC was chosen for this study due to the initiatives that indirectly target this nexus. The initiatives that have the most potential to generate a positive impact in this area are the incentives for innovation (such as the recently built innovation center and the municipal law to encourage innovation), the protected areas of the municipality, and the programs for environmental compensation.

16.2.2.2 São José dos Campos Case Study

The municipality of São José dos Campos has some relevant examples of developing programs and initiatives that explore and enhance its geospatial potentials, aligning conservation with technological development in public programs undertaken regarding FWEN elements. Also, the city has invested in projects that employ the potential of reforestation and conservation as GBI, an innovative strategy within the public arena in Brazil. Among these strategies existent in SJC toward a self-sufficient, technological, and sustainable city are portrayed below, two are included in this chapter, and were selected for deeper analysis within the IFWEN project through semi-structured exploratory interviews and analysis with the city technical focal points, due to their innovative approaches in the nexus (if not possible to all three thematic with at least one as strong components), related to GBI positive impacts, and associated with spatial planning and governance in the city: (I) the Alluvial Plains of the Rivers Paraíba do Sul and Jaguari Protected Area and (II) the Municipal Program for Payment of Environmental Services – *Programa Mais Água* (More Water Program).

I. *The Alluvial Plains of the Rivers Paraíba do Sul and Jaguari Environmental Protected Area.*

This municipal protected area was established in 1984 by the São José's Municipal Master Plan (Prefeitura de São José dos Campos, 1984) and currently is also known as the Banhado Environmental Protection Area (Banhado EPA), which is a sustainable use category under the *Sistema Nacional de Unidades de Conservação da Natureza* (SNUC) – the official Brazilian set of rules and procedures to create, implement, and manage protected areas. This is an innovative policy that supported conservancy in a context of growing urbanization trends (Fantin, 2005) and that guaranteed the protection of this area of crucial ecological importance for the maintenance and provision of essential environmental ecosystem services and quality of life of citizens, within this urban center: retention and processing of sediments, water quality, erosion and silting up control, filtering of toxins and sediments, recharge and discharge of aquifers, and buffering zone for floods (Figueroa, 1996). Moreover, this is an important area for the SJC population's quality of life, promoting key ecosystem services such as being an important feature of the city's landscape and improving air quality.

Therefore, restrictions on land use in these areas directly preserve these functions, improving water, soil, and air quality and improving flood mitigation capacity. Moreover, the Banhado EPA has been categorized as a "Zone One environmental protected area" by the SJC master plan, which is a category aimed at protecting areas that occupy the margins of rivers with hydromorphic formation and that are highly and naturally susceptible to floods. This legislation also permitted that the land use in this area can be partially destined for livestock and agricultural research and production (Prefeitura de São José dos Campos, 2019), reinforcing the food pillar of the FWEN and its correlation with GBI and water.

II. *Municipal Program for Payment of Environmental Services – Programa Mais Água.*

A municipal law[4] established in 2012 the *Programa Mais Água* (More Water Program). This program aims at promoting the conservation of areas with native vegetation in São José dos Campos and encourages sustainable practices in rural areas to protect water springs in the region, seeking increased water availability. Rural landowners need to commit to the program and follow the regulations regarding conservation of native vegetation and restoration of permanently protected areas (Áreas de Preservação Permanentes – APP) as well as the adoption of sustainable models of production to obtain monetary returns according to a standardized score (Prefeitura de São José dos Campos, 2014).

It is a premise to the program that the conservation of these green areas is essential for ensuring water provision safety and quality to the city of SJC (Prefeitura de São José dos Campos, 2015). Therefore, by offering monetary incentives to the landowners in the region, the program acknowledges the many ecosystem services pro-

[4] https://servicos2.sjc.sp.gov.br/legislacao/Leis/2012/8703.pdf

vided by the preserved areas, especially as water cycle regulation systems. Despite focusing on the potential of forests as green infrastructure for water regulation, the program also presents a clear relationship with food production in the region, as the majority of the areas contemplated by the program are destined for livestock production – 60% (Prefeitura de São José dos Campos, 2017). Properties with more sustainable practices (no pesticide use and organic production) are better evaluated by the standardized tests (Prefeitura de São José dos Campos, 2015) and get a higher reward.

The content from the case of São José dos Campos is mostly related to the water sector. The initiatives focus on the environmental management of protected areas and payment for ecosystem services related to water and springs conservation. The stronger identified link is between water and food sectors since there is a promotion of sustainable practices in rural areas to protect water springs in the region, seeking increased water availability. Another relevant observation is that none of the actions presented a direct link with all the three nexus sectors.

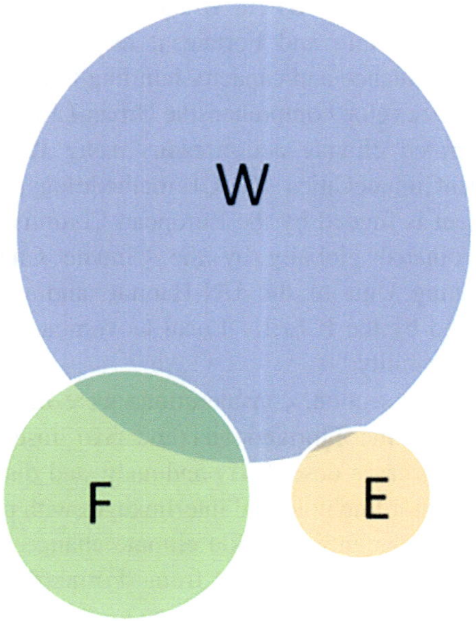

Fig. 16.3 Current integration and preponderance of nexus sectors from the São José dos Campos study case. (Source: Elaborated by the authors)

The image below represents the interlinkages and preponderance of nexus sectors related to the São José dos Campos analysis (Fig. 16.3).

16.3 Discussion

In his analysis of the WEF, Covarrubias (2021: 18) emphasizes that "the relationships for city authorities is different for either water, energy or food, because of the different geographic scales, and because of the division of governance responsibilities for each of these flows take place at different contexts." And as Covarrubias (2019) points out, this requires innovative and cross-sectorial systems of provisioning. What can be observed is that the prevailing logic is that city planning and governance still approach resources such as WEF as separate domains, ignoring their interconnectedness (Hoff, 2011; Howels & Rogner, 2014).

The nexus approach is also presented as a governance scheme that proposes a broader dialogue between stakeholders aiming to balance the trade-offs between water-energy-food sectors (Hoff, 2011). However, Brazilian municipalities still face many challenges developing, especially implementing, more integrated municipal plans and their corresponding actions. Moreover, it is crucial to find ways to improve local planning and governance processes (Benites-Lazaro & Giatti, 2020; Cabral, 2018).

This issue of local governance fragmentation can be noticed in the cases of Florianopolis and São José dos Campos. Despite being innovative initiatives, they still focus more strongly on one FWEN component and have more autonomous characteristics, not yet integrating a range of different municipal departments and stakeholders. A holistic approach to tackling global challenges could be a way to foster the integration between distinct challenges and opportunities at the local level.

The FWEN has become an important reference to urban planning and represents a new perception (Hoff, 2011) due to its potential to promote effective adaptation (Rasul & Sharma, 2015; Gondhalekar & Ramsauer, 2017) and innovative urban development. Considering the effects of climate change on increasing demands for natural resources and insecurities in water, energy, and food supply, it is imperative to avoid failures in management and governance strategies (Fernandes Torres et al., 2019).

Despite not being a legal obligation for Brazilian municipalities, local leaders have become more aware that urgent efforts are required to scale up climate action, dealing with climate change mitigation, adaptation, and resilience (Arikan et al., 2020), and are developing local instruments to tackle climate change.

So far, most efforts by cities to respond to climate change have focused on mitigation (i.e., reduction of greenhouse gas emissions (GHG)) and much less on adaptation (i.e., strategies to reduce exposure and susceptibility and improve the coping capacity of communities to hazards) as these strategies imply taking a precautionary and anticipatory approach (Broto & Bulkeley, 2012; Torres, 2020; Neder et al., 2021, Di Giulio et al., 2017; Di Giulio et al., 2018). However, the implementation of adaptation plans is urgent. Changes in global climate are already underway, and social, infrastructural, and economic costs of inaction are high (Geneletti & Zardo, 2015), directly affecting water-energy-food sectors.

Aiming to increase local climate actions, the Global Covenant of Mayors for Climate and Energy (GCoM) is one of the largest movements engaging municipalities in this agenda. This global alliance for city climate leadership has raised the commitment of over 10,000 cities and local governments that share a long-term vision of supporting voluntary action to combat climate change. Cities that join the GCoM should advance with their specific climate compliance: pledge to develop a greenhouse gas emission inventory and assess climate risks and set measurable emission targets, ambitious climate adaptation goals, and sustainable energy access goals in line with the Paris Agreement. Signatories agree to formally adopt plans and targets within 3 years of signing the commitment (GCoM, 2021). The Urban-LEDS project is one of the global initiatives that directly contributes to GCoM, collaborating to achieve climate compliance. And in Brazil, it provided specific support to Recife and Fortaleza to develop their current Local Climate Action Plans, connected to their respective municipal laws.

Besides the methods applied within the IFWEN project scope for both Brazilian cases, aiming to find instruments that can support Brazilian municipalities to enhance their planning and governance strategies, an exploratory question was elaborated: can municipal instruments to tackle climate change increase the interconnectivity of specific water-energy-food sectors at the local level?

For this specific purpose, additional literature review and qualitative analysis were conducted on the topics of climate change adaptation and mitigation, focusing on the experience of two Brazilian cities – Recife and Fortaleza – which used the same methodology to update their Local Climate Action Plan under the Urban-LEDS Phase II project.

The Urban Low Emissions Development Strategy (Urban-LEDS) addresses integrated low-emission and resilient development in more than 60 cities in 8 countries: Brazil, India, Indonesia, and South Africa (from Phase I) and Bangladesh, Colombia, Lao PDR, and Rwanda (added in Phase II). In Brazil, selected cities, such as Recife and Fortaleza, received tailor-made guidance and capacity building to consolidate or develop comprehensive Urban-LEDS and integrated climate action plans using ICLEI's GreenClimateCities (GCC) methodology. The project is funded by the European Commission, coordinated globally by the Climate Change Planning Unit of the UN-Habitat, and implemented by the ICLEI – Local Governments for Sustainability.[5]

In this section, considerations were made in terms of the Florianópolis and São José dos Campos cases described previously, and discussions regarding potential interlinkages with planning instruments to tackle climate change were raised, using examples from Fortaleza and Recife.

[5]https://urban-leds.org/

This combined process resulted in the analysis and discussion presented in this chapter.

16.3.1 Insights from Local Climate Action Plans as a Tool to Increase the Nexus Potential

Aware of the fact that climate change can affect different aspects of the municipal dynamics, especially emphasizing the existing challenges, Recife and Fortaleza updated throughout 2020 their Local Climate Action Plans (LCAPs), having as the main goal the neutralization of GHG emissions by 2050 and improvement of their citizens' life quality (ICLEI, 2020a, b).

The LCAPs demonstrate how the city strategically plans to reduce GHG emissions and adapt to the consequences of climate change. It aims to provide an alignment between sectors and municipal departments, existing legislation, planned projects, and actions. It is an executive, instrumental, and dynamic document that presents the city's level of ambition in planning climate change mitigation and adaptation (ICLEI, 2020a, b).

The process to develop the plans is part of the GreenClimateCities (GCC) methodology, composed of nine steps in three phases – analyze, act, and accelerate – outlining how climate risks and vulnerabilities can be assessed and options (to achieve low-to-no emission development and climate adaptive development) can be identified and integrated into urban development policies, plans, and instruments. It is also associated with measuring, reporting, and verification (MRV) logic and consists of a wide range of resources, tools, and guidance to support local and regional governments to deliver ambitious climate action.

The plan is further based on UN-Habitat's guiding principles which state that climate action plans should be ambitious, inclusive, fair, comprehensive, integrated, relevant, feasible, evidence-based, transparent, and verifiable (Un-Habitat, 2015).

Taking into account the guidelines of the GCC methodology and the UN-Habitat's principles for climate action plans, relevant actors from both cities' public sector and civil society were identified to contribute to the plan development directly.

The main steps for the content of both cities' plan development were the following:

(a) Analysis of legislation in place at the city, state, and national levels related to climate aspects.
(b) Identification of existing commitments, sectoral plans, instruments, and other documents that demonstrate the city's commitment to the climate agenda.
(c) Interviews with public managers and officers, representing the following sectors/departments: urban planning, infrastructure, finance, education, transport and mobility, sanitation, housing, and environmental and urban services.
(d) Evaluation of GHG emission scenarios through municipal emission inventories (using the Global Protocol for Community-Scale GHG Emissions – GPC methodology). The most recent emission inventory results were defined as the baseline for the projection of future emission scenarios. Evaluation of the municipal climate risk assessment, which identifies the main climate hazards and potential social impacts, is also presenting projection under future scenarios.
(e) Meeting with representatives of Recife and Fortaleza's youth leaders, to collect perceptions and suggestions from the different realities of the city, contributing to the refinement of principles and actions.
(f) Organization of participatory workshops and formal meetings with Sustainability and Climate Change Municipal Committees – COMCLIMA, GECLIMA, and FORCLIMA – in addition to representatives from academia, the third sector, civil society, and the private sector to present, prioritize, and validate the main actions to be considered by both plans (ICLEI, 2020a, b).

Four strategic axes were defined as priorities for mitigation and adaptation to climate change in both cities, detailing the goals, targets, actions, and results aiming to achieve the ambitious scenario outlined according to the baseline and strongly avoiding the business-as-usual (BAU) scenario. The four axes are energy, sanitation, mobility, and resilience (ICLEI, 2020a, b).

Analyzing these four main axes defined as priorities to tackle climate change at the municipalities' jurisdictions, its related goals, targets, and actions defined under both LCAPs present considerations for a specific question: can municipal instruments to tackle climate change increase the interconnectivity of specific water-energy-food sectors at the local level?

16.3.2 Can Municipal Instruments to Tackle Climate Change Increase the Interconnectivity of Specific Water-Energy-Food Sectors at the Local Level?

The information was organized from Fortaleza and Recife, under a table that brings together the main sectors of nexus (water-energy-food) and LCAP axes (energy, sanitation, mobility, and resilience).

The analysis only used the information available in the documents, and any interpretation of further reflections was made in terms of potential results or unfolding that could result from concrete implementation:

- In Table 16.1, the specific content of the goals, targets, and actions, specified under Recife and Fortaleza LCAPs, of energy, sanitation, mobility, and resilience is presented in the lines and located under the corresponding "nexus column." Merged cells represent a content that corresponds to two nexus sectors.

The organization of the content from the goals, targets, and actions established under the four axes of the LCAPs from Recife and Fortaleza (energy, sanitation, mobility, and resilience) allowed visualization of the linkages with the water-energy-food nexus sectors. Only the content related to at least one sector of the nexus was inserted in the table.

The outcomes show that most of the contents is related to the water sector. The majority of the identified linkages are between the water and energy sectors. This might be related to one of the main goals of the LCAPs which is the neutralization of GHG emissions by 2050. Another relevant observation is that none of the actions presented a direct link with all the three nexus sectors. It is important to note that the analysis only used the information written in the documents.

The image below represents the interlinkages and preponderance of nexus sectors related to Recife and Fortaleza analysis (Fig. 16.4).

Two actions that already present a link between two nexus sectors were chosen, from Recife and Fortaleza Local Climate Action Plans, to evaluate if possible results, or unfoldings related to concrete implementation, could provide further reflections on the three nexus sectors integration. The examples are shown here:

- Implementing ecological sanitation measures, such as biological treatment micro-station or constructed wetland, can generate organic waste/elements that could be used to produce compost for community gardens and agroforestry, among others, which would then link also with the food sector.
- Implementing low-impact development (LID) measures for water management in urban watersheds to reduce the speed of rainwater runoff, increase water infiltration into the soil, and provide alternative sources for non-primary uses. These could reduce the consumption of energy related to water or wastewater pumping and/or increase water availability in reservoirs for energy generation, which would then link also with the energy sector.

Following this exercise, one action was chosen from Florianópolis and São José dos Campos cases, which already present a link between two nexus sectors, to evaluate if potential results, or unfoldings related to concrete implementation,

Table 16.1 Analysis of Recife and Fortaleza regarding the FWEN sectors and Local Climate Action Plan axes

<table>
<tr><td colspan="5">Nexus sectors</td></tr>
<tr><td></td><td></td><td>Food</td><td>Water</td><td>Energy</td></tr>
<tr><td rowspan="9">Local Climate Action Plan Axes</td><td rowspan="4">Sanitation</td><td rowspan="4">Install, by 2024 (Rec and For), composting plant(s) with targeted products for community gardens and/or organic agriculture, and schools, associating with environmental education initiatives</td><td colspan="2">Universalize sewage services with the assurance that the treatment of wastewater is carbon neutral by 2050, also increasing the use of methane for energy generation (Rec and For)</td></tr>
<tr><td colspan="2">Implement and evaluate the benefits of other ecological sanitation measures, such as biological treatment micro-station, constructed wetland, evapotranspiration tanks, etc. (Rec and For).</td></tr>
<tr><td rowspan="2"></td><td>Reduce, by 2050, waste disposal in landfills up to 50% (Rec) and 60.6% (For).</td></tr>
<tr><td>Ensure that the methane emitted by the decomposition of waste in landfills is burned or used for energy in 60% by 2030 and 100% from 2037 (Rec) and 50% by 2030 and 100% by 2D40 (For)</td></tr>
<tr><td rowspan="2">Energy</td><td rowspan="2"></td><td colspan="2">Ensure, by 2037 (Rec) and by 2040 (For), 100% of electricity supplied is from renewable sources</td></tr>
<tr><td></td><td>Reduce energy consumption in all economic sectors by 20% (Rec) and 27,6% (For) by 2050, compared to the BAU scenario</td></tr>
<tr><td rowspan="3">Mobility</td><td rowspan="3"></td><td>Strengthen the resilience capacity of public transport systems upon extreme weather events (Rec)</td><td>Ensure that the public transport fleet is composed of 100% electric vehicles by 2050 (For and Rec)</td></tr>
<tr><td colspan="2">Assess, by 2023, the results of the solar boat project and the potential business model for scalability (Rec)</td></tr>
<tr><td>Promote, by 2037, the necessary interventions to enable the navigability of the Capibaribe and Beberibe rivers (Rec)</td><td></td></tr>
</table>

(continued)

Table 16.1 (continued)

Nexus sectors				
		Food	Water	Energy
	Resilience		Foster integrated water management to reduce climate risks and guarantee water security (Rec)	
			Ensure access and supply of drinking water to the entire population by 2025 (Rec).	
			Identify water bodies that can undergo revitalization and renaturalization processes by 2023 (Rec)	
			Create an Urban Water Committee to promote coordination between the institutions responsible for supply, sewage, and drainage (For)	
		Implement LIDs in urban watersheds to reduce the speed of runoff of rainwater, increase water infiltration into the soil, and provide alternative sources for non-primary uses, such as green roofs, rain gardens, bio-swales, cisterns, wetlands, permeable paving, etc. (For)		
		Promote systemic management of green areas, including afforestation, urban agriculture, municipal parks, and APPs (Rec and For)		
		Promote practices of ecological agriculture and agroforestry systems, providing technical support (Rec)		
		Recognize urban agriculture initiatives that value gender equity and cultural diversity (Rec)		

Source: Elaborated by the authors.
(For) content-specific from the city of Fortaleza; (Rec) content-specific from the city of Recife

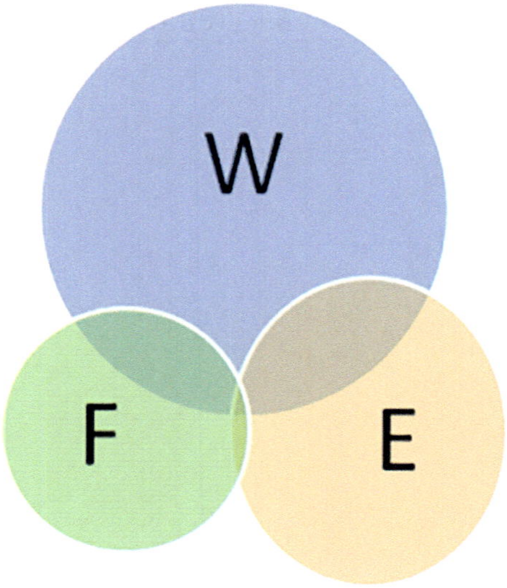

Fig. 16.4 Current integration and preponderance of nexus sectors considered under Recife and Fortaleza Local Climate Action Plans. (Source: Elaborated by the authors)

could provide further reflections on the three nexus sectors integration. The examples are shown here:

Florianópolis

- The use of organic compost for gardens and urban agriculture could also generate renewable energy through biogas and reduce waste in landfills that do not capture the methane, which would then link also with the energy sector.

São José dos Campos

- Promote sustainable agroecological practices in rural areas to protect water springs in the region; seeking increased water availability can also increase water availability in reservoirs for energy generation.

These examples above demonstrate that it might be possible to increase the integration between the FWEN sectors, through a specific GBI measure which could also be included under more cross-cutting municipal instruments, such as Local Climate Action Plans, therefore reinforcing the role of integrated instruments in promoting broader governance of municipal aspects.

The cases analyzed under this section present that the nexus logic can be enhanced in some aspects. According to the image below, one of the main potentials of the nexus approach is to increase the integration between sectors. The cases present one stronger sector – integrating with another sector, but no concrete integration between the three of them, as predicted in a high integration nexus circumstance, was identified (Fig. 16.5).

According to Covarrubias (2020), this lack of integration presents an important possibility to improve the overall sustainability performances of urban infrastructures and governance.

The innovation approaches, specifically focusing on GBI, from Florianópolis and São José dos Campos, already demonstrate some original improvements in more sustainable urban measures. These are also tackling other important specific aspects for local context such as waste management, in the case of Florianópolis, and economic mechanisms, such as the payment for environmental services to small-scale farmers in São José dos Campos.

These examples imply a clear understanding that an "urban nexus" approach assumes that socio-material flows interact and relate to one another in promoting cross-sectoral policies toward the attainment of a synergic urban sustainable development within a context of climate change. The main challenge is to overcome the prevailing logic based on addressing and formulating policies in silos that do not guarantee simultaneous attainment of WEF security as well as environmental sustainability (Bhaduri et al., 2015: 726).

Interdisciplinary approaches to urban nexus can support more collaborative efforts across sectors, scales, and jurisdictions, enabling new technical-scientific and societal capacities to anticipate changes and effectively respond to emerging risks (Sperling & Berke, 2017).

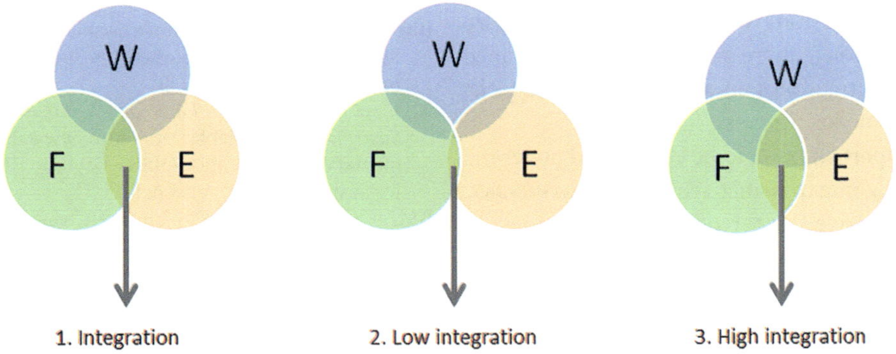

Fig. 16.5 General three types of nexus sectors integration. (Source: Adapted from Sperling and Berke 2017)

Especially with the analyses of the case from Recife and Fortaleza, the process to develop Local Climate Action Plans also presents a potential to use this multidisciplinary municipal instrument to foster this sectoral integration.

16.4 Conclusion

Municipal governments face complex challenges and can be directly or indirectly influenced by global circumstances, as explored under this chapter, especially by climate change or global pandemics, such as COVID-19. Also, facing an internal scenario in the public administration of lack of financial resources, technical capacity, mid-long-term strategic planning, traditional silo thinking, and operating, it is imperative to find better ways to improve existing dynamics.

The Brazilian cases presented in this article demonstrate improvements in parts of these complex approaches and also indicate potential ways to keep increasing more effective integration. The reflections arising from the analysis of the Local Climate Action Plan (LCAP) development methodology indicate a good potential to bring different actors, especially related to nexus sectors, to the same table to define strategic actions to tackle climate change at the local level. It is important to note that, currently, the main LCAP goals still have a stronger focus on GHG emission reduction. However, increasing the capacity to adapt to extreme weather events and improve the overall resilience of the territory, its citizens, and local systems is also an important highlight that is being considered in the recent versions of these plans. In future plans' versions, there might be more balance between mitigation and adaptation goals, tackling other nexus aspects.

An important point to be considered under the Brazilian context is that the concept associated with WEF does not recognize inequalities in access to water, energy, and food. These aspects should be considered as relevant components in building pathways in the sense of sustainability. Also, it maintains the importance associated with the demands of human needs, disregarding the pressures on ecosystems and their services that could put at risk biodiversity conservation or increase climate emergency.

Cities can be seen as nodes that demand increasing resources. It is crucial to articulate sectors, actors, resources, infrastructures, policies, and utility services to reach a satisfactory level for the provisioning of water, energy, and food. Finding ways to unite bottom-up, top-down, multilevel, multi-sectoral, and multi-actor efforts that together can improve local governance seems to be a common imperative from the nexus and climate approaches at the local level. As some cities have been experimenting, initiatives in this direction represent an innovative trend that depends very strongly on the role of local authorities. And their understanding of what is at stake by establishing an articulated and integrative network of socio-institutional agenda of transformations directed toward a more equitable and sustainable future.

References

Arikan, Y., Carreño, C., & van Staden, M. (2020, October). *ICLEI's Climate Neutrality Framework – Accelerating integrated climate action for sustainable urban development*. Version: 5.

Beck, U. (2018). Metamorfose do Mundo. Zahar, Rio de Janeiro. 1ª edição (22 março 2018) 279 pg.

Benites-Lazaro, L. L., & Giatti, L. L. (2020). O nexo água-energia-alimentos: Desafios da integração de políticas. In P. R. Jacobi (Ed.), *Diálogos socioambientais na macrometrópole paulista* (Vol. 03, n.° 09). Territórios da Energia. Available at: https://periodicos.ufabc.edu.br/index.php/dialogossocioambientais/issue/view/32/19

Bhaduri, A., Ringler, C., Dombrowski, I., et al. (2015). Sustainability in the water-energy-food nexus. *Water International, 40*, 723–732.

BID, et al. (2015). *Plano de Ação Florianópolis Sustentável*. Available at: http://www.ibam.org.br/media/arquivos/estudos/plano_de_acao_florianopolis_sustentavel_bid_caixa.pdf

BMZ (2014). Adaptation to Climate Change- Promising ways to tackle Climate Risks- Federal Ministry for Economic Development and Cooperation/Climate Policy Division. Bonn, Germany. 32pages. https: https://www.bmz.de/resource/blob/18754/5fc4c285ff5b178cfd307fd3e04d6ad5/Materilie325_Adaptation-to-Climate-Change.pdf

Broto, V. C., & Bulkeley, H. (2012). A survey of urban climate change experiments in 100 cities. *Global Environmental Change, 23*, 92–102. https://doi.org/10.1016/j.gloenvcha.2012.07.005

Bruno, A. P. (2020). *Desenvolvimento Urbano: o contexto de formulação da Política Nacional de Desenvolvimento Urbano*. Brasília. Available at: https://www.gov.br/mdr/pt-br/assuntos/desenvolvimento-urbano/SEINFRA_TCU_Desenvolvimento_Urbano_APB_final_para_publicacao_site_MDR.pdf

Cabral, W.S (2018). Water and Energy Nexusunder Climate Change Scenarios: Lessons from Brazil. New Water and PolicyPractice. https://onlinelibrary.wiley.com/doi/abs/10.18278/nwpp.4.2.2

Centro de Estudos e Promoção da Agricultura de Grupo – CEPAGRO. (2016). *TECNOLOGIA SOCIAL DA REVOLUÇÃO DOS BALDINHOS EM MANUAL PARA DOWNLOAD GRATUITO*. Available at: https://cepagroagroecologia.wordpress.com/2016/05/06/tecnologia-social-da-revolucao-dos-baldinhos-em-cartilha-para-download-gratuito/

CNM – Confederação Nacional dos Municípios. (2014). *Estudos Técnicos: Planos Municipais*. Brasília.

Costa, M. A., & Favarão, C. B. (2016). Institucionalidade e governança na trajetória recente da política urbana brasileira: legislação e governança urbanas. In IPEA (Ed.), *O Estatuto da Cidade e a Habitat III: um balanço de quinze anos da política urbana no Brasil e a nova agenda urbana*.

Dasgupta, P. (2021). *The economics of biodiversity: The Dasgupta review*. HM Treasury.

Deslatte, A., Hatch, M., & Stokan, E. (2020). How can local governments address pandemic inequities? *Public Administration Review, 80*. https://doi.org/10.1111/puar.13257

Di Giulio, G. M., Bedran-Martins, A. M., Vasconcellos, M. D. P., & Ribeiro, W. C. (2017). Mudanças climáticas, riscos e adaptação na megacidade de São Paulo, Brasil. Sustentabilidade em Debate. *Brasília, 8*(2), 75–87.

Di Giulio, G. M., Bedran-Martins, A. M., Vasconcellos, M. D. P., Ribeiro, W. C., & Lemos, M. C. (2018). Mainstreaming climate adaptation in the megacity of São Paulo, Brazil. *Cities, 72*, 237–244.

Douglas, I., Champion, M., Clancy, J., et al. (2020). The COVID-19 pandemic: Local to global implications as perceived by urban ecologists. *Socio-ecological Practice Research, 2*, 217–228. https://doi.org/10.1007/s42532-020-00067-y

Fantin, M. (2005). *Contribuição para a formulação de políticas públicas de desenvolvimento sustentável e gestão do meio ambiente em Áreas de Proteção Ambiental: um estudo de caso da várzea do rio Paraíba do Sul no Município de São José dos Campos-SP*. Dissertação de mestrado – Universidade do Vale do Paraíba, Instituto de Pesquisa e Desenvolvimento, São José dos Campos/SP. Available at: https://biblioteca.univap.br/dados/000001/0000013A.pdf

Fernandes Torres, C. J., Peixoto de Lima, C. H., Suzart de Almeida Goodwin, B., Rebello de Aguiar Junior, T., Sousa Fontes, A., Veras Ribeiro, D., Saldanha Xavier da Silva, R., Dantas Pinto Medeiros, Y. (2019). A Literature Review to Propose a Systematic Procedure to Develop "Nexus Thinking" Considering the Water–Energy–Food Nexus. Sustainability, 11/24. https://www.mdpi.com/2071-1050/11/24/7205

Figueroa, F. E. V. (1996). *Avaliação econômica de ambientes naturais – o caso das áreas alagadas – uma proposta para a represa do lobo (Broa)*. Dissertação Mestrado – UFSCar, Itirapina/SP.

Fortes, P. A. C., & Ribeiro, H. (2014). Saúde Global em tempos de globalização. *Saúde e Sociedade, 23*(2), 366–375.

GCoM – Global Covenant of Mayors for Climate & Energy. (2021). Home. Available at: https://www.globalcovenantofmayors.org/. Accessed 31 May 2021.

Geneletti, D., & Zardo, L. (2015). *Ecosystem-based adaptation in cities: An analysis of European urban climate adaptation plans*. Elsevier: Land Use Policy. http://www.sciencedirect.com/science/article/pii/S0264837715002732

Harari, Y. N. (2021, February). Lessons from a year of Covid. *Financial Times*. https://www.ft.com/content/f1b30f2c-84aa-4595-84f2-7816796d6841

Hezri, A. (2016) Embedding the environment sustainability for the 2030 Agenda in Malaysia. In:Persidangan Kebangsaan Mengenai Kependudukan dan Matlamat Pembangunan Lestari. Dewan Perdana Nur, KPWKM.

Hoff, H. (2011). *Understanding the Nexus. Background paper for the Bonn2011 conference: The Water, Energy and Food Security Nexus* (p. 52). Stockholm Environment Institute.

Howels, M. & Rogner, H.H. (2014). *Water-energy nexus: Assessing integrated systems. Nature Climate Change 4*(4) 246–247. https://pure.iiasa.ac.at/id/eprint/10960/

ICES-BID, et al. (2015a). *Estudo 1 de Mitigação*. Available at: http://www.pmf.sc.gov.br/arquivos/arquivos/pdf/27_08_2015_9.27.58.f256adc2f8bf21d6481e39eb1b350a0d.pdf

ICES-BID, et al. (2015b). *Relatório Final Estudo 3 Crescimento Urbano Florianópolis*. Available at: https://www.pmf.sc.gov.br/arquivos/arquivos/pdf/27_08_2015_9.30.19.2d57c5303b800097ab78796419b761af.pdf

ICLEI – Governos Locais pela Sustentabilidade. (2020a). *Plano Local de Ação Climática de Fortaleza*. São Paulo, Brasil. Available at: https://urbanismoemeioambiente.fortaleza.ce.gov.br/images/urbanismo-e-meio-ambiente/infocidade/mudancas-climaticas/plano_local_de_acao_climatica_de_fortaleza_2020.pdf

ICLEI – Governos Locais pela Sustentabilidade. (2020b). *Plano Local de Ação Climática do Recife*. São Paulo, Brasil. Available at: https://americadosul.iclei.org/wp-content/uploads/sites/78/2020/12/20-recife-acaoclimat-1.pdf

ICLEI – Local Governments for Sustainability. (2021). *Closing the loop: Innovation for increased efficiency in the food-water-energy nexus*.

Instituto Brasileiro de Geografia e Estatística – IBGE. (2010a). *CENSO DEMOGRÁFICO 2010*. Available at: https://cidades.ibge.gov.br/brasil/sp/sao-jose-dos-campos/panorama

Instituto Brasileiro de Geografia e Estatística – IBGE. (2010b). *CENSO DEMOGRÁFICO 2010 Características da população e dos domicílios: resultados do universo*. Available at: http://www.ibge.gov.br/home/estatistica/populacao/censo2010/caracteristicas_da_populacao/resultados_do_universo.pdf

IPCC. (2018). Summary for policymakers. In *Global Warming of 1.5°C. An IPCC Special Report on the impacts of global warming of 1.5°C above pre-industrial levels and related global greenhouse gas emission pathways, in the context of strengthening the global response to the threat of climate change, sustainable development, and efforts to eradicate poverty* (V. Masson-Delmotte, P. Zhai, H.-O. Pörtner, D. Roberts, J. Skea, P. R. Shukla, A. Pirani, W. Moufouma-Okia, C. Péan, R. Pidcock, S. Connors, J. B. R. Matthews, Y. Chen, X. Zhou, M. I. Gomis, E. Lonnoy, T. Maycock, M. Tignor, & T. Waterfield, Eds.). (in press).

Jacobi, P. R., Monteiro, R. & Souza, D.T. Caminhos para uma Nova Ética em tempos Pós COVID-19: o Desafio de Ampliar Diálogos e Fortalecer Aprendizagem Social In: Direitos Humanos e Educação- Calgaro, C., Pilau Sobrinho, L. & Rocha, L. UNIVALI, Itajai. https://www.file:///C:/Users/prjacobi/Downloads/Jacobi_Monteiro_Souza_Covid%2019%20e%20Dialogo.pdf 287-303

Keilmann-Gondhalekar, D. & Ramsauer, T. (2017). Nexus City: Operationalizing the urban Water-Energy-Food Nexus for climate change adaptation in Munich, Germany. Urban Climate – v. 19/2017- 28-48. https://www.sciencedirect.com/science/article/abs/pii/S2212095516300542

Klug, L., et al. (2016). Mudanças Climáticas e os Desafios Brasileiros para Implementação da Nova Agenda Urbana. In *O Estatuto da Cidade e a Habitat III: um balanço de quinze anos da política urbana no Brasil e a Nova Agenda Urbana*. IPEA.

Moran, E. (2011). *Meio ambiente e Ciências Sociais*. Cap. 5 – Análise multiescalar e multitemporal; Cap. 6 – Biocomplexidade nos sistemas ecológicos e Cap. 7 – Tomada de Decisão Ambiental.

Neder, E. A., de Araújo Moreira, F., Dalla Fontana, M., et al. (2021). Urban adaptation index: Assessing cities readiness to deal with climate change. *Climatic Change, 166*, 16.

OECD. (2020). *Building back better: A sustainable, resilient recovery after COVID-19*. Available at: https://www.oecd.org/coronavirus/policy-responses/building-back-better-a-sustainable-resilient-recovery-after-covid-19-52b869f5/

PBMC. (2016). *Mudanças Climáticas e Cidades. Relatório Especial do Painel Brasileiro de Mudanças Climáticas* (S. K. Ribeiro & A. S. Santos, Eds.). PBMC, COPPE – UFRJ.

Prefeitura de São José dos Campos. (1984). *Lei ordinária N° 2792 – Declara área de proteção ambiental a região do Banhado de São José dos Campos e dá outras providências, a respeito*. Available at: https://leismunicipais.com.br/a/sp/s/sao-jose-dos-campos/lei-ordinaria/1984/279/2792/lei-ordinaria-n-2792-1984-declara-area-de-protecao-ambiental-a-regiao-do-banhado-de-sao-jose-dos-campos-e-da-outras-providencias-a-respeito

Prefeitura de São José dos Campos. (2002). *LEI N° 11.262, DE 08 DE NOVEMBRO DE 2002*. Available at: https://www.al.sp.gov.br/repositorio/legislacao/lei/2002/lei-11262-08.11.2002.html

Prefeitura de São José dos Campos. (2014). *LEI N°. 8703/12 DE 21 DE MAIO DE 2012*. Available at: https://servicos2.sjc.sp.gov.br/legislacao/Decretos/2014/16086.pdf

Prefeitura de São José dos Campos. (2015). *Edital de Chamamento – Programa de Pagamento por Serviços Ambientais*. Available at: https://www.sjc.sp.gov.br/media/21903/edital-chamamento.pdf

Prefeitura de São José dos Campos. (2017). *Produto 3.1 – Projeto PSA Couves – Conservação de Água e Solo na Microbacia do Ribeirão das Couves*. São José dos Campos (SP). Available at: http://sigaceivap.org.br:8080/publicacoesArquivos/arq_pubMidia_Processo_093-2015-PMSJC.pdf

Prefeitura de São José dos Campos. (2019). *Lei complementar N. 623, de 9 de outubro de 2019*. Available at: https://servicos2.sjc.sp.gov.br/legislacao/Leis%20Complementares/2019/623.pdf

Rasul, G., & Sharma, B. (2015). The nexus approach to water–energy–food security: An option for adaptation to climate change. *Climate Policy*, 1–21. https://doi.org/10.1080/14693062.2015.1029865

Schulterbrandt-Gragg, R., Anandhi, A., Jiru, M., & Usher, K. M. (2018). A conceptualization of the urban food-energy-water nexus sustainability paradigm: Modeling from theory to practice. *Frontiers in Environmental Science, 6*, 133. https://doi.org/10.3389/fenvs.2018.00133

Seto, K. C., et al. (2014). Human settlements, infrastructure and spatial planning. In *Climate change 2014: Mitigation of climate change. Contribution of Working Group III to the Fifth Assessment Report of the Intergovernmental Panel on Climate Change*. Cambridge University Press.

Sperling, J. B., & Berke, P. R. (2017). Urban Nexus science for future cities: Focus on the energy-water-food-X Nexus. *Current Sustainable Renewable Energy Reports, 4*, 173–179. https://doi.org/10.1007/s40518-017-0085-1

Torres, P. H. C. (2020). Reaction or anticipation? The role of urban and regional planning in adapting to the climate emergency. In *Urban and regional responses to climate emergency*. Regional Studies Association. Available at: https://regions.regionalstudies.org/ezine/article/reaction-or-anticipation-the-role-of-urban-and-regional-planning-in-adapting-to-climate-emergency/#

UN-Habitat. (2015). *Guiding principles for city climate action planning*. Un-Habitat. https://unhabitat.org/sites/default/files/download-manager-files/English%20Publication.pdf

World Economic Forum (WEF). (2021). *The global risks report 2021*, 16th ed. http://wef.ch/risks2021

Zhang, W., Zhou, T., Zou, L., Zhang, L., & Xiaolong, C. (2018). Reduced exposure to extreme precipitation from 0.5 °C less warming in global land monsoon regions. Nat Commun. https://www.nature.com/articles/s41467-018-05633-3

17 Food Waste Redistribution and Implications for Achieving the Sustainable Development Goals: The Case of a Food Bank in the Municipality of São Paulo, Brazil

Luciana M. Vieira, Ísis Domingues, and Daniele Eckert Matzembacher

Abstract

Food loss and waste is a global problem that affects several countries on all continents. The closer to the demand side of the supply chain, the greater the impact of food losses and waste in economic, environmental, nutritional, and social terms. This occurs mainly because it has a significant impact from the embedded emissions, labor, water usage, financial investments, and other resources and process associated with food production, processing, transport, and retailing. This chapter describes the food bank operation and discusses the challenges faced by a food bank in the municipality of São Paulo to redistribute food that would be discarded. It focuses on efforts promoted to shorten and improve supply chain coordination and improve food and nutritional security to vulnerable urban communities, being an alternative to lessen the impacts on climate change and other environmental, economic, and social issues related to sustainability. Implications for achieving the SDGs and an agenda for future research are proposed. The relevance of this discussion lies in the fact that to achieve the objectives of the Agenda 2030 and to be able to promote changes in relation to climate change, it is necessary to know the possible interventions to reduce food waste, how they work, and the challenges faced and their impacts.

Keywords

Food waste · Food redistribution · Food banks · Emissions · SDGs

17.1 Introduction

Food loss and waste (FLW) can be defined as a decrease in the quantity or quality of food along the food supply chain. Empirically, it considers food losses as occurring along the food supply chain from harvest/slaughter/catchup to, but not including, the retail level. Food waste, on the other hand, occurs at the retail and consumption

L. M. Vieira (✉) · Í. Domingues
São Paulo School of Management (FGV EAESP), Fundação Getúlio Vargas (FGV), São Paulo, SP, Brazil
e-mail: luciana.vieira@fgv.com; idomingues@ashoka.org

D. E. Matzembacher
Post-graduation Program in Administration, Universidade Federal do Rio Grande do Sul (PPGA/UFRGS), Porto Alegre, RS, Brazil
e-mail: daniele.eckert@ufrgs.br

L. L. B. Lazaro et al. (eds.), *Water-Energy-Food Nexus and Climate Change in Cities*, Sustainable Development Goals Series, https://doi.org/10.1007/978-3-031-05472-3_17

level (FAO, 2019, p. 14). Both FLW have become a major global issue that threatens sustainable food systems and generates negative externalities in economic, environmental, nutritional, and social terms (Campoy-Muñoz et al., 2017; Kazancoglu et al., 2018; Papargyropoulou et al., 2014; Thyberg & Tonjes, 2016; United Nations Environment Programme, 2021).

Regarding the environmental aspect, the impacts of FLW on water resources and climate change are widely identified by academic literature. The interrelationship between these various elements is discussed in the literature as the food-water-energy nexus. This concept emerged in response to climate change and social changes including population growth, economic growth, globalization, and urbanization (Hoff, 2011) causing increased pressure on water, energy, and food resources. As these resources have complex interactions, these pressures lead to an increasing number of trade-offs among them. Demands for water, energy, and food are estimated to increase by 40%, 50%, and 35%, respectively, by 2030 (US NIC – United States National Intelligence Council, 2012). There are several studies seeking to better understand this complex phenomenon, the impact, and possible interventions. Endo et al. (2017) provide a good review of studies evaluating water-food nexus, water-energy nexus, and water-energy-food nexus. When it comes specifically to food waste analyses, for example, many studies calculate the greenhouse gas emissions due to wasted food using life cycle assessment (LCA) (Venkat, 2011) or the impact on water resources (Chapagain & James, 2013). It is possible to identify studies conducted in Sweden (Brancoli et al., 2020), the UK (Quested et al., 2011), and the United States (Venkat, 2011), among others. These studies found that the closer to the demand side of the supply chain, the greater the impact of FLW. This occurs mainly because it has a significant impact from the embedded emissions, labor, water usage, financial investments, and other resources and process associated with food production, processing, transport, and retailing (Garnett, 2008). For this reason, food waste has a more negative impact than food losses. Therefore, this chapter will focus on food waste.

The greater impact of food waste in the final stages of supply chains has a significant implication for large urban centers, as their supply chains are generally extended and food travel longer distances. This is the case of the city of São Paulo, the largest financial and economic center of the Brazil and South America. Managing food supply in large urban cities, such as São Paulo, is not a trivial task. According to Kawano et al. (2012), it involves aspects of large distances, perishability, and supply chain coordination. When the supply process is not efficient, it can generate high levels of food waste. São Paulo presents a situation that reflects the case of almost 50 megacities in the world, such as Tokyo (Japan), Jakarta (Indonesia), Shanghai (China), New York (United States), Mexico City (Mexico), Delhi (India), and Moscow (Russia), among others.

Actions aimed at reducing food waste in large urban centers can bring benefits related to climate change, water use, and other environmental, economic, and social aspects. A new agenda recently pulled together these efforts to reduce food waste in the 2030 Agenda for Sustainable Development, which must be adopted by all UN (United Nations) member countries. The 2030 Agenda addresses areas of crucial importance to humanity and to the planet, such as ensuring sustainable production and consumption patterns and achieving food security. The United Nations Sustainable Development Goal (SDG) Target 12.3 is "by 2030, halve per capita global food waste at the retail and consumer levels and reduce food losses along production and supply chains, including post-harvest losses" (UN General Assembly, 2015). One billion extra people could be fed if FLW is reduced by half (Kummu et al., 2012). Reducing FLW also has the potential to contribute to SDG 2 (to end hunger, the achievement of food security and improved nutrition), SDG 6 (sustainable water management), SDG 13 (climate change), SDG 14 (marine resources), SDG 15 (terrestrial ecosystems, forestry, biodiversity), and many other SDGs (FAO, 2019).

In this sense, food waste prevention generates the best positive impact regarding environmental, economic, and social aspects. The next preferred option to handle this problem would be food redistributing to vulnerable people, in the case of foods that still conserve quality and nutritional value for human consumption (USEPA, 2018). The Food Recovery Hierarchy elaborated by the US Environmental Protection Agency (USEPA, 2018) is available in Fig. 17.1.

Regarding food redistribution, Brazil is internationally recognized for its well-structured food bank network. The Brazilian Network of Food Banks was created to strengthen and integrate the performance of food banks and help the reduction and prevention of food waste. Currently, there are 249 food banks identified by the network. Food banks receive donations of food considered to be nonstandard for marketing but adequate for consumption. Food is passed on to nonprofit civil society institutions that produce and distribute free meals to people in situation of food vulnerability (Henz & Porpino, 2017). According to data provided last year by the Brazilian government, these establishments redistributed 59 thousand tons of food in the country, supplying 5,894,201 people in a situation of social and economic vulnerability, and helped 17,182 social institutions (CAISAN, 2021).

The role of managing a public food bank poses many challenges. Food banks are the main intermediary for receiving donations from producers, processors, distributors, and retailers. The food bank has a main role to select what is fit for consumption and redistribution among charities, nursing homes, schools, or low-income social groups. There are several barriers to improving the management of a food bank by a municipality such as regulation, urban logistics, assurance of nutrition quality of donations, and lack of resources, among many others.

This chapter describes food bank operation and discuss the challenges faced by a food bank in the municipality of São Paulo to redistribute food that would be discarded. It focuses on efforts promoted to shorten and improve supply chain coordination and improve food and nutritional security to vulnerable urban communities, being an alternative to lessen the impacts on climate change and other environmental, economic, and social issues related to sustainability. Implications for achieving the SDGs and an agenda for future research are proposed. The relevance of this discussion lies in the fact that to achieve the objectives of the Agenda 2030 and to be able to promote changes in relation to climate change, it is necessary to know the possible interventions to reduce food waste, how they work, and the challenges faced and their impacts.

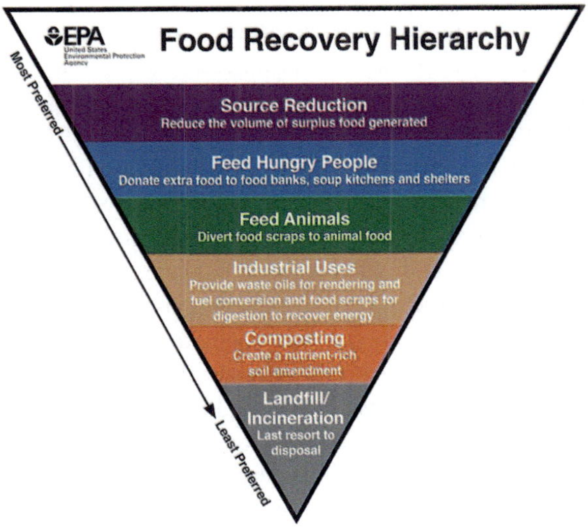

Fig. 17.1 Food recovery hierarchy. (Source: USEPA, 2018)

17.2 Literature Background

It is estimated that 25–33% of all the food produced in the world is either lost or wasted (Gustavsson et al., 2011). The waste of food impacts directly the supply and demand, increasing the value of food and consequently contributing to spread food poverty worldwide. There are other collateral effects that can be observed because of food waste. As an example, we have the increase of water consumption and the environmental exhaustion of resources, besides the increase of generation of carbon dioxide (CO_2) equivalent emissions with no counterpart (Baglioni et al., 2016). Almost 50% of organic resources captured are wasted, and when observing organic contributors to CO_2 emissions, food waste is the highest contributor (Brenes-Peralta et al., 2020). Consequently, the disregard for food waste causes economic, social, and environmental constraints.

Estimates suggest that 8–10% of global greenhouse gas emissions are associated with food that is not consumed (United Nations Environment Programme, 2021). It represents a waste of resources used in production, such as land, water, labor, energy, and inputs to produce food that is also wasted. It also leads to unnecessary CO_2 emissions and air pollution caused by farm machinery and trucks that transport food. Moreover, food waste represents a loss of economic value of the food produced (Gustavsson et al., 2011; Kummu et al., 2012; Lundqvist et al., 2008; Nellemann et al., 2009).

The food waste in Brazil has particularities when compared to other developing countries, as the lack of quantitative data in Brazil is an issue and the estimation of how much food the country wasted is scarce of precise explanations. Besides a considerable harvest lost, similar to other emerging countries, the amount of food that Brazilian households put away is high. To understant this context, we need to consider some cultural aspects, like cooking large portions more than necessary, which is a sign of wealth, which make families willing to have abundant food stocks which are highly valued (Porpino, 2016). The hyperinflation that Brazil faced in the early 1980s and 1990s forced people, especially low-income families, to maintain monthly food stockpiling in the house.

Strategies to tackle FLW are therefore needed to achieve environmentally sustainable food production (Foley et al., 2011) while ensuring food security (Godfray et al., 2010). A reduction of FLW would have a substantial positive effect on natural and societal resources. It would not only avoid pressure on scarce natural resources but would also decrease the need to raise food production to meet the 2050 population demand (FAO, 2013; Parfitt et al., 2010). In fact, one billion extra people could be fed if half reduced FLW. It is considered one of the most promising measures to improve food security in the coming decades (Kummu et al., 2012).

The Universal Declaration of Human Rights, proclaimed at the General Assembly of the United Nations in 1948, states the fundamental human rights to be universally protected and that includes the right to access adequate food (General Assembly of the United Nations, 1948). According to the Food and Agricultural Organization (FAO) of the United Nations, "food security exists when all people at all times have physical, social and economic access to sufficient, safe and nutritious food to meet dietary needs and food preferences for an active and healthy life" (FAO, 2009, p. 8).

When analyzing the Brazilian reality before COVID-19, the Brazilian Institute of Geography and Statistics (IBGE, 2013) reports that 22.6% of the Brazilian population were in food insecurity situation. When considering urban areas, IBGE evaluated that 6.8% of the residences were in the food insecurity spectrum. Paradoxically, food waste reaches 1.3 billion tons per year worldwide, and FAO (2020) estimates that the food production will need to increase about 50% by 2050 to feed the growing world population considering that the rates of food loss and waste will be maintained.

For this reason, Brazil promotes actions aimed at food insecurity, even when the problem is also very widespread in terms of environmental impacts. It is in this context that food banks emerge. In 1993, the food security and nutrition public policies started to be under the responsibility of the Ministry of Social Development. The ministry established the creation of the National Council for Food Security (CONSEA), and after that, several governmental initiatives were adopted to reduce food insecurity and improve access to food. In the early 2000s, CONSEA created the National System of Food Security and Nutrition (SISAN) and launched the National Policy of Food Security and Nutrition (PNSAN), two national plans, and held five national conferences (Machado, 2017). In June 2020, because of the pandemic crises, the Act #14.016 to establish the responsibility of food donors was approved, guarantying they will not be legally responsible for any problems, unless bad faith is proven.

In this sense, food banks are important instruments to reduce food waste. Food bank is defined as a food supply and security initiative that must capture and redistribute food from donation and, therefore, without commercial value, but safe for human consumption. Food banks can be considered the link between supply and demand in food supply chains. In Brazil, the food bank program is part of the integrated policies developed by the Operational Network of the National Food and Nutritional Security System.

The definition of Ordinance No. 17, of April 17, 2016, which establishes the Brazilian Network of Food Bank (BNFOOD BANKS), states the objective of food banks as (BRASIL, 2016):

> [...] supply initiative, which aims to combat hunger and food insecurity through the collection of donations of food that would be wasted along the production chain. Such initiatives can be Public – under the management and responsibility of subnational entities, in other words, states and municipalities – or they can be Private – under the management of non-profit social interest civil society organizations. (BRASIL, 2016)

Regarding the operational structure of the *food banks*, the Brazilian Food Bank Network presents the following definition:

> §1 [...] physical and / or logistical structures that offer the capture service and / or free reception and distribution of foodstuffs from donations from the sectors private and/or public and which are directed to public or private institutions characterized as providers of social assistance, protection and defense services education and justice units, health establishments and other health units food and nutrition.'§ 2 ° The logistical structures mentioned in § 1 refer to methodologies of the type 'Urban harvest', which are characterized by the immediate collection and delivery of food donated, excluding the need for a physical location for storage. (BRAZIL, 2016)

Since food banks are nonprofit organizations, the mechanisms developed by them to promote their mission can be rudimentary depending on the sponsors and local regulations involved. In the context of humanitarian supply chains, redistribution of food via food banks addresses environmental, economic, and social objectives (Facchini et al., 2018). Martins et al. (2019) consider the diversity of objectives may cause conflicts, since the food bank needs (a) to be efficient (maximizing the fundraising and redistribution at the lowest cost possible), (b) to promote social sustainability (reaching the greatest number of social assistance institutions and attending them in the fairest way possible), and (c) to mitigate the environmental impact (reducing CO_2 emissions through the reduction of food losses).

The food bank structure as humanitarian supply chain includes a strong correlation to logistic issues and components, what suggests the importance of operationalizing the logistic system in order to promote access to food for vulnerable population. As nonprofit systems, the resources (financial, logistical, and structural, among others) are often nonrecurring besides scarce; in other words, food banks are exposed to a considerable uncertainty and risk (Fianu & Davis, 2018). Additionally, as a humanitarian initiative, the resources acquired and managed by these organizations need to be well allocated, and these allocations need to be transparent to the civil society. Consequently, it is important to understand and monitor key drivers of the performance that can help to improve the operations (Martins et al., 2019).

Matzembacher et al. (2021) emphasize the importance of institutional stakeholders to lead the agenda whose objective is food loss reducing or food redistribution. The authors observed systematically the stakeholder's interactions, which involve deliberating and deciding how to solve operating problems. However, the processes to coordinate and legitimize multiple stakeholders can be costly, especially when the stakeholders have different compliance standards and interests. It is in this context that better understanding how food waste redistribution occurs through food banks, what challenges they face, and implications for achieving the Sustainable Development Goals is relevant: they are essential to reduce the negative environmental, social, and economic impacts of food waste.

17.3 Methods

The municipality of São Paulo's food bank case study was elaborated based on two in-depth interviews with the food bank manager carried out virtually during April and May 2020, a virtual meeting with the sustainability manager from one of the largest retail chains in Brazil and main donor to this food bank in June 2020, and a webinar with NGOs (nongovernment organizations) that receive the food bank donations in November 2020. The data analysis was exploratory, aiming to understand main challenges during the COVID-19, and was crossed with public data available (websites, reports, and articles in newspapers). Then, a narrative was elaborated and discussed with the food bank staff and is partially reproduced in the section below. The case presentation aims to illustrate the main theoretical discussion.

17.4 Case: The Operation and Challenges Faced by a Food Bank in the Municipality of São Paulo

The food bank of the city of São Paulo is a public institution that aims to purchase food from family farming and collect food from the food industries, retail chains, and wholesalers that are food considered to be nonstandard for marketing but adequate for consumption. These foods are donated to assistance entities, previously associated with the food bank. Their mission is to contribute to reduce hunger and food waste in the municipality of São Paulo. This food redistribution allows the nutritional complementation of meals served to thousands of children, youth, adults, and the elderly daily.

Any person, physical or legal, company, or public agency can become a food donor. The food bank team receives the food, selects those that are suitable for consumption, and delivers them to the assistance entities. These entities are in charge of distributing the food to the population, either through ready meals or direct transfer to low-income families. In return, the entities participate in training and educational activities related to food handling, preparation, etc.

The food bank manager understands that its value proposition is to help companies that produce and distribute food in social projects against hunger in the country and to preserve the environment by reducing organic waste. In addition to redistributing food, this food bank also works with companies and schools. The purpose is to raise awareness to generate a permanent culture change in relation to food waste, in addition to encouraging more people to donate food. The lectures have themes related to healthy eating, nutrition, full use of food, and dissemination of the work made by the food bank talking about social and food waste recovery indicators.

The food bank does not formally assess whether its operations are linked to a positive environmental impact. The data available informs that considering the year 2019, an average of 32 tons of food is collected per month. After evaluating products with sanitary conditions for consumption, the monthly donation average is 23 tons of food per month. They focus more on social issues, emphasizing that their action jointly addresses the problem of hunger and social exclusion. The distribution of food satisfies the basic need, while their educational actions promote greater use of the nutrients provided, enabling a more tasty and balanced diet. According to them, the impact of their action is that the hunger cycle is interrupted and the cycle of social inclusion is favored. Therefore, the focus of their action is to contribute directly to SDG 2 – to end hunger, the achievement of food security and improved nutrition – even though the impacts of their action permeate the goals related to, for example, food waste reduction, sustainable water management, climate change, terrestrial ecosystems, forestry, and biodiversity. Despite this, their action is crucial to improving the use of natural resources, being able to contribute directly to reducing greenhouse gas emissions.

An example of food redistribution made by the food bank that generates a shortening of the chain in a large urban town is described as follows. The municipality of São Paulo has been performing a campaign to combat food waste. Within this campaign, a pilot project has recently been initiated to redistribute surplus food from street markets that would otherwise have been thrown away, either because they were not sold due to lack of demand or because the appearance of the food was ugly even though it had good nutritional value. The city of São Paulo has almost 900 street markets that are organized by vendors in different streets every day supplying fresh food to citizens. It is a traditional distribution channel that provides food near urban households.

As of mid-April 2019, before the pandemic, eight street markets were involved, with the initial ambition to expand it. On average, 120–200 kg of food is collected at each street market, sometimes up to 250 kg depending on the day and situation. The food waste from the street market is collected, measured, and transported to the municipal food bank, where more than 300 charity institutions can retrieve and redistribute it to people in need.

The pilot project is financed by the municipality of São Paulo. The municipality initiated this project to redistribute the food that was previously being thrown away after the street markets close and driven to landfills. The environmental impact related to carbon footprint, water footprint, land use, and biodiversity, as well as making economic assessments, and social impact of throwing this food waste have never been measured.

Food bank managers mention that the cause of the problem in street fairs is related to supply chain coordination and logistics. In their perception, the street market vendors go to CEAGESP (main wholesaler in São Paulo and the largest one in Latin America) every day to buy the food they intend to sell and do not have any space to store it after the street market closes. Most of them are usually unable to bring the food from one street market to another unless they have cold storage or another place to sell the products that isn't a street market, like a small grocery store or other. Technology and food cooling problems are common in Brazil, especially in small producers, traders, and street vendors. This process might however differ depending on the price of the products sold. Vendors selling more expensive products have a greater incentive to retrieve their products, while vendors selling cheaper products – like lettuce – will probably opt to throw it away. After discussions with street vendors, the municipality of São Paulo found that food waste is considered by these individuals as a problem inherent to food commercialization, that is, they don't care about that since they understand that it's something that is expected to happen. They are aware at the time of purchase that they will have to throw some food away, but according to them, they are unable to buy less since a larger quantity presents better at the market and gives the customer an impression of quality. Moreover, the vendors have expressed that it is difficult to buy smaller quantities at CEAGESP. The wholesalers come to the terminal market expecting to sell their whole trucks, so if the vendors request smaller quantities, they will charge higher prices for less quantity of food.

According to the São Paulo municipality representatives, it has been easy to get the street vendors involved in the project. The food bank team has simply showed a promotional video about the project, and the street vendors have immediately been willing to cooperate and give their surplus food to the food banks. The video shows both other vendors donating food and thereafter the people who are being helped by their donations.

Because of the geographical constraints due to the location of the food bank, the project has only taken place in downtown São Paulo. The variety of products received might vary if the collections were to take place in the periphery, but it is believed that the street vendors' attitude toward redistribution would be the same.

The collection of the surplus food at the street markets is done by two employees on each street market by giving the vendors boxes in which they can collect food that is edible but that they will not be able to sell during the day. From what was observed, workers from the municipality arrive at 1 pm to deliver these boxes at the markets and thereafter collect them as the vendors successively close their stalls for the day. On the market observed, a large majority of the stall vendors choose to donate their surplus. The three vendors interviewed stated that they buy their fruits three times a week and store them in their trucks. Alike the vendors interviewed on previous markets that were not part of the pilot project, these vendors explained that they try to sell as much as possible by reducing the price but what they cannot sell and will not be fit for selling the following day is donated. The effort that they put on the clearance sale seems to vary however with one vendor stating that he gradually lowers the price and finally donates the rest whereas another vendor prefers to donate rather than to sell at a very low price. This also enables him to offer fresher products to his customers.

The project has increased the amount of food collected by charities to more than 400 during the pandemic. The connection between charity institutions and street markets is direct so that they can be in charge of collecting without involvement from the municipality. The initial proposal was for the fairs to supply the entities that receive the food directly to reduce processes and the risk of food loss, but these systems were not well accepted. The model that worked was the food bank receiving these products from the fair, sorting the quality of food for consumption and then donating to the entities. The negative point of this model is that it requires more dedication of the team's time and a greater need to move the food. Since there are over 800 street markets in São Paulo, the odds of there being one close to each of the 400 charities are very high. This would in time enable the redistribution of food to have a greater geographical spread since it is currently limited to the proximity of the sole municipal food bank that exists in the city. The participating charities are trained in relation to food cleaning and storing and utilization of the whole products but mainly the measurement of the collected food before they start operating. This is so that the municipality can still have control over how much food is collected and what is done with it.

The biggest challenge for the food bank is the variability of the volume of food available for the charities. The food that is collected from the street markets is brought to the location by the food bank employees and thereafter sorted, cleaned, and quality checked. The fruits and vegetables are thereafter stored in cold storage for the charities to collect. All the donated food is quantified by direct weighing. This is done both before and after it is sorted, creating data on both the edible and inedible food donated. The waste generated by the food collected from the market is only 1% of what is collected, much lower than the supermarkets' 25%. This is surprising due to the fact that all of the street markets' products are perishable while the supermarkets' products are not, but it is explained that the supermarkets' products have been displayed longer and handled more by the customer which results in more damages. The data the municipality currently considers to be lacking is of the food that isn't collected, how much of it is still edible but in worse quality, and how much is avoidable.

The food bank also faces other issues in its operation, such as the following:

(1) There is no system for the periodic distribution of donated food among the institutions – lack of predictability in donation.
(2) There is a difficulty for the associated institution to withdraw the donation, as they do not have financial resources for hiring transportation – this limits donations.
(3) Donations are made without considering the "profile" of the institutions.

(4) There is difficulty in disseminating and raising awareness of the purpose and importance of public facilities as is the case of this food bank to potential donors and the city hall itself.
(5) The reports with the data of the donations received are prepared without criteria of periodicity and visual standard – lack of visibility about donations and food waste reduction.
(6) There are overload and centralization of the team.

Some project proposals to improve the processes of the food bank that are being discussed and/or developed are the elaboration of a website to provide information and attract more donations, make a standardization of food donated according to the needs of the entities, create the routes and periodic collection of donations organized by proximity to the donor establishments, and establish partnerships with food delivery companies and the industry. However, in order to advance in these future projects, the food bank first needs to solve some of its operational problems. Overcoming these problems is essential to scale and increase the environmental, social, and economic impact of this food bank.

17.5 Implications for Achieving the SDGs and an Agenda for Future Research

The first step in the discussion between food redistribution and its impact on reaching SDGs (e.g., SDGs 2, 6, 12, 13, 14, and 15) is to understand how food banks can help achieve these goals. In addition to knowing the mode of operation and the challenges faced by these initiatives such as the one described, the case presented related to the food bank in the municipality of São Paulo brings some lessons related to the necessity to consider four important elements: (1) the replication of these initiatives, (2) cooperation between different stakeholders, (3) inclusion in public policies related to sustainability, and (4) measurement of the impact of this type of intervention.

The initiative by the food bank in the municipality of São Paulo could be replicated by other food banks in Brazil in relation to collaboration with street markets. It is a necessary action to promote change maximization. Hoolohan et al. (2018) agree on the importance of scale to promote changes that impact on the water-energy-food nexus and, consequently, on reaching SDGs. According to Perrini et al. (2010), replication of these initiatives increases impact and induces social change. Disseminating case studies helps to foster this type of initiative, but it is necessary that the managers of these food banks be attentive to this type of opportunity and develop cooperation with other stakeholders.

Cooperation between different stakeholders is an element of success in this Brazilian initiative. The literature emphasizes the importance of collaborative approaches between supply chain stakeholders and governments as essential for the implementation of the innovation processes that minimize the trade-offs when it comes to water-energy-food nexus (da Rosa et al., 2021). This logic is also valid for food banks. Kibler et al. (2018) also propose that a sustainable management of food waste involves varied mechanisms and actors at multiple levels of governance, since food waste influences the capacity to reach SDGs as collective decision-making at the societal level regarding methods of food production and consumption. However, as proposed by Matzembacher et al. (2021), the processes to coordinate and legitimize multiple stakeholders' relations can be costly, especially when they have different compliance standards and interests. Therefore, it is proposed that cooperation needs to be well planned and integrated in a way that is beneficial to everyone and manages to mitigate, as far as possible, conflicting interests, to be possible in the long term.

In this sense, the inclusion of public policies plays an important role. As proposed by Kibler et al. (2018), government is a key component, since it has a direct impact on the food-energy-water system through its regulatory powers over land use and through its allocation of resources for policy and programmatic initiatives. In the case analyzed, the initiative was possible because it arose from a cooperation between the municipality of São Paulo and different stakeholders. This type of action needs to be encouraged by the government, discussed, and included in a broader sustainability agenda, not only regulatory but incentive policies.

Actions that seek the replication of these initiatives through cooperation between different stakeholders and public policies need to consider that food waste is a complex problem, with an impact on climate, biodiversity, energy, and soil and water issues, and that solutions need to be holistic. De Laurentiis et al. (2016) give a few examples: food production requires water directly to grow crops. This water usually requires pumping and treating, which requires energy; in turn electricity production is dependent on water for cooling and steam generation. Energy and water are further required for processing, packaging, transport and storage, preparation from the end user, and ultimately final disposal of food. This is the water-energy-food nexus. According to Lal (2016), there has been a growing emphasis on the adoption of the nexus approach for addressing complex problems because of the widespread realization that the focus thus far has been on only one component, which has often aggravated the problems and exacerbated adverse consequences. The big goal is to maximize synergies and minimize trade-offs (De Laurentiis et al., 2016; Lal, 2016). One way to minimize these trade-offs is associated especially with environmental innovation (da Rosa et al., 2021).

One question that needs to be highlighted is that current studies on the food-energy-water nexus seeking to provide pathways to reach SDGs do not capture effects on human health (Slorach et al., 2020) and other social-related issues. The contradictions between food poverty and insecurity affecting a large section of the global population and the everyday wastage of food must also be considered as part of this nexus. In fact, the understanding that food waste is part of a more complex problem brings up questions about human health, since the initiative by the food bank in the municipality of São Paulo helps to solve the problem of food insecurity and hunger by redistributing food and promoting education initiatives. This is extremely relevant for a nation like Brazil and many other countries. For example, the 2018 report by the United Nations (UN) Food and Agriculture Organization (FAO) highlights the rise in global percentage of hunger for the third consecutive year, and according to the World Health Organization (2021), globally, nearly 80% of children live in countries where at least one data point on stunting, wasting, and overweight is less than five years old. The coronavirus 2019 disease (COVID-19) pandemic increased these global food insecurity alerts, since it led to food shortages, increased food prices, and loss of income (Paslakis et al., 2020). Therefore, food waste is more complex as it is part of food-energy-water-health nexus.

An issue that draws attention in relation to the operation of the food bank in the municipality of São Paulo is that, although there is some data regarding the operation of this initiative, there is no measurement of the environmental impact caused by them. To date, there is no knowledge of studies or actions that have measured the potential positive impact of these initiatives in relation to carbon footprint, water footprint, land use, and biodiversity, as well as making economic assessments, and social impact. This observation provided by our case is aligned with Kibler et al. (2018), who identified that little is known about the energy and water consumed in managing food waste. Some authors (e.g., see De Laurentiis et al., 2016; Eriksson et al., 2015; Matzembacher et al., 2020) understand that LCA is considered to be a major tool for measurement when it comes to food waste issues and to guide a shift toward sustainable food systems. The example presented showed evidence that the food bank itself does not perceive its potential to help Brazil to achieve the SDGs nor is not able to measure its environmental positive impact. Research funding agencies and academics may be aware of this gap. Making these measurements is essential to help achieve the objectives of the Agenda 2030 but also to analyze the feasibility of promoting such initiatives, investing resources in research, cooperation activities, and the formulation of public policies.

In addition, the environmental impact of throwing food waste into landfills, as it was before the pilot project, was never measured. This continues to occur at many other street markets, both in the municipality of São Paulo and in the country. Certainly, redistributing food waste in a large city such as São Paulo has a significant impact, as would be the case in large urban centers. Studies that correctly assess the impact of this are needed to assess the need to expand this initiative and support the government. Without measuring, it is not possible to know the size of the problem and the potential for possible interventions. To find out how the nations are progressing in relation to the SDGs, it is imperative to measure the starting point and the progress made. This is a lesson that this case brings to Brazil and maybe this issue also happens in other big cities and countries.

Finally, food waste redistribution and implications for achieving SDGs considering the food-energy-water (and health) nexus are an important topic of study and discussion. Future research should continue to examine these efforts to reduce food waste and deal with hunger and food insecurity. Research could focus on quantifying and measuring the impacts of redistribution actions and other interventions that seek to reduce food waste. Furthermore, studies of the food-energy-water nexus could integrate the dimension of health or other social impacts, which are closely related. Moreover, future studies could investigate how to promote the replication of these initiatives through public policies and how to make governance capable of reducing conflicting interests, generating mutual benefits, and being long-lasting. The sum between replication of interventions and long durability of these actions are essential elements on reaching SDGs.

Acknowledgment We are thankful to Cecile Bouju and Fanny Igergård, students at the Royal Institute of Technology in Sweden, for their insights on street markets in São Paulo during their internship at FGV (Fundação Getulio Vargas). This work was partially supported by the FGV's Applied Research and Knowledge Network and by the British Academy of Social Sciences.

References

Baglioni, S., De Pieri, B., & Tallarico, T. (2016). Surplus food recovery and food aid: the pivotal role of non-profit organizations. Insights from Italy and Germany. *International Society for Third-Sector Research, 28*, 2032–2052.

Brancoli, P., Bolton, K., & Eriksson, M. (2020). Environmental impacts of waste management and valorisation pathways for surplus bread in Sweden. *Waste Management, 117*, 136–145.

BRASIL. (2016). Lei n° 11.346, de 15 de setembro de 2006. Cria o Sistema Nacional de Segurança Alimentar e Nutricional (SISAN). Brasília. Presidência da República, [2006a]. Available at: http://www.planalto.gov.br/ccivil_03/_Ato2004-2006/2006/Lei/L11346.htm

Brenes-Peralta, L., Jiménez-Morales, M., Campos-Rodríguez, R., De Menna, F., & Vittuari, M. (2020). Decision-making process in the circular economy: A case study on University Food Waste-to-Energy Actions in Latin America. *Energies, 13*, 2291.

CAISAN. (2021). Câmara Interministerial de Segurança Alimentar e Nutricional. Available at: http://mds.gov.br/caisan-mds/rede-brasileira-de-bancos-de-alimentos#:~:text=Em%20um%20levantamento%20realizado%20em,principais%20regi%C3%B5es%20metropolitanas%20do%20pa%C3%ADs

Campoy-Muñoz, P., Cardenete, M. A., & Delgado, M. C. (2017). Economic impact assessment of food waste reduction on European countries through social accounting matrices. *Resources, Conservation and Recycling, 122*, 202–209.

Chapagain, A., & James, K. (2013). *Accounting for the impact of food waste on water resources and climate change. Food industry wastes: assessment and recuperation of commodities* (pp. 217–236). Academic Press, Elsevier.

da Rosa, F. S., Lunkes, R. J., Spigarelli, F., & Compagnucci, L. (2021). Environmental innovation and the food, energy and water nexus in the food service industry. *Resources, Conservation and Recycling, 166*, 105350.

De Laurentiis, V., Hunt, D. V., & Rogers, C. D. (2016). Overcoming food security challenges within an energy/water/food nexus (EWFN) approach. *Sustainability, 8*(1), 95.

Endo, A., Tsurita, I., Burnett, K., & Orencio, P. M. (2017). A review of the current state of research on the water, energy, and food nexus. *Journal of Hydrology: Regional Studies, 11*, 20–30.

Eriksson, M., Strid, I., & Hansson, P. A. (2015). Carbon footprint of food waste management options in the waste hierarchy – A Swedish case study. *Journal of Cleaner Production, 93*, 115–125.

Facchini, E., Iacovidou, E., Gronow, J., & Voulvoulis, N. (2018). Food flows in the United Kingdom: The potential of surplus food redistribution to reduce waste. *Journal of the Air and Waste Management Association, 68*(9), 887–899.

FAO – Food and Agriculture Organization of the United Nations. (2013). *Food wastage footprint: Impacts on natural resources*. Summary report.

FAO – Food and Agriculture Organization of the United Nations. (2018). *The State of Food Security and Nutrition in the World 2018*. Building climate resilience for food security and nutrition.

FAO – Food and Agriculture Organization of the United Nations. (2019). *The State of Food and Agriculture 2019*. Moving forward on food loss and waste reduction.

Fianu, S., & Davis, L. B. (2018). A Markov decision process model for equitable distribution of supplies under uncertainty. *European Journal of Operational Research, 264*(3), 1101–1115.

Foley, J. A., Ramankutty, N., Brauman, K. A., Cassidy, E. S., Gerber, J. S., Johnston, M., … Balzer, C. (2011). Solutions for a cultivated planet. *Nature, 478*(7369), 337.

Garnett, T. (2008). *Cooking up a storm: Food, greenhouse gas emissions and our changing climate*. Food Climate Research Network, Centre for Environmental Strategy, University of Surrey.

Godfray, H. C. J., Beddington, J. R., Crute, I. R., Haddad, L., Lawrence, D., Muir, J. F., & Toulmin, C. (2010). Food security: the challenge of feeding 9 billion people. *Science, 327*(5967), 812–818.

Gustavsson, J., Cederberg, C., Sonesson, U., van Otterdijk, R., & Meybeck, A. (2011). *Global food losses and food waste: Extent, causes and prevention*. FAO.

Henz, G. P., & Porpino, G. (2017). Food losses and waste: how Brazil is facing this global challenge? *Horticultura Brasileira, 35*(4), 472–482.

Hoff, J. (2011). Understanding the nexus. In *Background paper for the Bonn*. 2011 Conference: The Water, Energy and Food Security Nexus, Bonn, Germany, 16–18 November 2011, Stockholm Environment Institute (SEI).

Hoolohan, C., Larkin, A., McLachlan, C., Falconer, R., Soutar, I., Suckling, J., … Yu, D. (2018). Engaging stakeholders in research to address water–energy–food (WEF) nexus challenges. *Sustainability Science, 13*(5), 1415–1426.

IBGE. (2013). *Instituto Brasileiro de Geografia e Estatística. Pesquisa Nacional por Amostra de Domicílios: Segurança Alimentar – 2013*. IBGE, 2014.

Kawano, B. R., Mores, G. D. V., Silva, R. F. D., & Cugnasca, C. E. (2012). Estratégias para resolução dos principais desafios da logística de produtos agrícolas exportados pelo Brasil. *Revista de Economia e Agronegócio/Brazilian Review of Economics and Agribusiness, 10*(822-2016-54244), 71–88.

Kazancoglu, Y., Ozkan-Ozen, Y. D., & Ozbiltekin, M. (2018). Minimizing losses in milk supply chain with sustainability: An example from an emerging economy. *Resources, Conservation and Recycling, 139*, 270–279.

Kibler, K. M., Reinhart, D., Hawkins, C., Motlagh, A. M., & Wright, J. (2018). Food waste and the food-energy-water nexus: A review of food waste management alternatives. *Waste Management, 74*, 52–62.

Kummu, M., de Moel, H., Porkka, M., Siebert, S., Varis, O., & Ward, P. J. (2012). Lost food, wasted resources: Global food supply chain losses and their impacts on freshwater, cropland, and fertiliser use. *Science of the Total Environment, 438*, 477–489.

Lal, R. (2016). Global food security and nexus thinking. *Journal of Soil and Water Conservation, 71*(4), 85A–90A.

Lundqvist, J., de Fraiture, C., & Molden, D. (2008). *Saving water: from field to fork: curbing losses and wastage in the food chain*. Stockholm International Water Institute.

Machado, A. (2017). *Diagnóstico do tema: perdas e desperdícios de alimentos (PDA) no Brasil*. FAO Brasil. 72p. (FAO Brasil. Consultant Report).

Martins, C., Melo, M., & Pato, M. (2019). Redesigning a food bank supply chain network in a triple bottom line context. *International Journal of Production Economics, 214*(2019), 234–247.

Matzembacher, D. E., Brancoli, P., Maia, L. M., & Eriksson, M. (2020). Consumer's food waste in different restaurants configuration: A comparison between different levels of incentive and interaction. *Waste Management, 114*, 263–273.

Matzembacher, D., Vieira, L., & Barcellos, M. (2021). *An analysis of multi-stakeholder initiatives to reduce food loss and waste in an emerging country – Brazil*. Industrial Marketing Management. No MDS.

Nellemann, C., MacDevette, M., Manders, T., Eickhout, B., Svihus, B., Prins, A. G., & Kaltenborn, B. P. (2009). *The environmental food crisis – The environment's role in averting future food crises. A UNEP rapid response assessment* (pp. 1–104). United Nations Environment Programme, GRID-Arendal. Birkeland Trykkeri AS.

Papargyropoulou, E., Lozano, R., Steinberger, J. K., Wright, N., & Bin Ujang, Z. (2014). The food waste hierarchy as a framework for the management of food surplus and food waste. *Journal of Cleaner Production, 76*, 106–115.

Parfitt, J., Barthel, M., & Macnaughton, S. (2010). Food waste within food supply chains: quantification and potential for change to 2050. *Philosophical Transactions of the Royal Society, B: Biological Sciences, 365*(1554), 3065–3081.

Paslakis, G., Dimitropoulos, G., & Katzman, D. K. (2020). A call to action to address COVID-19–induced global food insecurity to prevent hunger, malnutrition, and eating pathology. *Nutrition Reviews*.

Perrini, F., Vurro, C., & Costanzo, L. A. (2010). A process-based view of social entrepreneurship: From opportunity identification to scaling-up social change in the case of San Patrignano. *Entrepreneurship and Regional Development, 22*(6), 515–534.

Porpino, G. (2016). Household Food Waste Behavior: Avenues for Future Research. *Journal of the Association for Consumer Research, 1*, 41–51.

Quested, T. E., Parry, A. D., Easteal, S., & Swannell, R. (2011). *Food and drink waste from households in the UK*. Blackwell Publishing.

Slorach, P. C., Jeswani, H. K., Cuéllar-Franca, R., & Azapagic, A. (2020). Environmental sustainability in the food-energy-water-health nexus: a new methodology and an application to food waste in a circular economy. *Waste Management, 113*, 359–368.

Thyberg, K. L., & Tonjes, D. J. (2016). Drivers of food waste and their implications for sustainable policy development. *Resources, Conservation and Recycling, 106*, 110–123.

UN General Assembly. (2015). *Transforming our world: the 2030 Agenda for Sustainable Development*, A/RES/70/1, 21 Oct 2015. Available at: http://www.un.org/ga/search/view_doc.asp?symbol=A/RES/70/1&Lang=E

United Nations Environment Programme. (2021). Food Waste Index Report 2021.

US NIC (United States National Intelligence Council). (2012). *Global trends 2030: Alternative worlds* (p. 137). US NIC.

USEPA. (2018). Sustainable management of food recovery hierarchy. https://www.epa.gov/sustainable-management-food/reducewasted-food-feeding-hungry-people. Accessed 29 Mar 2019.

Venkat, K. (2011). The climate change and economic impacts of food waste in the United States. *International Journal on Food System Dynamics, 2*(4), 431–446.

World Health Organization. (2021). *Levels and trends in child malnutrition*. UNICEF/WHO/World Bank Group Joint Child Malnutrition Estimates Key findings of the 2021 edition.

Index

L. L. B. Lazaro et al. (eds.), *Water-Energy-Food Nexus and Climate Change in Cities*, Sustainable Development Goals Series, https://doi.org/10.1007/978-3-031-05472-3

GPSR Compliance

The European Union's (EU) General Product Safety Regulation (GPSR) is a set of rules that requires consumer products to be safe and our obligations to ensure this.

If you have any concerns about our products, you can contact us on ProductSafety@springernature.com

In case Publisher is established outside the EU, the EU authorized representative is:

Springer Nature Customer Service Center GmbH
Europaplatz 3
69115 Heidelberg, Germany

Batch number: 10732809

Printed by Printforce, the Netherlands